Lecture Notes in Artificial Intelligence 9793

Subseries of Lecture Notes in Computer Science

More information about this series at http://www.springer.com/series/1244

Nathan F. Lepora · Anna Mura
Michael Mangan · Paul F.M.J. Verschure
Marc Desmulliez · Tony J. Prescott (Eds.)

Biomimetic and Biohybrid Systems

5th International Conference, Living Machines 2016
Edinburgh, UK, July 19–22, 2016
Proceedings

 Springer

Editors
Nathan F. Lepora
University of Bristol
Bristol
UK

Anna Mura
Universitat Pompeu Fabra
Barcelona
Spain

Michael Mangan
University of Lincoln
UK

Paul F.M.J. Verschure
Universitat Pompeu Fabra
Barcelona
Spain

Marc Desmulliez
Heriot-Watt University
Edinburgh
UK

Tony J. Prescott
University of Sheffield
Sheffield
UK

ISSN 0302-9743 ISSN 1611-3349 (electronic)
Lecture Notes in Artificial Intelligence
ISBN 978-3-319-42416-3 ISBN 978-3-319-42417-0 (eBook)
DOI 10.1007/978-3-319-42417-0

Library of Congress Control Number: 2016944482

LNCS Sublibrary: SL7 – Artificial Intelligence

Printed on acid-free paper

This Springer imprint is published by Springer Nature
The registered company is Springer International Publishing AG Switzerland

Preface

These proceedings contain the papers presented at Living Machines: The 5th International Conference on Biomimetic and Biohybrid Systems, held in Edinburgh, UK, during July 19–22, 2016. The international conferences in the Living Machines series are targeted at the intersection of research on novel life-like technologies inspired by the scientific investigation of biological systems, *biomimetics*, and research that seeks to interface biological and artificial systems to create *biohybrid* systems. The conference aim is to highlight the most exciting international research in both of these fields united by the theme of "Living Machines."

Biomimetics is the development of novel technologies through the distillation of principles from the study of biological systems. The investigation of biomimetic systems can serve two complementary goals. First, a suitably designed and configured biomimetic artifact can be used to test theories about the natural system of interest. Second, biomimetic technologies can provide useful, elegant, and efficient solutions to unsolved challenges in science and engineering. Biohybrid systems are formed by combining at least one biological component—an existing living system—and at least one artificial, newly engineered component. By passing information in one or both directions, such a system forms a new hybrid bio-artificial entity. The theme of the conference also encompasses biomimetic methods for manufacture, repair, and recycling inspired by natural processes such as reproduction, digestion, morphogenesis, and metamorphosis.

The following are some examples of *living machines* as featured at this and past conferences:

- Biomimetic robots and their component technologies (sensors, actuators, processors) that can intelligently interact with their environments
- Active biomimetic materials and structures that self-organize and self-repair
- Nature-inspired designs and manufacturing processes
- Biomimetic computers—neuromimetic emulations of the physiological basis for intelligent behavior
- Biohybrid brain–machine interfaces and neural implants
- Artificial organs and body parts including sensory organ–chip hybrids and intelligent prostheses
- Organism-level biohybrids such as robot–animal or robot–human systems

Five hundred years ago, Leonardo da Vinci designed a series of flying machines based on the wings of birds. These drawings are famous for their beautiful, lifelike designs, created centuries before the Wright brothers made their first flight. This inspiration from nature that Leonardo pioneered remains as crucial for technology today as it was many centuries ago.

Leonardo's inspiration was to imitate a successful biological design to solve a scientific problem. Today, this subject area is known as biomimetics. The American inventor Otto Schmitt first coined this term in the 1950s while trying to copy how nerve cells function in an artificial device. He put together the Greek words bios (life) and mimetic (copy) and the name caught on.

Why is nature so good at finding solutions to technological problems? The answer lies in Charles Darwin's theory of evolution. Life, by the process of natural selection, is a self-improving phenomenon that continually reinvents itself to solve problems in the natural world. These improvements have accumulated over hundreds of millions of years in plants and animals. As a result, there are a myriad natural design solutions around us, from the wings of insects and birds to the brains controlling our bodies.

Biomimetics and bio-inspiration has always been present in human technology, for example, making knives akin to the claws of animals. An exciting development, however, has been the dramatic expansion of the biomimetic sciences in the new millennium. The Convergent Science Network (CSN) of biomimetic and biohybrid systems, which organized the first Living Machines conference, has also completed a survey on *The State of the Art in Biomimetics* (Lepora, Verschure and Prescott, 2013). As part of the survey, we counted how much work on biomimetics is published each year. This revealed a surprising answer: from only tens of articles before the millennium, it has exploded since then to more than a thousand papers each year.

This huge investment in research inspired by nature is producing a wide variety of innovative technologies. Examples include artificial spider silk that is stronger than steel, super-tough synthetic materials based on the shells of molluscs, and adhesive patches mimicking the padded feet of geckos. Medical biomimetics is also leading to important benefits for maintaining health. These include bionic cochlear implants for hearing, fully functional artificial hearts, and modern prosthetic hands and limbs aimed at repairing the human body.

Looking to the future, one of the most revolutionary applications of biomimetics will likely be based on nature's most sophisticated creation: our brains. From our survey of biomimetic articles, we found that a main research theme is to take inspiration from how our brains control our bodies to design better ways of controlling robots. This is for a good reason. Engineers can build amazing robots that have seemingly human-like abilities. But so far, no existing robot comes close to copying the dexterity and adaptability of animal movements. The missing link is the controlling brain.

It is often said that future scientific discoveries are hard to predict. This is not the case in biomimetics. There are plenty of examples surrounding us in the natural world. The future will produce artificial devices with these abilities, from mass-produced flying micro devices based on insects to robotic manipulators based on the human hand to swimming robots based on fish. Less certain is what they will do to our society, economy, and way of life. Therefore the Living Machines conference also seeks to anticipate and understand the impacts of these technologies before they happen.

The main conference, during July 20–22, took the form of a three-day single-track oral and poster presentation program that included five plenary lectures from leading international researchers in biomimetic and biohybrid systems: Antonio Bicchi (University of Pisa) on robotics, haptics, and control systems; Frank Hirth (Kings College London, Institute of Psychiatry) on evolutionary neuroscience; Yoskiko Nakamura (University of Tokyo) on biomimetics in humanoids; Thomas Speck (Albert-Ludwigs-Universität, Freiburg) on plants and animals as concept generators for biomimetic materials and technologies; and Barbara Webb (University of Edinburgh) on perceptual systems and the control of behavior in insects and robots. There were also 20 regular talks and a poster session featuring approximately 40 posters. Session themes included: biomimetic robotics; biohybrid systems including biological-machine interfaces; neuromimetic systems; soft robot systems; active sensing in vision and touch; social robotics and the biomimetics of plants.

The conference was complemented with a further day of workshops and symposia, on July 19, covering a range of topics related to biomimetic and biohybrid systems: Our Future with Living Machines: Societal, Economic, and Ecological Impacts (Jose Halloy and Tony Prescott); Living Machines That Grow, Evolve, Self-Heal and Develop: How Robots Adapt Their Morphology to the Environment (Barbara Mazzolai and Cecilia Laschi); and The Emergence of Biological Architectures (Enrico Mastropaolo, Naomi Nakayama, Rowan Muir, Ross McLean, Cathal Cummins).

The main meeting was hosted at Edinburgh's Dynamic Earth, a five-star visitor attraction in the heart of Edinburgh's historic old town, next to the Scottish Parliament and Holyrood Palace. Dynamic Earth is a visitor experience that invites you to take a journey through time to witness the story of planet Earth through a series of interactive exhibits and state-of-the-art technology. Satellite events were held nearby at University of Edinburgh's School of Informatics in George Square. The Dynamics Earth experience, with its seamless integration of nature and technology, provided an ideal setting to host the 5th Living Machines Conference.

We wish to thank the many people that were involved in making LM2016 possible: Tony Prescott and Marc Desmulliez co-chaired the meeting; Nathan Lepora chaired the Program Committee and edited the conference proceedings; Paul Verschure chaired the international Steering Committee; Michael Mangan and Anna Mura co-chaired the workshop program; Anna Mura and Nathan Lepora co-organized the communications; Sytse Wierenga, Carme Buisan, and Mireia Mora provided additional administrative and technical support including organizing the website; and Katarzyna Przybcien and Lynn Smith provided administrative and local organizational support. We would also like to thank the authors and speakers who contributed their work, and the members of the Programme Committee for their detailed and considered reviews. We are grateful to the five keynote speakers who shared with us their vision of the future.

Finally, we wish to thank the sponsors of LM2016: The Convergence Science Network for Biomimetic and Neurotechnology (CSNII) (ICT-601167), which is funded by the European Union's Framework 7 (FP7) program in the area of Future Emerging Technologies (FET), and Heriot Watt University in Edinburgh, UK. Additional support was also provided by the University of Sheffield, the University of Bristol, the University of Pompeu Fabra in Barcelona, and the Institució Catalana de

Recerca i Estudis Avançats (ICREA). LM2016 was supported via a Santander Mobility Grant. Living Machines 2016 was also supported by the IOP Physics journal *Bioinspiration & Biomimetics*, who will publish a special issue of articles based on the best conference papers.

July 2016

Nathan F. Lepora
Anna Mura
Michael Mangan
Paul F.M.J. Verschure
Marc Desmulliez
Tony J. Prescott

Organization

Conference Chairs

Marc Desmulliez — Heriot Watt University, UK
Tony Prescott — University of Sheffield, UK

Program Chair

Nathan Lepora — University of Bristol, UK

Satellite Events Chairs

Michael Mangan — Lincoln University, UK
Anna Mura — Universitat Pompeu Fabra, Spain

International Steering Committee Chair

Paul Verschure — Universitat Pompeu Fabra and ICREA, Spain

Communications Chairs

Anna Mura — Universitat Pompeu Fabra, Spain
Nathan Lepora — University of Bristol, UK

Technical Support

Katarzyna Przybcien — Heriot Watt University, UK
Lynn Smith — Heriot Watt University, UK
Carme Buisan — Universitat Pompeu Fabra, Spain
Mireia Mora — Universitat Pompeu Fabra, Spain
Sytse Wierenga — Universitat Pompeu Fabra, Spain

Program Committee

Andrew Adamatzky — UWE, Bristol, UK
Robert Allen — University of Southampton, UK
Yoseph Bar-Cohen — JPL, USA
Federico Carpi — Queen Mary University of London, UK
Anders Christensen — University Institute of Lisbon, Portugal
Frederik Claeyssens — University of Sheffield, UK
Andrew Conn — University of Bristol, UK

Jorg Conradt	TU München, Germany
Holk Cruse	University Bielefeld, Germany
Mark Cutkosky	Stanford University, USA
Danilo De Rossi	Research Centre E. Piaggio, Italy
Marc Desmulliez	Heriot Watt University, UK
Sanja Dogramadzi	University of the West of England, UK
Stéphane Doncieux	ISIR, France
Volker Dürr	Bielefeld University, Germany
Wolfgang Eberle	Imec, Belgium
Maria Elena Giannaccini	University of Bristol, UK
Benoît Girard	CNRS and UPMC, France
Sabine Hauert	University of Bristol, UK
Helmut Hauser	University of Bristol, UK
Ivan Herreros	Universitat Pompeu Fabra, Spain
Koh Hosoda	Osaka University, Japan
Ioannis Ieropoulos	University of the West of England, UK
Cecilia Laschi	Scuola Superiore Sant'Anna, Italy
Nathan Lepora	University of Bristol, UK
Michael Mangan	Lincoln University, UK
Uriel Martinez-Hernandez	University of Leeds, UK
Ben Mitchinson	University of Sheffield, UK
Vishwanathan Mohan	Italian Institute of Technology, Italy
Anna Mura	Universitat Pompeu Fabra, Spain
Martin Pearson	Bristol Robotics Laboratory, UK
Hemma Philamore	University of Bristol, UK
Andrew Philippides	University of Sussex, UK
Tony Pipe	Bristol Robotics Laboratory, UK
Tony Prescott	University of Sheffield, UK
Roger Quinn	Case Western Reserve University, USA
Sylvian Saighi	University of Bordeaux, France
Thomas Schmickl	Karl-Franzens University Graz, Austria
Charlie Sullivan	University of the West of England, UK
Luca Tonin	University of Padova, Italy
Pablo Varona	Universidad Autonoma de Madrid, Spain
Eleni Vasilaki	University of Sheffield, UK
Benjamin Ward-Cherrier	University of Bristol, UK
Stuart Wilson	University of Sheffield, UK
Hartmut Witte	Technische Universität Ilmenau, Germany

Contents

Full Papers

The Natural Bipeds, Birds and Humans: An Inspiration for Bipedal Robots.... 3
Anick Abourachid and Vincent Hugel

Retina Color-Opponency Based Pursuit Implemented Through Spiking
Neural Networks in the Neurorobotics Platform 16
Alessandro Ambrosano, Lorenzo Vannucci, Ugo Albanese,
Murat Kirtay, Egidio Falotico, Pablo Martínez-Cañada, Georg Hinkel,
Jacques Kaiser, Stefan Ulbrich, Paul Levi, Christian Morillas,
Alois Knoll, Marc-Oliver Gewaltig, and Cecilia Laschi

A Two-Fingered Anthropomorphic Robotic Hand with Contact-Aided
Cross Four-Bar Mechanisms as Finger Joints 28
Guochao Bai, Jieyu Wang, and Xianwen Kong

Living Designs 40
Rina Bernabei and Jacqueline Power

iCub Visual Memory Inspector: Visualising the iCub's Thoughts 48
Daniel Camilleri, Andreas Damianou, Harry Jackson, Neil Lawrence,
and Tony Prescott

A Preliminary Framework for a Social Robot "Sixth Sense"............. 58
Lorenzo Cominelli, Daniele Mazzei, Nicola Carbonaro,
Roberto Garofalo, Abolfazl Zaraki, Alessandro Tognetti,
and Danilo De Rossi

A Bio-Inspired Photopatterning Method to Deposit Silver Nanoparticles
onto Non Conductive Surfaces Using Spinach Leaves Extract in Ethanol.... 71
Marc P.Y. Desmulliez, David E. Watson, Jose Marques-Hueso,
and Jack Hoy-Gig Ng

Leg Stiffness Control Based on "TEGOTAE" for Quadruped Locomotion ... 79
Akira Fukuhara, Dai Owaki, Takeshi Kano, and Akio Ishiguro

Wall Following in a Semi-closed-loop Fly-Robotic Interface............. 85
Jiaqi V. Huang, Yilin Wang, and Holger G. Krapp

Sensing Contact Constraints in a Worm-like Robot by Detecting Load
Anomalies 97
Akhil Kandhari, Andrew D. Horchler, George S. Zucker,
Kathryn A. Daltorio, Hillel J. Chiel, and Roger D. Quinn

Head-Mounted Sensory Augmentation Device: Comparing Haptic
and Audio Modality   107
    Hamideh Kerdegari, Yeongmi Kim, and Tony J. Prescott

Visual Target Sequence Prediction via Hierarchical Temporal Memory
Implemented on the iCub Robot   119
    Murat Kirtay, Egidio Falotico, Alessandro Ambrosano, Ugo Albanese,
    Lorenzo Vannucci, and Cecilia Laschi

Computer-Aided Biomimetics   131
    Ruben Kruiper, Jessica Chen-Burger, and Marc P.Y. Desmulliez

A Neural Network with Central Pattern Generators Entrained by Sensory
Feedback Controls Walking of a Bipedal Model   144
    Wei Li, Nicholas S. Szczecinski, Alexander J. Hunt, and Roger D. Quinn

Towards Unsupervised Canine Posture Classification via Depth Shadow
Detection and Infrared Reconstruction for Improved Image Segmentation
Accuracy   155
    Sean Mealin, Steven Howell, and David L. Roberts

A Bio-Inspired Model for Visual Collision Avoidance on a Hexapod
Walking Robot   167
    Hanno Gerd Meyer, Olivier J.N. Bertrand, Jan Paskarbeit,
    Jens Peter Lindemann, Axel Schneider, and Martin Egelhaaf

MIRO: A Robot "Mammal" with a Biomimetic Brain-Based Control
System   179
    Ben Mitchinson and Tony J. Prescott

A Hydraulic Hybrid Neuroprosthesis for Gait Restoration in People
with Spinal Cord Injuries   192
    Mark J. Nandor, Sarah R. Chang, Rudi Kobetic, Ronald J. Triolo,
    and Roger Quinn

Principal Component Analysis of Two-Dimensional Flow Vector Fields
on Human Facial Skin for Efficient Robot Face Design   203
    Nobuyuki Ota, Hisashi Ishihara, and Minoru Asada

Learning to Balance While Reaching: A Cerebellar-Based Control
Architecture for a Self-balancing Robot   214
    Maximilian Ruck, Ivan Herreros, Giovanni Maffei, Martí Sánchez-Fibla,
    and Paul Verschure

Optimizing Morphology and Locomotion on a Corpus of Parametric
Legged Robots   227
    Grégoire Passault, Quentin Rouxel, Remi Fabre, Steve N'Guyen,
    and Olivier Ly

Stick(y) Insects — Evaluation of Static Stability for Bio-inspired Leg
Coordination in Robotics 239
  Jan Paskarbeit, Marc Otto, Malte Schilling, and Axel Schneider

Navigate the Unknown: Implications of Grid-Cells "Mental Travel"
in Vicarious Trial and Error 251
  Diogo Santos-Pata, Riccardo Zucca, and Paul F.M.J. Verschure

Insect-Inspired Visual Navigation for Flying Robots 263
  Andrew Philippides, Nathan Steadman, Alex Dewar,
  Christopher Walker, and Paul Graham

Perceptive Invariance and Associative Memory Between Perception
and Semantic Representation USER a Universal SEmantic Representation
Implemented in a System on Chip (SoC) 275
  Patrick Pirim

Thrust-Assisted Perching and Climbing for a Bioinspired UAV 288
  Morgan T. Pope and Mark R. Cutkosky

The EASEL Project: Towards Educational Human-Robot Symbiotic
Interaction 297
  Dennis Reidsma, Vicky Charisi, Daniel Davison, Frances Wijnen,
  Jan van der Meij, Vanessa Evers, David Cameron, Samuel Fernando,
  Roger Moore, Tony Prescott, Daniele Mazzei, Michael Pieroni,
  Lorenzo Cominelli, Roberto Garofalo, Danilo De Rossi,
  Vasiliki Vouloutsi, Riccardo Zucca, Klaudia Grechuta, Maria Blancas,
  and Paul Verschure

Wasp-Inspired Needle Insertion with Low Net Push Force 307
  Tim Sprang, Paul Breedveld, and Dimitra Dodou

Use of Bifocal Objective Lens and Scanning Motion in Robotic Imaging
Systems for Simultaneous Peripheral and High Resolution Observation
of Objects 319
  Gašper Škulj and Drago Bračun

MantisBot Uses Minimal Descending Commands to Pursue Prey as
Observed in Tenodera Sinensis 329
  Nicholas S. Szczecinski, Andrew P. Getsy, Jacob W. Bosse,
  Joshua P. Martin, Roy E. Ritzmann, and Roger D. Quinn

Eye-Head Stabilization Mechanism for a Humanoid Robot Tested
on Human Inertial Data 341
  Lorenzo Vannucci, Egidio Falotico, Silvia Tolu, Paolo Dario,
  Henrik Hautop Lund, and Cecilia Laschi

Towards a Synthetic Tutor Assistant: The EASEL Project and its
Architecture   353
   Vasiliki Vouloutsi, Maria Blancas, Riccardo Zucca, Pedro Omedas,
   Dennis Reidsma, Daniel Davison, Vicky Charisi, Frances Wijnen,
   Jan van der Meij, Vanessa Evers, David Cameron, Samuel Fernando,
   Roger Moore, Tony Prescott, Daniele Mazzei, Michael Pieroni,
   Lorenzo Cominelli, Roberto Garofalo, Danilo De Rossi,
   and Paul F.M.J. Verschure

Aplysia Californica as a Novel Source of Material for Biohybrid Robots
and Organic Machines   365
   Victoria A. Webster, Katherine J. Chapin, Emma L. Hawley,
   Jill M. Patel, Ozan Akkus, Hillel J. Chiel, and Roger D. Quinn

A Soft Pneumatic Maggot Robot   375
   Tianqi Wei, Adam Stokes, and Barbara Webb

Short Papers

On Three Categories of Conscious Machines   389
   Xerxes D. Arsiwalla, Ivan Herreros, and Paul Verschure

Gaussian Process Regression for a Biomimetic Tactile Sensor   393
   Kirsty Aquilina, David A.W. Barton, and Nathan F. Lepora

Modulating Learning Through Expectation in a Simulated Robotic Setup. . . .   400
   Maria Blancas, Riccardo Zucca, Vasiliki Vouloutsi,
   and Paul F.M.J. Verschure

Don't Worry, We'll Get There: Developing Robot Personalities to Maintain
User Interaction After Robot Error   409
   David Cameron, Emily Collins, Hugo Cheung, Adriel Chua,
   Jonathan M. Aitken, and James Law

Designing Robot Personalities for Human-Robot Symbiotic Interaction
in an Educational Context   413
   David Cameron, Samuel Fernando, Abigail Millings, Michael Szollosy,
   Emily Collins, Roger Moore, Amanda Sharkey, and Tony Prescott

A Biomimetic Fingerprint Improves Spatial Tactile Perception   418
   Luke Cramphorn, Benjamin Ward-Cherrier, and Nathan F. Lepora

Anticipating Synchronisation for Robot Control   424
   Henry Eberle, Slawomir Nasuto, and Yoshikatsu Hayashi

MantisBot: The Implementation of a Photonic Vision System.   429
   Andrew P. Getsy, Nicholas S. Szczecinski, and Roger D. Quinn

Force Sensing with a Biomimetic Fingertip   436
  Maria Elena Giannaccini, Stuart Whyle, and Nathan F. Lepora

Understanding Interlimb Coordination Mechanism of Hexapod Locomotion
via "TEGOTAE"-Based Control   441
  Masashi Goda, Sakiko Miyazawa, Susumu Itayama, Dai Owaki,
  Takeshi Kano, and Akio Ishiguro

Decentralized Control Scheme for Myriapod Locomotion That Exploits
Local Force Feedback   449
  Takeshi Kano, Kotaro Yasui, Dai Owaki, and Akio Ishiguro

TEGOTAE-Based Control Scheme for Snake-Like Robots That Enables
Scaffold-Based Locomotion   454
  Takeshi Kano, Ryo Yoshizawa, and Akio Ishiguro

Modelling the Effect of Cognitive Load on Eye Saccades and Reportability:
The Validation Gate   459
  Sock C. Low, Joeri B.G. van Wijngaarden, and Paul F.M.J. Verschure

Mutual Entrainment of Cardiac-Oscillators Through Mechanical Interaction . . .   467
  Koki Maekawa, Naoki Inoue, Masahiro Shimizu, Yoshihiro Isobe,
  Taro Saku, and Koh Hosoda

"TEGOTAE"-Based Control of Bipedal Walking   472
  Dai Owaki, Shun-ya Horikiri, Jun Nishii, and Akio Ishiguro

Tactile Vision – Merging of Senses.   480
  Nedyalka Panova, Alexander C. Thompson,
  Francisco Tenopala-Carmona, and Ifor D.W. Samuel

Tactile Exploration by Contour Following Using a Biomimetic Fingertip   485
  Nicholas Pestell, Benjamin Ward-Cherrier, Luke Cramphorn,
  and Nathan F. Lepora

Towards Self-controlled Robots Through Distributed Adaptive Control   490
  Jordi-Ysard Puigbò, Clément Moulin-Frier, and Paul F.M.J. Verschure

Discrimination-Based Perception for Robot Touch.   498
  Emma Roscow, Christopher Kent, Ute Leonards, and Nathan F. Lepora

On Rock-and-Roll Effect of Quadruped Locomotion: From Mechanical and
Control-Theoretical Viewpoints.   503
  Ryoichi Kuratani, Masato Ishikawa, and Yasuhiro Sugimoto

Hydromast: A Bioinspired Flow Sensor with Accelerometers   510
  Asko Ristolainen, Jeffrey Andrew Tuhtan, Alar Kuusik,
  and Maarja Kruusmaa

Developing an Ecosystem for Interactive Electronic Implants   518
 Paul Strohmeier, Cedric Honnet, and Samppa von Cyborg

Gait Analysis of 6-Legged Robot with Actuator-Equipped Trunk and Insect
Inspired Body Structure   526
 Yasuhiro Sugimoto, Yuji Kito, Yuichiro Sueoka, and Koichi Osuka

Quadruped Gait Transition from Walk to Pace to Rotary Gallop
by Exploiting Head Movement   532
 Shura Suzuki, Dai Owaki, Akira Fukuhara, and Akio Ishiguro

Exploiting Symmetry to Generalize Biomimetic Touch   540
 Benjamin Ward-Cherrier, Luke Cramphorn, and Nathan F. Lepora

Decentralized Control Scheme for Centipede Locomotion Based on Local
Reflexes.   545
 Kotaro Yasui, Takeshi Kano, Dai Owaki, and Akio Ishiguro

Realization of Snakes' Concertina Locomotion
by Using "TEGOTAE-Based Control".   548
 Ryo Yoshizawa, Takeshi Kano, and Akio Ishiguro

Author Index   553

Full Papers

The Natural Bipeds, Birds and Humans: An Inspiration for Bipedal Robots

Anick Abourachid[1] and Vincent Hugel[2(✉)]

[1] National Museum of Natural History, Paris, France
abourach@mnhn.fr
[2] University of Toulon, Toulon, France
vincent.hugel@univ-tln.fr

Abstract. Despite many studies, the locomotion of bipedal legged robots is still not perfect. All the current robots are based on a humanoid model, which is not the unique one in Nature. In this paper we compare the natural bipedies in order to explore new tracks to improve robotic bipedal locomotion. This study starts with a short review of the historical bases of the biological bipedies to explain the differences between the structures of the human and bird bodies. The observations on the kinematics of bird walking describe a modular system that can be reproduced in robotics to take advantage of the bird leg versatility. For comparison purposes, a bird model is scaled up to have the same mass and the same height of the center of mass as a humanoid model. Simulation results show that the bird model can execute larger strides and stay on course, compared with the humanoid model. In addition the results confirm the functional decomposition of the bird system into the trunk and the thighs for the one part, and the distal part of the leg for the other part.

Keywords: Human bipedy · Bird bipedy · Kinematics · Robotics modeling

1 Introduction

All current operational bipedal robots are humanoid robots. The lower part is either a mobile wheeled base, or is composed of legs with classical kinematics inspired by human legs. Despite progress in locomotion ability, they are quite far from the performances of human locomotion. In this paper, we explore the bipedal systems used in nature in order to find new tracks for improving robots' bipedal locomotion.

The first part presents a short review of the natural bipedal systems, humans and birds, with a short summary of the history and a description of their structure. The second section focuses on the kinematics and dynamics of walking. Section 3 is dedicated to the robotic modeling of both bipeds that are tested on simulations, which are described in Sect. 4. Results are analyzed and discussed in Sect. 5.

© Springer International Publishing Switzerland 2016
N.F. Lepora et al. (Eds.): Living Machines 2016, LNAI 9793, pp. 3–15, 2016.
DOI: 10.1007/978-3-319-42417-0_1

The Natural Bipeds

Humans and birds are the only bipedal animals totaling 10001 strictly bipedal species: one species of primate and large diversity of birds. Primates are basically quadruped mammals. Apart from humans, some of them can walk bipedally, but typically in a transitory way, when carrying objects for instance, or for display. Even if humans are able to swim, it is a exceptional behavior, and human are basically specialized for one locomotor behavior: walking. Birds are specialized for the flight, but are also bipedal species. The wings are only used for flying, but the legs are versatile, used for hopping, walking, swimming, but also for take-off and landing. The differences between the natural bipeds arise from different histories. In the primate clade, the permanent bipedal posture appeared 4 million years ago, with the origin of the genus *Australopithecus*. Since then, a bipedal gait was used by seven species of *Homo* over the past 2 million years but is today used exclusively by the species *Homo sapiens*. In contrast, there are 10000 extant representatives of a group of cursorial bipedal dinosaurs, the theropods, that developed the ability to fly during the Mesozoic about 200 million years ago [1]. Some of these flying theropods survived the mass extinction of the dinosaurs and gave rise to the current diversity of birds. All the birds have inherited the same attributes typical of a flying animal and show stick-like forelimbs supporting feathers, a rigid trunk, and a bony short tail [2]. They all have three long bones in the legs: the femur, the tibiotarsus, and the tasometatarsus, and all of them are digitigrade, walking on their toes. The main variability in relation with the locomotor behaviors concerns the development of the sternal keel and the lengths of the wings for flight and the proportions of the leg bones. Depending on the species and their way of life, they vary from 10 to 32%, 37–56% and 13–45% of the total leg length for the femur, tibiotarsus and tasometatarsus respectively [3]. On the contrary, all the humans belong to the same species thus the leg proportions are very consistent with the tibia and femur varying up to 6% of the total leg length among extant populations of humans [4].

The human and bird bodies do not share the same design and geometry (Fig. 1). The human body posture is fully erected, the trunk vertical above the hip on the straight legs. The avian posture is crouched, the joints are flexed, the hips are located on the back, and the trunk is usually hanging horizontally between the thighs. These differences lead to differences in the walking kinematics.

2   Walking in Human and Birds

2.1   Kinematics

The kinematics of human walking can be approximated by an inverted pendulum gait with the moving thigh and the lower leg powering the forward motion of the upper body [5]. The center of mass (CoM) path is guided by adjusting the pelvic rotation, pelvic tilt, lateral pelvic displacement, the knee and hip flexion, and the knee and ankle interaction [6]. The motions of the trunk and the head

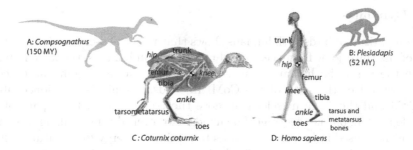

Fig. 1. CoMparison between the two biological bipedal systems. (A) a theropod, close to the birds' ancestor, and (B) a primate, close to the human ancestor, are represented backward. Parts of the body are shared between the birds (C) and the human (D), but the orientation of the segments are different, erected in the humans and crouched in the birds. The 11 bones of the plantar sole, tarsus and metatarsus, included in the human foot, are fused and form the tarsometatarsus bone of the leg in birds. The CoM is at the hip level in humans and at the knee level in birds.

are small and play a minor role in the kinematics compared to the pelvis and legs [6,7]. The arm swing balances the trunk torques induced by the legs [8].

The kinematics of the bird are more versatile as they use more than one locomotor mode, depending on the species. During walking, the contribution of the thigh to the forward body motion (femur motion) is far lower than that of the more distal segments, the tibiotarsus and the tarsometatarsus [9–11]. The thigh muscles that insert on the pelvis are mainly used to move the trunk that represents 80 % of the body mass, and to adjust the path of the CoM. In terrestrial birds such as the quail (*Coturnix coturnix*), during walking the trunk oscillates around the CoM, pitching, rolling and yawing being guided by the muscles of the thigh that cross the hip [10]. When a duck like the teal (*Callonetta leucophrys*) swims, the part distal to the knee is used as the paddle to propel the system. The thigh does not move at all, anchoring the leg on the trunk similar to a paddle on the hull of a boat. The hip is less flexible compared to that in terrestrial birds such as the quail. Consequently, the thigh is capable of moving the trunk less when walking. The foot is put medially on the ground, allowing medio-lateral translation of the trunk, keeping the CoM above the support polygon. These motions give the typical waddling of ducks [12]. When taking off, the birds hop before using the wings. All the acceleration is provided by the legs, the trunk being propelled first by an extension of the hip joints when the legs push on the substrate [13].

Thus, the birds bipedy may be considered as the association of two morphofunctional modules, the trunk plus thighs guiding the CoM in a different way depending on the species and the locomotor behavior, and the part of the legs distal to the knee, propelling the system. This functional modularity could participate in the versatility of the pelvic locomotor system.

2.2   Dynamics

The comparison of birds with humans shows that in both cases the CoM is at the level of the third joint from the ground up: the hip in humans, the knee in birds. Above the CoM the human trunk is erected and has no significant function in the control of the path of the CoM. In birds, the hip is positioned above the CoM and the thigh muscles can move the trunk to adjust the path of the CoM depending on the mechanical demands. However, during walking, the same mechanism is used in humans and birds to minimize energy expenditure, with a kinetic-gravitational energy transfer during each stride [14]. This mechanism can be modeled by the trajectory of the CoM during locomotion, successively going up and down at each stride like in an inverted pendulum. The relation between the metabolic energy consumption and the mechanical work, as function of speed and body size, is similar in birds, humans and all other mammals [15,16].

Since the two natural biped systems use a dynamic inverted pendulum model during walking, it is possible to use this property to model and to compare both systems. The next sections are dedicated to robotics modeling to test whether a bird-like bipedal robot could be an alternative to a humanoid robot.

3   Robotic Modeling

In order to compare the bird system with the human system, two robotic models were designed for simulation (Fig. 2). The objective is to make these models walk forward inside a dynamic simulator and to check the advantages of the bird system over the human system.

The bird model is inspired by the quail. The leg is composed of three segments – femur, tibiotarsus, and tarsometatarsus, and one main flat toe. The humanoid robotic model is the model used to design the ROMEO humanoid manufactured by the Aldebaran-Robotics company. The model has two segments – femur and tibia–, and one foot per leg. For both models, the CoM, namely G, is placed at the same height $h_G = 0.66$ m (humanoid in flexed position, see simulation Sect. 4). In the bird model, the CoM is located between the knees. The toe length is made larger than the humanoid foot. Actually the toe length was fixed as twice the longitudinal distance between the heel and the knee to ensure the projection of the CoM in the middle of the toe.

Both models have the same mass $m = 40.5$ kg. The mass distribution for the bird model was drawn from a real quail that was frozen and cut into pieces. Each piece was weighed and its CoM was calculated by the double suspension method [17]. In the humanoid model, the CoM is above the hips. The humanoid mass distribution was drawn from the mass distribution of a real human being (Table 1).

The size of the humanoid model is 1.4 m. The size of the bird model is 1.1 m. This size was obtained after scaling up the quail dimensions of the different parts using h_G.

Table 1. Mass rate distribution. For the bird the foot includes the tarsometatarsus and the toes.

	Quail	Humanoid
Head/Neck	9	7.6
Body	63	51
Wings/arms	4×2	5.3×2
Femurs	5×2	9.8×2
Tibias	4×2	4.3×2
Feet	1×2	1.3×2
Total	100	100

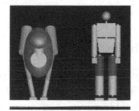 

Fig. 2. Robotic models of bird and humanoid.

3.1 Leg Kinematics

The leg of the bird model features two or three rotary joints at the hip, one at the knee, one at the ankle and two joints between the tibiotarsus and the toe, named the foot joint. The hip joints include a roll joint (longitudinal axis), a yaw joint (vertical axis), and a pitch joint (lateral axis) that can be inhibited when necessary to check specific inverse kinematics. The knee joint is a pitch joint (lateral axis) that is slightly rotated in the horizontal plane. The ankle joint is also a pitch joint but with a slight inclination about the longitudinal axis according to the analysis of the quail walk [18]. The foot joint includes a roll joint and a pitch joint.

The humanoid leg has a classical kinematics with three orthogonal rotary joints at the hip – yaw, roll, and pitch joints in this order starting from the pelvis –, a pitch joint at the knee, and a pitch and roll joint at the ankle in this order from the knee [19].

Table 2 summarizes the main characteristic differences between the bird model and the humanoid model. The number of DoF for the bird is set to 6 here since the pitch joint is considered to be inhibited (see simulations). The noticeable differences include the position of the CoM with respect to the hips, the shape of the legs and the distribution of the joint degrees of freedom along the leg.

Table 2. Main characteristic differences between bird and humanoid models.

Characteristics	Bird	Humanoid
Nb of segments/leg	3	2
DoF	6	6
Distribution of DoF (from pelvis)	2-1-1-2	3-1-2
Orientation of joint axes	inclined (knee+ankle)	straight
Legs in standing position	flexed and inclined (knees abducted / feet adducted)	vertically stretched
Location of CoM	middle of knees	above hips
Center-of-mass vertical	through middle of knees	close to frontal plane
Toe/Foot	1 flat main toe (rigid)	1 flat foot (rigid)

3.2   Dynamic Model and Walking Algorithm

The dynamic model adopted for the quail robot and the humanoid robot is the model of the 3D linear inverted pendulum (3D-LIP) where the total robot's mass m is concentrated at the top of a massless stick that joins the Zero Moment Point (ZMP) on the ground to the suspended mass [20,21]. The ZMP can be considered as the center of pressure (CoP) below the supporting foot. The stick is telescopic, i.e. the stick length varies according to the variation of the leg joint angles. The height of the CoM is kept constant during the walk. This means that the humanoid has to flex knees before starting to walk since feet are rigid and remain horizontal during the walk. The bird robot can walk directly without initial configuration. Its feet also remain horizontal. Actually the robotic feet used here are rigid and cannot roll unlike bird toes and human soles.

All the walks have instantaneous double support phases.

The walking algorithm adopted for both models is decomposed into the following steps:

Planning of Forward Foot/Toe Steps. A step is defined as the longitudinal distance between two consecutive feet touching down. The step length is the same along the walk and the first stride is a half step. For all models, the step length was set to 0.25 m for the first set of simulations, then it was set to 0.33 m. The ratio of knee flexion for the humanoid was set to 0.9, this means that the flexion motion leads to the humanoid vertically going down 10 % of the initial hip height. The step duration is the same for all the walks, $t_s = 0.4$ s.

Planning of ZMP. For each support leg, the ZMP is fixed in the middle of the foot/toe when the CoM velocity is constant (no acceleration). The acceleration phase at the beginning of the walk and the deceleration phase at the end were managed separately to calculate the ZMP automatically [21]. At start the ZMP is shifted to the heel, whereas at stop the ZMP is shifted to the front of the foot/toe. This is due to the longitudinal acceleration of the

CoM, which can be tuned by reducing the lengths of the first steps or by increasing the time to execute them.

CoM Trajectory. The horizontal coordinates of the CoM, namely x_G and y_G, are drawn from the equation that govern the linear inverted pendulum model [20]:

$$\ddot{x}_G = \frac{g}{h_G}(x_G - x_P), \quad \ddot{y}_G = \frac{g}{h_G}(y_G - y_P)$$

where g is the gravity and x_P, y_P are the horizontal coordinates of the ZMP. The analytic solution leads to a hyperbolic shape of the CoM trajectory.

Inverse Kinematics. This step is used to calculate the joint commands to be sent to the dynamic simulator. Here the CoM is assumed to be fixed with respect to the trunk of the model, which is a reasonable approximation. The inverse kinematics use the classical pseudo-inverse of the Jacobian matrix that gives the variations of the foot trajectory as a function of the joint angle variations.

4   Simulations

The simulator used is the Adams MSC-Software environment coupled with Matlab. Joint commands are calculated in Matlab and sent to Adams through Simulink using a periodic timer. Joint position and velocity commands are used as desired inputs to PID control blocks inside Adams to control the motors that drive the leg joints. There is no internal stabilizer, i.e. there is no closed-loop control with any inertial measurement unit feedback, because we want to check the suitability of the biped model for the walk using a classical walking algorithm used in robotics.

The interaction model of feet/toes with the ground is achieved through little spherical caps located below the soles of feet/toes. The contact of those caps with the ground result in normal forces that are modeled using penetration depth, penetration velocity, stiffness and damping coefficient [22].

Three kinds of simulations were carried out:

1. humanoid model flexing knees with a flexion ratio of 90 %, then walking 1 m forward with 0.25 m steps lasting 0.4 s each (Fig. 3A).
2. bird model walking 1 m forward with 0.25 m steps lasting 0.4 s each. All hip joints are controlled, and the inverse kinematics process tends to minimize the variations of all joints, and therefore to "distribute" the angle amplitude among all joints (Fig. 3B).
3. bird model walking 1 m forward with 0.25 m steps lasting 0.4 s each. The pitch joint at the hip was inhibited. Consecutively most of the thrusting is achieved by the part of the leg below the knee (Fig. 3C).

The same series of simulation is reproduced with a step length of 0.33 m.

Figure 4 presents the results relative to steps of 0.25 m. The top view figures show the theoretical and real foot(toe)prints, the theoretical trajectory of the

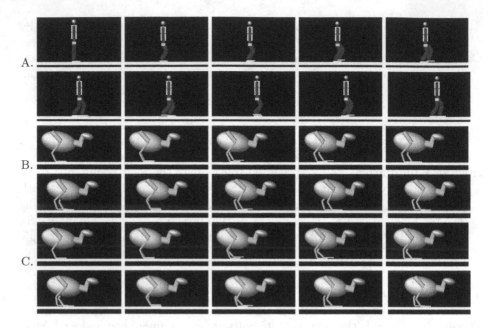

Fig. 3. Successive 0.1 s snapshots of a 1 m forward walking with step length of 0.25 m. A: humanoid model (First 0.7 s of walk after 2 s knee flexion); B: bird model with control of hip pitch joint and C: bird model with hip pitch joint inhibited (First 0.9 s of walk)

CoM, and the centers of pressure and the real CoM obtained from simulation in the dynamic environment. Figure 5 depicts the results relative to steps of 0.33 m. The CoP was calculated at each frame from the normal force sensors located on each spherical cap located on the corners of the feet/toes.

5   Results and Discussion

The results relative to walking steps of 0.25 m show that all models, namely the humanoid, the quail robot with control of the hip pitch joint for walking, and the quail model with the hip pitch joint inhibited, can follow the trajectory of the CoM that was planned offline (Fig. 4).

During the walk the centers of pressure (CoP) of the humanoid model are more scattered than the quail model (Fig. 4A). As long as the CoP remain inside the footprints this is not a problem, however it appears that it is more difficult to stabilize the CoP in the case of the humanoid structure. This observation can be related to the fact that the simulated humanoid can be considered as a double inverted pendulum (leg – trunk with CoM above the hips) actually whereas the simulated quail model can be considered as an inverted pendulum connected to a standard pendulum (leg – trunk with CoM below the hips), which is intrinsically more stable.

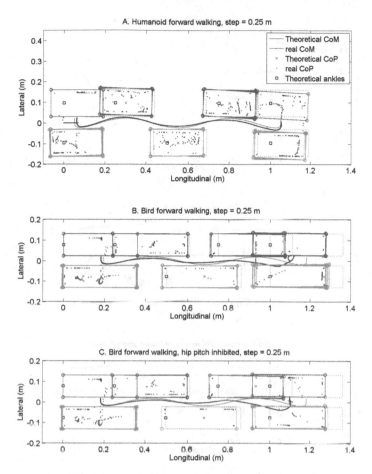

Fig. 4. Footprints of humanoid, and toeprints of bird model walking 1 m forward with step length of 0.25 m. (Color figure online)

It is noticeable that the deceleration phase in the case of the quails leads to little pitching oscillations as a significant number of centers of pressure accumulate on the fore and rear extremities of the toes in the final toeprints (Fig. 4B,C). These oscillations are responsible for a shorter path traveled compared with the path planned, i.e. 0.92 m instead of 1 m. The pitching oscillations can be explained by the fore and rear distributions of the mass of the trunk in the quail model than makes it more sensitive to pitching moves. The quail model with hip pitch joint inhibition is interesting here because the control of the hip pitch joints can be decoupled and used to counteract those pitching oscillations to enhance the walk balance during acceleration/deceleration phases. This matches the observation from biologists that consider that the hip pitch joints play the role of controlling the movement of the trunk with respect to the legs to adjust the posture of the bird.

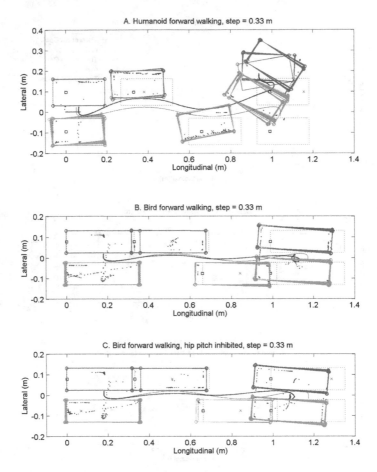

Fig. 5. Footprints of humanoid, and toeprints of bird model walking 1 m forward with step length of 0.33 m. (Color figure online)

The results relative to walking steps of 0.33 m are really selective. Actually the humanoid model cannot stay on course and suffer from disturbing yaw moves from the beginning (Fig. 5A). Centers of pressure accumulate on the lateral sides of the footprints and act as pivots that make the robot turn badly. This highlights the role of the arms in the human walk, whose back-and-forth swing counteracts the disturbing rotation about the vertical. On the contrary the quail model regularly stays on course (Fig. 5B and C) except at the end because the acceleration phase is too fast. The fact that bird legs only account for 20 % of the total mass while human legs account for 30 % diminishes the influence of the disturbing yaw moment due to leg swing in the case of the bird model. This series of simulations shows the superiority of the quail model to achieve larger steps in comparison with the humanoid model of the same mass, with the same height of the CoM. Here it must be noted that the distance between toes in the quail model is smaller than the distance between feet in the humanoid model.

This leads to a reduced amplitude of the oscillation of the CoM in the case of the quail model, which helps the quail model stay on course. Moreover, the real quail adopts a kind of catwalk by placing the landing toe in front of the other support toe, which constrains the amplitude of the CoM sway even more. When scaled up the quail model also features larger toes compared with the humanoid foot size. This difference also contributes to increase the stride range capabilities of the quail model.

The simulation results are interesting but the implementation of the robotic models has some limitations that must be taken into account in the analysis. The first limitation is the walking algorithm that assumes that feet and toes remain flat all along the walk. A consequence is that the humanoid leg and the set of tibiotarsus and metatarsus in the quail model tend to get stretched at the end of the stance phase when the step length is increased, which is not efficient as a starting position to raise the leg, and therefore limits the stride length. This can explain why the natural bipeds roll the foot/toe sole in the stance phase before raising it in order to avoid stretched legs in the rear position. In the case of the humanoid model, the rolling of feet can be useful to avoid the *flexed-knee* walk, and to set up human-like walks with folding and unfolding legs.

In addition the walking algorithms do not include a specific management of the starting and ending phases of the walk. In cruise mode the CoP is planned to stay in the middle of the footprint but in the starting phase the CoP is pushed backwards to initiate the walk, and in the end phase, the CoP is pushed forwards to halt the motion. At last the joints modeled in the dynamic simulator are rigid. There is no flexible joint, neither compliance to store/restore energy as occurs in the natural biped legs.

6   Conclusion

This study has described the biological differences between the human bipeds and the bird bipeds, from a kinematics and a dynamics point of view. Based on these observations, a robotics humanoid model and a robotic quail-like model were designed to compare the walking capabilities of both structures. Simulations carried out inside a dynamic environment have highlighted a major capability of the quail model over the humanoid model, namely the capability to increase the stride length while maintaining the course. Besides, the control of the hip pitch joint in the quail model can be decoupled and used to counteract the pitching oscillations of the trunk in the sagittal plane, the part of the bird leg below the knee being dedicated to thrusting the system. This confirms the mechanical advantage of the modular organization of the bird's bipedy with a trunk-thigh module and a distal module.

Future work will explore the mechanical consequences of the differences in the body proportions observed in birds. Up to now only the ratios of leg segments relative to the quail were used in the simulation study. The limb ratios of other birds could be more suited to the design of a robotic biped. Another research study will focus on the design of rolling toes in order to increase the walking efficiency of the robotic model.

References

1. Hope, S.: The Mesozoic radiation of Neornithes. In: Chiappe, L.M., Witmer, L.M. (eds.) Mesozoic Birds, Above the Head of Dinosaurs, pp. 339–388. University of California Press, Berkeley (2002)
2. Norberg, U.M.L.: Vertebrate Flight. Springer, Berlin (1990)
3. Gatesy, S.M., Middleton, K.M.: Bipedalism, flight and the evolution of theropod locomotor diversity. J Vertebr. Paleontol. **17**, 308–329 (1997)
4. Porter, A.M.W.: Modern human, early human and Neanderthal limb proportions. Int. J. Osteoarchaeology **9**, 54–67 (1999)
5. Pontzer, H., Holloway 3rd, J.H., Raichlen, D.A., Lieberman, D.E.: Control and function of arm swing in human walking and running. J. Exp. Biol. **212**, 523–534 (2009)
6. Hayot, C.: 3D biomechanical analysis of the human walk: comparison of mechanical models. PhD thesis, Poitiers (in French), France (2010)
7. Winter, D., Quanbury, A., Reimer, G.: Analysis of instantaneous energy of normal gait. J. Biomech. **9**, 253–257 (1976)
8. Li, Y., Wang, W., Crompton, R.H., Gunther, M.M.: Free vertical moments and transverse forces in human walking and their role in relation to arm-swing. J. Exp. Biol. **204**, 47–58 (2001)
9. Gatesy, S.M.: Guineafowl hind limb function. I cineradiographic analysis and speed effects. J. Morphol. **240**, 115–125 (1999)
10. Abourachid, A., Hackert, R., Herbin, M., Libourel, P.A., Lambert, F.O., Gioanni, H., Provini, P., Blazevic, P., Hugel, V.: Bird terrestrial locomotion as revealed by 3D kinematics. Zoology (Jena) **114**, 360–368 (2011)
11. Stoessel, A., Fischer, M.: Comparative intralimb coordination in avian bipedal locomotion. J. Exp. Biol. **215**, 4055–4069 (2012)
12. Provini, P., Tobalske, B.W., Crandell, K.E., Abourachid, A.: Transition from leg to wing forces during take-off in birds. J. Exp. Biol. **215**, 4115–4124 (2012)
13. Provini, P., Goupil, P., Hugel, V., Abourachid, A.: Walking, paddling, waddling: 3D kinematics of Anatidae locomotion (Callonetta leucophrys). J. Exp. Zool. **317**, 275–282 (2012)
14. Cavagna, G.A., Heglung, N.C., Taylor, R.: Mechanical work in terrestrial locomotion: two basic mechanisms for minimizing energy expenditure. Am. J. Physiol. **2**, 243–261 (1977)
15. Taylor, C.R., Heglund, N.C., Maloy, G.M.O.: Energetics and mechanics of terrestrial locomotion I Metabolic energy consumption as a function of speed and body size in birds and mammals. J. Exp. Biol. **97**, 1–21 (1982)
16. Fedak, M.A., Heglund, N.C., Taylor, C.R.: Energetics and mechanics of terrestrial locomotion II kinetic energy changes of the limbs and body as a function of speed and body size in birds and mammals. J. Exp. Biol. **79**, 23–40 (1982)
17. Abourachid, A.: Mechanics of standing in birds: functional explanation of lamness problems in giant turkeys. Br. Poult. Sci. **34**, 887–898 (1993)
18. Hugel, V., Hackert, R., Abourachid, A.: Kinematic modeling of bird locomotion from experimental data. IEEE Trans. Robot. **27**(2), 185–200 (2011)
19. Zorjan, M., Hugel, V.: Generalized humanoid leg inverse kinematics to deal with singularities. In: IEEE International Conference on Robotics and Automation, pp. 4791–4796 (2013)
20. Kajita, S.: Humanoid Robot. Ohmsha Ltd, 3-1 Kanda Nishikicho, Chiyodaku, Tokyo, Japan (2005)

21. Hugel, V., Jouandeau, N.: Walking patterns for real time path planning simulation of humanoids. In: IEEE RO-MAN, pp. 424–430 (2012)
22. Zorjan, M., Hugel, V., Blazevic, P., Borovac, B.: Influence of rotation of humanoid hip joint axes on joint power during locomotion. Adv. Robot. **29**(11), 707–719 (2015)

Retina Color-Opponency Based Pursuit Implemented Through Spiking Neural Networks in the Neurorobotics Platform

Alessandro Ambrosano[1]([✉]), Lorenzo Vannucci[1], Ugo Albanese[1],
Murat Kirtay[1], Egidio Falotico[1], Pablo Martínez-Cañada[6], Georg Hinkel[4],
Jacques Kaiser[2], Stefan Ulbrich[2], Paul Levi[2], Christian Morillas[6], Alois Knoll[5],
Marc-Oliver Gewaltig[3], and Cecilia Laschi[1]

[1] The BioRobotics Institute, Scuola Superiore Sant'Anna,
Viale R. Piaggio 34, 56025 Pontedera, Italy
`alessandro.ambrosano@sssup.it`
[2] Department of Intelligent Systems and Production Engineering (ISPE IDS/TKS),
FZI Research Center for Information Technology, Haidund-Neu-Str. 10-14,
76131 Karlsruhe, Germany
[3] Blue Brain Project (BBP), École polytechnique fédérale de Lausanne (EPFL),
Campus Biotech, Bâtiment B1, Ch. des Mines 9, 1202 Geneva, Switzerland
[4] Department of Software Engineering (SE), FZI Research Center for Information
Technology, Haid-und-Neu-Str. 10-14, 76131 Karlsruhe, Germany
[5] Department of Informatics, Technical University of Munich,
Boltzmannstrae 3, 85748 Garching, Germany
[6] Department of Computer Architecture and Technology,
CITIC, University of Granada, Granada, Spain

Abstract. The 'red-green' pathway of the retina is classically recognized as one of the retinal mechanisms allowing humans to gather color information from light, by combining information from L-cones and M-cones in an opponent way. The precise retinal circuitry that allows the opponency process to occur is still uncertain, but it is known that signals from L-cones and M-cones, having a widely overlapping spectral response, contribute with opposite signs. In this paper, we simulate the red-green opponency process using a retina model based on linear-nonlinear analysis to characterize context adaptation and exploiting an image-processing approach to simulate the neural responses in order to track a moving target. Moreover, we integrate this model within a visual pursuit controller implemented as a spiking neural network to guide eye movements in a humanoid robot. Tests conducted in the Neurorobotics Platform confirm the effectiveness of the whole model. This work is the first step towards a bio-inspired smooth pursuit model embedding a retina model using spiking neural networks.

1 Introduction

One of the most important characteristics of the primate visual system is represented by the space-variant resolution retina with a high-resolution fovea that

© Springer International Publishing Switzerland 2016
N.F. Lepora et al. (Eds.): Living Machines 2016, LNAI 9793, pp. 16–27, 2016.
DOI: 10.1007/978-3-319-42417-0_2

offers considerable advantages for a detailed analysis of visual objects. When light hits the retina, *photoreceptor cells* transform it in electrochemical signals, which get furtherly processed first by flowing through *bipolar cells* and then through *ganglion cells*, whose axons form the optic nerve. Two more types of cells, *horizontal* and *amacrine*, participate in this process by connecting multiple photoreceptors and bipolar cells respectively, acting as feedback channels.

All these cells cooperate in a complex "retina circuit", that can be split in many different "microcircuits" where around 80 types of identified neural cell types take part. These microcircuits process simultaneously the light and forward different information to the brain through dedicated pathways. Color information is carried by some of these pathways, combining signals from the cone photoreceptors, which are sensitive to coloured light [1].

There are three types of cone cells in the human retina, each one sensitive to a different wavelength range: S-cones are sensitive to short wavelengths, M-cones to middle wavelengths and L-cones to long wavelengths, corresponding roughly to blue, green and red light respectively. Both the density of photons on the cone and the color of the light determine the probability that an individual cone will capture a photon [2]. For this reason, signals from a single cone type can't carry any color clue, but at least two types of cone must be compared in order to get actual color information.

Two processes of this sort are known to happen inside the retina: the *red-green opponency*, where M-cones and L-cones responses are taken into account in order to get color information in the red-green axis, and *yellow-blue opponency*, where all three kinds of cone are considered to get color information in the yellow-blue axis [3]. They are called opponency mechanisms because in both cases some cone type contribute with a positive weight and some other with a negative weight. In the red-green case we have M-cones opposed to L-cones whereas in the yellow-blue case the joint effect of M-cones and L-cones is opposed to S-cones response, though the exact circuitry of both processes is still under investigation.

The space-variant resolution of the retina requires efficient eye movements for correct vision. Two forms of eye movements — saccades and smooth pursuit — enable us to fixate the object on the fovea. Saccades are high-velocity gaze shifts that bring the image of an object of interest onto the fovea. The purpose of smooth pursuit eye movements is to minimise the retinal slip, i.e. the target velocity projected onto the retina, stabilizing the image of the moving object on the fovea. Retinal slip disappears once eye velocity catches up to target velocity in smooth pursuit eye movements. In primates, with a constant velocity or a sinusoidal target motion, the smooth pursuit gain, i.e. the ratio of tracking velocity to target velocity, is almost 1.0. In the last years several models of robotic visual pursuit have been developed. Shibata and colleagues suggested a control circuits for the integration of the most basic oculomotor behaviours [4] including the smooth pursuit eye movement. A similar model of smooth pursuit and catch-up saccade [5] was implemented on the iCub robot. Also models based on artificial neural networks were developed for visual tracking tasks [6–9]. In this paper we present a first attempt of embedding a retina model inside a visual pursuit controller suitable for a robotic implementation. The controller, taking as input the output of the retina can generate an appropriate motor command to follow a moving target without needing for computer vision algorithm in extracting the target position from the incoming visual image. Moreover, the controller uses biologically inspired Spiking Neural Networks in order to implement parts of the controller, thus a proper framework that combines robotics and neural simulations has been used: the Neurorobotics Platform.

2   The Retina Simulation Platform

A great number of models have been proposed to reproduce different processing mechanisms of the retina [10–13]. However, these are often ad hoc models focused on fitting some specific retina functions rather than providing a general retina simulation platform. We chose the retina simulator COREM [14,15], which includes a set of computational retinal microcircuits that can be used as basic building blocks for the modeling of different retina functions. In addition, COREM implements an interface with Spiking Neural Networks (SNNs) that allows its integration with models of higher visual areas and the Neurorobotics Platform.

The computational retinal microcircuits that can be configured within COREM consist of one spatial processing module (a space-variant Gaussian filter), two temporal modules (a low-pass temporal filter and a single-compartment model), a configurable time-independent nonlinearity and a Short-Term Plasticity (STP) function. The simulation engine allows the user to create custom retina scripts and to easily embed the retina model in the neural simulator.

3   The Neurorobotics Platform

The Neurorobotics Platform (NRP) is developed as part of the Human Brain Project[1] to run coupled neural and robotics simulations in an interactive platform. Whereas there are multiple neural simulators (e.g. Neuron [16], NEST [17]), robotics and physics simulators (e.g. Gazebo [18]), the NRP aims at offering a platform that combines the neuroscientific and robotic fields, by providing coupled physics and neural simulations. A core part of the NRP is the Closed-Loop-Engine (CLE) that allows to specify the data exchange between the brain simulation and the robot in a programmatic manner and orchestrates the simulations.

The key concept of the NRP is offering scientists an easy access to a simulation platform using a state-of-the-art web interface. Scientists are relieved from the burdensome installation process of scientific simulation software and are able to leverage large-scale computing resources. Furthermore, support for monitoring and visualizing the spiking activity of the neurons or joint states of the robot is offered as well as the camera image perceived by the robot.

To give an impression on how the platform looks like, a screenshot of a visual pursuit experiment with the iCub [19] humanoid robot is depicted in Fig. 1.

---

[1] https://www.humanbrainproject.eu/.

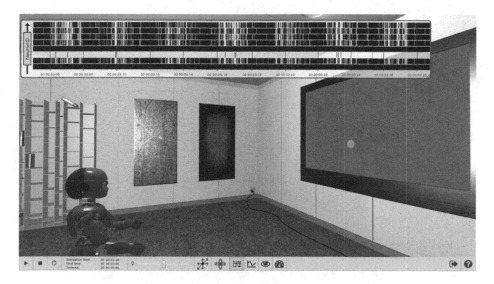

Fig. 1. Screenshot showing a pursuit experiment with the iCub robot in the Neurorobotics Platform (NRP). A plot at the top shows spiking activity of neurons.

The Closed-Loop-Engine (CLE), depicted in Fig. 2, implements the core of the NRP. The CLE controls the neural and the world simulation, and executes a lightweight simulation data interchange mechanism. Neural and robotics simulation are iteratively run in parallel for the same amount of time, after which the transfer functions (special functions that translate output from one simulation into input for the other) are executed periodically. Communication with the brain is realized through recording and injecting spike data. Interfacing with the robot simulation is done using the Robot Operating System middleware (ROS [20]). In order to ensure reproducibility, data exchange is conducted in a deterministic fashion.

3.1 Integrating the Retina Simulation Platform in the NRP

The original retina codebase [15] was implemented as a stand-alone program. So the first step towards the embedding in the NRP has been to isolate all the core functionalities of the model in a separate module, that could be used as a library. The original executable has been kept as a "frontend" application depending on such library.

In order to have a retina model involved in the robot control inside a NRP simulation, it is necessary to forward the camera image or the robot to the retina, get the retina output information so it can be processed in a SNN, and finally decode the SNN output to control the robot. All these steps must take place inside transfer functions.

The whole retina model is written in C++, whereas all the NRP backend, including transfer function related business logic, is written in Python.

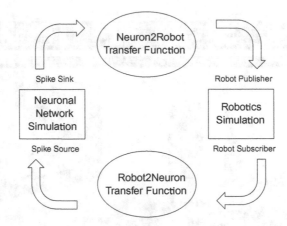

Fig. 2. A closed loop between a robotics simulation and a neural network

In particular, in order to easy the development of these transfer function modules, the part of the transfer functions that specify the connection to and from both simulations are implemented using a subset of python that defines a Domain Specific Language [21]. Thus, to allow the NRP to send input to the retina and get output from it, the functions in the retina library involved in these processes have been wrapped with Python bindings, making them accessible from Python code.

4   Retina Based Controller

We tested the retina-NRP interaction by creating a custom experiment on the platform performing a visual tracking task with a humanoid robot. Since our focus is on the red-green opponency, we created an environment in which a simulated iCub [19] humanoid robot will have to track a green ball moving horizontally on a red background.

In the experiment, image from one of the two cameras mounted on the robot eyes is sent to the retina simulator. Retinal output is then gathered and one horizontal stripe of the output image (320 pixels from a 320 × 240 image), intersecting the target, is forwarded to a neural network, where the spikes are filtered in order to avoid any undesired false detection. Finally, spike trains coming from the SNN are translated in a motion command for the robot eyes, making them following the target. An overview of the controller is shown in Fig. 3.

4.1   Retina Circuit

The retina circuit implemented with the framework described in Sect. 2 is composed of two symmetric subcircuits processing independently the input image. In the first layer we have simulated L-cones and M-cones, receiving the initial image. Output from the photoreceptors is then forwarded to the two different subcircuits, namely the L^+M^- circuit and the L^-M^+ circuit.

The L^+M^- circuit (L^-M^+ description can be derived by symmetry) sums the output signals from the two cones, giving a positive weight to L-cones and a negative weight to M-cones. Combined cones signal is processed by simulated horizontal, bipolar, amacrine and ganglion cells. Every simulated ganglion cell is then sensible to a red center on green background within its receptive field. The high level result is a sensitivity to borders on a static image, which is slightly accentuated around borders between red and green objects for both circuits. In case of moving images instead, ganglion cells in the L^+M^- are particularly sensitive, due to their temporal characteristics, to green objects appearing in receptive fields that were earlier impressed by red objects. We will exploit this peculiarity to infer the position of an object by combining responses from the two circuits.

4.2   Brain Model

The spiking neural network used in our experiment comprises 1280 integrate and fire neurons [22], organized in two layers. The first layer acts as a current to spike converter, it has been designed for taking current intensities value coming from the retina library to neural spikes. In the second layer, every neuron, except the "side" ones, gather information from 7 subsequent neurons on the first layer, acting as a local spike integrator.

Every layer embeds two independent populations of 320 neurons, processing separately output from the two different ganglion cell layers of the retina circuit, so there is a one to one correspondence between considered pixels and neurons in a single first layer population. Thus, neurons from 1 to 320 will get input from the first ganglion circuit and forward spikes to neurons from 641 to 960, and neurons from 321 to 640 will get input from the second ganglion circuit and forward spikes to neurons from 961 to 1280.

The summation occurring in the second layer, together with linearly decreasing synapse weights from the center of the image to the periphery, serves as a filter for avoiding undesired spikes in peripheral regions of sight.

4.3   Transfer Functions

The robot controller is implemented by means of two transfer functions (TFs), one robot to neuron and one neuron to robot.

In the robot to neuron TF, the image on the eye camera of the robot is collected and forwarded to the retina library, updating its status. Outcome from the ganglion layers of the retina circuit are processed, then current values for a strip of pixel containing the target are transmitted to the two populations in the first layer of the brain.

Output spikes from the second brain layer are then processed by the second TF with the following steps:

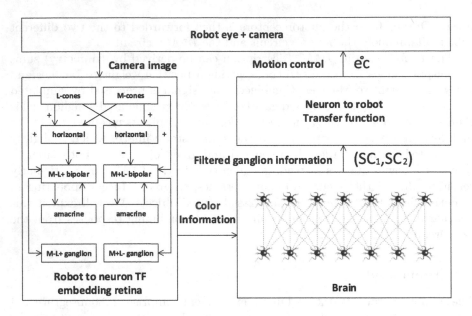

Fig. 3. Block diagram of the implemented robot controller. (Neuron image from Amelia Wattenberger, released under Creative Commons License 3.0 https://thenounproject. com/term/neuron/79860/)

- Spike counts for every neuron on the two output population are gathered in two collections SC_1 and SC_2, that we represent as two functions

$$SC_1, SC_2 : [1, 320] \to \mathbb{N}$$

where $[1, 320]$ is a discrete set, representing a neural population.
- As higher stimulation of ganglions correspond to higher spike frequency in the brain, we expect one clearly distinguished frequency maximum per population, the first one corresponding to the pixel where the ball enters the receptive field in the first ganglion layer and the other to the pixel where the ball leaves the receptive field in the second ganglion layer. Since we are using one neuron per pixel on an horizontal stripe of pixels, the two maxima will correspond to two column indexes of the original image, which we will call p_1 and p_2. For each population we obtain the said column index by computing first the maximum neural spike rate, then taking the first neuron index with that spike rate. In formulas, p_1 and p_2 are defined as

$$p_1 = \min\{\arg\max_{x}\{SC_1(x)\}\}$$

$$p_2 = \min\{\arg\max_{x}\{SC_2(x)\}\}$$

- We take as an estimate of the target center the value

$$\bar{p} = \frac{p_1 + p_2}{2}$$

– The estimated position of the ball $\bar{p}$ is then converted to an estimated angle $\mathring{p}$ with respect to the eye center with the equation

$$\mathring{p} = -\arctan\left(\frac{\bar{p} - 160}{160}\right)$$

– The eye position angle command $\mathring{e}c_{t+1}$, depending on the current eye position $\mathring{e}_t$ is computed as follows

$$\mathring{e}c_{t+1} = \mathring{e}_t + 0.2\mathring{p}$$

where the constant 0.2 is determined empirically and prevents quick eye movements that could result in the target loss from the robot sight.

5   Results

5.1   Target Detection

In this experiment the controller has been validated by testing only its target detection capabilities without any actuation on the robot. The experiment was

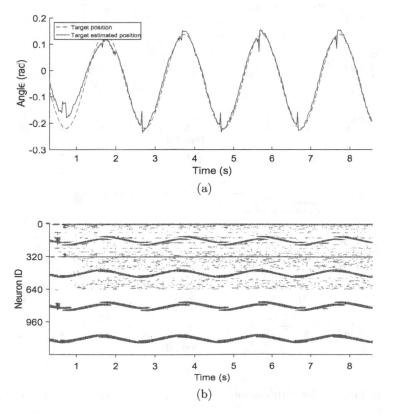

(a)

(b)

Fig. 4. The computed ball position (a) and the brain response (b) for a ball with sinusoidal motion moving at 0.5 Hz, robot eyes still. (Color figure online)

started with a slightly different neuron to robot transfer function which didn't send any control message back to the robot and with the target ball moving with a sinusoidal trajectory with a cycle frequency of 0.5 Hz, covering almost half the field of vision of the camera. During the experiment, data about both target perceived location and the neural network response was collected and it is reported in Fig. 4.

At the beginning of the simulation, the robot suddenly moves to a default position, while the controller responsible for the upright position starts. For this reason, the camera image during the first 2 seconds of simulation may be affected by this movement and thus its elaboration may provide wrong results. In Fig. 4a we can observe how the estimated target position, after the robot stabilization, follows a sinusoidal trend. The target position is plotted with a dashed line as a reference. Sensitivity of different ganglion inputs can be noticed in the first half of the spike plot of Fig. 4b, where the first population spikes in correspondence to the target entering the receptive field and the second spikes when the target

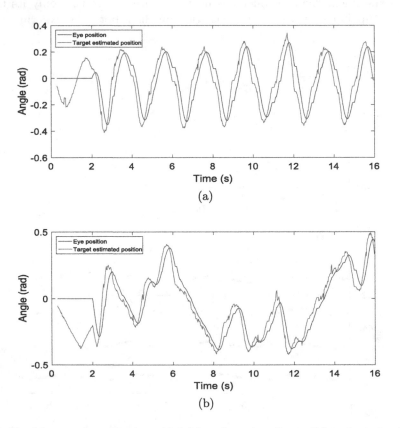

Fig. 5. Tracking results with sinusoidal (a) and random linear (b) trajectories (Color figure online)

leaves the receptive field. The filtering action of the second neural layer can be observed in the second half of the spike plot.

5.2 Target Pursuit

In these experiments we used the controller enabling motion commands with the same setup of the previous experiment. We tested the controller with a target moving with a sinusoidal trajectory (Fig. 5a) and with a random linear trajectory (Fig. 5b), namely a linear trajectory changing direction (left or right) randomly every second. For the same reasons explained in the previous test, we wait for the robot stabilization before starting actuating the eye. It can be observed how the target estimation is still effective even with a moving eye, which validates our choice of a retina simulator as a data source for the controller.

Comparing our approach with a purely image-processing based target detection [23], we can easily observe a greater estimate error on our controller, that can be noticed easily in Fig. 5b at seconds 5, 10 and 12. A one to one comparison, though, is imprecise, as in the image-processing approach all the pixels of the camera image are processed by the controller, whereas in our implementation we consider just one stripe of 320 pixels.

6 Discussion and Future Works

In this paper we propose a retinal red-green opponency based robot controller, processing image from the outside world through a retina model instead of using classic image processing methods. We set up a framework by integrating an existing retina framework in the NRP, implementing a custom retina circuit and a custom neural network and finally setting up an experiment in the NRP with suitable transfer functions. Two experiments allowed us to validate the effectiveness of this setup for both detection and pursuit of a moving green target on a red background, although with the limitations of a single retina pathway and a single image stripe processed by the brain model. This work represents a first attempt towards a bio-inspired visual pursuit controller embedding a retina model. In future works, we plan to extend the "field of view" of the controller to the whole image, after a careful design of the pursuit controller in order to simulate a brain function areas involved in the smooth pursuit eye movement through spiking neural networks. We will improve our retina circuit to simulate also space-variant resolution and try to model and combine more retinal pathways concurring in color detection or motion detection. In order to interface this controller model with a real robotic platform, the retina model could be interfaced with real-time neural simulations such a the ones provided by neuromorphic hardware like SpiNNaker [24].

Acknowledgements. The research leading to these results has received funding from the European Union Seventh Framework Programme (FP7/2007-2013) under grant agreement no. 604102 (Human Brain Project). The authors would like to thank the

Italian Ministry of Foreign Affairs, General Directorate for the Promotion of the "Country System", Bilateral and Multilateral Scientific and Technological Cooperation Unit, for the support through the Joint Laboratory on Biorobotics Engineering project.

References

1. Dacey, D.M.: Primate retina: cell types, circuits and color opponency. Prog. Retinal Eye Res. **18**(6), 737–763 (1999)
2. Baylor, D., Nunn, B., Schnapf, J.: Spectral sensitivity of cones of the monkey Macaca fascicularis. J. Physiol. **390**, 145 (1987)
3. Dacey, D.M., Packer, O.S.: Colour coding in the primate retina: diverse cell types and cone-specific circuitry. Curr. Opin. Neurobiol. **13**(4), 421–427 (2003)
4. Shibata, T., Vijayakumar, S., Conradt, J., Schaal, S.: Biomimetic oculomotor control. Adapt. Behav. **9**(3–4), 189–207 (2001)
5. Falotico, E., Zambrano, D., Muscolo, G., Marazzato, L., Dario, P., Laschi, C.: Implementation of a bio-inspired visual tracking model on the icub robot. In: Proceedings of IEEE International Workshop on Robot and Human Interactive Communication, pp. 564–569 (2010)
6. Vannucci, L., Cauli, N., Falotico, E., Bernardino, A., Laschi, C.: Adaptive visual pursuit involving eye-head coordination and prediction of the target motion. In: IEEE-RAS International Conference on Humanoid Robots, pp. 541–546 (2014)
7. Vannucci, L., Falotico, E., Di Lecce, N., Dario, P., Laschi, C.: Integrating feedback and predictive control in a bio-inspired model of visual pursuit implemented on a humanoid robot. In: Wilson, S.P., Verschure, P.F.M.J., Mura, A., Prescott, T.J. (eds.) Living Machines 2015. LNCS (LNAI), vol. 9222, pp. 256–267. Springer, Heidelberg (2015)
8. Zambrano, D., Falotico, E., Manfredi, L., Laschi, C.: A model of the smooth pursuit eye movement with prediction and learning. Appl. Bionics Biomech. **7**(2), 109–118 (2010)
9. Falotico, E., Taiana, M., Zambrano, D., Bernardino, A., Santos-Victor, J., Dario, P., Laschi, C.: Predictive tracking across occlusions in the icub robot. In: 9th IEEE-RAS International Conference on Humanoid Robots, HUMANOIDS 2009, pp. 486–491 (2009)
10. Benoit, A., Caplier, A., Durette, B., Hérault, J.: Using human visual system modeling for bio-inspired low level image processing. Comput. Vis. Image Underst. **114**(7), 758–773 (2010)
11. Wohrer, A., Kornprobst, P.: Virtual retina: a biological retina model and simulator, with contrast gain control. J. Comput. Neurosci. **26**(2), 219–249 (2009)
12. Hérault, J., Durette, B.: Modeling visual perception for image processing. In: Sandoval, F., Prieto, A.G., Cabestany, J., Graña, M. (eds.) IWANN 2007. LNCS, vol. 4507, pp. 662–675. Springer, Heidelberg (2007)
13. Morillas, C.A., Romero, S.F., Martínez, A., Pelayo, F.J., Ros, E., Fernández, E.: A design framework to model retinas. Biosystems **87**(2), 156–163 (2007)
14. Martínez-Cañada, P., Morillas, C., Pino, B., Ros, E., Pelayo, F.: A computational framework for realistic retina modeling. Int. J. Neural Syst. (Accepted for publication)
15. Martínez-Cañada, P., Morillas, C., Nieves, J.L., Pino, B., Pelayo, F.: First stage of a human visual system simulator: the retina. In: Trémeau, A., Schettini, R., Tominaga, S. (eds.) CCIW 2015. LNCS, vol. 9016, pp. 118–127. Springer, Heidelberg (2015)

16. Hines, M.L., Carnevale, N.T.: The NEURON simulation environment. Neural Comput. **9**(6), 1179–1209 (1997)
17. Gewaltig, M.O., Diesmann, M.: NEST (NEural Simulation Tool). Scholarpedia **2**(4), 1430 (2007)
18. Koenig, N., Howard, A.: Design and use paradigms for gazebo, an open-source multi-robot simulator. In: Proceedings of the IEEE/RSJ International Conference on Intelligent Robots and Systems (IROS 2004), vol. 3, pp. 2149–2154. IEEE (2004)
19. Metta, G., Natale, L., Nori, F., Sandini, G., Vernon, D., Fadiga, L., Von Hofsten, C., Rosander, K., Lopes, M., Santos-Victor, J., et al.: The iCub humanoid robot: an open-systems platform for research in cognitive development. Neural Netw. **23**(8), 1125–1134 (2010)
20. Quigley, M., Conley, K., Gerkey, B., Faust, J., Foote, T., Leibs, J., Wheeler, R., Ng, A.Y.: ROS: an open-source robot operating system. In: ICRA Workshop on Open Source Software, vol. 3, p. 5 (2009)
21. Hinkel, G., Groenda, H., Vannucci, L., Denninger, O., Cauli, N., Ulbrich, S.: A domain-specific language (DSL) for integrating neuronal networks in robot control. In: ACM International Conference Proceeding Series, pp. 9–15, 21 July 2015
22. Brette, R., Gerstner, W.: Adaptive exponential integrate-and-fire model as an effective description of neuronal activity. J. Neurophysiol. **94**(5), 3637–3642 (2005)
23. Vannucci, L., Ambrosano, A., Cauli, N., Albanese, U., Falotico, E., Ulbrich, S., Pfotzer, L., Hinkel, G., Denninger, O., Peppicelli, D., Guyot, L., Von Arnim, A., Deser, S., Maier, P., Dillman, R., Klinker, G., Levi, P., Knoll, A., Gewaltig, M.O., Laschi, C.: A visual tracking model implemented on the iCub robot as a use case for a novel neurorobotic toolkit integrating brain and physics simulation. In: IEEE-RAS International Conference on Humanoid Robots, pp. 1179–1184 (2015)
24. Painkras, E., Plana, L.A., Garside, J., Temple, S., Galluppi, F., Patterson, C., Lester, D.R., Brown, A.D., Furber, S.B.: SpiNNaker: A 1-w 18-core system-on-chip for massively-parallel neural network simulation. IEEE J. Solid-State Circuits **48**(8), 1943–1953 (2013)

A Two-Fingered Anthropomorphic Robotic Hand with Contact-Aided Cross Four-Bar Mechanisms as Finger Joints

Guochao Bai, Jieyu Wang, and Xianwen Kong[✉]

Heriot-Watt University, Edinburgh EH14 4AS, UK
{gb9,jw26,X.Kong}@hw.ac.uk

Abstract. This paper presents an anthropomorphic design of a robotic finger with contact-aided cross four-bar (CFB) linkages. Anatomical study shows that finger joints have a complex structure formed by non-symmetric surfaces and usually produce complex movement than a simple revolute motion. The articular system of human hand is firstly investigated. Kinematics of a CFB mechanism is then analyzed and computer aided design of fixed and moving centrodes of CFB mechanism is presented. Gripping analysis of human hand shows two easily ignored components of a finger, fingernail and soft fingertip. Based on the range of motion of the joints of the most flexible thumb finger, a two-joint anthropomorphic finger is developed by using contact-aided CFB linkages which can also be used for joint design of prosthetic knee. Prototype of a two-fingered hand is manufactured by using 3D printing technology and gripping of a wide range of objects is tested.

Keywords: Robotic finger · Anthropomorphic · Contact-aided · Cross four-bar

1 Introduction

It is known that investigating anthropomorphic robotic hands may help understand human hand. The grasping and manipulation abilities of current robotic hand are far less dexterous than the human hand, even though significant progresses have been made in the past four decades [1–3]. The design considerations include number of fingers, joints, DOF (degree-of-freedom), range of motion for each joint, speed of movements, and force generation capability [4]. Design of a versatile and robust robotic hand that has the similar dexterity as human hand is an undisputedly challenging task.

A finger of the human hand possesses several biological features which are hard to mimic simultaneously [5]. They include: (1) Unique shape of bones at the joints to determine DOF of joints; (2) Joint capsule formed by ligaments to limit the range of motion of joint; (3) Cartilage and synovial fluid for low-friction contact between bone surfaces; (4) Non-linear interactions between tendons and bones to dynamically determine the motion of finger.

In a human hand, the anatomical structure and nervous system have significant contribution to its ability. Anatomical study shows that finger joints have a complex

© Springer International Publishing Switzerland 2016
N.F. Lepora et al. (Eds.): Living Machines 2016, LNAI 9793, pp. 28–39, 2016.
DOI: 10.1007/978-3-319-42417-0_3

structure formed by non-symmetric surfaces and usually produce complex movement than a simple revolute motion.

Most of the existing robotic hands are connected with revolute joints between phalanges in order to achieve the required DOF and kinematic characteristics, such as hinges, gimbals, cables, gears or belts. Compliant materials are recently used as joints. Xu proposed one type of MCP joint whose biomechanics and dynamic properties are close to human counterparts [6]. The artificial joint is composed of a ball joint, crocheted ligaments, and a silicon rubber sleeve which is distinctive to the other finger joints.

Many robotic hands are driven by gears [7, 8] or tendon [9]. There are also smart motor [10] and air muscle [11] types. The function of these drive approaches is to make the trajectory of a fingertip similar to the typical human trajectory during reaching and grasping objects.

In this paper, a novel type of finger is proposed (see Fig. 1). Unlike most of the current finger design, fingernail and soft fingertip are considered for specific gripping. Firstly, the articular system of human hand including bones and joints is presented. Physical parameters of finger skeleton and approximate limits of joints' motion are summarized. Secondly, kinematics of cross four-bar (CFB) linkage is analyzed and then a contact-aided CFB mechanism is generated for the design of finger joint and knee. Thirdly, based on the limits of joints' motion referred in Sect. 2, a two-joint thumb finger is designed and the mechanism is analyzed. Then the developing process is summarized and design approach is proposed. Finally, a prototype of artificial finger considering function of fingernail and soft fingertip is fabricated using 3D printing and some grasping tests of finger are conducted to verify the performance of the finger.

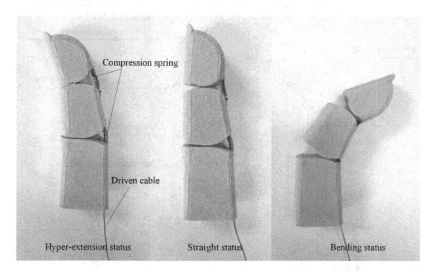

Fig. 1. Contact-aided joints and motion of a robotic finger.

2   Articular System of Human Hand

In order to mimic human hand for grasping objects, articular system of the human hand is investigated. The accurate models of the human finger have been proposed based on anatomical studies [12, 13]. The bone and articular structure of human hand are two key components to be imitated because the structures and their relative movements have essential effects on grasping objects. By analyzing and learning the articular system, similar mechanisms can be chosen. Human bones and articulations are introduced briefly. DOF and range of motion of each joint are summarized in literature [14, 15].

We use the subscripts I for the thumb, II for the index, III for the middle, IV four the ring, and V for the small fingers. Three types of hand bones, carpals, metacarpals and phalanges are shown in Fig. 2. The names of joints are nominated according to hand bones, such as carpalmetacarpal joint (CMC, for short) which locates between carpal and metacarpal. Hand fingers are separated into proximal, middle and distal phalanges, excepting thumb finger which has only two phalanges.

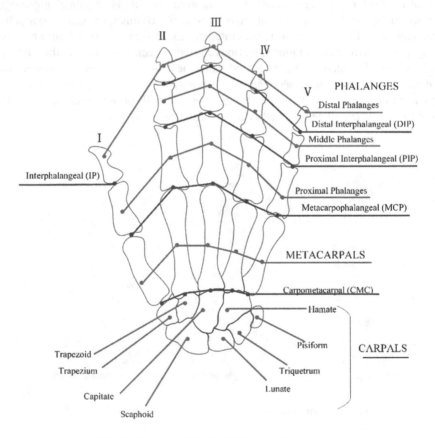

Fig. 2. Bones and joints (left hand anterior view).

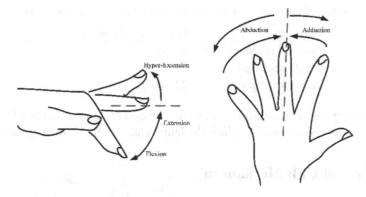

Fig. 3. Flexion/extension, abduction/adduction of middle finger [16].

The definition of movements of each finger in terms of extension/flexion, abduction/ adduction and hyper-extension which means the finger over pass the dotted line, shown in Fig. 3.

Fingers play an important role in grasping and manipulation. Research on movements of human fingers gives us targets to mimic. Tables 1 and 2 show the range of motion of joints and lengths of phalanges of the fingers.

Table 1. Range of movements of the finger joints(H refers to hyper extension) [16].

Fingers	Joints	Action	Ranges(in degree)
Thumb	CMC	Adduction/Abduction	0(contact)/60
		Extension/Flexion	25/35
	MCP	Adduction/Abduction	0/60
		Extension/Flexion	10H/55
	IP	Extension/Flexion	15H/80
Index	MCP	Adduction/Abduction	13/42
		Extension/Flexion	0/80
	PIP	Extension/Flexion	0/100
	DIP	Extension/Flexion	10H/90
Middle	MCP	Adduction/Abduction	8/35
		Extension/Flexion	0/80
	PIP	Extension/Flexion	0/100
	DIP	Extension/Flexion	10H/90
Ring	MCP	Adduction/Abduction	14/20
		Extension/Flexion	0/80
	PIP	Extension/Flexion	0/100
	DIP	Extension/Flexion	20H/90
Small	MCP	Adduction/Abduction	19/33
		Extension/Flexion	0/80
	PIP	Extension/Flexion	0/100
	DIP	Extension/Flexion	30H/90

Table 2. Length range of physical parameters of the finger skeleton.

Phalanges	Length(in millimeter)
Proximal phalanges	38–55
Middle phalanges	24–35
Distal phalanges	22–30

By analyzing the contained structure and movements of fingers, we intend to replace them with similar mechanisms to mimic the human hand in the following section.

3   Analysis of CFB Mechanism

Planar four-bar mechanism, which is widely used in many applications, is efficient at transferring motion and power. A planar four-bar linkage is composed of four revolute joints with parallel axes. Some biomechanics researchers have used four-bar linkage in the design of human body. For example, a four-bar linkage system is presented in [17] to replicate the polycentric motion of the knee that occurs during passive knee flexion-extension. A CFB mechanism is proposed in [18] for the knee design of bipedal robot. Applications on robotic hands for motion imitation also exists in [19–21]. In this paper, centrodes of the CFB mechanism are investigated and a contact aided CFB mechanism is proposed to mimic the complex rolling characteristics of finger joints. This design has an overconstrained characteristic to increase the structure stiffness.

3.1   Kinematics of CFB Mechanism

Centrode, as an important characteristic in planar kinematics, is the path traced by an instantaneous center of rotation of a rigid link moving in a plane [22]. The motion of the coupler link with respect to the ground link is pure rotation around the instantaneous center. According to Kennedy-Aronhold Theorem, the centrode can be found at the intersection of the crank link and follower link in a CFB mechanism, which means that CFB mechanism always has centrodes between the coupler link and the ground link. The fixed centrode can be found by drawing the trajectory of the intersection of crank link and follower link. The process of drawing the two trajectories of centrodes is shown in Fig. 4. The trajectories depend on lengths of the associated four links.

Partial trajectories of centrodes with the rotational angle of the crank link equaling to 35° are shown in Fig. 4. Random angles are available by using this approach. The motion of the coupler link with respect to the ground link is duplicated by making these two centrodes roll against one another without slipping. Because of the pure rolling of the two curves, they have the same lengths.

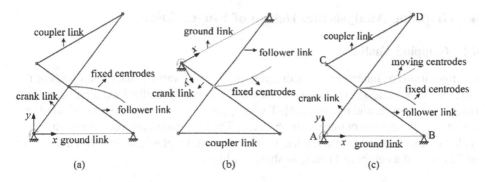

Fig. 4. CFB mechanism and trajectories of its centrodes. (a) Produced trajectory of fixed centrodes. (b) Produced trajectory of moving centrodes. (c) Trajectories integrated in one figure.

3.2   Contact-Aided CFB Mechanism

Due to the characteristics of the centrodes, we can design a contact-aided CFB mechanism which has a much higher stiffness than the CFM mechanism, by adding a high kinematic pair between the coupler and the ground link.

Contact-aided mechanism was first introduced in [23]. Cannon and Howell proposed a novel design of compliant rolling-contact element capable of performing the functions of bearing and a spring [24]. The use of contact surfaces is to enhance the functionality of compliant mechanism to be capable of performing certain kinematic tasks similar to rigid body. Contact surfaces integrating with rigid links as an overconstrained mechanism have certain kinematics and limited motion range. As shown in Fig. 5, a contact-aided CFB mechanism with limited motion range of 90°. An application on artificial joints will be introduced in Sect. 5.1.

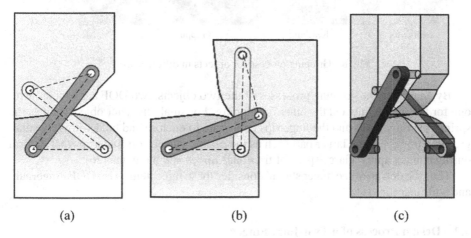

Fig. 5. Contact-aided CFB mechanism. (a) Initial position. (b) Final position. (c) Side view of the assembly.

4   Gripping Analysis and Design of Finger Joints

4.1   Gripping Analysis

Configurations of human hand are different in gripping variant shapes and sizes of the objects. Nevertheless, gripping processes are similar, typically including searching, reaching, gripping, and moving [25]. Two gripping examples are now considered and various configurations of fingers are observed. The two objects are a compression spring with a diameter of 2.5 mm and a length of 15 mm and a plastic wheel with a diameter of 22 mm and a height of 11 mm, as shown in Fig. 6.

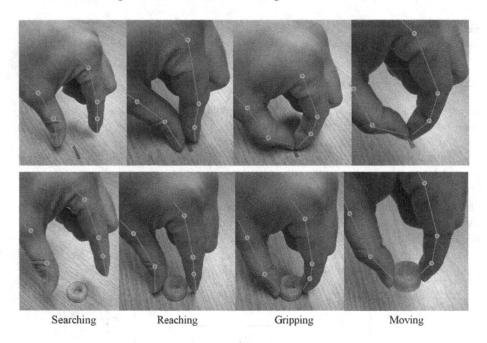

Searching   Reaching   Gripping   Moving

Fig. 6. Gripping process for objects in different sizes.

By analyzing the gripping processes of the two objects, two DOF are adopted with one for force applying and the other for gipping. For small diameter object such as the spring, the fingernails and the fingertips cooperate to enclose and hold with very small gripping force. For the larger part such as the plastic wheel, soft fingertips that deform during contact apply a large space of frictional forces and moments [26].

Thus the design of the finger should consider these three items: two DOF, fingernail, and soft fingertip.

4.2   Design Process of a Two-Joint Finger

Taking the thumb finger as an example, the MCP and IP joints of this finger are considered for design. The action ranges of extension/flexion as to MCP and IP joints are 10H/55 and 15H/80 respectively as shown in Fig. 7. Thumb finger is the only one which has hyper extension at MCP joint which means that it is the most flexible one in five fingers.

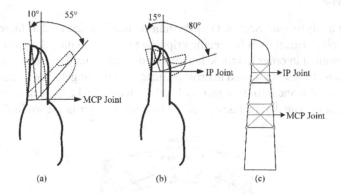

Fig. 7. Extension/flexion ranges of two joints of thumb finger and equivalent CFB mechanisms of the joints.

In order to mimic the ranges of motion of the two joints, two CFB linkages should be used with one for mimicking IP joint and the other for MCP joint.

According to the approach proposed in Sect. 3, the extension/flexion ranges of these two CFB mechanisms determine the range of motion of the mechanisms, thus the fixed and moving centrodes of the mechanisms can be calcualted. Meanwhile, two-joint finger will have a fixed motion range, as shown in Fig. 8.

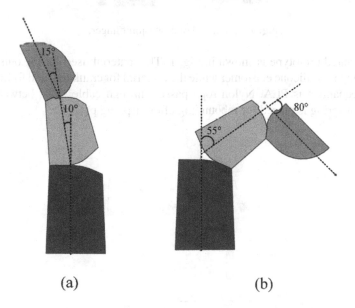

Fig. 8. Side view of two-joint finger connecting with fixed and moving centrodes.

5   Prototype and Testing

5.1   Prototype

The gripping analysis (see Sect. 4.1) of the human hand indicates that fingernails and soft fingertip play critical roles in specific gripping tasks. A 3D model considering these details are designed in order to mimic human hand's structure. Two joints of the finger are decoupled and a tendon is used to drive phalanges of the finger. The finger is divided by its symmetrical plane into front and back halves which make the manufacturing and assembly efficient. The detailed design of the two-joint finger is shown in Fig. 9.

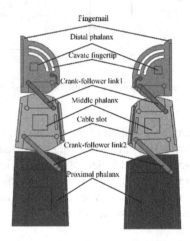

Fig. 9.   3D model of a two-joint finger.

A 3D printed prototype is shown in Fig. 1. The material used for the outside layer of the phalanges is silicone elastomer while the material for crank link and follower links of CFB mechanism is PLA. Nylon rope passes through cable slots, between which compression spring is placed for rebounding after gripping process.

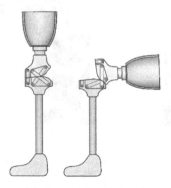

Fig. 10.   Prosthetic leg using contact aided CFB linkage.

Based on the above approach and design process, a prosthetic leg using a contact-aided CFB linkage is developed with the bending angle of 90° as shown in Fig. 10.

5.2 Testing

A two-fingered hand is developed for manually gripping test as shown in Fig. 11. The differential palm contains a movable pulley which drives the two fingers with underactuation. The objects used for grasping are of a wide range, including plastic cup, tiny spring and pinecone, etc. as shown in Fig. 12. Fingernail plays a significant role for gripping tiny spring. The cooperation of fingernail and fingertip ensures a successful gripping of a ball pen. The soft fingertips provide a large friction force and moment for

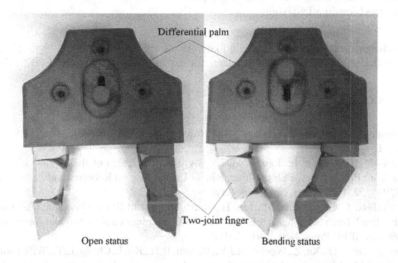

Fig. 11. Two-fingered hand at open and bending statuses.

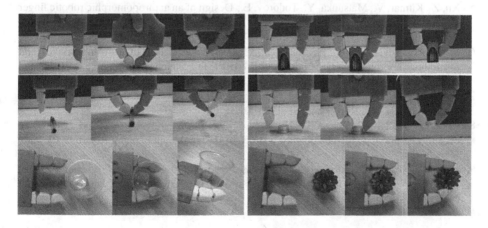

Fig. 12. Gripping testing on a variety of objects.

the cuboid-shaped battery. The differential drive contributes to the irregular non-centered pinecone gripping. The testing on these universal objects verified the feasibility of the design.

6   Conclusions and Discussions

The kinematics of contact-aided CFB mechanism was analyzed and a design approach to mimic human finger joints was proposed. The gripping process carried out by human hand was analyzed and the structures of fingernail, soft fingertip were designed for prototyping. A design example of two-joint thumb finger was conducted by using the proposed design approach. The prototype was manufactured by 3D printing technology with specific materials. The testing of gripping of a variety of objects verified the feasibility of the biomimetic design.

Acknowledgments. The authors would like to thank the Engineering and Physical Sciences Research Council (EPSRC), United Kingdom, for the support under grant No EP/K018345/1.

References

1. Salisbury, J., Craig, J.: Articulated hands: force control and kinematic issues. Int. J. Robot. Res. **1**, 4–17 (1982)
2. Jacobsen, S., Iversen, E., Johnson, D.R., Biggers, K.: Design of the Utah/M.I.T. dextrous hand. In: Proceedings IEEE the International Conference on Robotics and Automation, pp. 1520–1532. IEEE Press, New York (1986)
3. Butterfass, J., Grebenstein, M., Liu, H., et al.: DLR-hand II: next generation of a dextrous robot hand. In: Proceedings IEEE International Conference on Robotics and Automation, pp. 109–114. IEEE Press, New York (2001)
4. Deshpande, A.D., Xu, Z., Weghe, M.J.V., Brown, B.H., Ko, J., Chang, L.Y., Wilkinson, D.D., Bidic, S.M., Matsuoka, Y.: Mechanisms of the anatomically correct testbed hand. IEEE/ASME Trans. Mechatronics. **18**, 238–250 (2013)
5. Xu, Z., Kumar, V., Matsuoka, Y., Todorov, E.: Design of an anthropomorphic robotic finger system with biomimetic artificial joints. In: 4th IEEE RAS & EMBS International Conference, pp. 568–574. IEEE Press, New York (2012)
6. Xu, Z., Todorov, E., Dellon, B., Matsuoka, Y.: Design and analysis of an artificial finger joint for anthropomorphic robotic hands. In: 2011 IEEE International Conference Robotics and Automation (ICRA), pp. 5096–5102. IEEE Press, New York (2011)
7. Townsend, W.: The barrett hand grasper—programmably flexible part handling and assemble. Int. J. Ind. Robot. **27**, 181–188 (2000)
8. Kyberd, P.J., Light, C., Chappell, P.H., Nightingale, J.M., Whatley, D., Evans, M.: The design of anthropomorphic prosthetic hands: a study of the southampton hand. Robotica **19**, 593–600 (2001)
9. Liu, H.: DSP and FPGA-based joint impedance controller for DLR/HIT II dexterous robot hand. In: Proceedings IEEE/ASME International Conference on Advanced Intelligent Mechatronics, pp. 1594–1599 (2009)

10. Yamano, I., Maeno, T.: Five-fingered robot hand using ultrasonic motors and elastic elements. In: Proceedings IEEE International Conference on Robotics and Automation, pp. 2673–2678 (2005)
11. Shadow robot company. http://www.shadowrobot.com/hand/
12. Buchner, H., Hines, M., Hemami, H.: A dynamic model for finger interphalangeal coordination. J. Biomech. **21**, 459–468 (1988)
13. Dennerlein, J., Diao, E., Mote Jr., C.: Tensions of the flexor digitorum superficialis are higher than a current model predicts. J. Biomech. **31**, 295–301 (1998)
14. Sancho-Bru, J., Perez-Gonzalez, A., Vergara-Monedero, M.: A 3-D dynamic model of human finger for studying free movements. J. Biomech. **34**, 1491–1500 (2001)
15. Buchcholz, B., Armstrong, T., Goldstein, S.: Anthropometric data for describing the kinematics of the human hand. Ergonomics **35**, 261–273 (1992)
16. Peña, E., Yang, J., Abenoza, M., Tico, N., Abedel-Malek, K.: Virtual human hand: autonomous grasping strategy. In: Proceedings of the 17th IASTED International Conference on Applied Simulation and Modelling, pp. 609–618. Felice, F. de, Corfu (2008)
17. Goodfellow, J., O'Connor, J.: The Knee, W. Norman Scott Mosby-Year Book, Inc. (1994)
18. Hamon, A., Aoustin, Y.: Cross four-bar linkage for the knees of a planar bipedal robot. In: 10th IEEE-RAS International Conference Humanoid Robots (Humanoids), pp. 379–384. IEEE Press, New York (2010)
19. Birglen, L., Gosselin, C.M.: Geometric design of three-phalanx underactuated fingers. J. Mech. Design. **128**, 356–364 (2006)
20. Wu, L., Carbone, G., Ceccarelli, M.: Designing an underactuated mechanism for a 1 active DOF finger operation. Mech. Mach. Theory **44**, 336–348 (2009)
21. Moon, Y.M.: Bio-mimetic design of finger mechanism with contact aided compliant mechanism. Mech. Mach. Theory **42**, 600–611 (2007)
22. Eckhardt, H.D.: Kinematic Design of Machines and Mechanisms, pp. 249–283. McGraw-Hill, New York (1998)
23. Mankame, N.D., Ananthasuresh, G.K.: Contact aided compliant mechanisms: concept and preliminaries. In: ASME 2002 International Design Engineering Technical Conferences and Computers and Information in Engineering Conference. American Society of Mechanical Engineers, pp. 109–121 (2002)
24. Cannon, J.R., Lusk, C.P., Howell, L.L.: Compliant rolling-contact element mechanisms. In: ASME 2005 International Design Engineering Technical Conferences and Computers and Information in Engineering Conference, pp. 3–13. American Society of Mechanical Engineers (2005)
25. Niebel, B. W.: Motion and Time Study, pp. 197–205. Richard D Irwin (1993)
26. Akhtar Khurshid, A.G., Malik, M.A.: Robotic Grasping and Fine Manipulation Using Soft Fingertip. Adv Mechatronics, pp. 155–174 (2011)

Living Designs

Rina Bernabei[1]([✉]) and Jacqueline Power[2]

[1] University of New South Wales, Sydney, Australia
R.Bernabei@unsw.edu.au
[2] University of Tasmania, Launceston, Australia
Jacqueline.Power@utas.edu.au

Abstract. This paper will outline why product designers are exploring making processes of the natural world and how this is of benefit to traditional product design practice. This type of experimental design work pushes the boundaries of conventional product design in which mass-manufacture efficiency drives the design and production process. Products, which use growth processes as fundamental to the making process, are increasingly becoming more feasible for end-user acquisition. This paper will provide two case study examples. These case studies contextualise these products and how they co-exist and contribute to the well-established design approaches of digital fabrication and co-creation.

Keywords: Biodesign · Object design · Material technology · Emotional design

1 Introduction

Product designers are increasingly exploring the opportunities afforded by using living and artificial systems in material technologies and manufacture-related processes. This type of experimental work pushes the boundaries of conventional product design, in which mass-manufacture efficiency drives the design and production process. This paper provides two case study examples of bio-design - one is a product, a chair, and one a piece of jewelry, a ring. These designed artefacts demonstrate the use of natural growth processes for the creation of a finished product or artefact.

This paper questions how the emerging field of bio-design is of benefit to traditional design practice, particularly product design. This discourse rests on a rich foundation of design theory that has been developed since modernism. Bio-design crosses over and intersects some current paradigms in very interesting and innovative ways. The authors, researchers in the field of both product design and interior design, see the potential of bio-design to co-exist and contribute to a series of well-established design approaches including digital fabrication and co-creation approaches.

2 Defining Bio-design

Within the design disciplines, the term bio-design is used in reference to designed works that incorporate living organisms. The term bio-design is a broad one, capturing in its use of the word 'design' many disciplines including architecture, fashion, and products.

© Springer International Publishing Switzerland 2016
N.F. Lepora et al. (Eds.): Living Machines 2016, LNAI 9793, pp. 40–47, 2016.
DOI: 10.1007/978-3-319-42417-0_4

A clarification of the prefix 'bio-' is required. William Myers, author of the 2012 publication BioDesign: nature + science + creativity and curator of the exhibition BioDesign: on the cross pollination of nature, science and creativity, explains that:

> "Biodesign goes further than other biology-inspired approaches to design and fabrication. Unlike biomimicry, cradle to cradle, and the popular but frustratingly vague 'green design', biodesign refers specifically to the incorporation of living organisms as essential components, enhancing the function of the finished work. It goes beyond mimicry to integration, dissolving boundaries and synthesizing new hybrid typologies. The label is also used to highlight experiments that replace industrial or mechanical systems with biological processes." [1]

Such a definition enables many design approaches to fall under the umbrella of bio-design. Rather than explore products that incorporate living organisms to varying degrees and for various purposes, this paper seeks to instead consider products or artefacts that use actual growth processes as fundamental to the making process. In other words, without the biological processes, there would be no tangible or useable product or artefact – the product is quite literally grown. This requires a rethink of traditional product manufacturing systems.

Bio-design products and artefacts remain in their infancy, many as conceptual works or prototypes rather than available for user acquisition. The two pieces selected as case study examples below are either one-off prototypes or concept designs. [2] The case study examples will not be positioned to engage with ethical dilemmas that arise with such modes of production and the utilisation of biological processes. This is however, an important aspect of bio-design that demands in the future greater attention in the literature. Design and art historian Christina Cogdell has commented that, "it appears that none of the architects or designers who uphold their bioart as a prototype for biodesign has seriously engaged with their critical intent or with the ethical problems". [3] This is not to say that designers have not considered these issues as part of their practice but may have instead not articulated them, focusing on other facets of the bio-design process, such as material application, aesthetic values and collaborative processes.

3   Bio-design and Digital Fabrication

Digital fabrication is a well-established product design manufacturing practice and new making opportunities are presented when coupled with bio-design processes. According to computer scientist Albrecht Schmidt "Historically, the separation of the physical and digital worlds has been very clear, but digital fabrication is blurring that dividing line." [4] In design, digital fabrication is definable as:

> "a process whereby an object design is created on a computer, and the object is then automatically produced by a machine. Digital fabrication machines can be roughly sorted into two categories: subtractive and additive. Subtractive approaches use drill bits, blades or lasers to remove material from an original material source, thus shaping the three-dimensional object. Additive processes deposit progressive layers of a material until a desired shape is achieved." [5]

Digital fabrication includes technologies such as 3D printing, laser cutting and computer-numerically controlled (CNC) routing. Additive digital fabrication processes such as 3D printing, is relatively new. Like bio-design, its integration with and overlap with existing design paradigms, is still being explored.

Eric Klareenbeek's MyceliumChair (Fig. 3) is a bio-design example integrating both biological elements and digital fabrication technologies - "a chair in which 3D printing and growing material are combined." [6] Klareenbeek collaborated with a number of scientists on the project. [7] The final chair, exhibited at Dutch Design Week in 2013, has been printed from a mixture of materials, including water, straw, biodegradable plastic and mycelium, which Klareenbeek describes as "the threadlike network in fungi". [8] The mycelium itself grows inside the chair, giving it strength - the chair is effectively alive.

Fig. 1. Studio Eric Klarenbeek's, *MyceliumChair* complete with living oyster mushrooms. http://www.ericklarenbeek.com/

Although the growth of the mycelium within the chair is not visible, remarkably the chair supports a colony of oyster mushrooms actually grown from the chair (Fig. 3). Unlike the presence of the mycelium, which provides strength to the chair, the oyster mushrooms themselves are purely decorative. [9] With the mushrooms in full bloom, their continued growth was halted by drying the structure, and coating it in biodegradable plastic. [10] The materials used in the project have a raft of implications for production. For instance, Eric Klareenbeek has said:

"its all living material; meaning its growing and multiplying, so if you treat the organism well, you'll never run out of 'glue'. In other words, it's theoretically inexhaustible!!"

Fig. 2. Digital fabrication of Studio Eric Klarenbeek's, *myceliumchair.* http://www.ericklarenbeek.com/

Bio-design when used with digital fabrication, such as 3D printing, presents diverse possibilities for material use and its assembly. In the case of the MyceliumChair, 3D printing facilitated the application of a new composite material that could generate a desired complex form (Fig. 4). Digital fabrication technologies enable designers to experiment with new materials and with increasing home-based platforms, have the capacity in the future to extend the making into the hands of the users themselves.

4   Bio-design and Co-creation

Bio-design involves designing with nature. In this way, it could be considered a form of co-creation, which is defined by Elizabeth Sanders and Pieter Jan Stappers who are leading practitioners in the field of co-creation and co-design, as;

"any act of collective creativity, i.e. creativity that is shared by two or more people. Co-creation is a very broad term with applications ranging from the physical to the metaphysical and from the material to the spiritual..." [11]

We propose that bio-design presents an important new avenue for undertaking co-creation. Not only does bio-design require collaboration between different disciplines, including those beyond the design disciplines, but it also facilitates the creation and use of multi-, cross-, and inter- disciplinary knowledge. [12]

A bio-design artefact example that reveals alignments with co-creation is the *Bioje-wellery* project (Fig. 1). This project brought together researchers and designers Tobie Kerridge and Nicki Stott at the Royal College of Art, and Ian Thompson a bioengineer from the Kings College London. This project arguably provided a groundbreaking new approach to jewellery design. The dedicated project website provides the following overview of the project;

"Biojewellery started out by looking for couples who wanted to donate their bone cells. Their cells were seeded onto a bioactive scaffold. This material encouraged the cells to divide and grow rapidly, and the resulting tissue took on the form of the scaffold, which was a ring shape. The couple's cells were grown at Guy's Hospital, and the final bone tissue was taken to a studio at the Royal College of Art to be made into a pair of rings. The bone was combined with tradi-tional precious metals so that each has a ring made with the tissue of their partner." [13]

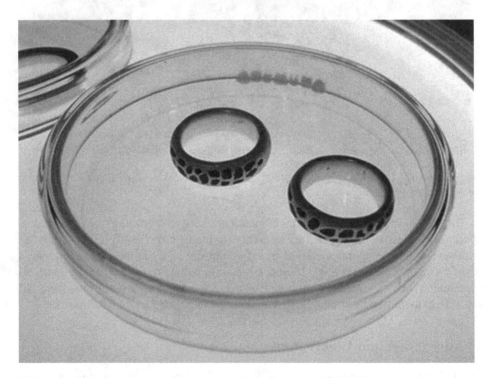

Fig. 3. Final rings grown for two of the *Biojewellery* project participants. http://www.biojewellery.com/project6.html

The *Biojewellery* rings were grown from the cells of a loved one, enabling the wearer to adorn their bodies literally with a part of someone else. As described in the exhibition catalogue accompanying the project, "each person wears the body of their lover on their hand." [14]

The success of the project required interdisciplinary involvement harnessing various disciplinary expertise and material knowledge (Fig. 2). In the approach taken to the coming together of various disciplines for the *Biojewellery* project:

"instead of attempting to smooth over tensions between the disciplines involved, we actively sought to highlight areas of conflict. Our aim was to provoke debate – not only between special-isms but amongst the public at large. In other words, by confounding expectations of the nature of this kind of multi-disciplinary work, we hoped to surprise both ourselves and others." [15]

In this project the users were also significantly involved, donating cells to grow the rings and providing designs of decorative patterns for the metalwork. The ability to grow the rings from bone provided an additional layer of potency to the meaning embedded in these rings. It also forms a strong emotional bond between the product and the user. The emotional connection differs from that of mass produced products, as the bio-product described is both co-created, and is made up from tangible user bio-products, forming an intimate extension of user with product.

Designer Toby Kerridge has said that "my aim was to reflect on how we (Ian, Nikki, myself) executed project work, including working with organisations and the function of these designs once they get out into the world, for example in accomplishing some form of public engagement." [16] This further suggests that the co-creation aspect of the project was extremely important, as well as the dialogue that exists in relation to the project outcomes.

Fig. 4. Examples of some of the biomedical materials used in the *Biojewellery* project. http://www.biojewellery.com/project6.html

5 Conclusion

Bio-design continues to grow and attract interest in the product design discipline. However little seems to have been said regarding how bio-design fits within the existing established design paradigms. Through co-creation, bio-design has the potential to create strong emotional connections between products and users, by having users intimately involved in the creation of the product. Digital fabrication processes and in particular 3D printing, when teamed with bio-design presents the opportunity to increase the designer's palette of materials and open new form possibilities.

Bio-design presents more than the promise of product manufacturing splintering from industrial production processes of old and incorporating biological processes. The case studies here of Tobie Kerridge and Eric Klareenbeek present a tantalising glimpse of how bio-design is being incorporated in the design methodologies of practicing designers and researchers.

References

1. Myers, W.: Bio Design, pp. 7–8. The Museum of Modern Art, New York (2012)
2. These definitions are provided in line with Christina Cogdell who commented that the exhibition *Design and the Elastic Mind* failed to differentiate between; *"imagined visions –* virtual pieces, if you will, materialized for the exhibition through digitally manipulated photographs or videos – *one-off prototypes* seemingly ready for production, and *post-production designs.*" Christina Cogdell, 'Design and the Elastic Mind, Museum of Modern Art (Spring 2008). Design Issues 25(3), 95 (Summer 2009)
3. Cogdell, C.: BioArt to BioDesign. Am. Art **25**(2), 28 (2011). Summer
4. Schmidt, A.: Where the physical and digital worlds collide. Computer **45**(12), 77 (2012)
5. Zoran, A., Buechley, L.: Hybrid reassemblage: an exploration of craft, digital fabrication and artifact uniqueness. Leonardo **46**(1), 6 (2013)
6. 'MyceliumChair,' Eric Klarenbeek. http://www.ericklarenbeek.com/. Accessed 9 Feb 2014
7. The Wageningen UR Plant Breeding Mushroom Research Group, CNC Exotic Mushrooms and Beelden op de Berg. 'MyceliumChair,' Eric Klarenbeek. http://www.ericklarenbeek.com/. Accessed 9 Feb 2014
8. 'MyceliumChair,' Eric Klarenbeek. http://www.ericklarenbeek.com/. Accessed 9 Feb 2014
9. Mycelium Chair by EricKlareenbeek is 3D-printed with Living Fungus, Dezeen. http://www.dezeen.com/2013/10/20/mycelium-chair-by-eric-klarenbeek-is-3d-printed-with-living-fungus/. Accessed 7 Jan 2014
10. Mycelium Chair by EricKlareenbeek is 3D-printed with Living Fungus, Dezeen. http://www.dezeen.com/2013/10/20/mycelium-chair-by-eric-klarenbeek-is-3d-printed-with-living-fungus/. Accessed 7 Jan 2014
11. Sanders, E.B.N., Stappers, P.J.: Co-creation and the new landscapes of design. CoDesign: Int. J. CoCreation Des. Arts **4**(1), 6 (2008)

12. Julieanna Preston defines multidisciplinary as referring to "knowledge shared by more than one discipline in which each is tackling a common challenge using its specialized tool kit – essentially, the whole is the sum of its part. Crossdisciplinary refers to one discipline using the knowledge set of another, i.e., importation across discipline boundaries." Julieanna Preston, "A Fossick for Interior Design Pedagogies," in After Taste: expanded practice in interior design, ed. Kent Kleinman, Joanna Merwood-Sailsbury and Lois Weinthal, p. 105. Princeton Architectural Press, New York (2012)
13. Biojewellery: designing rings with bioengineered bone tissue. http://www.biojewellery.com/project.html. Accessed 4 Feb 2014
14. Kerridge, T., Stott, N., Thompson, I.: Biojewellery: Designing Rings with Bioengineered Bone and Tissue, p. 8. Oral & Maxillofacial Surgery, King's College London, London (2006)
15. Kerridge, T., Stott, N., Thompson, I.: Biojewellery: Designing Rings with Bioengineered Bone and Tissue, p. 9. Oral & Maxillofacial Surgery, King's College London, London (2006)
16. Kerridge, T.: Personal communication, 12 February 2014

iCub Visual Memory Inspector:
Visualising the iCub's Thoughts

Daniel Camilleri[✉], Andreas Damianou, Harry Jackson,
Neil Lawrence, and Tony Prescott

Psychology Department, University of Sheffield, Western Bank, Sheffield, UK
d.camilleri@sheffield.ac.uk
http://www.sheffield.ac.uk

Abstract. This paper describes the integration of multiple sensory recognition models created by a Synthetic Autobiographical Memory into a structured system. This structured system provides high level control of the overall architecture and interfaces with an iCub simulator based in Unity which provides a virtual space for the display of recollected events.

Keywords: Synthetic autobiographical memory · Unity simulator · Yarp · Deep gaussian process

1 Introduction

Human episodic and autobiographical (or event) memory can be considered as an attractor network operating in a latent variable space, whose dimensions encode salient characteristics of the physical and social world in a highly compressed fashion [1]. The operation of the perceptual systems that provide input to event memory can be analogised to a deep learning process that identifies psychologically meaningful latent variable descriptions [2]. Instantaneous memories then correspond to points in this latent variable space and episodic memories to trajectories through this space. Deep Gaussian Processes (DGP) [3] are probabilistic, non-parametric equivalents of neural networks and have many attractive properties as models of event memory; for example, the ability to discover highly compressed latent variable spaces, to form attractors that encode temporal sequences, and to act as generative models [4].

As part of the WYSIWYD FP7 project to develop social cognition for the iCub [5] humanoid robot, we are exploring the hypothesis that an architecture formed by suitably configured DGPs can provide an effective synthetic analogue to human autobiographical memory, a system that we call SAM. Work so far has focused on the development of models for separate sensory modalities that demonstrate useful qualities such as compression, including identification of psychologically-meaningful latent variables, pattern completion, pattern separation, and uncertainty quantification. The next phase focuses on the integration of different sensory modalities using multiple sub-models which are cast within a

© Springer International Publishing Switzerland 2016
N.F. Lepora et al. (Eds.): Living Machines 2016, LNAI 9793, pp. 48–57, 2016.
DOI: 10.1007/978-3-319-42417-0_5

single and coherent framework. However, with the use of multiple sensory modalities modelled together, one requires firstly a single point of entry for easy and intuitive communication with all models. This interface controls access to the separate sub models. Secondly, with the availability of multiple sub models, one also requires a method for visualising and associating all the recollected events in a virtual environment. This offers an important window into the recollection process of multiple models by visually displaying all recollected sensory modalities. We refer to this idea as the "visual memory inspector" (VMI) environment, to highlight the fact that we can *actively* (i.e. in a user-driven manner) interact with it and explore the memory space. The unique generative properties of the SAM model which we employ significantly facilitate the deployment of the VMI.

The development of this interface is also interesting since studies of human autobiographical memory indicate that whilst episodic memories are recovered via a loop through the hippocampal system the outputs of that system generate activity within primary sensory areas that appears to encode a sensory experience of the recollected event [6]. These patterns can then be picked up for processing elsewhere in the brain, for instance, by systems that plan future actions, or that reflect on the implications of remembered experiences. This activity also feeds through to hippocampus (as part of the loop) and may play a role in the reconstruction of further memories. Whilst the brain architecture underlying human autobiographical memory is poorly understood, our hope is that the development of an integrative architecture for autobiographic memory in a humanoid robot could provide clues for unravelling the role of different brain areas in human memory, and provide a top-down functional description of how such a system could operate.

The rest of this paper first provides an introduction to the operation of SAM in Sect. 2 together with a brief description of the models that have been trained so far based on DGP in Sect. 3. Section 4 subsequently outlines the need for a supervisory process to interface with multiple SAM Models. Section 5 describes the implementation of the Visual Memory Inspector crucial to the understanding of how the iCub is analysing the situation and finally Sect. 6 provides a description of the upcoming work on this project.

2   SAM Backend

In this section we first explain the SAM architecture used as a backend and in the subsections that follow, we outline the SAM-based sub-models developed in our work.

The SAM [2] system is a probabilistic framework which approximates *functional* requirements to Auto-biographical memory as have been identified in previous studies [1]. In detail, denote the N observed sensory data as D multi-dimensional vectors $\{\mathbf{y}_n\}_{n=1}^N$, i.e. $\mathbf{y}_n \in \Re^D$. Typically these vectors are noisy and high-dimensional, for example if the robotic agent is perceiving visual signals, each frame $\mathbf{y}_n$ will be a noisy image with D equal to the number of all the pixels composing it. In SAM, each $\mathbf{y}_n$ is modelled through $\mathbf{y}_n = f(\mathbf{x}_n) + \epsilon$, where

ϵ is Gaussian noise. Here, $\mathbf{x}_n \in \Re^Q, Q \ll D$ is a low-dimensional vector, and $\{\mathbf{x}_n\}_{n=1}^N$ forms the (compressed) memory space (called a *latent* space) which is learned through the agents experience by Bayesian inference. Moreover, f is a Gaussian process mapping which maps latent points back to the original observation space. This is a *generative* mapping and plays a key role to the VMI. Furthermore, this mapping is anchored on a user-defined number of "anchor" points $\mathbf{U}$, meaning that any output of the function f will be a combination of elements from $\mathbf{U}$. [2] explains how the combination of anchor and latent points form the final memory space, where high-level analogies to neurons and synapses can be defined. In SAM, one can stack multiple latent spaces to form a hierarchical (deep) memory space. Notice that thanks to the Bayesian framework of SAM, we have access to the (approximate) posterior distribution $q(\mathbf{x}|\mathbf{y})$, meaning that once the model is trained we can readily obtain the reverse mapping of f when new sensory outputs $\mathbf{y}_*$ need to be considered.

We now proceed to outline the specific SAM-based sub-models which handle different types of sensory modalities. We will see later how these sub-models can be handled within a central framework which we call SAM Supervisor.

3   SAM Models to Date

3.1   Face Recognition Model

Face recognition with SAM has been demonstrated in previous work [7] which used a Viola-Jones face detector [8]. This method has been improved with the application of facial landmark identification and tracking through the use of a more robust face tracker called the Cambridge Face Tracker [9]. The output of this face tracker, depicted in Fig. 1a provides an outline of the face together with a general direction of looking. This face outline is then extracted from the original image and processed into a rotationally invariant representation along the roll axis as shown in Fig. 1b.

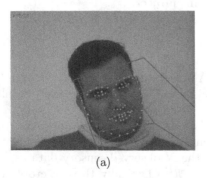

(a)   (b)

Fig. 1. (a) Cambridge Face Tracker with light blue facial landmarks and red box representing orientation of the head. (b) Augmented output with roll rotational invariance (Color figure online)

The image is subsequently resized, vectorised, labelled and then trained upon in the same manner as the previous work. Recollection is then carried out with the use of labels which returns a face extracted from the latent feature space that could either be a past observation or a fantasy face.

3.2 Tactile Model

The tactile model interfaces with the iCub's skin and collects the pressure reading over all texels of a specific body part, the arm, which are compiled into a single vector and trained upon to recognise four distinct types of touch which are:

1. Hard Touch
2. Soft Touch
3. Caress
4. Pinch

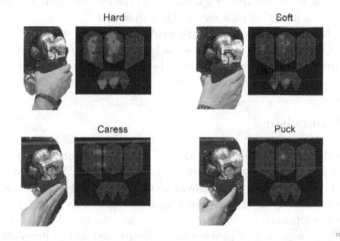

Fig. 2. Examples of the four types of touch on the iCub forearm classified by the Tactile Model

The recognition of these types of touch which can be seen in Fig. 2 is particularly important in social situations. The recollection of a fantasy instance as the inverse process describes the pressure which is required for the re-enactment of the recollected touch.

3.3 Emotion from Speech Model

Another important indication which guides social interaction is the detection of emotional state especially from voice. As such there is currently work being carried out on the use of Mel-Frequency Ceptral Coefficients (MFCC) [10] paired with a Gaussian Mixture Model (GMM) to construct classification vectors for training with SAM.

In the first stage of feature extraction mel-frequency ceptral coefficients (MFCCs) are created for each frame of the utterance. These are a standard in speech processing and have shown great success in a number of tasks including speaker recognition and emotion recognition - as well as their ubiquitous use in speech recognition systems. MFCCs are approximations to the Fourier transform of the power spectrum of a frame of audio.

These MFCCs are then made into supervectors. Firstly, for each speaker, a GMM is trained on every feature from each of their utterances. These make up the speaker-specific Gaussian mixture models (GMMs). The feature vector is then generated using the method described in [11] which extracts the MFCC features for each utterance combined with the posterior probability of the mixture of Gaussians and this vector is used for training with SAM.

On the other hand, extraction of MFCCs from the raw waveform loses much of the original information necessary to recreating sound waves, such as intonation and other long-term features of speech. Thus the current state of the system does not allow the conversion of a recollection to sound.

However, this will be tackled in future work through the extraction of different features from sound which do not abstract the original audio signal as highly as MFCC features. One such feature that is being researched is the use of power spectra.

4   SAM Supervisor

The current challenge with multiple models of separate sensory modalities is the requirement to launch and interface with each individually. This hampers the development of more complex hierarchical models that link multiple modalities and this issue has led to the development of a streamlined system through the use of a supervisory process.

The aim of this supervisory process is, as mentioned in the introduction, to provide a single framework where all models that are developed with SAM can interact with the rest of the modules developed for WYSIWYD. As such the role of SAM Supervisor is fourfold:

1. Provide a single point of contact with all external modules accessing the models which greatly facilitates external interfacing. This exposes two valid commands for each model which are *ask_modelName_label* which returns the label given an instance of data or *ask_modelName_instance* which returns a fantasy memory instance given a label.
2. Initialise all models as subprocesses of the supervisor to ensure parallel operation and perform routine checks on the status of the loaded models.
3. Check that all models are up to date with respect to the available data (experiences) in the ABMSql database. This ensures that the iCub is current with respect to the conglomeration of its experiences to date.

4. Allow for the specification of model configurations that specifically describe which model configurations are to be loaded thus allowing custom memory layouts to be design and implemented easily.

Moreover, the presence of multiple models brings to light another challenge and that is understanding what is currently happening within the memory system which leads to the requirement of a Visual Memory Inspector whose implementation is detailed in the next section.

5   Visual Memory Inspector

The aim of the Visual Memory Inspector, as stated before, is to understand better what is currently occurring within the reasoning and recollection processes of the iCub in a visual manner. Thus this requires, first of all, a virtual world that behaves similar to the real world as it is understood by the iCub, a model of the iCub himself as the protagonist of this world as well as a means of communication with Yarp [12] for the transmission of information and motor commands.

As such this virtual world requires a platform with a physics engine to govern interaction between objects while also offering flexibility in interfacing with external libraries, cross-platform execution for both Windows and Linux and finally an easy way of developing applications within this platform.

The four candidates considered as a platform for the VMI were the Unity Game Engine [13], V-Rep [14], Webots [15] and Gazebo [16]. On one hand, Gazebo offers a versatile environment for the simulation of robots but requires the installation of ROS. V-Rep and Webots are also oriented towards the simulation of robots but are both difficult to interface to with Yarp and also have licensing restrictions and a small niche user base making development challenging.

Unity on the other hand satisfies all the requirements set for the VMI. It has an advanced GPU accelerated physics engine for simulating collisions and motion, allows multiplatform compilation because it derives from a .NET programming paradigm and furthermore facilitates the inclusion of external libraries in C# and/or JavaScript. Moreover Unity also has a vast user base which facilitates development and has no licensing requirements for the basic version which is versatile enough for the requirements of the project. Consequently after the choice of a development platform, the next step is to set up communication with Yarp from within Unity.

5.1   Unity-Yarp Integration

In order to integrate Yarp libraries within Unity, a common language is required to bridge the two platforms. Unity on one hand, can be developed using two languages, C# or JavaScript, of which JavaScript is easy to use but C# provides a higher level of control over the execution of code. On the other hand, Yarp is developed in C but it also provides language bindings for a variety of languages through the use of SWIG (Simplified Wrapper Interface Generator) [17] of which one of these languages is C#.

Thus with C# as the chosen language, the integration of Yarp with Unity is carried out by adding the Swig generated .dll (Windows) or .so (Linux) and .cs files that are generated through the compilation of Yarp to the Plugins folder of a Unity project and it is then imported as a library within the code.

With Unity capable of implementing Yarp ports the following crucial step for integration looked into the implementation of bottle and image conversion functions that allow decoding and encoding of Yarp information into a more Unity friendly format. Finally, since a call to yarp read is a blocking call, a threaded class was employed for the communication processes so that they can run in parallel to the visualisation. This results in higher frame rates and more time efficient processing of events. The next section describes the implementation of a virtual iCub as the protagonist of the memory inspector.

5.2   iCub Simulation

The VMI is currently targeted towards the visualisation of memories and as such does not require a sensory interface within the virtual environment. Nonetheless, future work will look into expanding the scope of the VMI to a simulation space where future planning can also be virtually carried out.

This is why the current implementation of VMI pairs itself with iCub_SIM to allow motion control of the iCub within the VMI and also provides a stereo stream of the iCub's current point of view. This is accessible separately from the VMI interface which by itself provides a game like environment with the capability of walking around the 3D scene in the iCub's memory to change viewing angles.

6   Future Work

6.1   Recollection Visualisation

The current state of the project allows the placement of the protagonist within a previously saved environment that could be obtained from a .obj model which includes Kinect generated 3D models. The VMI also has the functionality to dynamically load pre-existing 3D objects within the environment and assign a given label and position. An example of this can be seen in Fig. 3 where the VMI has dynamically generated a person and two objects with specific locations within the environment.

The next major step in the implementation of the VMI is to integrate with the Language Reservoir developed by INSERM [18]. This module within WYSIWYD generates a Predicate-Action-Object-Recipient (PAOR) description of a previous memory retrieved through ABMSql. Each part of this concise description received by the VIM is then transmitted to SAM as an *ask_modelName_instance* request which returns a fantasy memory from the corresponding latent space.

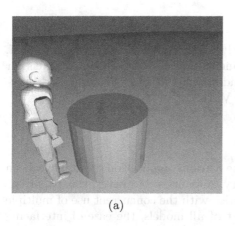
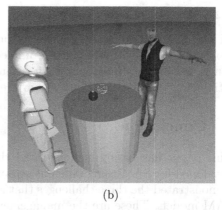

(a)   (b)

Fig. 3. Demonstration of VMI dynamic object loading. (a) Depicts the initial state of the VMI which starts off with just the iCub and an environment (b) Depicts the state of the VMI with the dynamic addition of a person and two objects within the loaded environment

Subsequently, after all constituents have been parsed and an instance received, the information is displayed as a scene within the VMI. A demonstrative example of such an interaction can be seen in Fig. 4 where the face instance recovered from SAM for the label '*Daniel*' has been embodied within a generic body.

Fig. 4. Demonstration of the addition of a face recalled from SAM to the generic body of a dynamically instantiated person within the VMI

6.2   Virtual Sensing for Planning

Of the iCub's four senses, currently only sight is available within the VMI. Upcoming work will focus on the implementation of depth cameras, texels for touch and binaural microphones for sound which will allow planning and simulating the outcome of actions within the VMI.

7   Conclusion

This paper has briefly demonstrated the various applications that have been developed for SAM using different sensory modalities. Moreover, this paper has demonstrated the three challenges that arise with the concurrent use of multiple SAM models. These are the management of all models, the ease of interfacing and finally the challenge of visualising what is happening within these memory models in an interactive manner. As such we proposed the use of two modules: Sam Supervisor the Visual Memory Inspector (VMI) which provide a solution to this systems problem. Finally we lay out a plan for the continued development of these modules into an easily expandable software system upon which the development of a synthetic human autobiographical memory can be based.

References

1. Evans, M.H., Fox, C.W., Prescott, T.J.: Machines Learning - Towards a New Synthetic Autobiographical Memory. In: Duff, A., Lepora, N.F., Mura, A., Prescott, T.J., Verschure, P.F.M.J. (eds.) Living Machines 2014. LNCS, vol. 8608, pp. 84–96. Springer, Heidelberg (2014)
2. Damianou, A., Ek, C.H., Boorman, L., Lawrence, N.D., Prescott, T.J.: A Top-Down Approach for a Synthetic Autobiographical Memory System. In: Wilson, S.P., Verschure, P.F.M.J., Mura, A., Prescott, T.J. (eds.) Living Machines 2015. LNCS, vol. 9222, pp. 280–292. Springer, Heidelberg (2015)
3. Damianou, A., Lawrence, N.: Deep gaussian processes. In: Carvalho, C., Ravikumar, P. (eds.) Proceedings of the Sixteenth International Workshop on Artificial Intelligence and Statistics (AISTATS). AISTATS 2013, JMLR W&CP, vol. 31, pp. 207–215 (2013)
4. Damianou, A.: Deep gaussian processes and variational propagation of uncertainty. PhD thesis, University of Sheffield (2015)
5. IIT: iCub: an open source cognitive humanoid robotic platform. http://www.icub.org/. Accessed 1 Mar 2016
6. Gelbard-Sagiv, H., Mukamel, R., Harel, M., Malach, R., Fried, I.: Internally generated reactivation of single neurons in human hippocampus during free recall. Science 322(5898), 96–101 (2008)
7. Martinez-Hernandez, U., Boorman, L., Damianou, A., Prescott, T.: Cognitive architecture for robot perception and learning based on human-robot interaction
8. Viola, P., Jones, M.J.: Robust real-time face detection. Int. J. Comput. Vis. 57(2), 137–154 (2004)
9. Baltrusaitis, T., Robinson, P., Morency, L.P.: Constrained local neural fields for robust facial landmark detection in the wild. In: Proceedings of the IEEE International Conference on Computer Vision Workshops, pp. 354–361 (2013)

10. Zheng, F., Zhang, G., Song, Z.: Comparison of different implementations of MFCC. J. Comput. Sci. Technol. **16**(6), 582–589 (2001)
11. Loweimi, E., Doulaty, M., Barker, J., Hain, T.: Long-Term Statistical Feature Extraction from Speech Signal and Its Application in Emotion Recognition. In: Dediu, A.-H., Martın-Vide, C., Vicsi, K. (eds.) SLSP 2015. LNCS, vol. 9449, pp. 173–184. Springer, Heidelberg (2015). doi:10.1007/978-3-319-25789-1_17
12. YARP: Yet another robot platform. http://wiki.icub.org/yarpdoc/. Accessed 10 Feb 2016
13. Unity Technologies: Unity 5.2.2. https://unity3d.com/. Accessed 1 Mar 2016
14. Copelia Robotics: V-rep. http://www.coppeliarobotics.com/index.html. Accessed 1 Mar 2016
15. Cyberbotics Ltd.: Webots. https://www.cyberbotics.com/overview. Accessed 1 Mar 2016
16. Open Source Robotics Foundation: Gazebo. http://gazebosim.org/. Accessed 1 Mar 2016
17. Swig: Swig. http://www.swig.org/. Accessed 1 Mar 2016
18. Hinaut, X., Twiefel, J., Petit, M., Bron, F., Dominey, P., Wermter, S.: A recurrent neural network for multiple language acquisition: Starting with english and french

A Preliminary Framework for a Social Robot "Sixth Sense"

Lorenzo Cominelli[1(✉)], Daniele Mazzei[1], Nicola Carbonaro[1],
Roberto Garofalo[1], Abolfazl Zaraki[1], Alessandro Tognetti[1,2],
and Danilo De Rossi[1,2]

[1] Faculty of Engineering, Research Center "E. Piaggio",
University of Pisa, Pisa, Italy
lorenzo.cominelli@for.unipi.it
[2] Department of Information Engineering, University of Pisa, Pisa, Italy
http://www.faceteam.it

Abstract. Building a social robot that is able to interact naturally with people is a challenging task that becomes even more ambitious if the robots' interlocutors are children involved in crowded scenarios like a classroom or a museum. In such scenarios, the main concern is enabling the robot to track the subjects' social and affective state modulating its behaviour on the basis of the engagement and the emotional state of its interlocutors. To reach this goal, the robot needs to gather visual and auditory data, but also to acquire physiological signals, which are fundamental for understating the interlocutors' psycho-physiological state. Following this purpose, several Human-Robot Interaction (HRI) frameworks have been proposed in the last years, although most of them have been based on the use of wearable sensors. However, wearable equipments are not the best technology for acquisition in crowded multi-party environments for obvious reasons (e.g., all the subjects should be prepared before the experiment by wearing the acquisition devices). Furthermore, wearable sensors, also if designed to be minimally intrusive, add an extra factor to the HRI scenarios, introducing a bias in the measurements due to psychological stress. In order to overcome this limitations, in this work, we present an unobtrusive method to acquire both visual and physiological signals from multiple subjects involved in HRI. The system is able to integrate acquired data and associate them with unique subjects' IDs. The implemented system has been tested with the FACE humanoid in order to assess integrated devices and algorithms technical features. Preliminary tests demonstrated that the developed system can be used for extending the FACE perception capabilities giving it a sort of sixth sense that will improve the robot empathic and behavioural capabilities.

Keywords: Affective computing · Behaviour monitoring · Human-Robot Interaction · Social robotics · Synthetic tutor

© Springer International Publishing Switzerland 2016
N.F. Lepora et al. (Eds.): Living Machines 2016, LNAI 9793, pp. 58–70, 2016.
DOI: 10.1007/978-3-319-42417-0_6

1   Introduction

Nowadays, it is well-known that our emotional state influences our life and decisions [1]. Education and learning processes are not excluded from this claim and the review, written by Bower [2], about how emotions might influence learning, as well as the more specific research about the effects of affect on foreign language learning, done by Scovel [3], are two of several important studies confirming this theory. As a consequence, in the last years this topic has triggered also the interest of social robotics scientists [4,5]. Indeed, they develop humanoid robots destined to interact with people, and the emotional and psychological states of these androids' interlocutors have to be necessarily taken into account. This becomes even more important in case where robots have to interpret the role of synthetic tutors intended for teaching children and conveying pedagogical contents in scenarios like musea or schools [6].

In order to interpret the emotional states of the pupils who interact with robots, social robotics researchers have proposed many different solutions that can be divided in two main categories: *Visual-Auditory Acquisition* and *Physiological Signals Acquisition*. The processing of both kinds of information aims to an estimation of the interlocutors' affective states, and these social/emotional information is exploited, in turn, by the control architecture of the robots, to modulate their behaviour, or completely change the actions they were planning. This improves the empathy and facilitates the dialogue between robots who are teaching and children who are learning. In any case, the main problem of these two approaches is that, up to date, there is not a good integration of data delivered by both the acquisitions. In most of the cases, the perception systems are designed to use only one kind of acquisition, and this is not sufficient to determine a stable perception of the human emotional state, but just a temporary assessment that can't be used in a long-term interaction [7].

The visual-auditory acquisition has the peculiarity to be more stable and well functioning thanks to a lot of available devices, which are easy to use but also easy to mislead (e.g., a smiling face has not to be always interpreted as an happy person); while the acquisition of physiological parameters has the advantage to reveal hidden emotional states and involuntary reactions that are relevant for human behaviour understanding, but this approach has the major outstanding problem to be an obtrusive, if not even invasive method.

In this paper, we present a novel architecture that supports the acquisition of both visual-auditory and physiological data. It is composed of the Scene Analyzer, a software for audio and video acquisition and social features extraction [8]; and the TouchMePad, consisting of two electronic patches and a dedicated signal processing software for the users physiological monitoring and affective state inference. TouchMePad is thought to be integrated with the other half of the perception system to become the sixth sense of a social robot. Moreover, the dedicated software, as well as the whole acquisition framework, has been designed to collect data in a sporadic and unobtrusive way in order to minimise the stress for the subjects and reach a high level of naturalness during the HRI. This is highly beneficial considering that Scene Analyzer is able to identify different persons by means of a QR codes reader module. Thanks to this recognition capability, once a subject is recognised, the physiological signals acquired by the Touch-MePad can be stored in a database that will permanently associate a unique ID with the recognised subject. This information is compared with other meaning-ful social information (e.g., facial expression, posture, voice direction), providing the robot with the capability to change or modulate the tutoring according to different persons and their emotional state both in real time and accordingly to past interactions (e.g., previous lessons).

2 The Perception Framework

The perception framework is composed of two main parts: the Scene Analyzer, our open-source visual-auditory acquisition system that we continuously upgrade and deliver on Github[1]; the TouchMePad, composed of two hardware patches for physiological parameters acquisition and a dedicated software for signal process-ing and affective computing.

2.1 Scene Analyzer

We designed Scene Analyzer (SA) as an out-of-the-box human-inspired percep-tion system that enables robots to perceive a wide range of social features with the awareness of their real-world contents. SA enjoys of a peculiar compatibility, indeed, due to its modular structure, it can be easily reconfigured and adapted to different robotics frameworks by adding/removing its perceptual modules. Therefore, it can be easily integrated to any robot irrespective of the working operating system.

SA consists of four distinct layers (shown in Fig. 1) data acquisition, low-level and high-level features extraction, structured data creation and communication. SA collects raw visual-auditory information and data about environment and, through its parallel perceptual sub-modules, extracts higher level information that are relevant from a social point of view (e.g., subject detection and tracking, facial expression analysis, age and gender estimation, speaking probability, body postures and gestures). All this information is stored in a dynamic storage called meta-scene. The SA data communication layer streams out the created meta-scene through YARP (Yet Another Robot Platform) middleware [9].

SA, supports a reliable, fast, and robust perception-data delivering engine specifically designed for the control of humanoid synthetic tutors. It collects visual-auditory data through Kinect ONE 2D camera, depth IR sensor, and microphone array with the highest level of precision. For example, it detects and keeps track of 25 body-joints of six subjects at the same time (Fig. 2), which is a fundamental capability for crowded scenarios. With such an accuracy it is possible to perceive gestures that are very important in educational contexts (e.g., head rubbing, arm crossed, exulting etc.). For the same purpose, extraction

---

[1] https://github.com/FACE-Team/SceneAnalyzer-2.0.

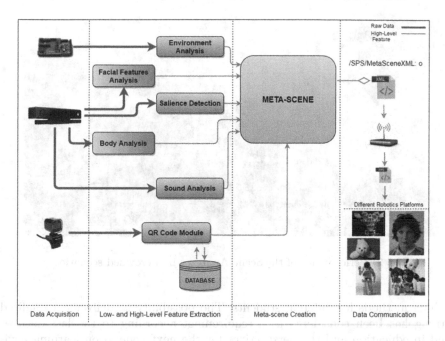

Fig. 1. Framework of the Scene Analyzer, the visual-auditory part of the perception system.

of data about joint coordinates of the lower part of the body has been added to the coordinates of the upper torso. These added coordinates makes the system capable to detect also the body pose of a person. Accordingly, we can distinguish between a seated and a standing position. For further information about meta-scene data structure and SA visual-auditory acquisition please refer to [10].

In order to endow the robot with a stable and reliable ID recognition capability, a new module has been implemented: the *QR Code Module*. It works in parallel analysing the image provided by a separated IID camera, and it is conceived for detecting and extracting information from QR codes. These bar-codes matrices can be printed and applied to the user's clothes or attached to the objects with which the robot has to interact. SA exploits an SQL database in which a list of QR codes are associated with permanent IDs and names. However, any further information can be added to any stored subject or object. Every time the synthetic agent will be able to recognise in the FOV a QR code that is known, because saved in the internal database, it will retrieve this information and automatically assign a known ID and name to that entity. Moreover, assuming that the entity is a person who has already interacted with the robot, once the subject has been recognised by the means of the QR code, the association between the permanent ID and that subject will continue even if the QR code will be no more visible by the camera.

This solution for assigning, storing and retrieving permanent information about interlocutors is fully automatic and mandatory in order to customise

Fig. 2. Performance of the Scene Analyzer in a crowded scenario.

teaching methods for different students. The possibility to deliver personalised learning has been considered as a requirement according to the last decades trend in education and the perspectives for the next generation learning environment [11,12]. Another important feature of SA is processing of environmental data streamed by Zerynth Shield[2], an electronic board for environmental analysis (e.g., light intensity, temperature, audio level), which is useful to determine potential influences of the environment on interaction and learning engagement.

2.2 TouchMePad

In order to estimate and recognise human emotions from physiological parameters, several techniques have been developed in the last years, and most of them exploit wearable sensors (e.g., [13]). Since our system is intended to be used in crowded environments involving pupils or young users, the usage of sensorized clothes such as gloves or shirts is a considerable complication. Furthermore, a continuous and permanent contact would invalidate the naturalness of the interaction which is already difficult enough since a humanoid robot is implicated in it. Last but not least, an unceasing acquisition of multiple data from many subjects, including who is not currently involved in a real interaction with the robot, would be useless as well as overwhelming for the data processing phase.

For all these reasons, we opted for a user-centred solution that is non-invasive, unobtrusive and keeps the naturalness of a social interaction. Indeed, it is conceived to prevent discomfort for the user who has not to be permanently attached to sensors. On the contrary, two electronic patches are attached to the synthetic tutor shoulders and the subjects are asked to touch sporadically the shoulders of the robot in order to acquire physiological signals only in some key moments of

---

[2] http://www.zerynth.com/zerynth-shield/.

the interaction. This facilitates both the user and the acquisition system, reducing the number of contacts with the sensors, as well as the amount of gathered data, to the strictly necessary.

Therefore, TouchMePad (TMP) can be considered as the other half of the perception framework providing the robot with a sort of sixth sense. TMP is conceived to monitor the variation of the physiological parameters that are correlated to human affective state. As shown in Fig. 3 the system is composed of the electronic patches, acting as electrodes, and a central electronic unit for power supply, elaboration and transmission of user physiological parameters. The electronic unit is designed to conveniently combine ECG analog front-end with EDA one by means of a low-power micro controller. The developed ECG block is a three leads ECG system that samples signals at the frequency of 512 Hz. The front-end is based on the INA321 instrumentation amplifier and on one of the three integrated operational amplifiers available in the micro controller (MSP family made by Texas Instruments, MSP430FG439), to reach the total 1000x amplification. Regarding the EDA circuit, a small continuous voltage (0.5 V) is applied to the skin and the induced current is measured through two electrodes positioned in correspondence with the middle and the ring finger. The ratio between voltage drop and induced current represents the skin electric impedance, the EDA signal. An ad-hoc finger electrodes patches were designed and developed allowing the acquisition of user physiological parameters in a natural way. Therefore, the user does not have to wear any type of band, shirt etc. but simply touching the patches with the fingers it is possible to calculate the *Inter-Beat Interval* (IBI) parameter and the *Electro Dermal Activity* (EDA) signal. Finally, the system evaluates the robustness of user contact to identify physiological signal segments that are affected by artifacts and have to be discarded in further analysis. The detection of an artifact can also trigger an automatic request, by the robot to the user, to modify fingers position on the electrodes or to place them on the patches once again for a better acquisition.

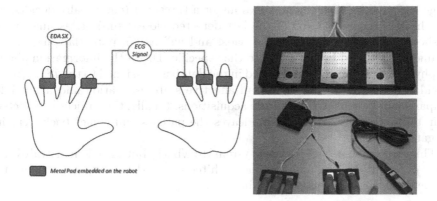

Fig. 3. Schema and photos of the acquisition patches.

For further information about physiologic signal acquisition and the methods we use for processing this data, please refer also to our previous works involving the acquisition of ECG [14,15] and EDA [16].

3   The ID Tracking Module

Since the perception system is composed of two separated main acquisition systems working in parallel (i.e., SA for visual-auditory acquisition and TMP for physiological parameters acquisition), we needed to find a solution to unify the data extracted from subjects in order to have a unique classification associated with IDs. The issue about having a link between information delivered by the SA and the one delivered by TMP depended on a very practical problem: every time a subject had to place their fingers on the electrodes, the camera got covered by that subject, and the SA was no longer able to detect any person in the field of view. This entails that the entire perception system, as it was initially conceived, was not able to assign the physiologic parameters to any person, because no person was detected while signals were acquired. Therefore, we developed a dedicated sub-module, which provided a workaround for this 'identity problem', ensuring the assignment of both SA and TMP data to a univocal ID.

We named it the *ID Tracking Module* (IDTm) and it works as a bridge between the TMP and the SA. This module continuously calculates the standard deviation of the pixels extracted by a cropped, re-sized, central part of the image acquired by the Kinect camera. Every time a subject approaches the patches covers most of the image field, as well as all the other people present in the scene. Nevertheless, we have another effect: the image becomes almost the same, especially the central part, regardless light conditions, contrast and brightness. Furthermore, the SA delivers information about the distance in meters of any subject until is detected. Considering that only one subject at a time has the possibility to put their fingers on the patches, we retrieve the information about the last detected closest subject and, when the standard deviation of the central image gathered by the camera comes under a threshold (empirically decided at 50), the IDTm saves the ID of the last detected closest subject, assumes that he/she is the one approaching the pads, and will assign potential physiologic parameters detected by the TMP, to that specific ID. All the information about the physiological parameters is stored in the database, assigned to the $ID_{Touching}$ provided by the IDTm, ready for an affective state estimation and a possible comparison between past and future acquisitions. Finally, the subject is detected again and recognised as soon as he leaves the patches and comes back clear in the gathered image.

This strategy not only solves a system drawback, but also demonstrates the power of a multi-tasking and modular architecture as the one we are presenting. A schema of this solution is shown in Fig. 4.

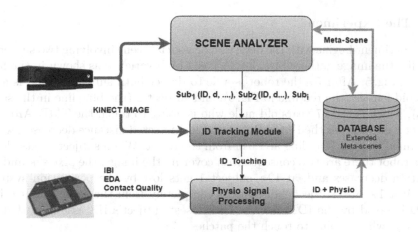

Fig. 4. The ID tracking schema.

4 Evaluation of the Perception System

4.1 Materials and Methods

The use case is based on the installation of the SA and the TMP on the FACE Robot body [17].

Data collected by the SA are stored in an SQL database together with the one provided by the TMP, thanks to the ID Tracking module. Once a contact on the TMP is detected, the quality of the signal is analysed and the information about contact quality is shown on the screen in which the interlocutor can read instructions. Projected instructions are: *"Please put your fingers on the patches"*, *"Bad contact! Please place your fingers again"* or *"Good Contact! Please keep your fingers on the patches"*. In this latter case, a 60 s countdown immediately starts on the screen. This is the duration that we set for having a reliable acquisition of physiological data. As a consequence, if the contact is maintained at a sufficient quality level, at the end of the countdown it does appear the following message: *"Thank you, you can remove your hands now"*. Acquired data are pre-filtered by the *Physio Signal Processing Module* shown in Fig. 4, which discards unreliable values, then calculates the IBI mean value, the EDA, asks for the $ID_{Touching}$ to the IDTm, then finally sends through yarp all this information to the database, adding the physio data to the proper ID.

Several tests were performed in order to verify the capability of the TMP to monitor user physiological state. A 27 years old male was selected for the experiment whose heart beat frequency in rest condition is known and in value of 80 bpm. This information has been useful for the comparison with the value calculated from the IBI values extracted by the sensor patches. The test consisted of different sessions of 1 min in which the subject was asked to maintain a rest condition and to put his fingers of both hands in direct contact with the TMP.

4.2   The Experiment

The experiment begins with the detection of a social scene involving two subjects. Initially, the image gathered by the SA has a high variance as shown in the first chart of Fig. 5. After 5 s the robot is able to detect both subject 1, who is a 26 years old female (the red line in Fig. 5), and subject 2 (the blue line in the same chart), which is the 27 years old male who is going to touch the TMP. After 8 s subject 2 approaches the FACE Robot and his detected distance decreases, while the female remains standing at 1.6 m from the robot. When subject 1 gets closer to the robot there are two consequences: covering the image, the pixels' standard deviation decreases and, at 10 s, subject 1 gets lost by the perception system. After just 1 s, also subject 2 disappears completely from the scene ($d = 0$), but his ID is saved by the IDTm as the last closest subject's ID, assuming that he is the one who is going to touch the patches.

Therefore, the acquisition system starts looking for a potential contact, checking and analysing the contact quality of the finger electrodes. This can be noticed in the first 2 s of Fig. 6, that shows the trend of the *Contact Quality* (CQ) parameter in correspondence with the two acquired physiological parameters (i.e., IBI and EDA). CQ allows to distinguish not only between bad and good sensor contact, but also to determine which fingers are not well positioned by means of the elaboration of the relative physiological signal. In fact, each of the six fingers involved in the acquisition has some peculiarities: for example the right hand medium finger is used as the reference electrode for the ECG circuit, while the ring and the medium finger of the left hand allow to track the EDA signal, as previously shown in the schema of Fig. 3. Considering Table 1, in which all the possible combination of CQ values are reported, we can claim that the system is working correctly when the CQ parameter is equal to 50, while in other condition the validity of physiological data is not guaranteed and fingers should be repositioned.

The physiological parameters gathered during all the 60 s of acquisition are shown in Fig. 6. At the beginning CQ is 60, representing that the circuit for hand detection is activated and ready for the elaboration. Then the IBI and EDA values are tracked and stored. At $t = 9s$ the CQ value decreases to 30,

Table 1. CQ value definition

CQ value	Meaning
0	*No Contact*
10	*IBI No Contact, EDA Good Contact*
20	*IBI & EDA Bad Contact*
30	*IBI Bad Contact, EDA Good Contact*
40	*IBI Good Contact, EDA Bad Contact*
50	*IBI & EDA Good Contact*
60	*Contact Circuit Active*

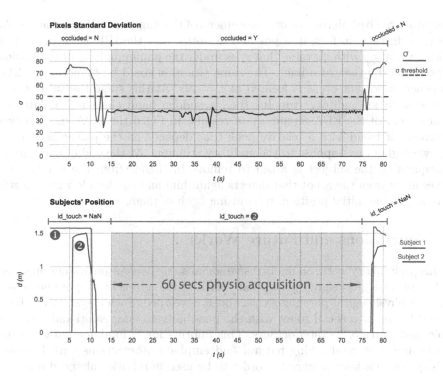

Fig. 5. The social robot sixth sense evaluation test - ID tracking module. (Color figure online)

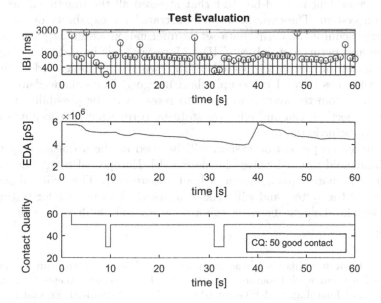

Fig. 6. Physiological Parameters extracted by the perception framework in real time during the 60 s of acquisition highlighted in Fig. 5.

likely due to a bad placement or a movement of the finger for IBI acquisition. At the same instant, in fact, it is possible to notice that the IBI signal presents a sudden variation with its value that goes under the minimum level of the session (represented by the green line). The same situation could be found at $t = 31s$. An other important event is located at $t = 40s$. At this stage the subject was elicited with an auditory stimulus (i.e., a sudden hand clap) to validate the contribution of the EDA circuit. As a results, in Fig. 6 the EDA signal leaves the constant trend with a relative peak that depends on the phasic activity of the sweat gland in response to the auditory stimulus. At the end of one minute of acquisition the subject is asked to remove his hands from the patches. He moves away from the robot that detects again him and the female who was still standing in the initial position, recognising both of them.

5   Conclusions and Future Works

In this preliminary work, an unobtrusive social scene perception system provided with a physiological signal perception "sixth sense" given by a sporadic acquisition of physiological parameters, has been developed. Such a system has been proved to endow a social robot with the possibility to infer emotional and psychological states of its interlocutors. This information is fundamental for social robots aimed at establishing natural and emphatic interactions with humans. This system has been designed in order to be used in robotic enhanced teaching contexts, where social robots assume the role of synthetic tutors for children and pupils. To evaluate the developed perception framework including all its features, we designed an ad-hoc test that stressed all the functionalities of the acquisition system. The experiment demonstrated the capability of the system to properly acquire the entire data-set of information assigning gathered data to unique and permanent subjects' IDs. Moreover, it is important to highlight that the fusion of the data collected by the two system integrated in the presented setup goes beyond the simple data merging. The enhanced meta-scene which results from the system will give to researchers the possibility to better infer the subject's social and affective state by correlating psycho-physiological signals with behavioural data.

The presented perception system will be used as the acquisition system of several humanoid robots involved in the EASEL European Project[3] in different educational scenarios (e.g., musea, school classrooms). This will validate the portability of the system and will make it a practical framework for testing how to improve the dialogue between robotic tutors and children as well as their learning.

Acknowledgment. This work was partially funded by the European Commission under the 7th Framework Program projects EASEL, Expressive Agents for Symbiotic Education and Learning, under Grant 611971-FP7- ICT-2013-10. Special thanks to Daniela Gasperini for her fundamental contribution in the experiments organization.

---

[3] http://easel.upf.edu/.

References

1. Damasio, A.: Descartes' Error: Emotion, Reason, and the Human Brain. Grosset/Putnam, New York (1994)
2. Bower, G.H.: How might emotions affect learning. Handb. Emot. Mem. Res. Theor. **3**, 31 (1992)
3. Scovel, T.: The effect of affect on foreign language learning: a review of the anxiety research. Lang. Learn. **28**(1), 129–142 (1978)
4. Hudlicka, E.: To feel or not to feel: the role of affect in human-computer interaction. Int. J. Hum. Comput. Stud. **59**(1), 1–32 (2003)
5. Fong, T., Nourbakhsh, I., Dautenhahn, K.: A survey of socially interactive robots. Robot. Auton. Syst. **42**(3), 143–166 (2003)
6. Causo, A., Vo, G.T., Chen, I.M., Yeo, S.H.: Design of robots used as education companion and tutor. In: Zeghloul, S., Laribi, M.A., Gazeau, J.-P. (eds.) Robotics and Mechatronics. Mechanisms and Machine Science, vol. 37, pp. 75–84. Springer, Switzerland (2016)
7. Yan, H., Ang Jr., M.H., Poo, A.N.: A survey on perception methods for human-robot interaction in social robots. Int. J. Soc. Robot. **6**(1), 85–119 (2014)
8. Zaraki, A., Mazzei, D., Giuliani, M., De Rossi, D.: Designing and evaluating a social gaze-control system for a humanoid robot. IEEE Trans. Hum. Mach. Syst. **44**(2), 157–168 (2014)
9. Metta, G., Fitzpatrick, P., Natale, L.: Yarp: yet another robot platform. Int. J. Adv. Robot. Syst. **3**(1), 43–48 (2006)
10. Zaraki, A., Giuliani, M., Dehkordi, M.B., Mazzei, D., D'ursi, A., De Rossi, D.: An rgb-d based social behavior interpretation system for a humanoid social robot. In: 2014 Second RSI/ISM International Conference on Robotics and Mechatronics (ICRoM), pp. 185–190. IEEE (2014)
11. Sampson, D., Karagiannidis, C.: Personalised learning: educational, technological and standardisation perspective. Interact. Educ. Multimedia (4) 24–39 (2010)
12. Brusilovsky, P.: Developing adaptive educational hypermedia systems: from design models to authoring tools. In: Murray, T., Blessing, S.B., Ainsworth, S. (eds.) Authoring Tools for Advanced Technology Learning Environments, pp. 377–409. Springer, Netherlands (2003)
13. Lisetti, C.L., Nasoz, F.: Using noninvasive wearable computers to recognize human emotions from physiological signals. EURASIP J. Adv. Sign. Process. **2004**(11), 1–16 (2004)
14. Tartarisco, G., Carbonaro, N., Tonacci, A., Bernava, G., Arnao, A., Crifaci, G., Cipresso, P., Riva, G., Gaggioli, A., De Rossi, D., et al.: Neuro-fuzzy physiological computing to assess stress levels in virtual reality therapy. Interact. Comput. (2015). iwv010
15. Carbonaro, N., Anania, G., Mura, G.D., Tesconi, M., Tognetti, A., Zupone, G., De Rossi, D.: Wearable biomonitoring system for stress management: a preliminary study on robust ECG signal processing. In: 2011 IEEE International Symposium on a World of Wireless, Mobile and Multimedia Networks (WoWMoM), pp. 1–6. IEEE (2011)

16. Carbonaro, N., Greco, A., Anania, G., Dalle Mura, G., Tognetti, A., Scilingo, E., De Rossi, D., Lanata, A.: Unobtrusive physiological and gesture wearable acquisition system: a preliminary study on behavioral and emotional correlations. In: Global Health, pp. 88–92 (2012)
17. Mazzei, D., Zaraki, A., Lazzeri, N., De Rossi, D.: Recognition and expression of emotions by a symbiotic android head. In: 2014 14th IEEE-RAS International Conference on Humanoid Robots (Humanoids), pp. 134–139. IEEE (2014)

A Bio-Inspired Photopatterning Method to Deposit Silver Nanoparticles onto Non Conductive Surfaces Using Spinach Leaves Extract in Ethanol

Marc P.Y. Desmulliez[✉], David E. Watson, Jose Marques-Hueso, and Jack Hoy-Gig Ng

School of Engineering & Physical Sciences, Nature Inspired Manufacturing Centre (NIMC), Heriot-Watt University, Earl Mountbatten Building, Edinburgh EH14 4AS, Scotland, UK
m.desmulliez@hw.ac.uk

Abstract. Densely packed silver nanoparticles (AgNPs) were produced as continuous films on non-conductive substrates in a site-selective manner. The formation of the AgNPs were directed by blue light using extract from spinach leaves acting as photo-reducing agent. This bio-inspired production of reduced ions nanofilms benefit applications where seed conductive layers are required for the manufacture of metal parts embedded into plastics.

Keywords: Nature-inspired manufacturing · Biomimetism · Biomimicry · Additive manufacturing · Artificial photosynthesis

1 Introduction

Intense research efforts have focused on the synthesis and deposition of silver nanoparticles (AgNPs) due to their usefulness in plasmonics, electronics, biological sensing and as an antimicrobial agent [1–3]. Some applications require these silver nanoparticles to be assembled in a reproducible manner as micro- or nano-scale periodical structures such as arrays or lines of specific width and space. Usually, AgNPs are synthesized in a colloidal solution and then consolidated onto a substrate using laser sintering [4], microwave [5] or plasma treatment [6]. They can also be prepared using the bottom-up approach whereby metal ions are molecularly linked to the substrate onto which they are deposited and then reduced to form nanoparticles within the molecular network of the linker layer. The latter approach can easily anchor metal nanoparticles onto a strong electrostatic adhesive such as dopamine [7] or ion-exchange resins [8].

In that respect, surface hydrolysed polyimide or polyetherimide films are ideal substrate materials for hosting metal nanoparticle thin films [9, 10]. These metal thin films can be produced *in situ* with a controllable thickness within a surface layer of ion-exchange resins of the R-COO⁻ type obtained as the result of alkaline hydrolysis of polyimide forming an amorphous layer of poylamic acid on its surface [8]. In addition, different reduction methods, including thermal, chemical or photo-induced, can be readily applied to such silver ion-exchange polyimide substrate (Ag⁺-PI).

© Springer International Publishing Switzerland 2016
N.F. Lepora et al. (Eds.): Living Machines 2016, LNAI 9793, pp. 71–78, 2016.
DOI: 10.1007/978-3-319-42417-0_7

Photo-induced reduction is the simplest route for replicating the desired patterns by regulating the shape of the light source such as in laser writing applications or by using a photomask as in photolithography. To the best of the authors' knowledge, reports regarding high-throughput photo-induced patterning of silver nanoparticle thin films that is compatible with large area substrate are still lacking due to the poor quantum yield of the photo-reducing agents or catalysts employed today.

Herein we demonstrate a nature-inspired strategy to reduce metal ions using photo-system 1 (PS I) found in spinach extract dissolved in ethanol. PS I is thought to be acting as the photo-reducing agent to produce patternable metal nanoparticles that meet the requirements of both production speed and process scalability. PS I comprises of light harvesting pigment proteins, amongst which chlorophyll is being most commonly referred to for photosynthesis.

2   Materials and Methods

For our purpose, the PS I and mineral agents contained in the leaves were simply extracted from spinach leaves by dissolving them in ethanol solvent, after using a blender to collapse the chloroplasts in the leaves which house the PS I. In a typical process, 100 g of fresh leaves was mixed with 500 ml of absolute ethanol in a blender and processed for 1 min. After passing the mixture through a 1:11 μm grade filter paper, 90 % of the liquid was then poured into a second flask and left to stand for some time for sedimentation. The process of filtration was then repeated again to avoid any debris in the extract solution. Next, 2 ml of the PS I and minerals extract solution were pipetted onto the surface of the Ag^+-PI substrate.

The substrate was originally prepared by a 2-step immersion: first into a 1 M KOH solution at 50 °C for 5 min, followed by a 0.01 M $[Ag(NH_3)_2]^+$ aqueous solution at room temperature for 5 min. As the preparation process involves only a couple of dipping process, substrates of large dimensions can readily be prepared depending on the size of the tanks that contain the two solutions described above. The photo-induced reduction step took up to five minutes using a 460 nm wavelength fibre delivered light emitting diode with an intensity of around 1270 mW/cm^2.

3   Results

Figure 1, top left, shows photographs of circular spots of silver patterns with an approximate diameter of 1 mm produced upon exposure to the blue light source at the tip of the optical fibre. The corresponding field emission scanning electron microscopy (FESEM) images where the growth of silver nanoparticles can be clearly observed in the same figure. Generally, particle size and density of the AgNPs increased with increasing exposure times. At 30 s exposure, AgNPs as small as 15 nm can be easily identified across the illuminated area. Some particles aggregated to sizes of about 70–80 nm. Within 60 s of exposure, the whole illuminated area was compactly covered with AgNPs. The smaller particles have an average size of about 30 nm and the larger aggregated particles about 70 nm. At 180 s exposure, a large amount of particles had aggregated to 40 nm or bigger, where, noticeably, some particles as large as 120 nm can be seen.

Fig. 1. Photographs and FESEM images of: (a) control background with no photo-induced reduction. (b–e) the location selective production of AgNPs on Ag⁺-PI upon exposure to the light source for 30, 60, 180 and 600 s, respectively. Scale bar is 500 nm.

In the latter case the AgNPs are not compactly connected as in the case of the 60 s exposed sample indicating that the silver ion source in the substrate supplying the reaction has been depleted and left a void behind when the smaller particles aggregated to form bigger particles.

At the prolonged exposure of 600 s, a dark mark can be clearly seen in the photograph at the centre of the circular pattern in the Fig. 1 (top left). The excess photon energy under this exposure time perhaps induced some thermal effects on the substrate resulting in polymer restructuring. The corresponding FESEM micrograph in Fig. 1(e) suggests that a reaction, as yet not explained, seemed to have taken place, which removed the larger particles seen in Fig. 1(d). The small amount of AgNPs observed in Fig. 1(a) with no light exposure could be the result of spontaneous reduction of Ag⁺ ions, which purely utilize the redox potential difference between impurity entities and the Ag + ions in the absence of a reducing agent.

4 Discussion

4.1 Short Exposure Time

The short exposure times required using a relatively low light intensity irradiation for the formation of high density AgNPs were surprising. Up to now, irradiation methods used for metal nanoparticle formation usually produce sparse particles from a dispersion in the liquid phase [11]. The enhancement reported here can be credibly contributed to the function of PS I extract. It is known that impurity of metal traces is present in the absolute ethanol solvent, however they did not contribute to the significantly faster photoreduction rate. Our previous work with the same Ag^+-PI substrate system using a synthetic photoreducing agent methoxy (ethylene glycol) and with the same absolute ethanol solvent required several hours of light exposure to obtain appreciable amount of AgNPs [9]. Although the PS I structure might have been altered from its natural membrane arrangements through the extraction procedure, it has definitely enabled a significant increase of the photoreduction rate. It would be interesting to study the performance in the photoreduction rate along side purified PS I, and the effects of different extraction methods and the different types of plant leaves as the source material.

The reason why the photo-induced reduction can be conducted in such a short time with a PS I coating on top of the Ag^+-PI is partly due to the efficient electron transfer within the PS I protein complex. Upon reception of the incident photon energy, rapid charge separation occurs within 10–30 ps, releasing an electron down an intra-protein energy cascade to an iron-sulfur complex called F_B^- [12]. When the PS I is in its natural environment, the soluble, iron-containing protein ferredoxin shuttles the electrons away from F_B^- to achieve a nearly perfect quantum yield [13]. In addition, the nucleation and

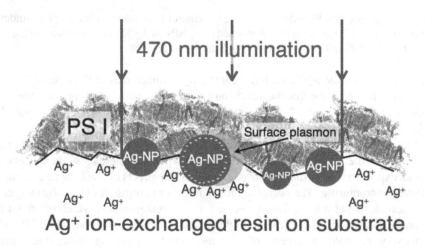

Fig. 2. Schematics of plasmon field enhancement and charge storage on the surface of AgNPs, which assist the photo-reduction of Ag^+ ions in the PS I/AgNPs hybrid system. The figure is a schematic representation of the assumed process with size of the nanoparticles not at the right scale. The substrate has a prior deposited Ag + ion-exchanged resin.

growth of the AgNPs in the present system benefit from a self-catalysed mechanism as illustrated in Fig. 2.

During the growth of the AgNPs, the PS I proteins are confined into the nano-cavities amongst the Ag aggregates. From this situation, two main driving forces can be attributed to the long range electron transfer from the PS I coating to the Ag^+ ions within the ion-exchange resins on the substrate. Firstly, the plasmon enhancement of photon fields within a hybrid system of PS I molecular complex and AgNPs aggregates has been modelled [14] and measured [15] to provide a several fold increase in the generation of excited state electrons. Secondly, the AgNPs act as ultra large surface area nanostructured electrodes for storing the excited electrons released from PS I as charges on the AgNPs surfaces [16] and thus minimising charge recombination. More recent studies carried out by the group seem to indicate that salts from the extract of spinach leaves might also contribute to the donation of electrons. This new information will be presented during the conference.

4.2 Three Potential Photoreduction Mechanisms

A cross-section FESEM image prepared by Broadband Ion Beam (BIB) machining is, shown in Fig. 3 and provides an insight into the photoreduction mechanisms. The process, developed and used by MCS Ltd, allows deformation and smear-free cross-sectioning without obscuring the AgNPs-polymer matrix. The BIB system allows representative sample sizes (up to 2 mm) to be cross-sectioned without mechanically touching the sample enabling absolute confidence that deformation or voids did not occur during the cross-sectioning procedures as in the FIB system.

There are three key features to note:

(i) The AgNPs produced by the photoreduction effectively formed a uniform nano-composite layer within the modified polyimide substrate surface. No gradient of AgNPs distribution was observed across the layer. The electrons generated from the PS-I coating and the natural mireal salts from the leaves were effectively transferred into the surface-modified polymer substrate. This suggests that some electron relay mechanisms must be involved as suggested earlier, since there would inevitably be an illumination gradient as the light impinges into the surface of the substrate due to both the polymer molecules and the *in situ* growth of the AgNPs.

(ii) The uniform boundary of the AgNPs nanocomposite illustrates the diffusion-controlled and well-defined nature of the 2-step immersion surface modification process where the polyimide molecules were first hydrolysed by KOH followed by Ag^+ ion-exchange. The line of shade underneath the AgNPs nanocomposite layer represents a small amount of unreacted Ag^+ ions that were not reduced. An explanation could be that the rate of Ag^+ ion diffusion within the polymer matrix is much slower than the rate of the electron transfer.

(iii) A relatively high thickness of approximately 1.5 μm layer of AgNPs nanocomposite was produced. Clusters of AgNPs were created through their nucleation and growth, however they were unable to form chains of continuous network within the volume of the nanocomposite to render the layer electrically conductive.

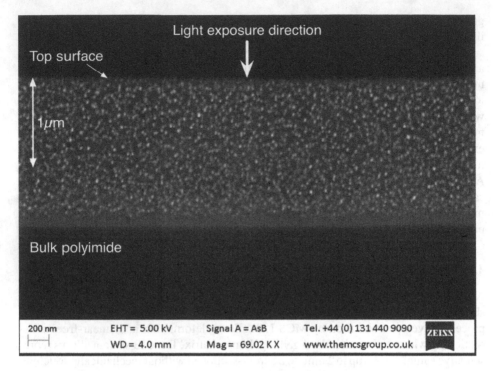

Fig. 3. FESEM image of a cross-section of the AgNPs nanocomposite within the surface-modified layer of the polyimide substrate.

5   A Scalable Nature-Inspired Manufacturing Process

Our approach in the synthesis and patterning of the AgNPs is easily scalable to large area industrial production. Regular methods commonly involve producing the AgNPs in a synthesis matrix first, followed by the separation of the particles and their consolidating onto desired locations on a substrate according to the following sequence:

1. Homogeneous mixing of the capping agent and the stabilizing agent with the metal ions.
2. Reduction of the capped metal ions.
3. Extraction of metal nanoparticles.
4. Purification to remove organic residues.
5. Consolidation and patterning onto a surface.

This procedure can take hours to perform. By first incorporating the metal ions into the ion-exchange resin, the molecular networks of the polymer served the role of a capping agent when the AgNPs were grown in situ within the surface layer of the substrate. As a result, significant production time is reduced in the present process:

1. Hydrolysis of polyimide by KOH immersion. 5 min.
2. Ion-exchange in a silver salt solution. 5 min.

3. Extraction of PS I and of the mineral salts form the leaves. The extract can be prepared separately and stored frozen prior to reaction. 30 min.
4. Exposure to light source for photo-reduction of silver ions. 1 min or less.

The requirement of the light intensity (around 1270 mW/cm^2) is also low enough such that a flood exposure lamp can be used for large area parallel production of AgNP patterns through a photomask. However ambient room lighting or daylight does not provide a large enough energy dose for any accelerated photoreduction of Ag$^+$ ions in the system here.

6 Conclusions

In summary, we have reported in this article a rapid strategy to photopattern macro-scale compact thin films consisting of AgNPs.

We propose some mechanisms involved in the outstanding electron transfer phenomenon, which resulted in the charge separation and transfer within the PS I coating. Furthermore, some electron relay systems must be present within the silver/polymer matrix to allow the electron transfer to proceed further and produce an almost 1.5 μm thickness layer metal matrix composite.

The whole production process itself can be as short as 11 min, if the PS I and mineral salts extract is prepared in advance. The absence of any toxic reducing agents or complex organic ligands in the whole production process also makes this patternable AgNPs nanocomposite fabrication method suitable for biological and sustainable applications.

Acknowledgements. The authors of this article would like to acknowledge the financial support of the UK Engineering & Physical Sciences Research Council (EPSRC) through the funding of the two research grants entitled Photobioform I (EP/L022192/1) and Photobioform II (EP/N018222/1).

References

1. Luo, X., Morrin, A., Killard, A.J., Smyth, M.R.: Application of nanoparticles in electrochemical sensors and biosensors. Electroanalysis **18**(4), 319–326 (2006)
2. Cobley, C.M., Skrabalak, S.E., Campbell, D.J., Xia, Y.: Shape-controlled synthesis of silver nanoparticles for plasmonic and sensing applications. Plasmonics **4**(2), 171–179 (2009)
3. Sondi, I., Salopek-Sondi, B.: Silver nanoparticles as antimicrobial agent: a case study on *E. coli* as a model for Gram-negative bacteria. J. Colloid Interf. Sci. **275**, 177–182 (2004)
4. Yung, K., Wu, S., Liem, H.: Synthesis of submicron sized silver powder for metal deposition via laser sintered inkjet printing. J. Mater. Sci. **44**(1), 154–159 (2009)
5. Perelaer, J., de Gans, B., Schubert, U.: Ink-jet printing and microwave sintering of conductive silver tracks. Adv. Mater. **18**(16), 2101–2104 (2006)
6. Reinhold, I., Hendriks, C., Eckardt, R., Kranenburg, J., Perelaer, J., Baumann, R., Schubert, U.: Argon plasma sintering of inkjet printed silver tracks on polymer substrates. J. Mater. Chem. **19**(21), 3384–3388 (2009)

7. Cheng, C., Nie, S., Li, S., Peng, H., Yang, H., Ma, L., Sun, S., Zhao, C.: Biopolymer functionalized reduced graphene oxide with enhanced biocompatibility *via* mussel inspired coatings/anchors. J. Mater. Chem. B **1**, 265–275 (2013)

8. Akamatsu, K., Ikeda, S., Nawafune, H., Deki, S.: Surface modification-based synthesis and microstructural tuning of nanocomposite layers: monodispersed copper nanoparticles in polyimide resins. Chem. Mater. **15**(13), 2488–2491 (2003)

9. Ng, J.H.-G., Watson, D.E.G., Sigwarth, J., McCarthy, A., Prior, K.A., Hand, D.P., Yu, W., Kay, R.W., Liu, C., Desmulliez, M.P.Y.: On the use of silver nanoparticles for direct micropatterning on polyimide substrates. IEEE Trans. Nano. **11**(1), 139–147 (2012)

10. Watson, D.E., Ng, J.H.-G., Aasmundtveit, K.E., Desmulliez, M.P.Y.: *In-situ* silver nanoparticle formation on surface-modified polyetherimide films. IEEE Trans. Nano. **13**(4), 736–742 (2014)

11. Balan, L., Schneider, R., Turck, C., Lougnot, D., Morlet-Savary, F.: Photogenerating silver nanoparticles and polymer nanocomposites by direct activation in the near infrared. J. Nanomaterials **2012**, 1–6 (2012)

12. Chitnis, P.R.: Photosystem I: function and physiology. Annu. Rev. Plant Biol. **52**(1), 593–626 (2001)

13. Noy, D., Moser, C.C., Dutton, P.L.: Design and engineering of photosynthetic light-harvesting and electron transfer using length, time, and energy scales. Biochim. Biophys. Acta **1757**(2), 90–105 (2006)

14. Govorov, A.O., Carmeli, I.: Hybrid structures composed of photosynthetic system and metal nanoparticles: plasmon enhancement effect. Nano Lett. **7**(3), 620–625 (2007)

15. Carmeli, I., Lieberman, I., Kraversky, L., Fan, Z., Govorov, A.O., Markovich, G., Richter, S.: Broad band enhancement of light absorption in photosystem I by metal nanoparticle antennas. Nano Lett. **10**(6), 2069–2074 (2010)

16. Lioubashevski, O., Chegel, V.I., Patolsky, F., Katz, E., Willner, I.: Enzyme-catalyzed bio-pumping of electrons into au-nanoparticles: a surface plasmon resonance and electrochemical study. J. Am. Chem. Soc. **126**(22), 7133–7143 (2004)

Leg Stiffness Control Based on "TEGOTAE" for Quadruped Locomotion

Akira Fukuhara[1,2]([⊠]), Dai Owaki[1], Takeshi Kano[1], and Akio Ishiguro[1,3]

[1] Research Institute of Electrical Communication, Tohoku University,
2-1-1 Katahira, Aoba-ku, Sendai 980-8577, Japan
{a.fukuhara,owaki,tkano,ishiguro}@riec.tohoku.ac.jp
[2] JSPS, 5-3-1 Kojimachi, Chiyoda-ku, Tokyo 102-0083, Japan
[3] CREST, Japan Science and Technology Agency,
4-1-8 Honcho, Kawaguchi, Saitama 332-0012, Japan
http://www.cmplx.riec.tohoku.ac.jp

Abstract. Quadrupeds exhibit adaptive limb coordination to achieve versatile and efficient locomotion. In particular, the leg-trajectory changes in response to locomotion speed. The goal of this study is to reproduce this modulation of leg-trajectory and to understand the control mechanism underlying quadruped locomotion. We focus primarily on the modulation of stiffness of the leg because the trajectory is a result of the interaction between the leg and the environment during locomotion. In this study, we present a "TEGOTAE"-based control scheme to modulate the leg stiffness. TEGOTAE is a Japanese concept describing the extent to which a perceived reaction matches the expected reaction. By using the presented scheme, foot-trajectories were modified and the locomotion speed increased correspondingly.

Keywords: Quadruped locomotion · Autonomous distributed control · Foot trajectory · Leg stiffness · TEGOTAE

1 Introduction

Efficient and adaptive locomotion of quadrupeds can be realized by flexible inter-limb and intra-limb coordination. For example, quadrupeds exhibit various types of gait depending on the locomotion speed, morphology, and the environment [1]. Although various quadruped robots with particular focus on inter-limb coordination have been reported [2,3], the mechanisms of intra-limb coordination without inverse-kinematics are still an open challenge. As regards to intra-limb coordination, cursorial mammals change their stroke length in response to increasing locomotion speed [4]. In addition, the period of the cyclic motion and the length of one step change adaptively with locomotion speed [5]. In order to reveal the

A. Fukuhara—This work was supported by JSPS KAKENHI Grant-in-Aid for JSPS Fellows (16J03825).

© Springer International Publishing Switzerland 2016
N.F. Lepora et al. (Eds.): Living Machines 2016, LNAI 9793, pp. 79–84, 2016.
DOI: 10.1007/978-3-319-42417-0_8

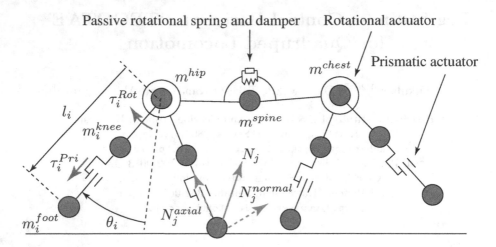

Fig. 1. Kinematic model of quadruped robot.

mechanism behind efficient and adaptive locomotion, both inter-limb and intra-limb coordination should be investigated in detail.

The goal of this study is to design adaptable and cursorial quadruped robots as well as to understand the mechanism of adaptive limb coordination. First, we focus on the modification of leg stiffness given that the foot trajectory is affected by the interaction between the leg and the environment during locomotion. Many studies on bio-inspired robots have shown the merits of compliant legs for stable locomotion [6]. However, the control schemes for adjusting stiffness were different depending on each study. In this study, we propose a new control scheme for quadruped locomotion to adaptively modulate the stiffness of different leg regions, based on the concept "TEGOTAE." In a 2D simulation, leg-stiffness control based on TEGOTAE leads to a change in the foot trajectory, and locomotion speed increases correspondingly.

2   Model

In order to reproduce the inter-limb coordination, we implemented the simple central pattern generator (CPG) model proposed in our previous study [7] into a simple 2D robot-model, as shown in Fig. 1. The robot is modeled as a mass-spring system. Each leg of the robot has 2 degrees of freedom (DOF) owing to the presence of rotational and prismatic actuators. These actuators are modeled using a virtual spring and damper. The stiffness of the virtual springs represents the leg stiffness. In addition, the body of the quadruped comprises 3 mass points; these mass points are connected using passive rotational spring and damper. In this study, we present a new TEGOTAE-based control method for adjustment of leg stiffness thereby enabling improved intra-limb coordination.

2.1   CPG Model

In our previous study [7], we proposed a CPG model that consists of four decoupled oscillators with local sensory feedback obtained from only a force sensor placed on each corresponding leg. Each oscillator of the CPG is described as

$$\dot{\phi}_i = \omega - \sigma N_i \cos \phi_i, \tag{1}$$

where ϕ_i is the phase of the oscillator of the i th leg, ω is the intrinsic angular velocity, σ is a weighting factor for sensory feedback, and N_i is sensory information about ground reaction force (GRF). Each i th leg tends to be in the swing phase for $0 < \phi_i < \pi$, while it tends to be in the stance phase for $\pi < \phi_i < 2\pi$. When $N_i = 0$, ϕ_i increases with ω and the leg repeats recovery-stroke and power-stroke motions depending on ϕ_i. When $N_i > 0$, ϕ_i is modulated to pull toward $3\pi/2$. Therefore, the leg supporting the body tends to keep stance phase in response to N_i. In spite of the simple rules of this CPG, our quadruped robots exhibited a variety of gaits in response to mass distribution and the frequency of locomotion [7].

2.2   Leg Stiffness Control Based on TEGOTAE

In order to model flexible intra-limb coordination, we deal with leg stiffness as the first step. In this study, while the basic foot trajectory is designed using simple sinusoidal curves, the stiffness of the leg is adjusted on the basis of TEGOTAE. The basic mapping of ϕ_i to foot trajectory is described as follows:

$$\bar{\theta}_i = C^{amp} \cos \phi_i, \tag{2}$$
$$\bar{l}_i = L^{offset} - L^{amp} \sin \phi_i, \tag{3}$$

where $\bar{\theta}_i$ and $\bar{l}_i$ are target angle and position for rotational and prismatic actuators, respectively, which constitute i th leg in Fig. 1, C^{amp} is amplitude value for the rotational motion, and L^{offset} and L^{amp} are offset and amplitude values for the prismatic motion, respectively. According to Eqs. (2) and (3), each i th leg tends to be in recovery-stroke when $0 < \phi_i < \pi$, while it tends to be in power-stroke when $\pi < \phi_i < 2\pi$. Based on the reference trajectory, the rotational and prismatic actuators are controlled by using PD control, which is given by the following equations:

$$\tau_i^{Rot} = K_i^{Rot}(\bar{\theta}_i - \theta_i) - D_i^{Rot}\dot{\theta}_i \tag{4}$$
$$\tau_i^{Pri} = K_i^{Pri}(\bar{l}_i - l_i) - D_i^{Pri}\dot{l}_i, \tag{5}$$

where τ_i^{Rot} is the control torque for the rotational actuator, τ_i^{Pri} is the control force for the prismatic actuator, K_i^{Rot} and K_i^{Pri} are P gains for rotational and prismatic actuator, respectively, and D_i^{Rot} and D_i^{Pri} are D gains for rotational and prismatic actuator, respectively. K_i^{Rot} and K_i^{Pri} also reflect the leg stiffness for rotational and prismatic joints. These K are adjusted in response to TEGOTAE.

In order to quantify TEGOTAE, we describe a function T as

$$T(x, \boldsymbol{S}) = f(x)g(\boldsymbol{S}), \tag{6}$$

where $f(x)$ is expectation of a controlled parameter x and $g(\boldsymbol{S})$ is a function of a sensory information vector $\boldsymbol{S}$ measuring the actual reaction from the environment, respectively. T is described as a product of the expectation f and the reaction g. There are various kinds of feedback to be considered for TEGOTAE depending on the kind of animal locomotion [8]. For the sake of simplicity, we define TEGOTAE with respect to propulsion and body-support. In addition, we regard that the rotational actuator is responsible for propulsion while the prismatic actuator is responsible for body-support and stability of locomotion. Based on this assumption, TEGOTAE for each actuator is described as follows:

$$T_i^{Rot} = \tau_i^{Rot} N_i^{normal}, \tag{7}$$
$$T_i^{Pri} = \tau_i^{Pri} N_i^{axial}, \tag{8}$$

where T_i^{Rot} and T_i^{Pri} are TEGOTAE functions for rotational and prismatic actuators, respectively, and N_i^{normal} and N_i^{axial} are normal and axial components of GRF, respectively, as shown in Fig. 1. Although, according to Eq. (6), the expectation part is a function of x, the control inputs for actuators (τ^{Rot} and τ^{Pri}) replace $f(x)$ in order to simplify TEGOTAE. Based on these TEGOTAE functions, the stiffness of rotational and prismatic actuators are adjusted as follows:

$$K_i^{Rot} = K_{def}^{Rot} + \sigma^{Rot} T_i^{Rot}, \tag{9}$$
$$K_i^{Pri} = K_{def}^{Pri} - \sigma^{Pri} T_i^{Pri}, \tag{10}$$

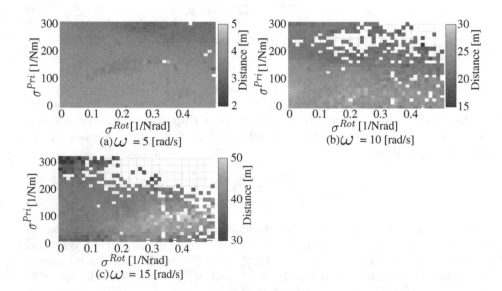

Fig. 2. Plots of distance traveled against each ω. (Color figure online)

where K_{def}^{Rot} and K_{def}^{Pri} are positive constants for default value of each joint, and σ^{Rot} and σ^{Pri} are the weighting factors for each joint. When $T_i^{Rot} > 0$, K_i^{Rot} increases so as to increase the leg stiffness in rotational motion. With increasing stiffness of rotational joints, the rotational actuator tends to obtain more TEGOTAE from environment. However, when $T_i^{Pri} > 0$, K_{Pri} decreased and the prismatic movement became softer; this was attributable to increased shock-absorption.

3   Preliminary Results

In the 2D simulation environment, we implemented the proposed TEGOTAE-based control into a quadruped robot shown in Fig. 1. Only ω was increased from 5 to 10, and then from 10 to 15. The robot moved for 20 s for each ω.

Figure 2 shows the distance traveled against each value of ω. According to Fig. 2(a), the effects of the TEGOTAE-based control are not observed with respect to distance traveled during locomotion with low ω. Whereas for medium

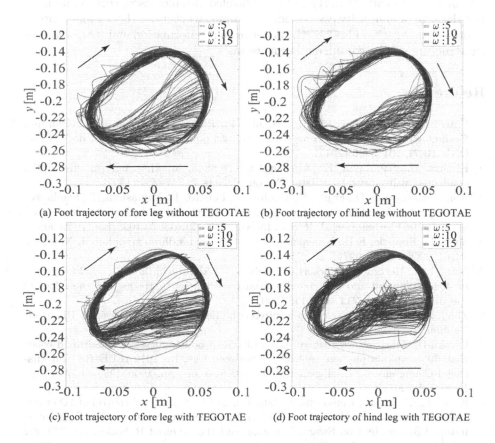

(a) Foot trajectory of fore leg without TEGOTAE   (b) Foot trajectory of hind leg without TEGOTAE

(c) Foot trajectory of fore leg with TEGOTAE   (d) Foot trajectory of hind leg with TEGOTAE

Fig. 3. Measured foot trajectories during the simulation. (Color figure online)

and high value of ω (Fig. 2(b) and (c)), the distance traveled were increased where both T_i^{Rot} and T_i^{Pri} affect.

Figure 3 shows the trajectory of m_i^{foot} during the simulation. In Fig. 3(a) and (c), foot trajectory of fore leg is plotted so as to make m_i^{chest} the origin, while the trajectory of the hind leg is plotted so as to make m_i^{hip} the origin in Fig. 3(b) and (d). Without TEGOTAE, the contact point with the ground changes randomly for high ω. On the other hand, with TEGOTAE, the contact point of each stride becomes closer. In addition, the bottom part of trajectories with TEGOTAE converged to a flat line because of contact with the ground, especially when $\omega = 15$.

4   Conclusion and Future Work

In this study, we focused on the adjustment of leg stiffness to investigate flexible foot trajectories in response to locomotion speed, and also presented a TEGOTAE-based control scheme. The proposed scheme changes the foot trajectory at high ω; in addition, the distance traveled also increased, correspondingly. For the sake of simplicity, we consider a simple structure of the leg and sensory information, N_i^{axial} and N_i^{normal}. Richer sensory information and more complex leg-structure will be considered in future work.

References

1. Catavitello, G., Ivanenko, Y.P., Lacqaniti, F.: Planar covariation of hindlimb and forelimb elevation angles during terrestrial and qquatic locomotion of dogs. PLoS ONE **10**(7), e0133936 (2015)
2. Kimura, H., Sakurama, K., Akiyama, S.: Dynamic walking and running of the quadruped using neural oscillator. Auton. Robots **7**, 247–258 (1999)
3. Aoi, S., Katayama, D., Fujiki, S., Tomita, N., Funato, T., Yamashita, T., Senda, K., Tsuchiya, K.: A stability-based mechanism for hysteresis in the walk-trot transition in quadruped locomotion. J. R. Soc. Interface (2012). doi:10.1098/rsif.2012.0908
4. Shen, L., Poppele, R.E.: Kinematic analysis of cat hindlimb stepping. J. Neurosci. **74**(6), 2266–2280 (1997)
5. Maes, L.D., Herbin, M., Hackert, R., Bels, V.L., Abourachid, A.: Steady locomotion in dogs: temporal and associated spatial coordination patterns and the effect of speed. J. Exp. Biol. **211**, 138–149 (2007)
6. Zhou, X., Bi, S.: A survey of bio-inspired compliant legged robot designs. Bioinspir. Biomim. **7**, 041001 (2012)
7. Owaki, D., Morikawa, L., Ishiguro, A.: Listen to body's message: quadruped robot that fully exploits physical interaction between legs. In: 2012 IEEE/RSJ International Conference on Intelligent Robots and Systems, pp. 1950–1955. IEEE Press, New York (2012)
8. Kano, T., Chiba, H., Umedachi, T., Ishiguro, A.: Decentralized control of 1D crawling locomotion by exploiting "TEGOTAE" from environment. In: The First International Symposium on Swarm Behavior and Bio-Inspired Robotics, pp. 279–282 (2015)

Wall Following in a Semi-closed-loop Fly-Robotic Interface

Jiaqi V. Huang[✉], Yilin Wang, and Holger G. Krapp

Department of Bioengineering, Imperial College London,
London SW7 2AZ, UK
j.huang09@imperial.ac.uk

Abstract. To assess the responses of an identified optic-flow processing interneuron in the fly motion-vision pathway, the H1-cell, we performed semi-closed-loop experiments using a bio-hybrid two-wheeled robotic platform. We implemented a feedback-control architecture that established 'wall following' behaviour of the robot based on the H1-cell's spike rate. The analysis of neuronal data suggests the spiking activity of the cell depends on both the momentary turning radius of the robot as well as the distance of the fly's eyes from the walls of the experimental arena. A phenomenological model that takes into account the robot's turning radius predicts spike rates that are in agreement with our experimental data. Consequently, measuring the turning radius using on-board sensors will enable us to extract distance information from H1-cell signals to further improve collision avoidance performance of our fly-robotic interface.

Keywords: Motion vision · Brain machine interface · Blowfly · Collision avoidance

1 Introduction

Avoiding collisions with obstacles or other moving objects is a fundamental requirement for any robotic platform moving in terrestrial, aerial, or aquatic environments. The same is true for most animals. Except for animals such as bats and whales, which avoid collisions based on self-generated acoustic signals, most species - especially flying insects, rely on visual cues [1, 2].

Several insects respond with escape behaviour and/or adjusting steering manoeuvers when facing retinal image expansion (looming stimuli) as demonstrated for instance in locusts [3] and hawk moths [4]. Based on models of the underlying neuronal circuits, bio-inspired control strategies have been adopted in engineering to support collision avoidance capabilities in technical applications [5, 6].

The blowfly, as one of the most manoeuvrable flying insects, is a well-established model system for studying visual guidance at the behavioural and neuronal level [7], with the potential to inspire novel technical solutions to collision avoidance in robotic systems. We have chosen to work on one of the blowfly's individually identified visual interneurons, the H1-cell, and explore its responses to directional motion as a collision avoidance sensor on a bio-hybrid robotic system under open- and closed-loop conditions.

© Springer International Publishing Switzerland 2016
N.F. Lepora et al. (Eds.): Living Machines 2016, LNAI 9793, pp. 85–96, 2016.
DOI: 10.1007/978-3-319-42417-0_9

The blowfly H1-cell has been studied for more than thirty years [8, 9]. It is excited by horizontal back-to-front motion and inhibited by motion in the opposite direction. In the absence of visual motion, the cell produces a spontaneous spike rate in the range of 25–50 Hz. Most previous electrophysiological studies have characterized H1-cell responses during rotational pattern motion [10–12], while translational motion has been less frequently applied [13].

Our research is focusing on both rotational and translational movement. Initially, we built and verified a mobile recording platform [14], then investigated a control strategy based on the H1-cell's temporal frequency tuning in lab environment [15], and implemented an autonomous fly-robot interface to test a temporal frequency tuning based collision avoidance control algorithm [16]. Here we study the distance-dependent response properties of the H1-cell, which will be exploited for collision avoidance.

For collision avoidance, distance estimation is crucial. It provides information on when to turn away from the potential obstacles and in which direction. In theory, optic flow induced by pure rotation does not contain any relative distance information. Distance information requires translational self-motion [17]. Blowflies generate saccadic rotations followed by translational drift phases during which the activity of two further directional selective interneurons, the HSE- and HSN-cell, encodes relative distance. This information is then used to generate the next saccade away from a potential obstacle [18]. The preferred directions of the HSE/N cells are opposite to those in the H1-cell, which enables them to respond during forward and slightly lateral translations, but their activity cannot be recorded using extracellular recording techniques.

By combining both rotational and translational movements, we may be able to stimulate the H1-cell and potentially obtain distance information from its extracellularly recorded responses [15]. Any trajectories may be described in terms of a circular trajectory with a turning radius ranging from zero to infinity. A zero turning radius is equivalent to a pure rotational movement, while an infinite turning radius would be equivalent to straight translational movement – any turning radius in between zero and infinity results in a circular trajectory.

To test whether the H1-cell responses contain distance information we characterised the H1-cell spike rate in an open-loop experiment with different turning radii. We implemented a semi-closed-loop bang-bang control algorithm on our fly-robot-interface with a turning radius, which was chosen based on earlier open-loop experiments. The algorithm alternatingly turns the robot towards and away from the wall to keep a certain average distance, and to avoid collision with the wall. This semi-closed-loop algorithm requires the control of three parameters: (i) the duration of the turn towards the wall, (ii) the duration of the turn away from the wall, (iii) the speed of the wheels, which are related to turning radius. The turns in opposite directions are performed at the same turning radius, but their execution is triggered either by a timer overflow interrupt in microprocessor under open-loop control, or by neural signal analysis, which counts the number of action potentials under closed-loop control. Both open-loop and closed-loop controls are outlined in the following sections.

2   Methods

2.1   Configuration of Experimental System

The fly-robot-interface includes three parts: (i) a blowfly, (ii) a mobile extracellular recording platform; right-hand side single unit recording only, (iii) a two-wheeled robot (Pololu© m3pi). These three parts were assembled as shown in Fig. 1.

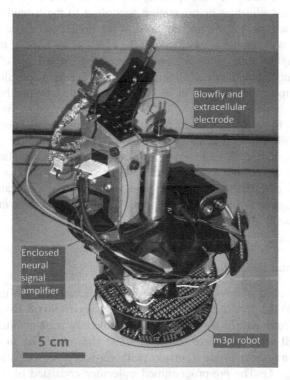

Fig. 1. The fly robot bio-hybrid system. The mobile extracellular recording platform with a single electrode is mounted on top of an m3pi robot. A blowfly is positioned in the centre of the chassis platform. H1-cell action potentials are pre-amplified with a customised amplifier enclosed by an aluminium case for electrical shielding.

H1-cell spiking activity was recorded using a tungsten electrode and was fed into the customised preamplifier with a gain of 10000. The amplified signals were fed into the robot controller for closed-loop control and a PC for off-line data analysis.

The neural data for closed-loop control were sent to the ADC port on the robot controller (ARM microprocessor NXP© LPC1768). The neural action potentials were sorted after digitisation and used to calculate the spike rate over a time interval of 50 ms as input to the robotic control algorithm.

A copy of the data was recorded to a computer hard disk by a data acquisition card (NI USB-6215, National Instruments Corporation, Austin, TX, USA) at a rate of 20 kHz.

2.2 Blowfly Preparation

The preparation of blowfly is the same as in previous series of experiments [14–16], which were performed on the same platform. Female blowflies, *Calliphora vicina*, aged 4–11 days, were chosen for the experiments. Wings were immobilised by adding a small droplet of bee wax on the wing hinges. Legs and proboscis were removed to reduce any movements that could degrade the quality of the recordings. All wounds were sealed with bee wax. The head of the fly was attached to a custom-made fly holder using bee wax after adjusting its orientation according to the 'pseudopupil methods' [19]. The thorax was bent down and attached to the holder to expose the back of the head. The cuticle was cut and removed along the edges at the back of the head capsule under optical magnification using a stereo microscope (Stemi 2000, Zeiss©). After carefully removing fat and muscle tissue with fine tweezers, physiological Ringer solution (for recipe see Karmeier et al. [20]) was added to the brain to prevent desiccation.

The dissected fly was then moved to the recording platform. Tungsten electrodes (3 MΩ tungsten electrode, product code: UEWSHGSE3N1M, FHC Inc., Bowdoin, ME, USA) were mounted on a micromanipulator (MM-1-CR-XYZ, National Aperture, Inc., Salem, NH, USA) and placed in close proximity to the H1-cell. Acceptable recording quality required a signal-noise-ratio > 2:1, i.e. the peak amplitude of the recorded H1-cell spikes was at least twice as high as the largest amplitude of the background noise.

2.3 Experimental Procedure

Experiments consisted of two parts: (i) the neural characterisation under open-loop conditions and (ii) the control algorithm verification under closed-loop conditions.

For the open-loop part of the experiment, the blowfly was mounted on the robot, where the H1-cell responses were recorded while the robot was moving along the pre-programmed trajectory inside a corridor with vertically striped patterns attached to the walls (see Fig. 4). The pre-programmed trajectory consisted of a series of alternating semi-circular turns with increasing turning radius. The sequence of robotic instructions was programmed as: robot initialisation (left/right 90° spin/return for infrared sensor initialisation); stop for 2 s; starting semi-circular turns with turning radii of 5, 10, 15, 20, 25, and 30 cm.

For the closed-loop experiments, the robot was programmed with a combined open-loop and closed-loop control algorithm, as only the left H1-cell was used as a sensor for detecting the distance to the wall. The control architecture is shown as a flow chart in Fig. 2.

The purpose of the experiment was to test if the robot would adjust its average distance towards the wall while moving forward, and keep the distance without collision until reaching the end of the experimental arena.

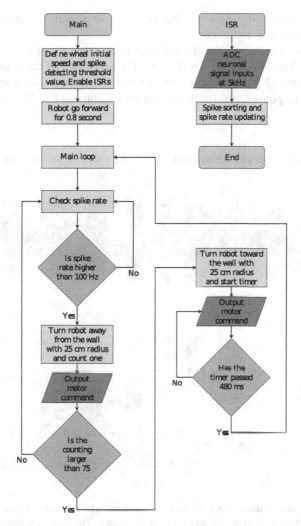

Fig. 2. Control algorithm flow chart of the bio-hybrid robotic firmware for wall following. (ADC: Analog to Digital Converter, ISR: Interrupt Service Routine)

3   Results

3.1   Open-Loop Neural Recording

The neural activities of left H1-cell were recorded by the NI-DAQ for 10 s and the data were stored on a computer hard disk.

The off-line data analysis included: spike sorting, ISI (inter-spike interval-based) spike rate calculation, and Kaiser filtering (window length = 10001, alpha = 128) of the time-continuous spike rate. Because of the high signal-to-noise-ratio, which was >> 2:1, the spikes could be sorted by a single threshold algorithm. The spike occurrence was recorded whenever the potential exceeded the threshold voltage (magenta band in Fig. 3, top trace).

From the plot, we can see the H1-cell responses when the robot starts to perform alternating turns from around t = 4 s (Fig. 3). The spike rate of the H1-cell decreased with increasing turning radius, which is in agreement with the result from a model that predicts the cell responses as a function of turning radius and wall distance (see next section).

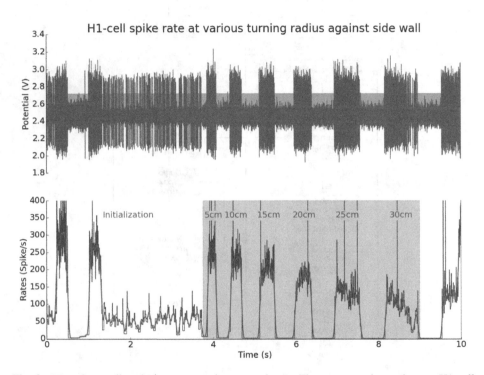

Fig. 3. Neural recording during an open-loop experiment. The top trace shows the raw H1-cell extracellular recording with its threshold (shown in magenta). The bottom subplot gives the spike rate calculated from the data presented in the top trace (blue curve is the ISI spike rate and red curve is the ISI rate filtered by Kaiser filtering (window length = 10001, alpha = 128). The robot initialization generates two activity peaks at the beginning. After 2 s of a stationary period, the robot starts to turn with different radii, which were 5, 10, 15, 20, 25, and 30 cm, shown in the grey area. Note that the average spike rate decreases with increasing turning radius. (Color figure online)

3.2   Verification of Closed-Loop Control Algorithm

We recorded a video from a ceiling-mounted camera, which showed that the robot was able to adjust the distance to the left wall, and avoid collision until the end of the corridor (Fig. 4).

The video can be found at https://www.youtube.com/watch?v=IEgyMUs3_QU.

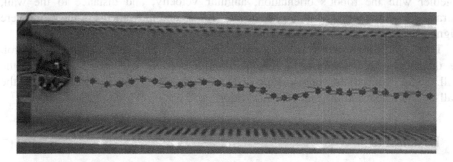

Fig. 4. The robot trajectory. The experimental arena was a tunnel with vertical stripes (gratings) attached to the walls (spatial wavelength: 30 mm or 16.2° from initial point). The green lines are the orientations (posterior) of the robot; the red dots are the positions of the robot along its trajectory recorded every 0.18 s. (Color figure online)

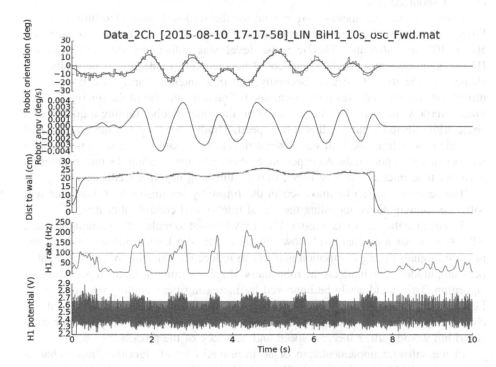

Fig. 5. The data from robot video and H1 neural recording. (1st trace) Robot orientation calculated from each frame of the video recording. The red curve is the robot orientation. The blue curve is the filtered robot orientation. (2nd trace) Robot angular velocity after filtering. (3rd trace) The distance between robot and wall calculated from each video frame. The blue curve shows the distance after filtering. (4th trace) H1-cell spike rate after smoothing with a filter kernel (Kaiser, window length = 10001, alpha = 128). (5th trace) The raw H1-cell extracellular recording data, with magenta band which indicates the threshold voltage for spike detection. (Color figure online)

The H1-cell activities were sampled using a data acquisition card. They are plotted together with the robot's orientation, angular velocity, and distance to the wall, extracted from the video recording. Neuronal signals and movements of the robot were aligned in time by a synchronisation signal (Fig. 5).

This control algorithm was successful in 5/12 trials under the bang-bang control, i.e. the robot reaching the end of the experimental corridor without colliding with the wall. Some of the trials were showing distance adjustment, but the robot touched the wall before it reached the end of the corridor.

4   Discussion

4.1   Wall Following as a First Approximation of Collision Avoidance

The control algorithm shown in Fig. 2 enabled the robot in several experiments to successfully follow the wall during oscillatory forward movement at an average distance of about 20 cm.

Three essential parameters were crucial for the wall-following algorithm to work. Firstly, a turning radius had to be chosen so that the H1-cell response just exceeded above 100 spikes/second. This response level was sufficiently different from the H1-cell's spontaneous activities and could be increased or decreased depending on changes of the robot trajectory. Secondly, the time interval during which the robot turned towards the wall was pre-programmed. The time interval of the microprocessor timer overflow interrupt was chosen based on the time of robot turning a quarter of a circle with selected turning radius at full speed. Thirdly, the length of the time interval of turning away from the wall was dependent on the H1-cell generating a pre-defined number of action potentials. As expected, higher spike rates during the turn resulted in a shorter time interval and *vice versa*, thus establishing negative feedback.

The success rate can be improved in the future by optimising both hardware and software configurations, regarding the signal fidelity and control efficiency.

To advance the hardware, a faster ADC may be used to replace the current on-board ADC, which has a comparatively low sampling rate of 5 kHz. Undersampling of the neuronal signals during digitisation is likely to reduce the detection of action potential peak amplitudes, which affects the robustness of spike sorting using a simple threshold algorithm. This could easily be improved by increasing the sampling rate to 20 kHz. Furthermore, spike sorting was performed on the same microprocessor used for closed-loop control. A separate spike sorting microprocessor with adaptive threshold algorithm would further increase speed and accuracy of the process.

In the software implementation of an improved control algorithm, more characterisations would be helpful to make the control algorithm more efficient. For example, multiple thresholds can be used to replace single threshold in bang-bang control, which brings finer adjustment during driving.

4.2 H1-cell Wall Distance Dependence and Its Model

The H1-cell does not directly encode any distance information. However, considering a third parameter, i.e. the turning radius, it is possible to estimate the distance of obstacle indirectly, based on H1-cell spike rate. We developed a phenomenological model to study the relationship between turning radius, distance and H1-cell spike.

Similar to retrograde and prograde movement of planets in Ptolemy's geocentric model of solar system, during self-motion the direction of optic flow generated by an optical system of an observer depends on the position of visual structures relative to the centre of the turning radius. If described in egocentric coordinates, visual structures positioned within a sphere that connects the observer with the centre of the turning radius move progressively (front-to-back; blue sphere in Fig. 6 left), while structures outside the sphere move regressively (back-to-front; red area in Fig. 6 left). We may transfer Ptolemy's model to the directional optic flow the H1-cell is confronted with while the fly engages on a trajectory that consists of a combination of translational and rotational self-motion described by a given turning radius, Rt. Relative motion of structures within the blue sphere (Fig. 6 left) would be in the anti-preferred direction of the cell while the motion of structures within the red area (Fig. 6 left) would be in the cell's preferred direction. Note that the receptive field of the H1-cell, i.e. the area within which motion affect the activity of the cell, comprises an entire visual hemisphere. In the following section, we will refer to the blue sphere in Fig. 6 (left) as the "inhibitory zone", and the red area as the "excitatory zone". The spike rate of H1-cell depends on the ratio between the inhibitory and excitatory zones and therefore on the turning radius of the robot.

The underlying principle becomes clear when considering the two extreme cases: A turning radius of zero is equivalent to the fly rotating on the spot, in which case the blue

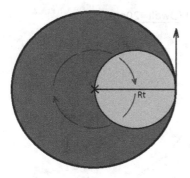
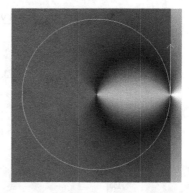

Fig. 6. (Left) Directional map of optic flow during turns of the robot at a finite turning radius, Rt. The red area indicates back-to-front optic flow, while the blue sphere indicates front-to-back optic flow. A cross indicates the turning centre. For further explanation, see text. (Right) Optic flow sensitivity field of H1-cell based on relative motion. The yellow arrow indicates the orientation of the robot. The yellow line describes the trajectory of the robot. The red zone is the excitatory region; the blue zone is the inhibitory region. The black zone is the region where there is no optic flow caused by relative rotation. The gradients indicate the scalar of the optic flow projected towards the observer. (Color figure online)

sphere completely vanishes and the cell is confronted only with motion in its preferred direction (for clockwise rotations – and only with motion in its anti-preferred direction for counter-clockwise rotations). An infinite turning radius, on the other hand, is equivalent to the robot moving on a straight line in the forward direction in which case the red area will vanish and the cell experiences only motion in its anti-preferred direction. This normally results in a perfect inhibition of H1-cell spiking. From this consideration, it follows that H1-cell response only contains information about distance when the robot is travelling on a finite turning radius.

After analysis of the open-loop experiment data, the H1-cell responses could be plotted as a function of distance to the visual surroundings (Fig. 7, red curve). However, when a wall partially reaches the inhibitory zone, H1-cell experiences both excitatory and inhibitory optic flow. Further explanation of signal integration at the level of optic flow processing interneurons can be found in publications from the Borst lab [21].

We developed a simple phenomenological model to estimate the H1-cell responses during robot movements along a given turning radius based on the directional motion of the surroundings as Eq. 1.

$$S(x,y) = \sin\left(\frac{\pi}{2} - \left|\cos^{-1}\left(\frac{(x-x_o)^2 + (y-y_o)^2 + (x-x_f)^2 + (y-y_f)^2 - (x_o-x_f)^2 - (x_o-y_f)^2}{2\sqrt{(x-x_o)^2+(y-y_o)^2}\sqrt{(x-x_f)^2+(y-y_f)^2}}\right)\right|\right)$$

(1)

Where $S(x,y)$ is the sensitivity to an object at a given location during relative rotation; (x_o, y_o) is the turning centre; (x_f, y_f) is the blowfly location.

An optic flow sensitivity map is plotted in Fig. 6 (right).

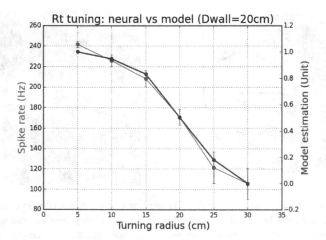

Fig. 7. The sensitivity to wall based optic flow during turns with various turning radius. The red curve shows experimental data obtained from H1-cell recordings and the blue curve gives the normalized output from the model simulation. In the experiments, the robot distance to the wall was 20 cm and spatial wavelength of the vertical bar pattern on the wall was 30 mm (as seen by the lateral eye of the fly). (N = 1, n = 2) (Color figure online)

In addition, a series of points along a line was specified in the model to simulate the wall in the experimental arena with a pattern of a certain spatial wavelength. The simulated wall was rotated from an azimuth angle of 135° to 45° (according to blowfly anterior direction) around a vertical axis that coincides with the ventral-dorsal axis of the fly, simulating rotation of a blowfly on a certain turning radius. When integrating the local motion (optic flow), we took into account the spatial resolution on the inter-ommatidial angle of the H1-cell receptive field [22].

The simulation result (Fig. 7) shows that the estimated sensitivity of the cell to a wall in relative motion was modulated by the turning radius; the smaller the turning radius, the higher the predicted H1-cell responses. The result (Fig. 7, blue trace) is in good agreement with neural recording data obtained under open-loop conditions (Fig. 7, red trace).

5 Summary

We were able to demonstrate that spike rate of the H1-cell provides spatial information about the visual surroundings under specific circumstances. At a fixed distance to the wall, the spike rates are modulated by the turning radius of the robot.

The current model estimates the spike rate as a function of two parameters: the turning radius and the distance to the wall. In the future, by fixing the turning radius in the model, the spike rate could be used to estimate the distance to the wall –information the robot needs to avoid a collision in a more generic environment.

In future experiments, we will analyse the integration of simultaneous excitatory and inhibitory local optic flow more systematically to derive ideal closed-loop control architectures for a collision-free operation of our fly-robot-interface. Further improvements may be achieved by including a forward model to calculate turning radius, in which the parameters such as the wheel speeds and the distance between wheels, can be locally measured, to predict the wall distance for collision avoidance control.

Acknowledgments. The authors would like to thank Caroline Golden for improving the proofreading the manuscript. This work was partially supported by US AFOSR/EOARD grant FA8655-09-1-3083 to HGK.

References

1. Fox, J.L., Frye, M.: Animal behavior: fly flight moves forward. Curr. Biol. **23**, R278–R279 (2013)
2. Reiser, M.B., Dickinson, M.H.: Visual motion speed determines a behavioral switch from forward flight to expansion avoidance in Drosophila. J. Exp. Biol. **216**, 719–732 (2013)
3. Krapp, H.G., Gabbiani, F.: Spatial distribution of inputs and local receptive field properties of a wide-field, looming sensitive neuron. J. Neurophysiol. **93**, 2240–2253 (2005)
4. Wicklein, M., Strausfeld, N.J.: Organization and significance of neurons that detect change of visual depth in the hawk moth Manduca sexta. J. Comp. Neurol. **424**, 356–376 (2000)

5. Blanchard, M., Rind, F.C., Verschure, P.F.M.J.: Collision avoidance using a model of the locust LGMD neuron. Robot. Auton. Syst. **30**, 17–38 (2000)
6. Bertrand, O.J.N., Lindemann, J.P., Egelhaaf, M.: A bio-inspired collision avoidance model based on spatial information derived from motion detectors leads to common routes. PLoS Comput. Biol. **11**, e1004339 (2015)
7. Kern, R., Boeddeker, N., Dittmar, L., Egelhaaf, M.: Blowfly flight characteristics are shaped by environmental features and controlled by optic flow information. J. Exp. Biol. **215**, 2501–2514 (2012)
8. Hausen, K.: Functional characterization and anatomical identification of motion sensitive neurons in the lobula plate of the blowfly Calliphora erythrocephala. Z. Naturforsch. **31c**, 629–633 (1976)
9. Krapp, H.G., Hengstenberg, R., Egelhaaf, M.: Binocular contributions to optic flow processing in the fly visual system. J. Neurophysiol. **85**, 724–734 (2001)
10. Longden, K.D., Krapp, H.G.: Octopaminergic modulation of temporal frequency coding in an identified optic flow-processing interneuron. Front. Syst. Neurosci. **4**, 153 (2010)
11. Maddess, T., Laughlin, S.B.: Adaptation of the motion-sensitive neuron H1 is generated locally and governed by contrast frequency. Proc. R. Soc. Lond. B Biol. Sci. **225**, 251–275 (1985)
12. Lewen, G.D., Bialek, W., de Ruyter van Steveninck, R.R.: Neural coding of naturalistic motion stimuli. Netw. Bristol Engl. **12**, 317–329 (2001)
13. Lindemann, J.P., Egelhaaf, M.: Texture dependence of motion sensing and free flight behavior in blowflies. Front. Behav. Neurosci. **6**, 92 (2013)
14. Huang, J.V., Krapp, H.G.: Miniaturized electrophysiology platform for fly-robot interface to study multisensory integration. In: Lepora, N.F., Mura, A., Krapp, H.G., Verschure, P.F.M.J., Prescott, T.J. (eds.) Living Machines 2013. LNCS, vol. 8064, pp. 119–130. Springer, Heidelberg (2013)
15. Huang, J.V., Krapp, H.G.: A predictive model for closed-loop collision avoidance in a fly-robotic interface. In: Duff, A., Lepora, N.F., Mura, A., Prescott, T.J., Verschure, P.F.M.J. (eds.) Living Machines 2014. LNCS, vol. 8608, pp. 130–141. Springer, Heidelberg (2014)
16. Huang, J.V., Krapp, H.G.: Closed-loop control in an autonomous bio-hybrid robot system based on binocular neuronal input. In: Wilson, S.P., Verschure, P.F.M.J., Mura, A., Prescott, T.J. (eds.) Living Machines 2015. LNCS, vol. 9222, pp. 164–174. Springer, Heidelberg (2015)
17. Koenderink, J.J., van Doorn, A.J.: Facts on optic flow. Biol. Cybern. **56**, 247–254 (1987)
18. Lindemann, J.P., Kern, R., van Hateren, J.H., Ritter, H., Egelhaaf, M.: On the computations analyzing natural optic flow: quantitative model analysis of the blowfly motion vision pathway. J. Neurosci. **25**, 6435–6448 (2005)
19. Franceschini, N.: Pupil and pseudopupil in the compound eye of Drosophila. In: Wehner, R. (ed.) Information Processing in the Visual Systems of Anthropods, pp. 75–82. Springer, Heidelberg (1972)
20. Karmeier, K., Tabor, R., Egelhaaf, M., Krapp, H.G.: Early visual experience and the receptive-field organization of optic flow processing interneurons in the fly motion pathway. Vis. Neurosci. **18**, 1–8 (2001)
21. Mauss, A.S., Pankova, K., Arenz, A., Nern, A., Rubin, G.M., Borst, A.: Neural circuit to integrate opposing motions in the visual field. Cell **162**, 351–362 (2015)
22. Petrowitz, R., Dahmen, H., Egelhaaf, M., Krapp, H.G.: Arrangement of optical axes and spatial resolution in the compound eye of the female blowfly Calliphora. J. Comp. Physiol. **186**, 737–746 (2000)

Sensing Contact Constraints in a Worm-like Robot by Detecting Load Anomalies

Akhil Kandhari[1]([✉]), Andrew D. Horchler[1], George S. Zucker[1],
Kathryn A. Daltorio[1], Hillel J. Chiel[2], and Roger D. Quinn[1]

[1] Department of Mechanical and Aerospace Engineering,
Case Western Reserve University, Cleveland, OH 44106-7222, USA
{axk751, rdq}@case.edu
[2] Departments of Biology, Neurosciences, and Biomedical Engineering,
Case Western Reserve University, Cleveland, OH 44106-7080, USA
hjc@case.edu

Abstract. In earthworms, traveling waves of body contraction and elongation result in soft body locomotion. This simple strategy is called peristaltic locomotion. To mimic this kind of locomotion, we developed a compliant modular worm-like robot. This robot uses a cable actuation system where the actuating cable acts like the circumferential muscle. When actuated, this circumferential cable contracts the segment diameter causing a similar effect to the contraction due to the circumferential muscles in earthworms. When the cable length is increased, the segment diameter increases due to restoring forces from structural compliance. When the robot comes in contact with an external constraint (e.g., inner walls of a pipe) continued cable extension results in both slack in the cable and inefficiency of locomotion. In this paper we discuss a probabilistic approach to detect slack in a cable. Using sample distributions over multiple trials and naïve Bayes classifier, we can detect anomalies in sampled data which indicate the presence of slack in the cable. Our training set included data samples from pipes of different diameters and flat surfaces. This algorithm detected slack within ±0.15 ms of slack being introduced in the cable with a success rate of 75 %. We further our research in understanding reasons for failure of the algorithm and working towards improvements on our robot.

Keywords: Robot · Worm · Soft · Probability · Naïve Bayes · Cable · Actuation

1 Introduction

During the crawling movements of the earthworms, sensory feedback provides the animal with an ability to adapt to different types of environmental perturbations that may occur [16]. The importance of sensory feedback for maintaining rhythmic crawling motion has been established [7]. Mechanosensory organs and stretch, touch, and pressure receptors are the feedback sources in earthworms [15]. Due to the flexibility in the earthworm's body, it is unable to sense its posture from only stretch receptors [16]. However, the sensory input activities from the setae allows it to adapt to its environment and crawl smoothly on rough surfaces.

© Springer International Publishing Switzerland 2016
N.F. Lepora et al. (Eds.): Living Machines 2016, LNAI 9793, pp. 97–106, 2016.
DOI: 10.1007/978-3-319-42417-0_10

Many robots use cable actuation methods, like the Softworm [1–3], exoskeleton-type robots [19], cable-actuated parallel manipulators [8] and cable-suspended robots [18]. In our cable actuated robot, slack occurs in the cable when the robot comes in contact with an external constraint. It is undesirable as it reduces the efficiency during peristaltic loco-motion. Detection of slack allows us to sense external constraints, which allows the robot to locomote efficiently. In this paper, our question is: can we recognize cable slack by searching for anomalies in feedback measurements from the actuators?

Anomaly detection techniques traditionally detect irregularities in new data after training on clean data [6]. Probabilistic approaches in predicting these anomalies have been studied widely [6, 13]. We use a probabilistic model to detect if slack has been introduced in our actuation cable to produce more efficient peristaltic locomotion [1] in our worm-like robot.

2   Slack Detection in a Cable-Actuated Robot

A significant issue faced when using cables is that they allow actuation in only one direction, i.e., a cable cannot be pushed. Initially, our Compliant Modular Mesh Worm (CMMWorm) robot had a bi-directional actuation system [9]. There were two sets of cables, one for circumferential actuation, (elongates the segment while making the diameter smaller) and one for longitudinal actuation (reduces segment length while increasing segment diameter). Due to nonlinear kinematics, the rate at which the longitudinal cables are spooled in is not equal to the rate at which the circumferential cables are spooled out. This difference in rates results in excess circumferential cable being spooled out. To prevent this excess cable from tangling or causing uneven wear, circumferential cable tensioning springs were required to keep the cable taut. Thus, cable slack was avoided, but as a result of the antagonistic cables, the design had high body stiffness.

For CMMWorm, Dynamixel MX-64T actuators are used. These actuators use PID (proportional-integral-derivative) control. Load feedback measures the torque applied by the actuator internally. We developed an open source library, DynamixelQ [12], for the OpenCM9.04 microcontroller used by CMMWorm for communication with AX and MX series Dynamixel actuators. It has been shown that the ability for a soft robot to respond to sensed load of its surroundings can permit efficient navigation, e.g., in narrowing pipes [4]. In the previous bi-directionally actuated iteration of CMMWorm [9], we used load measurements from the actuators to detect the walls of the pipe. Load measurements were sampled and then smoothed to filter noise using exponential smoothing. The data was then compared to a threshold value based on load data sampled. When a segment came in contact with the wall of a pipe the filtered load increased and exceeded the threshold. This allowed the algorithm to stop any further expansion of a segment.

In the current version of the CMMWorm (Fig. 1), only circumferential cables are used for actuation. The longitudinal cables have been replaced by six springs along the length of each segment such that, as the circumferential cable spools out, the springs passively return the segment to the maximum diameter allowed by the circumferential cable. In this simpler design, the spring stiffness need only be sufficient to restore the

Fig. 1. The six-segment Compliant Modular Mesh Worm (CMMWorm) robot crawling through a 20.3 cm inner diameter pipe. The linear springs that passively return the robot to its maximum diameter are visible along the length of each segment.

segments to their initial maximum diameter. The design is more robust, however, without the longitudinal cable, obstacle detection relies on detecting when the circumferential cable goes slack, rather than detecting an increase in longitudinal cable tension.

By reconfiguring the robot to use only one cable (circumferential), we can improve the mechanics of the motion (reducing cable friction losses, plastic deformation of the mesh, and limitations on range of motion), but detecting contact becomes a challenge. As the circumferential cable is pulled onto the spool, the diameter of a segment is constricted due to the length of cable available. In the other direction, when the cable is unspooled, the longitudinal springs expand the segment diameter as much as the increased cable length allows. However, when a segment comes in contact with an external constraint, such as that from the interior wall of a pipe, the segments stop expanding in proportion to the cable length that is unspooled and the cable can become slack. This causes reduced efficiency in responsive peristalsis that depends on accurate detection of contact [4]. Thus, a new way to control the segments of the robot was required.

It was observed that when the cable did experience slack, there was a small spike in the load values being recorded by the actuator. These load values are noisy and change based on the PID responses to cable tension. The small peaks due to slack can be thought of as anomalies in the load data being obtained. Hence, we approached this problem using a statistical probabilistic model. The problem then becomes to find the probability of slack occurring in the system given the load value that has been measured. If sampled values are not within the measured distribution, there is a higher probability that the segment has come in contact with an external force and the cable has gone slack.

3   Methods

When working with the robot, the actuators are configured to use a position control scheme. Position limits are specified and the actuator will rotate between these preset limits for a specified number of rotations (three rotations allow each segment to move from maximum to minimum diameter). The actuators are run at their maximum speed and with no torque limits. Load measurements gathered from an actuator, although noisy, tend to spike when slack is introduced in the cable. This is analogous to lifting an object using a cable; if the cable is cut, a short jerk is felt due to the sudden change in tension. If it is possible to detect this spike within the noisy data, then it is possible to determine that the segment has come in contact against an external obstacle and should stop further expansion in order to avoid introducing slack.

To do so, a probabilistic approach is used to capture the occurrence of slack in each segment of CMMWorm by measuring the distribution of load values in the cases of slack and no slack. Load data is gathered from the actuators by expanding each segment in different diameter pipes and visually observing when the cable becomes slack. Slack is visually determined when the circumferential cable tensioning springs return to their equilibrium state. We then ensured that the segment was anchored against the walls of the pipe. Measurement of load values were continued after slack was introduced in order to collect data for the case of slack. The noisiness of the data and structural variability [10] leads to inconsistent training sets between segments. Due to nonlinear frictional properties, fraying of cables over time, and minor repairs being made over various runs, an individual segment also has variability in the load measurements recorded. This means that the distributions of load values (given slack or no slack) will be different for each segment and will change every time an adjustment is made to a segment.

To overcome this problem, we used a moving median filter for each actuator. Then the filtered median of the previous ten load measurements is subtracted from the current load measurement. This is referred to as the "relative median load" (RML). The absolute value of this difference is used such that the relative median load is always positive (Fig. 2). This helps filter variability in the range of load values among different segments, as well as within a segment, yielding a constant range of data for the entire robot.

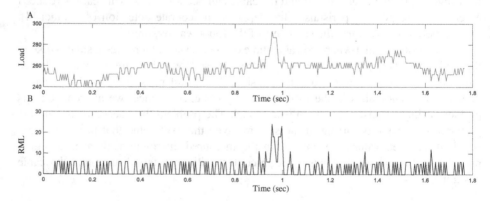

Fig. 2. (A) Load measurements from the actuator during expansion of a segment in a 20.3 cm inner diameter pipe. (B) The median filter measures the median of the previous ten load measurements. The output from this filter is then subtracted from the current load measurement. The absolute value of this is the relative median load (RML). This helps normalize the load data from different segments and use a single probability distribution for the entire robot. The spike in the above graphs at 0.95 s indicates that the cable has gone slack.

Based on the RML values, we build our distributions (Fig. 3). Data was gathered through ten runs for each of the six segments in 17.4 cm and 20.3 cm inner diameter pipes and on a flat surface. On flat surfaces the segments could expand to their maximum diameter without obstruction. Load values from these data sets were divided into two parts: load given slack in the cable and load given no slack in the cable. From these distributions it is observed that higher load values are more prominent when there is slack in the cable as compared to when there is no slack in the cable. Using naïve Bayes classification the possibility of slack occurring given a particular load value can be predicted based on probabilities, P, obtained from the training data.

The probability of a load measurement V_t at time t given slack in the cable, S is

$$P(V_t \mid S) = \frac{P(V_t)P(S \mid V_t)}{P(S)} \rightarrow P(S \mid V_t) = \frac{P(S)P(V_t \mid S)}{P(V_t)} \tag{1}$$

Similarly, the probability of a load measurement V_t at time t given no slack in the cable, $\bar{S}$ is

$$P(V_t \mid \bar{S}) = \frac{P(V_t)P(\bar{S} \mid V_t)}{P(\bar{S})} \rightarrow P(\bar{S} \mid V_t) = \frac{P(\bar{S})P(V_t \mid \bar{S})}{P(V_t)} \tag{2}$$

Dividing (1) by (2), the ratio of likelihood of slack versus no slack given a particular load is

$$\frac{P(S \mid V_t)}{P(\bar{S} \mid V_t)} = \frac{P(S)P(V_t \mid S)}{P(\bar{S})P(V_t \mid \bar{S})} \tag{3}$$

To prevent the algorithm from identifying noise as slack, the previous five load measurements are used to find the product of probabilities of these measurements. This product is a ratio of probability of slack to no slack given a load value and is called the slack confidence (SC), where

$$SC = \prod_{i=0}^{4} \frac{P(S)P(V_{t-i} \mid S)}{P(\bar{S})P(V_{t-i} \mid \bar{S})} \tag{4}$$

This helps not classify a noisy measurement for slack. Since the SC can be a large value, its natural logarithm, ln(SC), is used for plotting purposes throughout this paper. Through testing it was found that it took at least five probability values to localize a peak caused by slack in the cable.

A peristaltic controller needs to coordinate the simultaneous expansion and contraction of segments in order to minimize slip. At any given point, there are two active segments, one contracting and the other expanding in diameter. In between the two active segments there is a contracted inactive segment referred to as the "spacer segment" [10]. The remaining three inactive segments are expanded to the permissible limit and help in anchoring.

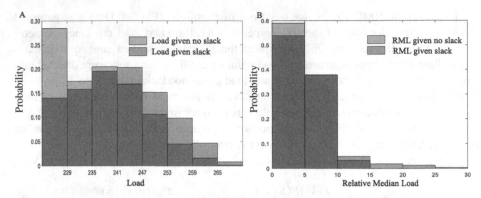

Fig. 3. Distributions showing the probability of load measurements (A) and relative median load measurements (RML) (B) given no cable slack (blue) and given cable slack (orange). The distributions show that the probability of obtaining higher load values decreases when there is no slack in the cable and the probability of obtaining higher load values increases when there is slack in the cable. The same trend is observed in the RML values. The RML distributions (B) are used to calculate slack confidence (SC) on the microcontroller by assigning probabilities to the load values obtained from the actuators. (Color figure online)

Once the wave is initiated, load measurements are recorded from the expanding segment after 150 ms to ignore transients that occur when the actuator starts from rest. After the actuator stabilizes, load measurements are recorded from the actuator of the expanding segment at 200 Hz. The slack confidence (SC) is then calculated by the microcontroller using Eq. (4). When slack occurs the SC has a large value compared to when there is no slack in the cable. At this point a distinct peak is observed (Fig. 4) and expansion of the segment is stopped and the control transitions to the next expanding segment. The wave travels down the length of the robot repeating the algorithm in each expanding segment. For the contracting segments a preset minimum position allows it to return to its initial configuration.

4   Results

Twenty runs each were done in 15.2 cm, 17.4 cm, and 20.32 cm inner diameter pipes and on a flat surface. The results shown in Table 1 summarize the success rate of slack being detected when the segment came in contact with the wall of the pipe or, in the

Table 1. $N = 20$ runs each were done in 15.2 cm, 17.4 cm, and 20.3 cm inner diameter pipes and on a flat surface. The percentage of successful runs, false positive errors, and false negative errors for the four different cases is shown.

	Successful runs	False positive errors	False negative errors
15.2 cm Pipe	70 %	25 %	5 %
17.4 cm Pipe	70 %	20 %	10 %
20.3 cm Pipe	75 %	25 %	0 %
Flat surface	85 %	15 %	0 %

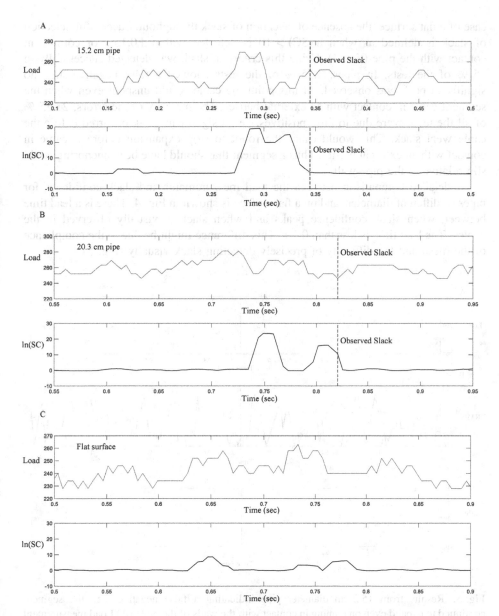

Fig. 4. Load measurements and natural logarithm of slack confidence, ln(SC), from the actuator in three different cases, (A) 15.2 cm diameter pipe, (B) 20.32 cm diameter pipe, and (C) flat surface. In (A) and (B) the dashed vertical line depicts when slack is visually observed on the robot. The slack confidence has a significant peak (ln(SC) > 10) close to when slack is introduced in the cable in both cases. In (C), minor peaks are observed, but are not significant compared to when slack is introduced in the cable due to the walls of the pipe.

104   A. Kandhari et al.

case of a flat surface, the absence of detection of slack throughout. Successful detection of slack is defined as when ln(SC) > 10 is detected within ±0.15 s of coming in contact with the pipe (Fig. 4). Using this criterion, slack was detected successfully in 75 % of the tests. In pipes, 3.75 % of the tests showed a false negative, i.e., no significant peak was observed. At this point the cable would unspool even when the segment came in contact with an external barrier. The majority of the errors, 21.25 % of all the tests, were due to false positives, i.e., a significant peak occurred before the cable went slack. This would cause the robot to stop expansion before it came in contact with an external barrier. Thus a segment that should have been anchoring could slip relative to the pipe wall.

A detailed comparison between the load measurements and slack confidence for pipes of different diameters and on a flat surface is shown in Fig. 4. There is a lead time between when slack confidence peaks and when slack is visually observed in the cables. The lead time is less than 0.1 s. This difference might be due to the compliance of the mesh and the difficulty of precisely detecting slack visually (Fig. 5).

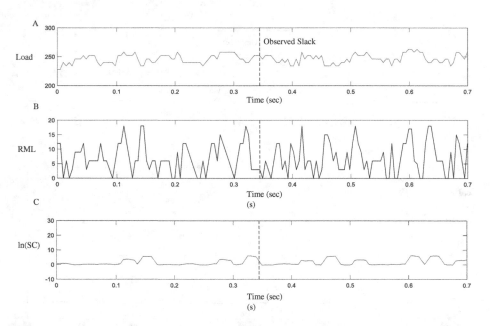

Fig. 5. Results from 15.2 cm diameter pipe indicating a false negative, i.e., the segment continued to expand even on coming in contact with the walls of the pipe. (A) Load measurement from the actuator, where no significant peak is observed. (B) The relative median load (RML) shows no significant peak throughout the test to quantify a point where slack is introduced. (C) The logarithm of SC shows a small spike when slack is introduced in the cable but it is not large enough to indicate that slack has been introduced.

False positive and false negative errors were observed in 25 % of runs. False positive errors cause inefficiency as it might cause the segment to slip. These errors might occur because of fraying of cables, high frictional loads on the cable, and damaged components. False negative errors also cause inefficiency in the locomotion, as the robot is not able to adapt to its environment. False negative errors also cause excess cable being spooled out. This cable can get tangled with other components and might damage or break itself. This error might be due to the high amount of noise causing interference with the sampled data when slack occurs. False negative errors might also be an indication of plastic deformation of the mesh, where the loads are not uniformly transferred along the mesh of the segment.

5 Conclusions

The predictive model of detecting slack using probability distributions has allowed us to redesign our robot using a single cable for actuation. We can now detect slack in different diameter pipes along different segments using this algorithm. Using the sensory capabilities of our current actuators, we might be able to eliminate the need for additional sensors on the robot. Our future work will include achieving higher accuracy within a smaller time interval using a more dynamic method of predicting load measurements, such as Kalman filters [14]. There have also been studies on detecting anomalies using Dynamic Bayesian Networks [5, 13, 17]. We also want to apply a similar feedback algorithm in conjunction with dynamic oscillator networks, such as stable heteroclinic channels [4, 11] that could help achieve a more adaptive gait pattern.

Currently our robot can react to static environments, by controlling the expansion of each segment on coming in contact with an external constraint, but what if the environment was to change suddenly? If the robot slipped while locomoting, would it be able to adjust to such a change? In the future, the worm-like robot could continuously check for an external load and not just during expansion. In this way any change in the environment will trigger it to expand or contract in order to comply to its surroundings.

Acknowledgements. This work was supported by NSF research grant No. IIS-1065489. The authors would like to thank Dr. Soumya Ray and Kenneth Moses for their help during the course of this project.

References

1. Boxerbaum, A.S., Chiel, H.J., Quinn, R.D.: A new theory and methods for creating peristaltic motion in a robotic platform. In: Proceedings of IEEE International Conference on Robotics and Automation, Anchorage, pp. 1221–1227 (2010)
2. Boxerbaum, A.S., Shaw, K.M., Chiel, H.J., Quinn, R.D.: Continuous wave peristaltic motion in a robot. Int. J. Robot. Res. **31**, 302–318 (2012)
3. Boxerbaum, A.S., Horchler, A.D., Shaw, K.M., Chiel, H.J., Quinn, R.D.: A controller for continuous wave peristaltic locomotion. In: Proceedings of IEEE International Conference on Intelligent Robots and Systems, San Francisco, pp. 197–202 (2011)

4. Daltorio, K.A., Boxerbaum, A.S., Horchler, A.D., Shaw, K.M., Chiel, H.J., Quinn, R.D.: Efficient worm-like locomotion: slip and control of soft-bodied peristaltic robots. Bioinspir. Biomim. **8**, 035003 (2013)
5. Dean, T., Kanazawa, K.: A model for reasoning about persistence and causation. Comput. Intell. **5**(3), 142–150 (1990)
6. Eskin, E.: Anomaly detection over noisy data using learned probability distributions. In: Seventeenth International Conference on Machine Learning, pp. 255–262 (2000)
7. Gray, J., Lissmann, J.W.: Locomotory reflexes in the earthworm. J. Exp. Biol. **15**, 5006–5017 (1938)
8. Hassan, M., Khajepour, A.: Analysis of bounded cable tensions in cable-actuated parallel manipulators. IEEE Trans. Robot. **27**(5), 891–900 (2011)
9. Horchler, A.D., Kandhari, A., Daltorio, K.A., Moses, K.C., Andersen, K.B., Bunnelle, H., Kershaw, J., Tavel, W.H., Bachmann, R.J., Chiel, H.J., Quinn, R.D.: Worm-like robotic locomotion with a compliant modular mesh. In: Wilson, S.P., Verschure, P.F., Mura, A., Prescott, T.J. (eds.) Living Machines 2015. LNCS, vol. 9222, pp. 26–37. Springer, Heidelberg (2015)
10. Horchler, A.D., Kandhari, A., Daltorio, K.A., Moses, K.C., Ryan, J.C., Stultz, K.A., Kanu, E.N., Andersen, K.B., Kershaw, J., Bachmann, R.J., Chiel, H.J., Quinn, R.D.: Peristaltic locomotion of a modular mesh-based worm robot: precision, compliance, and friction. Soft Robotics **2**(4), 135–145 (2015)
11. Horchler, A.D., Daltorio, K.A., Chiel, H.J., Quinn, R.D.: Designing responsive pattern generators: stable heteroclinic channel cycles for modeling and control. Bioinpir. Biomim. **10**(2), 026001 (2015)
12. Horchler, A.D.: DynamixelQ Library, version 1.2. https://github.com/horchler/DynamixelQ
13. Liang, K., Cao, F., Bai, Z., Renfrew, M., Cavusoglu, M.C., Podgurski, A, Ray, S.: Detection and prediction of adverse and anomalous events in medical robots. In: Proceedings of Twenty-Fifth IAAI Conference, pp. 1539–1544 (2013)
14. Manfredi, V., Mahadevan, S., Kurose, J.: Switching Kalman filters for prediction and tracking in an adaptive meteorological sensing network. In: Proceedings of Second Annual IEEE Communications Society Conference on Sensor and AdHoc Communications and Networks (SECON), pp. 197–206 (2005)
15. Mill, P.J.: Recent developments in earthworm neurobiology. Comp. Biochem. Physiol. **12**, 107–115 (1982)
16. Mizutani, K., Shimoi, T., Ogawa, H., Kitamura, Y., Oka, K.: Modulation of motor patterns by sensory feedback during earthworm locomotion. Neurosci. Res. **48**(4), 457–462 (2004)
17. Pavlovic, V., Rehg, J.M., Murphy, K.P.: A dynamic Bayesian network approach to figure tracking using learned dynamic models. In: Proceedings of the Seventh IEEE International Conference on Computer Vision, vol. 1, pp. 94–101 (1999)
18. Roberts, R., Graham, T., Lippitt, T.: On the inverse kinematics, statics, and fault tolerance of cable-suspended robots. J. Robot. Syst. **15**(10), 581–597 (1998)
19. Veneman, J.F.: A series elastic- and Bowden-cable-based actuation system for use as torque actuator in exoskeleton-type robots. Int. J. Robot. Res. **25**(3), 261–281 (2006)

Head-Mounted Sensory Augmentation Device: Comparing Haptic and Audio Modality

Hamideh Kerdegari[1(✉)], Yeongmi Kim[2], and Tony J. Prescott[1]

[1] Sheffield Robotics, University of Sheffield, Sheffield, UK
{h.kerdegari,t.j.prescott}@sheffield.ac.uk
[2] Department of Mechatronics, MCI, Innsbruck, Austria
yeongmi.kim@mci.edu

Abstract. This paper investigates and compares the effectiveness of haptic and audio modality for navigation in low visibility environment using a sensory augmentation device. A second generation head-mounted vibrotactile interface as a sensory augmentation prototype was developed to help users to navigate in such environments. In our experiment, a subject navigates along a wall relying on the haptic or audio feedbacks as navigation commands. Haptic/audio feedback is presented to the subjects according to the information measured from the walls to a set of 12 ultrasound sensors placed around a helmet and a classification algorithm by using multilayer perceptron neural network. Results showed the haptic modality leads to significantly lower route deviation in navigation compared to auditory feedback. Furthermore, the NASA TLX questionnaire showed that subjects reported lower cognitive workload with haptic modality although both modalities were able to navigate the users along the wall.

Keywords: Sensory augmentation · Haptic feedback · Audio feedback · Classification algorithm

1 Introduction

Sensory augmentation is an exciting domain in human-machine biohybridicity that adds new synthesized information to an existing sensory channel. The additional senses provided by sensory augmentation can be used to augment the spatial awareness of people with impaired vision [1–5] or for people operating in environments where visual sensing is compromised such as smoked-filled buildings [6–8].

The sensitive tactile sensing capabilities supported by facial whiskers provide many mammals with detailed information about local environment that is useful for navigation and object recognition. Similar information could be provided to humans using a sensory augmentation device that combines active distance sensing of nearby surfaces with a head-mounted tactile display [6, 7]. One of the attempts to design such a device was the 'Haptic Radar' [7] that linked infrared sensors to head-mounted vibrotactile displays allowing users to perceive and respond simultaneously to multiple spatial information sources. In this device, several sense-act modules were mounted together on a band wrapped around the head. Each module measured distance from the user to nearby surfaces, in the direction of the sensor, and transduced this information into a

© Springer International Publishing Switzerland 2016
N.F. Lepora et al. (Eds.): Living Machines 2016, LNAI 9793, pp. 107–118, 2016.
DOI: 10.1007/978-3-319-42417-0_11

vibrotactile signal presented to the skin directly beneath the module. Users intuitively responded to nearby objects, for example, by tilting away from the direction of an object that was moving close to the head, indicating that the device could be useful for detecting and avoiding collisions. Marsalia [9] has evaluated the effectiveness of a head-mounted display in improving hazard recognition for distracted pedestrians using a driving simulator. Results showed that response hit rates improved and response times were faster when participants had a display present.

The above studies indicate the value of head-mounted haptic display for alerting wearers to possible threats. The 'Tactile Helmet' [6] was a prototype sensory augmentation device developed by the current authors that aimed to be something more than a hazard detector—a device for guiding users within unsafe, low-visibility environments such as burning buildings. We selected a head-mounted tactile display as this facilitates rapid reactions, can easily fit inside a modified fire-fighter helmet, and leaves the hands of the firefighters free for tactile exploration of objects and surfaces. Our first generation device (see Fig. 1) comprised a ring of eight ultrasound sensors on the outside of a firefighter's safety helmet with four voice coil-type vibrotactile actuators fitted to the inside headband. Ultrasound distance signals from the sensors were converted into a pattern of vibrotactile stimulation across all four actuators. One of the goals of this approach was to have greater control over the information displayed to the user, and, in particular, to avoid overloading tactile sensory channels by displaying too much information at once.

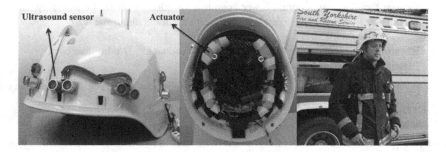

Fig. 1. The first generation 'Tactile Helmet' [6] was composed of a ring of ultrasound sensors and four actuators inside the helmet and was designed to help firefighter's navigate inside smoked-filled buildings.

Auditory guidance in the form of non-verbal acoustic sound or synthetic speech is another means for providing augmented navigation information for people with visually impairments or for rescue workers [3–5].

The effectiveness of haptic and audio modalities have been compared in a number of augmented navigation tasks with mixed results. For example, in [3], audio and haptic interfaces were compared for way finding by blind pedestrians and it was found that haptic guidance resulted in closer path-following compared to audio feedback. Marston et al. [10] also evaluated nonvisual route-following with guidance from audio and haptic display. Their results showed that haptic feedback produced slightly faster path completion time and shorter distance, however, there was no significant difference between audio and haptic modality. In [11], multimodal feedback strategies (haptic, audio and combined) were compared. Whilst there were no significant differences between modalities in navigation performance, subjects reported that the audio guidance was less comfortable than others. Kaul et al. [12] have evaluated audio and haptic guidance in a 3D virtual object acquisition task using HapticHead (a cap consisting of vibration motors) as a head-mounted display. User study indicated that haptic feedback is faster and more precise than auditory feedback for virtual object finding in 3D space around the user. Finally, in [13] haptic and audio modalities were compared in terms of cognitive workload, in a short-range navigation task, finding that workload was lower in haptic feedback compared to audio for blind participants.

The aim of the current paper is to evaluate and compare audio and haptic guidance for navigation using a head-mounted sensory augmentation device. We designed a second-generation vibrotactile helmet as a sensory augmentation device for fire fighters' navigation that sought to overcome some of the limitations of our first prototype (Fig. 1) [6] such as low-resolution tactile display. We previously investigated how to design our tactile interface worn on the forehead [14] to present useful navigational information as a tactile language [15]. Here, we use this tactile language to generate haptic guidance signals and compare this to audio guidance in the form of synthetic speech. In order to simulate a wall-following task similar to that faced by fire-fighters exploring a burning building, we constructed temporary walls made of cardboard in the experimental room and asked subjects to follow these walls using the two alternative guidance systems. The vibrotactile helmet uses ultrasound sensors to detect the user's distance to the walls and then a neural network algorithm to determine appropriate guidance commands (Go-forward/Turn right/Turn left). We evaluated the effectiveness of haptic and audio guidance according to the objective measures of task completion time, distance of travel and route deviation, and subjective measure of workload measurement using NASA TLX questionnaires.

2 Method

2.1 Subjects

Ten participants - 4 men and 6 women, average age 25 - voluntarily took part in this experiment. All subjects were university students or staff. The study was approved by the University of Sheffield ethics committee, and participants signed the informed consent form before the experiment. They did not report any known abnormalities with haptic perception.

2.2 Vibrotactile Helmet

The second generation vibrotactile helmet (Fig. 2) consists of an array of twelve ultrasound sensors (I2CXL-MaxSonar-EZ2 by MaxBotic), a tactile display composed of 7 tactors (Fig. 2 (b)) [14], a sound card, a microcontroller unit and two small lithium polymer batteries (7.4 V) to provide the system power. Furthermore, five reflective

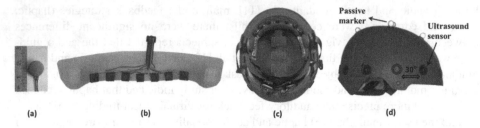

Fig. 2. (a) Eccentric rotating mass vibration motor (Model 310-113 by Precision Microdrives). (b) Tactile display interface. (c) Tactile display position inside the helmet. (d) Vibrotactile helmet.

passive markers were attached to the vibrotactile helmet surface (Fig. 2 (d)) to enable us to track the user's position and orientation using Vicon motion capture system.

Twelve ultrasound sensors were mounted with approximately 30 degrees separation to the outside of a skiing helmet (Fig. 2 (d)). The ultrasound sensors are employed sequentially one at a time. A minimum pulse-pause time of 50 ms is maintained between consecutive readings to make measurements more stable against ultrasound reflections. Using a 50 ms pulse-pause time, a complete environmental scan is accomplished every 0.6 s. The practical measuring range by this ultrasound sensor is between 20 cm and 765 cm with 1 cm resolution. The tactile display consists of seven eccentric rotating mass (ERM) vibration motors (Fig. 2 (a)) with 3 V operating voltage and 220 Hz operating frequency at 3 V. These vibration motors are mounted on a neoprene fabric and attached on a plastic sheet (Fig. 2 (b)) with 2.5 cm inter-tactor spacing which can easily be adjusted inside the helmet. Furthermore, a sound card was connected to the microcontroller to produce the synthetic speech for audio modality. The ultrasound sensors data are sent to the microcontroller through I2C BUS. The microcontroller in the helmet reads the sensors values and sends them to the PC wirelessly using its built-in WiFi support. The PC receives the sensor values and performs the required processing then generates commands, sending them back to the microcontroller wirelessly for onward transmission to the tactile display/sound card.

2.3   Haptic and Audio Guidance

In low visibility environments, firefighters navigate using the existing infrastructure such as walls and doors. These reference points help them to stay oriented and make a mental model of the environment [16]. To facilitate this form of navigation behavior we used a wall-following approach inspired by algorithms developed in mobile robotics that maintain a trajectory close to walls by combining steering-in, steering-out and moving forward commands [17]. Specifically, to navigate the user along the wall, we utilized three commands: turn-left, turn-right, and go-forward. The turn-left/right commands are intended to induce a rotation around the user (left/right rotation) in order to control the orientation of the user; the go-forward command is intended to induce forward motion. These three commands are presented to users in the form of haptic and audio feedback.

Haptic feedback in the form of vibrotactile patterns is used to present the commands to the user through the vibrotactile display. Figure 3 illustrates the positions of tactors in the vibrotactile display and vibrotactile patterns for presenting different commands. Note that tactor 4 is placed in the center of forehead. The turn-left command starts from tactor 3 and ends with tactor 1 while turn-right starts from tactor 5 and finishes with tactor 7. Go-forward command starts from tactor 3 and tactor 5 simultaneously and ends with tactor 4. We already investigated the utility and user experience of these commands as our tactile language using the combination of two command presentation modes— continuous and discrete and two command types— recurring and single [18]. Results showed that "recurring continuous (RC)" tactile language improved the performance better than other commands.

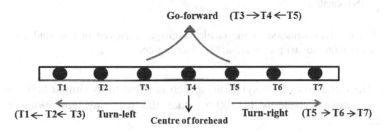

Fig. 3. Vibrotactile patterns for presenting turn-left, turn-right and go-forward commands in the tactile display.

The continuous presentation mode takes advantage of the phenomena of tactile apparent movement [19]. Specifically when two or more tactors are activated sequentially within a certain time interval, subjects experience the illusionary sensation of a stimulus travelling continuously from the first stimulation site to the second. The two main parameters that control the feeling of apparent motion are the duration of stimulus (DoS) and the stimulus onset asynchrony (SOA) [20]. In the current study, a DoS of 400 ms and a SOA of 100 ms were utilized respectively. This results in a total rendering time of 600 ms for turn right/left commands and 500 ms for go-forward command. However, in the discrete presentation mode the tactors are activated sequentially with no stimulus overlap that creates the experience of discrete motion across the forehead for all three commands. As command type, recurring condition presents the tactile command to the user's forehead repeatedly with interval between patterns of 500 ms until a new command is received; while for the single condition the tactile command is presented just once when there is a change in the command. A schematic representation of continuous command presentation and recurring command type for the turn-left command is presented in Fig. 4.

An alternative modality to haptic is audio modality through spoken direction [3]. Similar to the haptic modality, our audio modality also uses three commands to navigate the user along the wall. However, rather than using tactile language for presenting these commands, the following synthetic speech is applied: Go-forward, Turn-right and

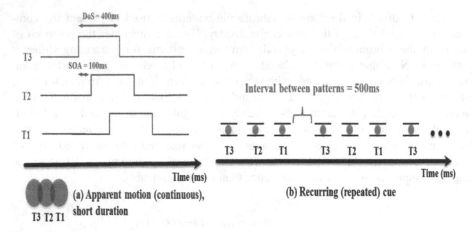

(a) Apparent motion (continuous), short duration

(b) Recurring (repeated) cue

Fig. 4. Schematic representation of the tactile language employed in this study. (a) Tactile apparent motion (continuous) presentation, (b) recurring cue.

Turn-left. The duration of each synthetic speech is equal to its similar haptic one and the interval between patterns is 500 ms like the recurring condition in haptic commands.

2.4 Procedure

We made a path consisting of several cardboard walls in the experiment room to navigate the subjects along it (Fig. 5 (a)). In order to track the subject's position and orientation during the navigation, we used a Vicon motion capture system in the experiment room. At the beginning of the experiment, each subject was invited into the experiment room and asked to wear the tactile helmet and a blindfold. They were not be able to see the experiment set-up and cardboard walls before starting the experiment. Participants were told that haptic/audio feedback would assist them to follow the walls either by turning to the left or right or by maintaining a forward path. Subjects were also asked to put on headphone playing white noise to mask any sounds from tactors during navigation with haptic feedback. Furthermore, subjects were asked to keep their head oriented in the direction of travel and to avoid making unnecessary sideways head movements. A short training session was then provided to familiarize subjects with the tactile language, audio feedback and with the experimental set-up. Once the participant felt comfortable, the trial phase was started. We considered two starting points (1 and 2 as shown in Fig. 5 (a)) to not let the subjects to memorize the paths. Blind-folded subjects (Fig. 5 (b)) started the first trial from position 1 and the second trial from position 2 and repeated it for the third and fourth trial. When each trial finished, subjects were stopped by the experimenter. Subjects were allowed to rest after each trial and started the next trial whenever they were ready. The maximum duration of the experiment was approximately 20 min. In total, each subject performed 4 trials including 2 feedback types (haptic and audio), each for two times in a pseudo-random

Fig. 5. (a) Overhead view of the experimental set-up consisting of cardboard walls and motion capture cameras, position 1 and 2 show the trial stating points. The length of the walls from the start point to the end is 20 m. (b) Subject is navigating along the wall.

order. Task completion time, travel distance and route deviation as objective measures for each trial were measured.

After finishing the experiment, subjects were asked to complete a paper and pencil version of the NASA task load index (TLX) [21] to measure subjective workload. It consists of six dimensions including mental demand, physical demand, temporal demand, performance, effort and frustration with 21 graduations. Additionally, subjects were asked to rate their preference for completing the task with audio and haptic modality.

3 Classification Algorithm

The wall-following task as a pattern classification problem is nonlinearly separable which is in favor of multilayer perceptron neural network [22]. In this work, multilayer perceptron neural network algorithm was utilized to guide the user along the wall as one of common methods used for robot navigation using ultrasound sensors [23]. As a classification algorithm, it associates the ultrasound data to the navigation commands (go-forward and turn right/left) in the form of haptic or audio modality. In order to collect data for training the classification algorithm, the experimenter wore the helmet and kept the laptop in her hands and followed the cardboard walls in the experiment room without wearing a blindfold (Fig. 5 (a)). The datasets are the collection of ultrasound readings when the experimenter follows the walls in a clockwise and anti-clockwise direction, each for 8 rounds. The data collection was performed at a rate of 1 sample (from 12 ultrasound sensors) per 0.6 s and generated a database with 4051 samples. Data were labeled during data collection by pressing the arrow key on the laptop keyboard when turning or going forward is intended (pressing left/right arrow key button for turn left/right and up arrow key button for go-forward). Ultrasound data in every scan were saved with a related label in a file. Three classes were considered in all the files: (1) Go-forward (2) Turn-right and (3) Turn-left and were used to train the classifier. We used Multilayer Perceptron (MLP) neural network to classify ultrasound data into three navigation commands. Our MLP (as shown in Fig. 6) consists of 12 input nodes (distance measurement form 12 ultrasound sensors), 1 hidden layer with 15

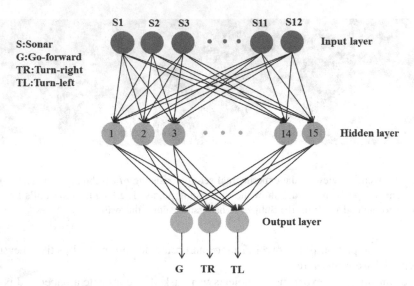

Fig. 6. The structure of the proposed MLP. It consists of 12 input nodes, 15 hidden nodes and 3 outputs.

nodes and 3 nodes in output layer (three navigation commands). Back propagation algorithm was used to train the data and evaluation was done using 10 times 10-Folds cross-validation. Sensitivity and specificity of the MLP algorithm for recognizing go-forward (G), turn-right (R) and turn-left (L) commands are defined as:

$$\text{Sensitivity} = \frac{TP_{G,R,L}}{TP_{G,R,L} + FN_{G,R,L}} \tag{1}$$

$$\begin{aligned}
\text{Specificity (G)} &= \frac{TP_R + TP_L}{TP_R + TP_L + FP_G}, \\
\text{Specificity(R)} &= \frac{TP_L + TP_G}{TP_L + TP_G + FP_R}, \\
\text{Specificity (L)} &= \frac{TP_R + TP_G}{TP_R + TP_G + FP_L}
\end{aligned} \tag{2}$$

where $TP_{G,R,L}$ (True Positive) corresponds to successfully classified Go-forward, Turn-right and Turn-left commands, $FP_{G,R,L}$ (False Positive) corresponds to erroneously classified Go-forward, Turn-right and Turn-left commands and $FN_{G,R,L}$ (False Negative) corresponds to missed Go-forward, Turn-right and Turn-left commands [24].

The overall accuracy of the MLP is 94.9 %. Table 1 presents the results of sensitivity and specificity of the MLP algorithm for recognizing the go-forward and turning commands. Finally, the proposed trained MLP algorithm was used to navigate the subjects along the walls.

Table 1. Sensitivity and specificity for recognizing go-forward and turning commands.

	Go-forward	Turn-right	Turn-left
Sensitivity (%)	95.9	95.3	90.6
Specificity (%)	92.7	98.6	98

4 Results

An alpha value of 0.05 was chosen as the threshold for statistical significance, all reported p-values are two-tailed. Shapiro-Wilk test showed that data are normally distributed. We measured task completion time (minute), travel distance (meter) and route deviation (meter) for audio and haptic modality as our objective measures. Task completion time was recorded as the time that subject took to navigate along the wall from start point to the end point. Task completion time for audio and haptic modality in Fig. 7 (a) shows that subjects navigated faster with haptic modality than audio modality. However, paired t-test showed no significant difference between audio and haptic modality in task completion time (t = −1.287, p = 0.33). Travel distance as a distance that subjects have walked along the wall was measured using motion capture system. As shown in Fig. 7 (b), subjects traveled shorter distance with haptic modality. A paired t-test revealed no significant difference between audio and haptic modality in travel distance (t = 2.024, p = 0.074). We further measured route deviation using motion capture system when navigating with audio and haptic modality. It shows subjects' position deviation relative to the walls during the navigation. Subjects had lower route deviation (Fig. 7 (c)) when navigating with haptic modality. A paired t-test showed a significant difference in route deviation between audio and haptic modality (t = 2.736, p = 0.023).

After completing the experiment, we subjectively measured workload for each modality by asking subjects answer the NASA TLX questionnaire. As shown in Fig. 8, physical and temporal demand did not vary much between two modalities which shows both of them were able to navigate the subjects. However, subjects rated that mental demand and effort are higher when navigating with audio feedback. These higher mental workload and effort are because subjects had to concentrate more to process audio feedback to navigate successfully along the wall. Subjects also rated better performance and lower frustration with haptic modality, which shows the capability of haptic modality for navigation along the wall consistent with our objective measure. Furthermore, subjects were asked to rate their preference for navigation with audio and haptic modality. This preference was rated on a scale of 1–21 to keep continuity with our NASA TLX, where (1) represents a strong preference for navigation with haptic feedback and (21) represents strong preference for navigation with audio feedback. The average preference rate of 3.4 as illustrated in Fig. 8 indicated subjects' preference for navigating with haptic modality.

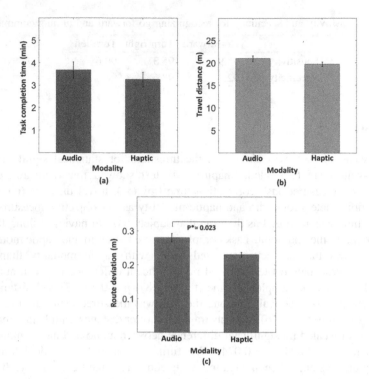

Fig. 7. Objective measures. (a) Task completion time, (b) Travel distance, (c) Route deviation. The unit of task completion time is in minute and unit of travel distance and route deviation is in meter. Error bars show standard error.

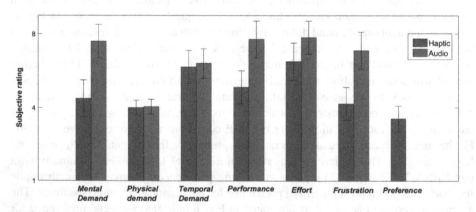

Fig. 8. Questionnaire feedback. The first six bar plots represent the NASA TLX score for audio and haptic modality. The rating scale is 1–21, where 1 represents no mental, physical and temporal demand, best performance, no effort required to complete the task and, no frustration. The last bar plot shows subjects' preference for navigation with haptic modality. The error bars indicate standard error. (Color figure online)

5 Conclusion

This paper compares and investigates haptic and audio modalities as non-visual interfaces for navigation in low visibility environment using the vibrotactile helmet as a sensory augmentation device. The haptic modality utilizes our tactile language in the form of vibrotactile feedback while audio modality applies synthetic speech to present navigation commands. The objective measure showed that haptic feedback leads to lower route deviation significantly. We also measured task completion time and travel distance. Although subjects had faster task completion time and lower travel distance with haptic feedback, no significant difference was found between these two modalities. Unlike [13] which blindfolded users had higher cognitive workload in navigation with haptic modality than with audio modality, our analysis using NASA TLX questionnaire indicated that haptic modality had lower workload on the subjects. The results of this study show the effectiveness of haptic modality for guided navigation without vision. Future work will use a local map of the environment estimated with the ultrasound sensors to generate the navigation commands in place of the MLP algorithm. Furthermore, it would be interesting to conduct this experiment with visually impaired people to investigate the potential of haptic and audio modality as a communication channel for assisted navigation devices.

Acknowledgment. We would like to thank the subjects for their help in data collection for this study. This work was supported by the University of Sheffield Cross-Cutting Directors of Research and Innovation Network (CCDRI), Search and Rescue 2020 project.

References

1. Kärcher, S.M., Fenzlaff, S., Hartmann, D., Nagel, S.K., König, P.: Sensory augmentation for the blind. Front. Hum. Neurosci. **6**, 37 (2012)
2. Kim, Y., Harders, M., Gassert, R.: Identification of vibrotactile patterns encoding obstacle distance information. IEEE Trans. Haptics **8**(3), 298–305 (2015)
3. Flores, G., Kurniawan, S., Manduchi, R., Martinson, E., Morales, L.M., Sisbot, E.A.: Vibrotactile guidance for wayfinding of blind walkers. IEEE Trans. Haptics **8**(3), 306–317 (2015)
4. Shoval, S., Borenstein, J., Koren, Y.: Auditory guidance with the NavBelt-a computerized travel aid for the blind. IEEE Trans. Syst. Man Cybern. C Appl. Rev. **28**(3), 459–467 (1998)
5. Holland, S., Morse, D.R., Gedenryd, H.: AudioGPS: spatial audio navigation with a minimal attention interface. Pers. Ubiquitous Comput. **6**(4), 253–259 (2002)
6. Bertram, C., Evans, M.H., Javaid, M., Stafford, T., Prescott, T.: Sensory augmentation with distal touch: the tactile helmet project. In: Lepora, N.F., Mura, A., Krapp, H.G., Verschure, P.F., Prescott, T.J. (eds.) Living Machines 2013. LNCS, vol. 8064, pp. 24–35. Springer, Heidelberg (2013)
7. Cassinelli, A., Reynolds, C., Ishikawa, M.: Augmenting spatial awareness with haptic radar. In: 10th IEEE International Symposium on Wearable Computers, pp. 61–64 (2006)
8. Carton, A., Dunne, L.E.: Tactile distance feedback for fire-fighters: design and preliminary evaluation of a sensory augmentation glove. In: Proceedings of the 4th Augmented Human International Conference, pp. 58–64 (2013)

9. Marsalia, A.C.: Evaluation of vibrotactile alert systems for supporting hazard awareness and safety of distracted pedestrians, Master Thesis at Texas A&M University (2013)
10. Marston, J.R., Loomis, J.M., Klatzky, R.L., Golledge, R.G.: Nonvisual route following with guidance from a simple haptic or auditory display. J. Vis. Impair. Blind. 101(4), 203–211 (2007)
11. Hara, M., Shokur, S., Yamamoto, A., Higuchi, T., Gassert, R., Bleuler, H.: Virtual environment to evaluate multimodal feedback strategies for augmented navigation of the visually impaired. In: IEEE Engineering in Medicine and Biology Society, pp. 975–978 (2010)
12. Kaul, O.B., Rohs, M.: HapticHead: 3D guidance and target acquisition through a vibrotactile grid. In: Proceedings of the 2016 CHI Conference Extended Abstracts on Human Factors in Computing Systems, pp. 2533–2539 (2016)
13. Martinez, M., Constantinescu, A., Schauerte, B., Koester, D., Stiefelhagen, R.: Cognitive evaluation of haptic and audio feedback in short range navigation tasks. In: Miesenberger, K., Fels, D., Archambault, D., Peňáz, P., Zagler, W. (eds.) ICCHP 2014, Part II. LNCS, vol. 8548, pp. 128–135. Springer, Heidelberg (2014)
14. Kerdegari, H., Kim, Y., Stafford, T., Prescott, T.J.: Centralizing bias and the vibrotactile funneling illusion on the forehead. In: Auvray, M., Duriez, C. (eds.) EuroHaptics 2014, Part II. LNCS, vol. 8619, pp. 55–62. Springer, Heidelberg (2014)
15. Kerdegari, H., Kim, Y., Prescott, T.: Tactile language for a head-mounted sensory augmentation device. In: Wilson, S.P., Verschure, P.F., Mura, A., Prescott, T.J. (eds.) Living Machines 2015. LNCS, vol. 9222, pp. 359–365. Springer, Heidelberg (2015)
16. Denef, S., Ramirez, L., Dyrks, T., Stevens, G.: Handy navigation in ever-changing spaces: an ethnographic study of firefighting practices. In: Proceedings of the 7th ACM Conference on Designing Interactive Systems, pp. 184–192 (2008)
17. Ando, Y., Yuta, S.: Following a wall by an autonomous mobile robot with a sonar-ring. In: Proceedings 1995 IEEE International Conference on Robotics and Automation, vol. 3, pp. 2599–2606 (1995)
18. Kerdegari, H., Kim, Y., Prescott, T.: Head-mounted sensory augmentation device: designing a tactile language. IEEE Trans. Haptics PP(99), 1 (2016)
19. Sherrick, C.E., Rogers, R.: Apparent haptic movement. Percept. Psychophys. 1(6), 175–180 (1966)
20. Kirman, J.H.: Tactile apparent movement: the effects of interstimulus onset interval and stimulus duration. Percept. Psychophys. 15(1), 1–6 (1974)
21. Hart, S.G., Staveland, L.E.: Development of NASA-TLX (Task Load Index): Results of Empirical and Theoretical Research. In: Hancock, P.A., Meshkati, N. (eds.) Human Mental Workload, pp. 239–250. Elsevier, North Holland Press, Amsterdam (1988)
22. Freire, A.L., Barreto, G.A., Veloso, M., Varela, A.T.: Short-term memory mechanisms in neural network learning of robot navigation tasks: a case study. In: 6th Latin American Robotics Symposium (LARS), no. 4 (2009)
23. Zou, A.-M., Hou, Z.-G., Fu, S.-Y., Tan, M.: Neural networks for mobile robot navigation: a survey. In: Wang, J., Yi, Z., Żurada, J.M., Lu, B.-L., Yin, H. (eds.) ISNN 2006. LNCS, vol. 3972, pp. 1218–1226. Springer, Heidelberg (2006)
24. Ando, B., Baglio, S., Marletta, V., Valastro, A.: A haptic solution to assist visually impaired in mobility tasks. IEEE Trans. Hum. Mach. Syst. 45(5), 641–646 (2015)

Visual Target Sequence Prediction via Hierarchical Temporal Memory Implemented on the iCub Robot

Murat Kirtay[✉], Egidio Falotico, Alessandro Ambrosano, Ugo Albanese,
Lorenzo Vannucci, and Cecilia Laschi

The BioRobotics Institute, Scuola Superiore Sant'Anna, 56025 Pontedera, Pisa, Italy
{m.kirtay,e.falotico,a.ambrosano,u.albanese,l.vannucci,c.laschi}@sssup.it
http://sssa.bioroboticsinstitute.it

Abstract. In this article, we present our initial work on sequence prediction of a visual target by implementing a cortically inspired method, namely Hierarchical Temporal Memory (HTM). As a preliminary test, we employ HTM on periodic functions to quantify prediction performance with respect to prediction steps. We then perform simulation experiments on the iCub humanoid robot simulated in the Neurorobotics Platform. We use the robot as embodied agent which enables HTM to receive sequences of visual target position from its camera in order to predict target positions in different trajectories such as horizontal, vertical and sinusoidal. The obtained results indicate that HTM based method can be customized for robotics applications that require adaptation of spatiotemporal changes in the environment and acting accordingly.

Keywords: Hierarchical Temporal Memory · Sequence prediction · Neurorobotics Platform

1 Introduction

Understanding mammalian brain functions and mechanisms have been attractive research field for robotics and artificial intelligence (AI). These functions and mechanisms provide proof of concept solutions for known problems in robotics and AI such as operating massively parallel processes with relatively low power consumption, processing noisy sensory information for action execution, decision making, etc. Thus, brain-inspired approaches have been investigated to derive principles of brain functions and mechanisms in order to propose solutions for various existing challenges. The several known examples are image recognition in large databases by means of deep neural networks [1], artificial neural network based path planning [2], to mention a few. Although these methods perform well in a specific task with notable accuracy, they lack in biological plausibility and generalization of the solution in different domains.

To build more biologically realistic models with generalization capabilities, a method based on operating principles of Neocortex has been proposed in [3] and

© Springer International Publishing Switzerland 2016
N.F. Lepora et al. (Eds.): Living Machines 2016, LNAI 9793, pp. 119–130, 2016.
DOI: 10.1007/978-3-319-42417-0_12

named as Hierarchical Temporal Memory (HTM). HTM based solutions[1] have been successfully implemented in a considerable number of applications and in a wide range of areas where sequential learning and generative prediction aimed at anomaly detection, image classification, rogue behavior detection, geospatial tracking. In robotics HTM can be used to anticipate the world future states in order to properly control motion and deal with the continuously changing environment. Humans appear to solve this problem by predicting the changes in their sensory system as a consequence of their actions [5]. The predictions are obtained using internal models that represent their own bodies and the external objects dynamics [6–8]. There are three main types of internal models [7]: the forward models, the environment models, and the inverse models. The forward models allow predicting the future data from the past perceptions and the planned actions; the environment models predict the dynamics of external objects or agents, and the inverse models find the actions needed to obtain the desired state starting from the actual one.

So far in robotics there have been a number of implementations of anticipatory sensory-motor systems based on internal models. Examples of control systems for mobile robots that predict the visual sensory data using forward models are [9,10], while [11–13] proposed systems that anticipate the dynamics of moving objects in order to accomplish catching tasks. Other models based on prediction have been proposed to improve performances of humanoid robots in different tasks: visual pursuit and prediction of the target motion [14–17], anticipation in reaching [18] or manipulation tasks [19–21].

In this study, we focus on sequence prediction of a visual target using HTM. Despite promising features of HTM, there are only limited applications in the robotics field. In study [22], authors used HTM to create and edit various postures in order to enable an inexperienced user to get familiarity with a humanoid robot's motion authoring concepts. The study in [23] proposed perception approach based on receiving images of various categories such as doors, plants and bookcases. Then, authors performed HTM as classification method to label these images and based on the output of image category, the predefined actions of a conceptual robot determined which are opening a door and move forward. Instead of presenting continuous prediction and online learning which can be considered as a core component of HTM, these robotics related studies concerned with image classification and action execution applications. On the other hand, Nguyen et al. investigate a spatiotemporal sequence learning architecture for autonomous navigation [24]. Although the proposed architecture contains basics from short and long-term memory mechanisms (STM and LTM) which are partially similar to HTM, the authors used a robot vision data set to extract sequence properties of the navigated environment. As described in this literature review, there are only limited robotics applications and research involving HTM. In this paper, we perform HTM for visual target position prediction of the iCub humanoid robot in a simulated environment.

---

[1] http://numenta.com/#applications.

The rest of paper organized as follows; in the Sect. 2, we highlight operating principles of Neocortex and biological plausibility of HTM. The experimental setup and implementation details of the HTM framework are presented in the Sect. 3. The obtained results and discussions are explained in the Sect. 4 while concluding remarks and future works are emphasized in the Sect. 5.

2 Biological Inspiration Behind HTM

In this section, the biological functions and structure of mammalian neocortex will be introduced and the biological plausibility of HTM will be addressed. Hierarchical temporal memory was proposed in study [25]. It is a biologically constrained model inspired by operating principles of the neocortex. Neocortex is responsible for visual pattern recognition, understanding speech, object recognition, manipulation by touch, etc. Biological organization of the neocortex consists of columnar structure repeated in all over the mammalian brain [26]. This structure of repeating pattern led to an idea of the neocortex performing same computational mechanism to process different inputs from sensory organs (e.g. eye, ear) and produce motor commands to execute necessary actions [3]. These biological insights provide support for HTM in order to have a layered architecture which performs same algorithms for different inputs. The aforementioned sensorimotor features of neocortex attract researchers to build Artificial Neural Networks (ANNs) which are able to implement computational and learning systems that imitate the behavior of neurons (e.g. pyramidal cells). Despite considerable success of the ANN based methods in specific applications, these approaches lack in biological plausibility and show performance degradation in different domains [27]. For instance, these methods require considerable amount of data and computational resources for training via batch processing and may not perform well if data have a streaming characteristic.

Unlike classical ANN based approaches, HTM provides more biologically realistic computation concepts to overcome existing difficulties against invariant and streaming data. By doing so, HTM employs an encoding phase for preprocessing data, spatial pooling to generate sparse distributed representation of the encoders, and temporal pooling (or memory) to capture time-based changes from the outputs of spatial pooling. These concepts will be re-emphasized and implementation details will be described in the Sect. 3.

3 Experimental Setup and Implementation of Hierarchical Temporal Memory

3.1 Neurorobotics Platform (NRP) Setup

In this study we used the Neurorobotics platform (NRP)[2] which is being developed as part of Human Brain Project (HBP)[3]. The aim of NRP is to provide

---

[2] http://neurorobotics.net.
[3] http://www.humanbrainproject.eu.

Fig. 1. iCub humanoid robot in Neurorobotics Platform

physically realistic simulation environment which enables neuroscientists to test their models on different robots such as iCub, Husky and Lauron. Furthermore, the platform can also be used by roboticists to build and customize biologically inspired robot control architectures for specific tasks. The software and functional details of the NRP with an integration use case is described in study [4]. The NRP is being developed for spiking neuron models, yet it also allows users to extend and integrate new models. We integrate HTM to derive target position sequence prediction while processing iCub's camera data. As can be seen in Fig. 1, iCub is used as an embodied agent receiving camera data to feed HTM implementation pipeline (see Fig. 2) in order to predict target position. A circular target moves on the screen along predefined paths which are horizontal, vertical and sinusoidal. After collecting images from a camera, the task is restarted and one step ahead prediction is performed using HTM.

We set two distinct goals to conduct simulation experiments with the iCub robot. In that, we firstly aimed to investigate the NRP and HTM capabilities to assess target sequence prediction in online mode. For the latter studies, we are in the process of extending HTM model to integrate sensorimotor transition modules for eye movements and grasping while continuously predicting target sequence.

3.2 Implementation Procedure

In order to exploit HTM for sequence prediction of a visual target, we customized an existing open-source software platform, Numenta Platform for Intelligent Computing (NuPIC)[4], that has already been used in various applications where data had streaming charateristics. As a first step of this study, we used

---

[4] http://www.numenta.org.

NuPIC, with its default parameters, to predict scalar values of different periodic functions. Afterward, we tuned NuPIC parameters in order to predict target sequences streamed from the simulated environment.

The sequence prediction mechanism works as follows: in a first phase, data received from the robot camera is collected until the target has reached the end of the screen (see Figs. 1 and 4). During this phase, the images are processed by extracting the region of interest, using prior knowledge of the target motion, and then by applying a binary threshold to obtain a black and white image. The thresholded image is converted into a vector where ones correspond to white pixels and zeroes to black ones. It should be noted that extracting the region of interest from the image reduces computation times for preprocessing, but the same results could be achieved by using the whole camera image data. Moreover, in most HTM applications, the encoding phase is usually employed for getting the binary representation of the input data before passing it to the HTM implementation pipeline, as depicted in Fig. 2.

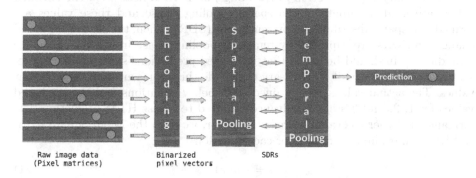

Fig. 2. HTM implementation pipeline: inputs and outputs of each phase

The generated binary representation is used as input of spatial pooling phase in order to obtain a sparse distributed representation (SDR) of the received input in the available dataset. The SDR is a cortically inspired representation of input data where only a small percentage of the bits are active [29]. Similarly, temporal pooling takes a series of SDR inputs from the spatial pool and generates predictions by learning sequences. This learning phase enables HTM to form stable sequences of SDRs while capturing temporal relations for prediction.

After the data collecting phase is over, i.e. when the target motion restarts, prediction are started to being produced by the HTM, while the new images are still being processed and fed into the spatial and temporal pools. Such predictions can be from one to several steps ahead, depending on the application requirements. In this work, the model is used to perform single and multi-step predictions for periodic functions and single step predictions of the target image sequences. A more comprehensive neuroscientific and mathematical background about the aforementioned HTM implementation phases can be found in the following studies: [27–29].

4   Results and Discussions

The obtained results will be presented in twofold which are HTM error analysis with periodic functions and HTM implementation in NRP. We consider that the error interpretation will give rise to an idea that whether HTM can be implemented in visual sequence prediction.

4.1   HTM-Prediction Analysis Using Periodic Functions

In this step, we aim to derive prediction errors while performing HTM on two periodic functions with different prediction steps. To obtain prediction values by performing HTM pipeline, 3000 sequential iteration steps are generated and converted to binary representation in encoding phase. As a next step, the encoded values are used as inputs of the spatial pooling phase and these values are formed into sparse distributed representations (SDR). Then, the sequence related changes in SDRs are captured by temporal pooling phase in order to generate a prediction. It should be noted that generated predictions share similar representation. That is why it is necessary to decode this representation to get scalar values. The actual values are obtained from below equations and the predicted values for 1, 2, and 3 steps in advance are used to assess HTM prediction performance. In order to evaluate prediction error, we used the Mean Square Error (MSE) as an evaluation metric for each function.

$$y = sin(x) \tag{1}$$

$$y = sin(2 \times x) + cos(4 \times x) \tag{2}$$

Figure 3(a) and (b) show that the obtained one and three steps prediction which are generated by using the Eq. 1 with iteration range between 2440 and 2600. Similarly, Fig. 3(c) and (d) illustrate single and multi-step prediction by performing Eq. 2 with iteration range between 2440 and 2550. As can be seen from these figures HTM can make a single and multi-step prediction for a given function. It should be noted that the predefined iteration ranges represent prediction trends for both equations.

Table 1. Mean square error results of the performed functions

Prediction step	MSE (Eq. 1)	MSE (Eq. 2)
1	$2.7514e-05$	0.0233
2	$5.3602e-05$	0.0263
3	$1.2051e-04$	0.0254

Moreover, the calculated MSE values in Table 1 are derived for the iteration ranges and this values also indicate that HTM can adapt to changes in streaming data to generate either single or multi-step prediction. We also observed that an increasing the prediction steps will lead to an increment in the MSE values. To sum up, we interpret HTM predictions using periodic functions and we observe that obtained results can lead to implementing the same method for visual target sequence prediction in a simulated environment.

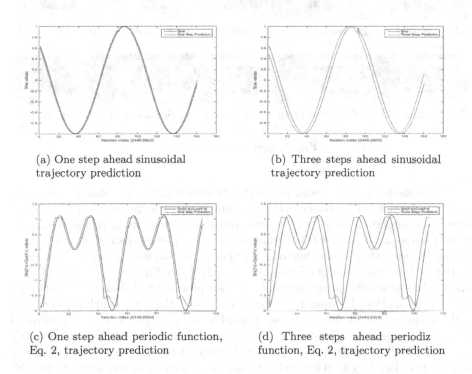

(a) One step ahead sinusoidal trajectory prediction

(b) Three steps ahead sinusoidal trajectory prediction

(c) One step ahead periodic function, Eq. 2, trajectory prediction

(d) Three steps ahead periodiz function, Eq. 2, trajectory prediction

Fig. 3. One step and three steps ahead periodic functions' trajectory predictions

4.2 HTM-NRP Implementation for Visual Target Sequence Prediction

After error analysis step with periodic functions, we consider performing HTM for visual target sequence prediction in a simulated environment. To implement this task, we customized HTM to process perceived image data from iCub's camera in the NRP. Instead of generating input data via predefined functions, we use a visual target which continuously moves along a horizontal, a vertical and a sinusoidal path. A subset of obtained image sequence for the target are shown in Fig. 4. Images in the first row are shown in Fig. 4 and recorded during horizontal movement, the second-row images consist of vertical movement and last row images belong to the sinusoidal movement.

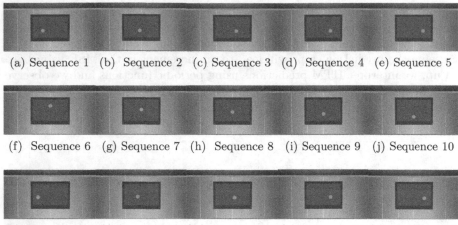

(a) Sequence 1   (b) Sequence 2   (c) Sequence 3   (d) Sequence 4   (e) Sequence 5

(f) Sequence 6   (g) Sequence 7   (h) Sequence 8   (i) Sequence 9   (j) Sequence 10

(k) Sequence 11 (l) Sequence 12 (m) Sequence 13 (n) Sequence 14 (o) Sequence 15

Fig. 4. iCub camera images

The considered experiment in NRP starts with target movement along one of the above mentioned trajectories (e.g. sinusoidal) till reach end of the virtual screen. Throughout this movement, each received camera image is stored in order to be used as an input for HTM implementation. The number of stored images during learning step for the horizontal sequence is 101, for the vertical sequence is 100 and for the sinusoidal sequence is 64. As outlined in the Sect. 3, raw pixel values of stored images preprocessed for pipeline and their SDR forms are used for prediction. In order to make a prediction, the same pipeline procedure will be applied to received images. The SDR form of input will be generated for next sequence prediction. To obtain similarity value between input SDR and stored SDRs, it is needed to decode prediction from available similarity scores. The decoding step for scalar values can be easily constructed by NuPIC framework. HTM has currently no probabilistic model to reconstruct predicted image from SDR. To overcome this difficulty, we implement a method to get overlapping scores based on active bit positions in among learned SDRs and predicted SDR values. If same overlapping score exists in learned SDRs, we refer the last pattern in the sequence as predicted pattern since we assume that the target will continuously move in the forward direction. Instead of performing classification and regression methods (e.g. K-nearest neighbors), we noticed that this simple method will lead to understanding HTM dynamics in detail.

However, applying this method brings about predicting forward jumps in the sequence or getting the same prediction more than once as a next step prediction. That behavior illustrated with the data obtained from Neurorobotics Platform in Fig. 5 where colored * refers to movement trajectory of the target center. For instance, red * characters are obtained prediction values while the target moves in a vertical path. Based on overlapping method 21 different predictions are generated during vertical movement. As can be seen from Fig. 5, predicted

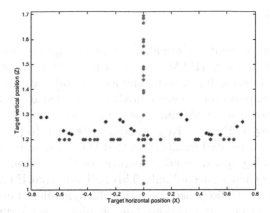

Fig. 5. Predicted path trajectories in NRP experiment (Blue - Horizontal, Red- Vertical, Black - Sinusoidal) (Color figure online)

sequence are consecutive with gaps and HTM can still capture spatiotemporal changes in these experiments.

To test overlapping score strategy in a trajectory where one point is far to another, we construct a sequence which is a combination of horizontal, vertical and sinusoidal, respectively. Such sequence is composed of a number of captured images which is lower compared to the previous experiment. In that, we randomly select eight patterns from horizontal, nine patterns from vertical and ten patterns from sinusoidal paths and convert them to a single sequence (see Fig. 6(a)). Then we employ the overlapping method and extracting a prediction as one of the patterns in the selected sequence which are shown in Fig. 6(b). It is observed that HTM can successfully predict sequences without having above mentioned problems.

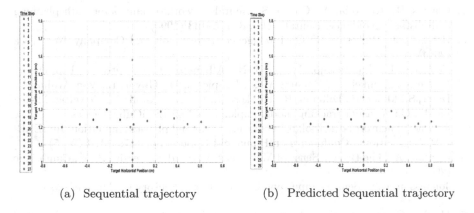

(a) Sequential trajectory   (b) Predicted Sequential trajectory

Fig. 6. Prediction path trajectories (Color figure online)

5   Conclusion

In this study we implement a cortically inspired computational method, Hierarchical Temporal Memory (HTM), to predict target position sequence (e.g. traversed path) via extracting target's spatiotemporal features. To do this, we firstly analyze HTM performance while predicting sequence in advance on different periodic functions. After obtaining promising results with these predefined functions, we employed HTM in Neurorobotics Platform to predict target position sequences. The presented results show that HTM paves the way for sequence prediction by learning spatiotemporal features of the environment and adopting these statistical changes accordingly. This indicates that HTM has promising dynamics for robotics studies where exploring the dynamics of unknown environment required to take necessary actions. Our ongoing research directions focused on developing HTM based algorithms on simulated and real humanoid robot in order to achieve a task which requires sensorimotor coordination and control such as predictions of the eye movements while tracking a visual object and continuously learning grasped object dynamics.

Acknowledgements. The research leading to these results has received funding from the European Union Seventh Framework Programme (FP7/2007-2013) under grant agreement no. 604102 (Human Brain Project). The authors would like to thank the Italian Ministry of Foreign Affairs, General Directorate for the Promotion of the "Country System", Bilateral and Multilateral Scientific and Technological Cooperation Unit, for the support through the Joint Laboratory on Biorobotics Engineering project.

References

1. Krizhevsky, A., Sutskever, I., Hinton, G.E.: ImageNet classification with deep convolutional neural networks. In: Neural Information Processing Systems (NIPS), pp. 1106–1114 (2012)
2. Glasius, R., Komoda, A., Gielen, S.: Neural network dynamics for path planning and obstacle avoidance. Neural Netw. **8**, 125–133 (1995)
3. Hawkins, J., Blakeslee, S.: On Intelligence. Henry Holt and Company, New York (2004)
4. Vannucci, L., Ambrosano, A., Cauli, N., Albanese, U., Falotico, E., Ulbrich, S., Pfotzer, L., Hinkel, G., Denninger, O., Peppicelli, D., Guyot, L., Von Arnim, A., Deser, S., Maier, P., Dillman, R., Klinker, G., Levi, P., Knoll, A., Gewaltig, M.-O., Laschi, C.: A visual tracking model implemented on the iCub robot as a use case for a novel neurorobotic toolkit integrating brain and physics simulation. In: IEEE-RAS International Conference on Humanoid Robots, pp. 1179–1184 (2015)
5. Berthoz, A.: The Brain's Sense of Movement. Harvard University Press, Cambridge (2002)
6. Johansson, R.S.: Sensory input and control of grip. Sensory guidance of movement (1998)
7. Miall, R.C., Wolpert, D.M.: Forward models for physiological motor control. Neural Netw. **9**, 1265–1279 (1996)

8. Nguyen-Tuong, D., Peters, J.: Model learning for robot control: a survey. Cogn. Process. **12**, 319–340 (2011)
9. Gross, H.-M., Heinze, A., Seiler, T., Stephan, V.: Generative character of perception: a neural architecture for sensorimotor anticipation. Neural Netw. **12**, 1101–1129 (1999)
10. Hoffmann, H.: Perception through visuomotor anticipation in a mobile robot. Neural Netw. **20**, 22–33 (2007)
11. Bauml, B., Birbach, O., Wimbock, T., Frese, U., Dietrich, A., Hirzinger, G.: Catching flying balls with a mobile humanoid: system overview and design considerations. In: 2011 11th IEEE-RAS International Conference on Humanoid Robots (Humanoids), pp. 513–520 (2011)
12. Kim, S., Billard, A.: Estimating the non-linear dynamics of free-flying objects. Robot. Auton. Syst. **60**, 1108–1122 (2012)
13. Kober, J., Glisson, M., Mistry, M.: Playing catch and juggling with a humanoid robot. In: 2012 12th IEEE-RAS International Conference on Humanoid Robots (Humanoids), pp. 875–881 (2012)
14. Falotico, E., Zambrano, D., Muscolo, G.G., Marazzato, L., Dario, P., Laschi, C.: Implementation of a bio-inspired visual tracking model on the iCub robot. In: Proceedings of the 19th IEEE International Symposium on Robot and Human Interactive Communication (ROMAN 2010), pp. 564–569. IEEE (2010)
15. Vannucci, L., Falotico, E., Di Lecce, N., Dario, P., Laschi, C.: Integrating feedback and predictive control in a bio-inspired model of visual pursuit implemented on a humanoid robot. In: Wilson, S.P., Verschure, P.F.M.J., Mura, A., Prescott, T.J. (eds.) Living Machines 2015. LNCS, vol. 9222, pp. 256–267. Springer, Heidelberg (2015)
16. Falotico, E., Taiana, M., Zambrano, D., Bernardino, A., Santos-Victor, J., Dario, P., Laschi, C.: Predictive tracking across occlusions in the iCub robot. In: Proceedings of the 9th IEEE-RAS International Conference on Humanoid Robots (Humanoids 2009), pp. 486–491 (2009)
17. Zambrano, D., Falotico, E., Manfredi, L., Laschi, C.: A model of the smooth pursuit eye movement with prediction and learning. Appl. Bionics Biomech. **7**(2), 109–118 (2010)
18. Cauli, N., Falotico, E., Bernardino, A., Santos-Victor, J., Laschi, C.: Correcting for changes: expected perception-based control for reaching a moving target. IEEE Robot. Autom. Mag. **23**, 63–70 (2016)
19. Datteri, E., Teti, G., Laschi, C., Tamburrini, G., Dario, P., Guglielmelli, E.: Expected perception: an anticipation-based perception-action scheme in robots. In: Proceedings of the 2003 IEEE/RSJ International Conference on Intelligent Robots and Systems (IROS 2003), vol. 1, pp. 934–939 (2003)
20. Laschi, C., Asuni, G., Teti, G., Carrozza, M.C., Dario, P., Guglielmelli, E., Johansson, R.: A bio-inspired neural sensory-motor coordination scheme for robot reaching and preshaping. In: The First IEEE/RAS-EMBS International Conference on Biomedical Robotics and Biomechatronics, BioRob 2006, pp. 531–536 (2006)
21. Laschi, C., Asuni, G., Guglielmelli, E., Teti, G., Johansson, R., Konosu, H., Wasik, Z., Carrozza, M.C., Dario, P.: A bio-inspired predictive sensory-motor coordination scheme for robot reaching and preshaping. Auton. Robots **25**, 85–101 (2008)
22. Seok, K.H., Kim, Y.S.: A new robot motion authoring method using HTM. In: International Conference on Control, Automation and Systems, pp. 2058–2061 (2008)

23. Mai, X., Zhang, X., Jin, Y., Yang, Y., Zhang, J.: Simple perception-action strategy based on hierarchical temporal memory. In: 2013 IEEE International Conference on Robotics and Biomimetics (ROBIO), pp. 1759–1764, 12–14 December 2013
24. Nguyen, V.A., Starzyk, J.A., Tay, A.L.P., Goh, W.: Spatio-temporal sequence learning of visual place cells for robotic navigation. In: The 2010 International Joint Conference on Neural Networks (IJCNN), pp. 1–8, 18–23 July 2010
25. Hawkins, J., George, D.: Hierarchical temporal memory: Concepts, theory and terminology. Technical report, Numenta (2006)
26. Mountcastle, V.B.: The columnar organization of the neocortex. Brain **120**(4), 701–722 (1997)
27. Hawkins, J., Ahmad, S.: Why Neurons Have Thousands of Synapses. A Theory of Sequence Memory in Neocortex. Frontiers in Neural Circuits (2015)
28. Cui, Y., Surpur, C., Ahmad, S., Hawkins, J.: Continuous online sequence learning with an unsupervised neural network model (2015). arXiv preprint: arXiv:1512.05463
29. Hawkins, J., Ahmad, S.: Properties of Sparse Distributed Representations and their Application to Hierarchical Temporal Memory (2015). arXiv preprint: arXiv:1503.07469v1

Computer-Aided Biomimetics

Ruben Kruiper$^{(\boxtimes)}$, Jessica Chen-Burger, and Marc P.Y. Desmulliez

Heriot-Watt University, Edinburgh EH14 4AS, Scotland, UK
rk22@hw.ac.uk

Abstract. The interdisciplinary character of Bio-Inspired Design (BID) has resulted in a plethora of approaches and methods that propose different types of design processes. Although sustainable, creative and complex system design processes are not mutually incompatible they do focus on different aspects of design. This research defines areas of focus for the development of computational tools to support biomimetics, technical problem solving through abstraction, transfer and application of knowledge from biological models. An overview of analysed literature is provided as well as a qualitative analysis of the main themes found in BID literature. The result is a set of recommendations for further research on Computer-Aided Biomimetics (CAB).

Keywords: Bio-Inspired Design (BID) · Biomimicry · Biomimetics · Bionics · Design theory · Innovation · Invention · Computer Aided Design (CAD)

1 Introduction

Bio-Inspired Design (BID) is associated with the application of *"nature's design principles"* to *"create solutions that help support a healthy planet"* [1]. Vandevenne (2011) added that the premise of bio-inspired design allows the finding and use of existing, optimal solutions. Additional factors include the sustainable image, association to an organism and 'high' probability of leapfrog innovations [2]. Although the technology that evolved in nature is not always ahead of man-made technology, the assumption that organisms have ways of implementing functions more efficiently and effectively than we do is assumed to be true in many cases [3, 4].

However, the search for biological systems and transfer of knowledge is non-trivial. Most Bio-Inspired Design (BID) methods use function, many in terms of the *'functional basis'* to model biological functions and flows, as the analogical connection between biology and engineering [5]. This paper identifies the insufficient definition of function throughout BID approaches as one of the main obstacles for knowledge transfer. The Biomimicry 3.8 Institute for example refers to a function as *"the role played by an organism's adaptations or behaviours that enable it to survive. Importantly, function can also refer to something you need your design solution to do"* [6].

This paper identifies the main areas of focus for research on Computer-Aided Biomimetics (CAB). Firstly, a definition of biomimetics is given that reflects its focus on technical problem-solving. Secondly, important notions from existing literature on BID are outlined based on themes for qualitative analysis. Finally, the results of the thematic analysis provide recommendations for further research on computational design tools for biomimetics.

© Springer International Publishing Switzerland 2016
N.F. Lepora et al. (Eds.): Living Machines 2016, LNAI 9793, pp. 131–143, 2016.
DOI: 10.1007/978-3-319-42417-0_13

2 Definitions

Pahl and Beitz (2007) noted that the analysis of biological systems can lead to useful and novel technical solutions, referring to bionics and biomechanics as fields that investigate the connection between biology and technology [7]. Biomimetics is often regarded as a synonym for BID, bionics and biomimicry, and refers to the transfer of biological knowledge from nature to technical applications [8–10]. As BID is an umbrella term we adopt the following definitions, based on Fayemi et al. (2014) [11]:

> **Biomimetics**: Interdisciplinary creative process between biology and technology, aiming to solve technospheric problems through abstraction, transfer and application of knowledge from biological models.
> **Biomimicry/Biomimesis**: Philosophy that takes up challenges related to resilience (social, environmental and economic ones), by being inspired from living organisms, particularly on an organizational level.

Bio-inspiration can be useful in early design stages, e.g. the fuzzy front end, when the design process has no clear direction. In later stages the search for functional, biological analogies becomes less urgent and the focus lies on transferring quantitative knowledge from biological systems to technical problems. Nachtigall (2002) introduced the term technical biology as a field in biology that describes and analyses structures, procedures and principles of evolution found in nature using methodological approaches from physics and engineering sciences [12]. According to Speck et al. (2008) technical biology is the basis of many biomimetic projects. It *"allows one to understand the functioning of the biological templates in a quantitative and technologically based manner"* [13]. However, according to Julian Vincent (personal communication, March 31, 2016) technical biology has been known as biomechanics for decades.

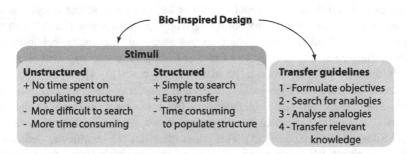

Fig. 1. Approaches of methods to 'enhance' BID (based on [5])

Martone et al. (2010) stated that, although biological models provide multifunctional properties with high potential for biomimetic applications, they have hardly been studied quantitatively in terms of their form-structure-function to transfer knowledge [14]. Figure 1 gives an overview of how methods approach BID. The next section is a literature review that aims to summarise important notions in a structure that is loosely based on seven themes for qualitative analysis. The themes represent areas that require attention in research on CAB: the transfer guidelines, the notion of function in literature and our definition of biomimetics – abstraction, transfer and application of knowledge from biological models.

3 Bio-Inspired Design Literature Review

3.1 Direction

Similar to Biomimicry 3.8, Helms et al. (2009) and Speck et al. (2008) indicated two possible directions of a biomimetics process [13, 15, 16]. The first is a problem driven, top-down process. If the knowledge of identified biological solutions is too little, Speck et al. propose an extension that involves further research on the biological system. The second is a solution driven, bottom-up process that searches for technical applications of a specific biological solution. Main characteristics are [13]:

Bottom-up (biological solution to technical problem)

- Abstraction often proves to be one of the most important as well as difficult steps
- Often several iterative loops
- 3–7 years for bottom-up to final product
- Possibly multiple implications of technology
- Potentially highly inventive

Top-down (technical problem to biomimetics)

- Existing product: Initially define problem and constraints, then search for possible biological solutions
- One or two most appropriate selected for further analysis
- 6–18 months top-down to functional prototype
- Usually not very innovative

Extended top-down

- Existing knowledge of biological model is too low, further research is required
- Can be as highly inventive as bottom-up
- Smaller range of implications of technology compared to bottom-up
- 1–5 years typically, in between top-down and bottom-up

Helms et al. (2009) noted that the steps in these processes in practice do not necessarily occur in the prescribed sequence. Once a biological solution is selected in the problem-driven process, the design process tends to be fixated around this one solution.

Furthermore, they identified several common errors and practice patterns that emerged during classroom projects on bio-inspired design [15]. Both are listed in Table 1.

3.2   Formulate Objectives

Lindemann et al. (2004), Stricker (2006), Inkermann et al. (2011) proposed a procedure that starts with a goal definition, based on the the Müncher Vorgehensmodell (MVM) by Lindemann (2003) [17–20]. The goal, as well as the later search for alternative solutions, are at a relatively abstract level. Stricker (2006) identified several problems during BID processes listed in Table 1 [18].

By our definition, the goal in biomimetics is solving technical problems. Stricker (2006) noted that knowing the type of problem you are dealing with, helps planning a solution route (based on Dorner 1987, Badke-Schaub et al. 2004, Erhlenspiel 2003) [18]:

Table 1. Mistakes and trends that often occur during a bio-inspired design process (based on [15, 18]).

Stricker (2006) errors	Helms et al. (2009) errors	Helms et al. (2009) trends
• Over-reduction of context • Neglecting form and processes by focusing mainly on structure • Generalising where different functions originate and simply copying existing foreknowledge • Structure is expected to be directly transferable (same elements, same relations) • Neglecting of constraints	• Vaguely defined problems • Poor problem-solution pairing • Oversimplification of complex functions • Using 'off-the-shelf' biological solutions • Simplification of optimisation problems • Solution fixation • Misapplied analogy • Improper analogical transfer	• Mixing problem-driven versus solution-driven approach • Usually focus on structure instead of function • The focus on function, usually problem-driven • Solution-driven generated multi-functional designs • Partial problem definition leads to compound solutions for new-found sub-problems. • Not many problems are framed as optimisation problems • Tendency to choose known biological solutions • Choice of biological solutions hard, too many or not enough

Synthesis barrier: goal is known, but the means to achieve it are not known
Dialectical barrier: goal is not known, ambiguous, multi-faceted, interrelated or too generic
Synthesis and dialectical barrier: combination of above, but knowledge not sufficient
Interpolation problem: means and goal known, but not clear how to achieve the goal

In BID both problem and solution decompositions are transferred [21, 22]. *"BID often involves compound analogies, entailing intricate interaction between problem decomposition and analogical reasoning"* [21].

3.3   Search for Analogies

Analogies in BID can be useful for solution generation, design analysis and explanation [5, 10]. To validate applicability of analogies through similarity, Inkermann et al. (2011) distinguish four types of similarities [19]:

Formal similarity: same rules and physical principles
Functional similarity: similar function
Structural similarity: similar structural design
Iconic similarity: similar form or shape

Databases are a common approach to store biological analogies, usually indexed by function. According to Yen et al. *"a designer can compare the functions of what they are designing and also compare the structures and behaviours of their design to biological systems"* [23]. Hill (1997) classified 191 biological systems into 15 descriptive technical and biological abstractions on basis of 5 general functions and 3 types of transactions: energy, information and matter [18, 20]. DANE, SAPPhIRE and the more recent Biologue system are databases indexed using the Structure-Behaviour-Function (SBF) model [23–25]. Wilson et al. (2008, 2009) and Liu et al. (2010) used an ontology based knowledge modelling approach to reuse strategies for design [26–28].

Natural language approaches are another way to search for relevant biological analogies. Shu et al. (2014) proposed a semi-automatic search method using functional keywords [2, 29]. According to Vandevenne (2011) these studies indicate that representation of analogues in natural language format should be considered as input for filtering, analysis and transfer. *"Automated characterization of biological strategies, and of the involved organisms, enables a scalable search over large databases"* [2].

Yen et al. (2014) noted that the search strategy for biological systems is a problem area. To make the search process more efficient, Vandevenne (2011) proposed searching on basis of function and further specification of behaviour and structure before commencing knowledge transfer. Other areas that require attention are methods for teaching analogical mapping, evaluating good analogies, good designs and good design problems [23].

An example of research on better understanding and supporting the use of biological analogies is the work by Linsey et al. (2014). According to them design by analogy is powerful, but a difficult cognitive process. To overcome cognitive bias and challenges they propose several principles and design heuristics. An example is providing uncommon examples to overcome design fixation [30]. The use of analogies and heuristics may be useful for biological inspiration and even supporting transfer of knowledge. However, *"databases can only record history and cannot deduce new relationships"* [31]. Therefore, Vincent (2002, 2006, 2009) proposes the use of TRIZ to facilitate the comprehension of biological systems [8, 9, 32]. Lindemann et al.

Gramann (2004) provide a structured, associative checklist to support deduction of technical analogies from biological systems that is loosely based on TRIZ [17]. Vincent et al. developed BioTRIZ, a reduced form of TRIZ that inventive principles to biological dialectical problems [8, 33]. *"Studying 5,000 examples, the conflict matrix was reduced from 39 conflict elements to 6 elements that appear in both biology and engineering, and a 6 by 6 contradiction matrix that contains all 40 of the inventive principles was created. These 6 conflict elements are substance, structure, time, space, energy/field, and information/regulation"* [5].

3.4 Function

Fratzl (2007) noted biomimetics studies start with the study of structure-function relationships in biology; the mere observation of nature is not sufficient [34]. Vincent (2014) noted defining problems in functions is key to knowledge transfer from biology to technology [4]. According to Stone et al. (2014) natural systems have to be modelled using normal function modelling techniques to use function as an analogical connection [5]. An example is using the functional basis that defines function as *"a description of a device or artefact, expressed as the active verb of the sub-function"* [35]. However, Deng (2002) identified two types of functions in literature [36]:

Purpose function: is a description of the designer's intention or the purpose of a design (not operation oriented).
Action function: is an abstraction of intended and useful behaviour that an artefact exhibits (operation oriented).

Furthermore, Deng (2002) stated that action functions can be described semantically or syntactically. Chakrabarti et al. (2005) adopted this view, an overview of definitions is given in Table 2. For SAPPhIRE Chakrabarti et al. (2005) view function as *"the intended effect of a system (Chakrabarti & Bligh, 1993) and behaviour as the link between function and structure defined at a given level. Thus, what is behaviour is specific to the levels at which the function and structure of a device are defined"* [25].

Table 2. Action functions from the perspective of semantic and syntactic formulation - based on [25, 36].

Semantic views	Syntactic views
Functions as input/output of energy, material and information	Functions using informal representation (e.g. verb-noun transformation)
Functions as a change of state of an object or system.	Functions using formal representation (e.g. mathematical transformation).

Nagel (2014) used a semantic view and noted that a function represents an operation performed on a flow of material, signal or energy [37]. According to Nagel (2014) the use of functional design methods for BID offers several advantages [38]:

- archival and transmittal of design information
- reduces fixation on aesthetic features or a particular physical solution
- allows one to define the scope or boundary of the design problem as broad or narrow as necessary
- encourages one to draw upon experience and knowledge stored in a
- database or through creative methods during concept generation

Vattam et al. (2010) adopted the formal definition for function used in Structure-Behaviour-Function (SBF) models, which was developed in AI research on design to support automated, analogical design [24]. This is a semantic view on functions for SBF: "*A function is represented as a schema that specifies its pre-conditions and its post-conditions*" [39].

Goel (2015) noted that BID presents a challenge for Artificial Intelligence (AI) fields such as knowledge representation, knowledge acquisition, memory, learning, problem solving, design, analogy and creativity [40]. Vandevenne (2011) noted that the instantiation of functional models is a labour-intensive task that requires biology and engineering knowledge [2].

3.5 Abstraction

Abstraction, the reduction of context, is an important aspect in all BID methodologies. According to Vattam et al. (2010) successful BID requires rich and multimodal representations representations of systems during design. Such representations are organised at different levels of abstraction. They "*explicitly capture functions and mechanisms that achieve those functions on the one hand, and the affordances and constraints posed by the physical structures for enabling the said mechanisms on the other hand*" [41]. Table 3 is an adaption of the lists by Fayemi et al. (2014) on requirements for abstraction from both a theoretical and a practical perspective.

Table 3. Requirements for abstraction - based on [11]

Theoretical	*Practical*
• Ability to model simple as well as complex problems	• Fast process
o Integrate different systemic levels	• Intuitive process
• Effective selection of significant data	• Allow combination with other tools
o Maintain specific constraints of problem	• Applicable over various industrial/scientific domains
• Ease of translation	• Allow for collaborative design
o Determine the solution in generics terms	

Abstraction eases the implementation of biological solutions [4, 20]. Chakrabarti (2014) found that exploration at higher levels of abstraction has a greater impact on novelty of solutions generated [22]. Lindemann et al. (2004) noted that, if one finds no technical analogies, the level of abstraction as well as the feasibility of solving the problem should be reconsidered [17].

Diagrammatic representations of biological systems lead to generation of more and better design ideas than textual representations [42]. Descriptive accounts of design lead to more effective educational techniques and computational tools for supporting design, advantages include: realism, accuracy of predictions and accuracy of design behaviour [15].

3.6   Transfer

In order to deduce new relationships, Vincent (2014) suggested using a descriptive approach of technology to ease knowledge transfer. *"There is very little indication of how one can take a concept from its biological context and transfer it to an engineering or technical context"* [31]. Differences in context, high complexity, high amount of interrelated and integrated multifunctional elements make transfer the most difficult part of BID. Lack of biological training increases the difficulty of transferring knowledge. Chakrabarti (2014) defined transfer as *"the reproduction of information from a model of a biological system in a model or prototype for a technical system"* [22]. There are four kinds of exchanges in bio-inspiration [22, 42]:

1. Use of analogy as idea stimulus
2. Exchange of structures, forms, materials
3. Exchange of functions, and processes
4. Exchange of knowledge about processes, information, chaos

With the aim of initiating work in systematic biomimetics Chakrabarti et al. (2005) developed a generic model for representing causality of natural and artificial systems [25, 42]. Using this model they analysed existing entries of biological systems and corresponding technical implementations based on similarities and identified several mechanisms of transfer [42]. Only the transfer of physical effects and the transfer of state changes of processes were however considered.

Vandevenne (2011) notes that DANE forces designers to understand the systems of candidate biological strategies in detail, as well as the obtained abstraction. This facilitates knowledge transfer and communication [2].

3.7   Application

Helms et al. (2010) sketched a macro-cognitive information processing model of BID to support the development of computational tools that support this form of design. They noted the complex interplay between problem definition and analogy mapping for knowledge transfer. Salgueiredo (2013) applied C-K theory to BID in early design phases and showed that biological knowledge is usually implemented once the traditional path is blocked. In this case the acquisition of biological knowledge is required to find 'unexpected properties'. She noted that the systematic implementation of BID in companies requires a different form of knowledge management. In essence, C-K theory is a theory of design knowledge management [10, 43]:

- The theory captures the iterative expansion of knowledge and enrichment of design concepts.
- Concept-space can only be partitioned into restrictive and/or expanding.
- partitions, relating to constraints to existing knowledge and addition of new knowledge.
- The theory is domain-independent, which is useful in supporting/analysing multi-disciplinary design.

These findings are in line with the co-evolution of problems and solutions mentioned by Chakrabarti (2014) and the generation of compound solutions through iterative analogy and problem decomposition mentioned by Vattam et al. (2010) [22, 24]. Salgueiredo (2013) concludes that *"biological knowledge does not offer solutions, it stimulates the reorganization of knowledge bases, creating bridges between different domains inside the traditional knowledge. This conclusion is important for reducing the risks of idolizing nature processes and systems"* [10].

4 Results Thematic Analysis

Based on the seven themes 40 features were identified that are of interest for CAB. These features repeatedly occurred in the literature described in the previous section. 190 relationships between them were documented. Relationships were weighted with factor 1, 3 or 5, based on their relative importance. Relations weighted with factor 3 or 5 were annotated. Figure 2 shows a selection of visualisations, features are represented as nodes and relationships between them as edges. In the software package Gephi a radial axis layout was used to create these visualisations, which supports qualitative analysis of similarity for the determination of main themes [44, 45]. As one might expect, nodes like *Transfer* and *Analogy* are highly inter-connected with the network. Less expected are *Holistic approach, Validation* and *Confusion of terms*. Other factors of interest are *Transfer-impediment, For computational purposes* and the differences in the sub-groups of abstraction. The annotated edges were used to determine feature influence and overlap. Main areas of focus we found and specific points of attention are:

- Holistic approach
  - Iteratively improve problem/goal and expand knowledge supports validation
  - Alternating problem-driven and solution-driven approach
  - Attention to heuristics and principles, e.g. [30]
  - Attention to macro-cognitive model, e.g. [21, 24]
- Abstraction
  - Abstraction should maintain relevant constraints and affordances
  - Use various levels of abstraction for complexity and multi-functional problems
  - Attention to problem definition and problem types
  - Attention to validation of abstracted entries in knowledge base
- Analogies
  - Representation of relevant knowledge for transfer
  - Attention to analogy mapping and analogical reasoning
  - Attention to analogy representation: natural language (descriptive), diagrams

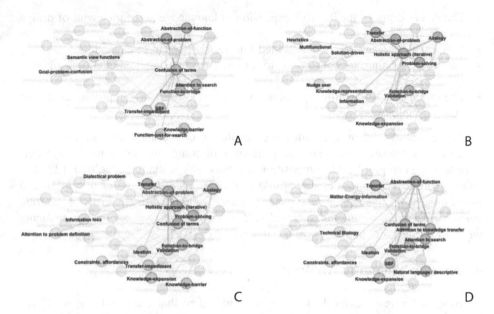

Fig. 2. Visualisations of features and relationships found during thematic analysis. The groups display a selected node and their most influential neighbours. The highlighted nodes in A-D are in alphabetical order: *Confusion of terms, Holistic approach, Abstraction of problem, Abstraction of function.*

- Confusion of terms and computational approach
  - Function is key and needs accurate definition for:
    - Problem decomposition
    - Wrong interpretation of interrelated terms (SBF)
    - Improve search for analogies and transfer
  - Automated characterisation improves scalability of knowledge base
- Transfer (impediments) and validation
  - Descriptive biological knowledge support realism and accuracy
  - Attention to supporting possible lack of biological knowledge
  - Attention to validation of analogies, e.g. through similarity

5 Discussion and Outlook

Some attention points for Computer Aided Biomimetics (CAB) were briefly introduced in the previous section. These are derived from a simplified qualitative analysis of the literature addressed in Sect. 3. The focus during analysis was on finding generic guidelines for the development of computational tools that support a BID process. The usefulness of these findings is left to the reader to decide.

In general, more attention should be paid to definitions, constraints and affordances, optimisation problems, the holistic approach and the validation mechanism. Validation for example is only possible when the required, relevant knowledge is accessible. However, the extended process described by Speck et al. (2008) implies that knowledge expansion drastically affects development time [13]. Investing additional time may not be desirable and Salgueiredo (2013) is right in concluding that BID should not be idolised [10].

BioTRIZ has, based on a non-systematic review of the literature on biological processes, changed the usual classification into '*matter, energy and information*' to '*substance, structure, time, space, energy/field, and information/regulation*' to allow better abstraction of biological phenomena. Further research is to aim at using TRIZ tools for CAB and Inventive Design [46] in cooperation with Julian Vincent and Denis Cavallucci.

Acknowledgements. The authors thank Julian Vincent and Denis Cavallucci for their advice.

References

1. The Biomimicry Institute Toolbox. http://toolbox.biomimicry.org/. Accessed 14 Mar 2016
2. Vandevenne, D., Verhaegen, P.-A., Dewulf, S., Duflou, J.R.: A scalable approach for the integration of large knowledge repositories in the biologically-inspired design process. In: Proceedings 18th International Conference on Engineering Design (ICED 2011), vol. 6, Lyngby/Copenhagen, Denmark (2011)
3. Fish, F.E., Beneski, J.T.: Evolution and bio-inspired design: natural limitations. In: Goel, A. K., McAdams, D.A., Stone, R.B. (eds.) Biologically Inspired Design, Chap. 12, pp. 287–312. Springer, London (2014)
4. Vincent, J.F.V.: Biomimetics in architectural design. In: Intelligent Buildings International, pp. 1–12 (2014)
5. Stone, R.B., Goel, A.K., McAdams, D.A.: Chartering a course for computer-aided bio-inspired design. In: Goel, A.K., McAdams, D.A., Stone, R.B. (eds.) Biologically Inspired Design, Chap. 1, pp. 1–16. Springer, London (2014)
6. The Biomimicry Institute Toolbox – Learn more. http://toolbox.biomimicry.org/core-concepts/function-and-strategy/. Accessed 14 Mar 2016
7. Pahl, G., Beitz, W., Feldhusen, J., Grote, K.-H.: Engineering Design: A Systematic Approach, 3rd edn. Springer, Berlin (2007) ISBN 978-1-84628-319-2
8. Vincent, J.F.V., Bogatyreva, O.A., Bogatyreva, N.R., Bowyer, A., Pahl, A.-K.: Biomimetics: its practice and theory. J. Roy. Soc. Interface 3(9), 471–482 (2006)
9. Vincent, J.F.V.: Biomimetics–a review. Proc. Inst. Mech. Eng., Part H: J. Eng. Med. 223(8), 919–939 (2009)
10. Salgueiredo, C.F: Modeling biological inspiration for innovative design. In: 20th International Product Development Management Conference, Paris, France, June 2013
11. Fayemi, P.-E., Maranzana, N., Aoussat, A., Bersano, G.: Bio-inspired design characterisation and its links with problem solving tools. In: Design Conference, Dubrovnik, Croatia, May 2014
12. Nachtigall, W.: Bionik: Grundlagen und Beispiele für Naturwissenschaftler und Ingenieure. Springer, Heidelberg (2002). ISBN 978-3-642-18996-8

13. Speck, T., Speck, O.: Process sequences in biomimetic research. Des. Nat. **4**, 3–11 (2008)
14. Martone, P.T., Boiler, M., Burgert, I., Dumais, J., Edwards, J., Mach, K., Rowe, N., Rueggeberg, M., Seidel, R., Speck, T.: Mechanics without muscle: biomechanical inspiration from the plant world. Integr. Comp. Biol. **50**, 888–907 (2010)
15. Helms, M., Vattam, S.S., Goel, A.K.: Biologically inspired design: process and products. Des. Stud. **30**, 606–622 (2009)
16. The Biomimicry Institute. http://toolbox.biomimicry.org/methods/integrating-biology-design/. Accessed 21 Mar 2016
17. Lindemann, U., Gramann, G.: Engineering design using biological principles. In: International Design Conference, Dubrovnik, Croatia, May 2004
18. Stricker, H.M.: Bionik in der Produktentwicklung unter der Berücksichtigung menschlichen Verhaltens. Ph.D. thesis, Technical University München (2006)
19. Inkermann, D., Stechert, C., Löffler, S., Victor, T.: A new bionic development approach used to improve machine elements for robotics applications. In: Proceedings of IASTED (2011)
20. Gramann, J.: Problemmodelle und Bionik als Methode. Ph.D. thesis, TU München (2004)
21. Goel, A.K., Vattam, S.S., Wiltgen, B., Helms M.: Information-processing theories of biologically inspired design. In: Goel, A.K., McAdams, D.A., Stone, R.B. (eds.) Biologically Inspired Design, pp. 127–152. Springer, London (2014)
22. Chakrabarti, A.: Supporting analogical transfer in biologically inspired design. In: Goel, A. K., McAdams, D.A., Stone, R.B. (eds.) Biologically Inspired Design, pp. 201–220. Springer, London (2014)
23. Yen, J., Helms, M., Goel, A.K., Tovey, C., Weissburg, M.: Adaptive evolution of teaching practices in biologically inspired design. In: Goel, A.K., McAdams, D.A., Stone, R.B. (eds.) Biologically Inspired Design, pp. 153–200. Springer, London (2014)
24. Vattam, S., Wiltgen, B., Helms, M., Goel, A.K., Yen, J.: DANE: fostering creativity in and through biologically inspired design. In: Taura, T., Nagai, Y. (eds.) International Conference on Design Creativity, Japan, pp. 115–122. Springer, London (2010)
25. Chakrabarti, A., Sarkar, P., Leelavathamma, B., Nataraju, B.S.: A functional representation for aiding biomimetic and artificial inspiration of new ideas. AIE EDAM **19**, 113–132 (2005)
26. Yim, S., Wilson, J.O., Rosen, D.W.: Development of an ontology for bio-inspired design using description logics. In: International Conference on PLM (2008)
27. Wilson, J.O., Chang, P., Yim, S., Rosen, D.W.: Developing a bio-inspired design repository using ontologies. In: Proceedings of IDETC/CIE (2009)
28. Liu, X., Rosen, D.W., Yu, Z.: Ontology based knowledge modeling and reuse approach in product redesign. In: IEEE IRI, Las Vegas, Nevada, USA, August 2010
29. Shu, L.H., Cheong, H.: A natural language approach to biomimetic design. In: Goel, A.K., McAdams, D.A., Stone, R.B. (eds.) Biologically Inspired Design, pp. 29–62. Springer, London (2014)
30. Linsey, J.S., Viswanathan, V.K.: Overcoming cognitive challenges in bioinspired design and analogy. In: Goel, A.K., McAdams, D.A., Stone, R.B. (eds.) Biologically Inspired Design, pp. 221–245. Springer, London (2014)
31. Vincent, J.F.V.: An ontology of biomimetics. In: Goel, A.K., McAdams, D.A., Stone, R.B. (eds.) Biologically Inspired Design, pp. 269–286. Springer, London (2014)
32. Vincent, J.F.V., Mann, D.L.: Systematic technology transfer from biology to engineering. Philos. Trans. Roy. Soc. Lond. A **360**, 159–173 (2002)
33. Bogatyrev, N., Bogatyrev, O.A.: TRIZ-based algorithm for Biomimetic design. Procedia Eng. **131**, 377–387 (2015)

34. Fratzl, P.: Biomimetic materials research: what can we really learn from nature's structural materials? J. Roy. Soc. Interface **4**(15), 637–642 (2007)
35. Stone, R.B., Wood, K.L.: Development of a functional basis for design. J. Mech. Design **122** (4), 359–370 (2000)
36. Deng, Y.-M.: Function and behavior representation in conceptual mechanical design. Artif. Intell. Eng. Des. Anal. Manuf. **16**, 343–362 (2002)
37. Nagel, J.K.S.: A thesaurus for bioinspired engineering design. In: Goel, A.K., McAdams, D. A., Stone, R.B. (eds.) Biologically Inspired Design, pp. 63–94. Springer, London (2014)
38. Nagel, J.K.S., Stone, R.B., McAdams, D.A.: Function-based biologically inspired design. In: Goel, A.K., McAdams, D.A., Stone, R.B. (eds.) Biologically Inspired Design, pp. 95–126. Springer, London (2014)
39. Goel, A.K., Rugaber, S., Vattam, S.S.: Structure, Behavior and Function of Complex Systems: The SBF Modeling Language. https://home.cc.gatech.edu/dil/uploads/SBF2.pd
40. Goel, A.K.: Biologically inspired design: a new paradigm for AI research on computational sustainability? In: Computational Sustainability, Workshop Papers (2015)
41. Vattam, S.S., Helms, M., Goel, A.K.: Biologically inspired design: a macrocognitive account. In Proceedings of the ASME IDETC/CIE (2010)
42. Sartori, J., Pal, U., Chakrabarti, A.: A methodology for supporting "transfer" in biomimetic design. AI Eng. Des. Anal. Manuf. **24**, 483–506 (2010)
43. Hatchuel, A., Weil, B.: C-K design theory: an advanced formulation. Res. Eng. Des. **19**, 181–192 (2009)
44. Gephi, Open Graph Viz Platform. https://gephi.org/
45. Braun, V., Clarke, V.: Using thematic analysis in psychology. Qual. Res. Psychol. **3**(2), 77–101 (2006)
46. Cavallucci, D., Rousselot, F., Zanni, C.: An ontology for TRIZ. Procedia Eng. **9**, 251–260 (2009)

A Neural Network with Central Pattern Generators Entrained by Sensory Feedback Controls Walking of a Bipedal Model

Wei Li[✉], Nicholas S. Szczecinski, Alexander J. Hunt, and Roger D. Quinn

Department of Mechanical and Aerospace Engineering, Case Western Reserve University,
Cleveland, OH 44106-7222, USA
wxl155@case.edu

Abstract. A neuromechanical simulation of a planar, bipedal walking robot has been developed. It is constructed as a simplified musculoskeletal system to mimic the biomechanics of the human lower body. The controller consists of a dynamic neural network with central pattern generators (CPGs) entrained by force and movement sensory feedback to generate appropriate muscle forces for walking. The CPG model is a two-level architecture, which consists of separate rhythm generator (RG) and pattern formation (PF) networks. The presented planar biped model walks stably in the sagittal plane without inertial sensors or a centralized posture controller or a "baby walker" to help overcome gravity. Its gait is similar to humans' with a walking speed of 1.2 m/s. The model walks over small obstacles (5 % of the leg length) and up and down 5° slopes without any additional higher level control actions.

Keywords: Biologically inspired · Central pattern generator · Sensory feedback · Bipedal walking

1 Introduction

It is generally accepted that basic rhythmic motor signals driving walking and other forms of locomotion in various animals are generated centrally by the neural networks in the spinal cord (reviewed by [1–3]). These neural networks, capable of producing coordinated and rhythmic locomotor activity without any input from sensory afferents or from higher control centers are referred to as central pattern generators (CPGs). Cats with transection of spinal cord and removal of major afferents can still produce rhythmic locomotor patterns [4, 5]. Despite lacking clear evidence of CPGs in humans, observations in patients with spinal cord injury provide some support [6]. Sensory input also plays an important role in locomotion. Locomotion is the result of dynamic interactions between the central nervous system, body biomechanics, and the external environment [7]. As a closely coupled neural control system, sensory feedback provides the information about the status of the biomechanics and the relationship between the body and the external environment to make locomotion adaptive to the real environment. Sensory input could reinforce CPG activities, provide signals to ensure correct motor output for all body parts, and entrain the rhythmic pattern to facilitate phase transition when a

© Springer International Publishing Switzerland 2016
N.F. Lepora et al. (Eds.): Living Machines 2016, LNAI 9793, pp. 144–154, 2016.
DOI: 10.1007/978-3-319-42417-0_14

proper body posture is achieved [8, 9]. There is strong evidence [10, 11] to suggest that sensory feedback affects motoneuron activity through common networks, rather than directly through reflex pathways acting on specific motoneurons. It implies sensory feedback and CPGs are highly integrated networks in the spinal cord.

Several 2D bipedal walking models have been previously developed with controllers containing CPGs and reflexes. Some simulation models [12, 13] had over-simplified multi-link structures driven by motors at joints, which lack the viscoelasticity of musculoskeletal structures. This lack provides poor mechanical properties that may reduce stability. Some models [14, 15] adopt more realistic musculoskeletal structures, but their CPG activities are not affected by sensory feedback. Instead, sensory afferents are connected directly with motoneurons. This is contradictory to findings that sensory feedback could strongly affect CPG activities [11, 12]. Geyer et al. [16] produced realistic human walking and muscle activities in a 2D model. They claim that their model only relies on muscle reflexes without CPGs or any other neural networks. It is realized by two sets (swing phase and stance phase) of coupled equations for every muscle to produce human walking and muscle activities. However, these equations could be viewed as abstract mathematical expressions of neural networks controlling muscle activities and the transition between two phases works like a finite-state machine. More recently, CPGs were added to a similar muscle-reflex model to control muscles actuating the hip joint [17]. Both of these works rely on precisely tracking muscle and other biomechanical parameters and provide limited insight into how the central nervous system is structured in humans. CPGs have also been applied to physical bipedal robots [18, 19], but usually these CPGs are abstract mathematical oscillators tuned to generate desirable trajectories of joints angles or torques, and the phases of the oscillators could simply be reset by reflexes. Klein and Lewis [20] claimed that their robot is the first that fully models human walking in a biologically accurate manner. But it has to rely on a baby walker-like frame to prevent it from falling down in the sagittal plane. Its knees remain bent distinctly before the foot touches the ground and its heels are always off the ground. It looks like it is pushing a heavy box, fundamentally different from a human's normal gait.

In this paper we present a biologically rooted neuromechanical simulation model of a planar, bipedal walker. It has a musculoskeletal structure and is controlled by a two-level CPG neural network incorporating biological reflexes via sensory afferents. The model is developed in Animatlab, which can model biomechanical structures and biologically realistic neural networks, and simulate their coupled dynamics [21]. We demonstrate that natural-looking, stable walking can be achieved by this relatively simple but biologically plausible controller combined with a human like morphological structure.

2   Structure of the Model

We based the biomechanical structure of our model on the human lower body, which has been shaped and selected by millions of years' evolution, and is energetically close to optimal [22]. The planar skeleton is made up of rigid bodies, and is actuated by antagonistic pairs of linear Hill muscles. The morphology structure employed signifi-cantly reduces the control effort with its inherent self-stability and provides a basis for comparison to the real human gait. The simulated biped (Fig. 1) is modeled as a pelvis supported by two legs. Each leg has 3 segments (thigh, shank and foot), which are connected by 3 hinge joints (hip, knee and ankle) and actuated by 6 muscles. The foot is composed of two parts connected by one hinge joint, spanned by a passive spring with damping (spring constant = 16 kN/m, damping coefficient = 20 N·m/s). This makes the foot flexible, which is the basis of the toe-off motion. The foot has a compliance of 10 μm/N to mimic the soft tissue of a human foot. The mass and length ratio of different segments (except the foot length) are based on human data [23]. The total mass of the model is 50 kg and its height is approximately 1 m (leg length is 0.84 m). The mass of the pelvis, thigh, shank and foot are 25 kg, 8 kg, 4 kg and 0.5 kg. The lengths of the thigh, shank and foot are 0.42 m, 0.42 m and 0.24 m.

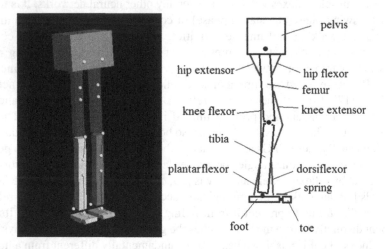

Fig. 1. Left: biomechancal model in Animatlab. Right: the muscle and skeleton system of the simulated biped.

3   Neural Control System

3.1   Neural Model

The neurons are leaky "integrate-and-fire" models. Each neuron is modeled as a "leaky integrator" of currents it receives, including synaptic current and injected current:

$$C_m \frac{dv(t)}{dt} = I_{leak}(t) + I_{syn}(t) + I_{inj}(t) \tag{1}$$

where $v(t)$ is the membrane potential, C_m is the membrane capacitance, $I_{leak}(t)$ is negative current due to the passive leak of the membrane, $I_{syn}(t)$ is the current from total synaptic input to the neuron, $I_{inj}(t)$ is the current externally injected into the neuron. The leak current is:

$$I_{leak}(t) = g_{leak} \cdot \left(E_{reset} - v(t)\right) \tag{2}$$

where g_{leak} is the leak conductance, E_{reset} is the resting potential. When the membrane potential reaches the threshold value the neuron is said to fire a spike, and is reset to E_{rest}.

Transmitter-activated ion channels are modeled with a time-dependent postsynaptic conductance $g_{syn}(t)$. The current that passes channels $I_{syn}(t)$ depends on the reversal potential E_{syn} of the synapse and the postsynaptic neuron membrane potential $v(t)$:

$$I_{syn}(t) = g_{syn}(t) \cdot \left(E_{syn} - v(t)\right) \tag{3}$$

If $E_{syn} < E_{rest}$, the synapse is inhibitory and it induces negative inhibitory postsynaptic current. If $E_{syn} > E_{rest}$, the synapse is excitatory and it induces positive excitatory postsynaptic current. The postsynaptic conductance $g_{syn}(t)$ increases to a value with a time delay to the arrival of the presynaptic spike and then decays exponentially to zero:

$$g_{syn}(t) = g_{\max} \cdot e^{-\dfrac{t - t_0 - \delta t_{delay}}{\tau_{syn}}} \quad , \ t \geq t_0 + \delta t_{delay} \tag{4}$$

where $g_{\max}$ is the initial amplitude of the postsynaptic conductance increase, t_0 denotes the arrival time of a presynaptic spike, δt_{delay} is the time delay between the presynaptic spike and the postsynaptic response, and τ_{syn} is the time constant of the decay rate.

3.2 Half-Center Oscillator

A basic neural pattern generator is a half-center neural oscillator composed of two reciprocally inhibited neurons. When one neuron spikes, it inhibits the other. Due to the facilitation of the synapse [24] the repeated spiking of the presynaptic neuron gradually hampers the synaptic transmission and reduces the postsynaptic conductance so the amplitude of the postsynaptic current gradually declines. The postsynaptic neuron would spike again as soon as the total current input recovers to the threshold value. When it starts to spike it would inhibit the other neuron in the same way.

3.3 CPG

Rybak et al. [25] proposed a two-level CPG architecture composed of a central rhythm generator (RG) network controlling several coupled unit pattern formation (UPF) modules in the pattern formation (PF) level. Sensory afferents could access and affect RG and PF networks separately. In this model the CPG network adopts a similar architecture as shown in Fig. 2. The RG is a half-center oscillator that generates the basic periodic stance signal for both legs. The PF network contains half-center oscillators as

UPF modules and each joint of each leg (hip, knee, and ankle) has one. Each half-center in the UPF module is connected to the corresponding extensor and flexor motoneurons to drive the extensor and flexor muscles for each joint. All half-center oscillators consist of two reciprocally inhibited "integrate-and-fire" neurons as described previously.

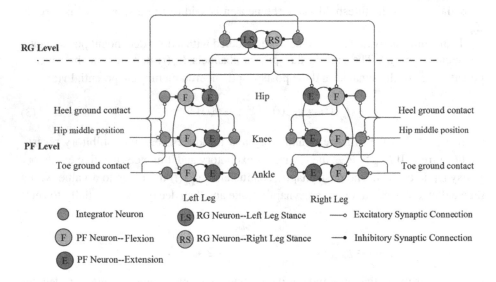

Fig. 2. The architecture of the CPG network. Black lines are synaptic connections between CPG neurons. Blue lines are synaptic connections from sensory neurons. (Color figure online)

Each neuron in the RG has an excitatory connection with a hip extension neuron and the other leg's hip flexion neuron. It encourages the legs to step in antiphase. At the PF level the hip flexion neuron has an excitatory connection with the ankle dorsiflexion neuron and an inhibitory connection with knee extension neuron. The hip extension neuron has an excitatory connection with the knee extension neuron. This controller uses a layer of interneurons to combine and filter afferent inputs from different receptors. These interneurons have inhibitory connections to the interneurons in the RG and PF networks.

3.4 Sensory Afferents

It is generally agreed that sensory input affects CPG timing directly since stimulation of corresponding afferents entrains or resets the locomotor rhythm [26]. In legged animals, there are three main categories of afferents that satisfy this requirement; group I muscle afferents in extensor muscles, cutaneous afferents in feet, and muscle afferents around the hip joint signaling its position. The first two largely depend on leg loading information, and the last is based on movement.

1. Ground Contact Afferents

Studies show that increasing the load on an infants' feet while they walk with support significantly increases the duration of the stance phase. It suggests that load receptor reflex activity can strongly modify and regulate gait timing and resides within the spinal circuits [27]. In our model, separate force sensors located on the plantar surface of the heels and toes separately detect and measure the ground contact force. As shown in Fig. 2, heel ground contact sensors excite afferent neurons which make inhibitory connections to the hip flexion, knee flexion, and ankle plantarflexion neurons at the PF level. They also apply inhibition to the other leg's stance neuron in the RG. Toe ground contact sensors inhibit the knee flexion and ankle dorsiflexion neurons at the PF level.

2. Hip Middle Position Afferents

During normal walking the knee extends in swing before the foot touches the ground. The knee reaches its peak flexion between 25 % and 40 % of the swing phase before extending. As shown in Fig. 2 in our model the hip middle position afferent is connected to the knee flexion neuron at the PF level and inhibits it when hip movement passes the predetermined hip angle during hip flexion.

4   Experiment and Results

In software simulation experiments, we confined the movement of the body to the sagittal plane, but it has no gravitational or pitching support, so it can fall forward or backward.

4.1   Normal Walking

The model walks at a moderate speed (1.2 m/s) on a rigid surface. Figure 3(a) shows a frame by frame comparison to that of a human. It is interesting to observe that they appear similar at corresponding phases of the cycle. Figure 3(b) compares hip, knee, and ankle angles between the model and human data during one gait cycle. The model's motion matches a human's in several aspects. The range of motion and general shapes of hip, knee and ankle angle curves are similar to a human's, although the model's knee motion lacks initial flexion at early stance phase, and the ankle dorsiflexion is insufficient. The knee reaches maximum flexion at about 75 % of the cycle and begins to extend before the hip reaches the anterior extreme position to prepare for ground contact. The transition from stance to swing is at about 60 % of the cycle, when the hip and knee begins to flex, and ankle reaches maximum plantarflexion. Figure 3(a) also compares the motion of the model's foot with a human's foot at early, middle and late stance phases.

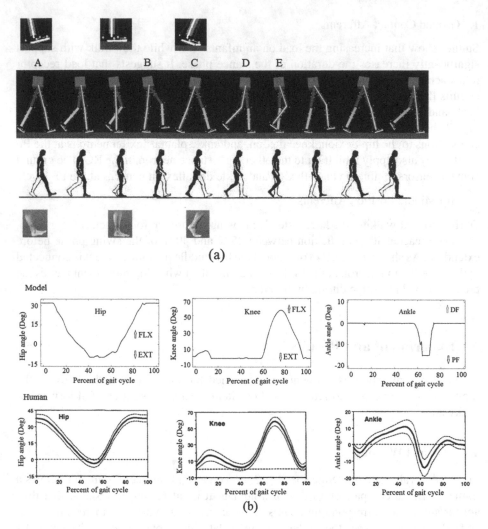

Fig. 3. Comparison of gait between the model and human. (a) Series of frames of the model and human at the same points in the gait cycle. (b) Model's hip, knee, and ankle angles compared with human's in the sagittal plane [28].

4.2 Interaction Between Sensory Feedback and CPGs

When the right leg enters stance (frame A in Fig. 3(a)) the right heel load sensor causes the heel ground contact sensor neuron to spike. This triggers the right leg stance RG neuron, causing right leg hip and knee UPF extension neurons to spike and initiate stance (time A in Fig. 4(c)). At time B in Fig. 3(a) the right heel just leaves ground and right heel ground contact sensor neuron stops spiking (time B in Fig. 4(a)). The right ankle UPF plantarflexion neuron is released from inhibition and begins to spike, causing right ankle plantarflexion (time B in Fig. 4(c)). The double support phase can be seen just following time C in Fig. 3(a). At this point, both the right toe and the left heel are in contact with the ground. The left heel ground contact sensor neuron begins to spike (time C in Fig. 4(b)) and inhibit the right leg stance RG neuron (time C in Fig. 4(c)). The left leg enters stance and begins to take load from the right leg. Loading of the left leg triggers right leg swing and further encourages stance in the left leg. At time D in Fig. 3(a) the right foot toe part leaves the ground. Right toe sensor neuron stops spiking (time D in Fig. 4(a)), releasing right knee UPF flexion neuron and right ankle UPF dorsiflexion neuron from inhibition (time D in Fig. 4(c)). Right knee further flexes and right ankle goes into dorsiflexion. Toward the end of the right leg's swing phase (time E in Fig. 3(a)) the right hip middle position sensor neuron spikes (time E in Fig. 4(a)), inhibiting the right knee UPF flexion neuron from spiking (time E in Fig. 4(c)). This causes the knee to extend and prepare for ground contact.

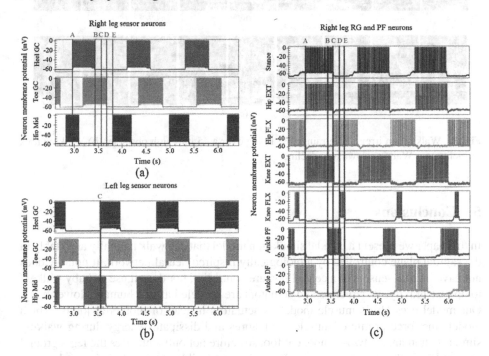

Fig. 4. Neuron membrane potentials in the gait cycle. (a) Right leg sensor neuron membrane potentials. GC: ground contact, Mid: middle position. (b) Left leg sensor neuron membrane potentials. (c) Right leg RG and PF neuron membrane potentials. EXT: extension, FLX: flexion, PF: plantarflexion, DF: dorsiflexion.

4.3   Walking on Irregular Terrain

The model can walk on some irregular terrains with the CPG network alone. No higher control actions are needed. It can walk over a 45 mm high obstacle, approximately 5 % of the leg length (Fig. 5(a)). It can also walk up a 5° incline (Fig. 5(b)) and walk down a 5° decline (Fig. 5(c)).

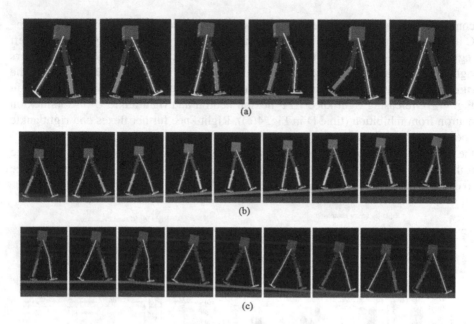

Fig. 5. Walking on irregular terrain (a) Walking over a 45 mm high obstacle. (b) Walking up a 5° incline. (c) Walking down a 5° decline.

5   Conclusions

In this paper we present a bipedal simulation model that can walk naturally and stably in the sagittal plane controlled by a biologically inspired neural network. It notably does not have inertial sensors or a central posture controller, nor does it use a "baby walker" to oppose gravity. Its structure and actuators are modeled after a human's lower body. Our model uses Hill's muscle model as actuators instead of motors at joints, which models the force profile of muscles and stores and dissipates energy during walking similar to humans. A two-component foot structure not only enhances the leg's ground contact late in the stance phase but also makes it possible to provide accurate dynamic sensory information of the relation between the foot and ground during stance by using individual sensors on toe and heel.

Using a neural network as a controller provides a more tangible and consistent imitation of the human central nervous system related to walking than switching between two sets of abstract equations to coordinate muscle reflexes. Compared to other bipedal models or robots adopting CPGs, our neural network has several more biologically plausible features. The rhythmic patterns are generated by oscillators composed of biological neuron models, not by mathematical oscillators tuned to reproduce the trajectory of joint movement. In our model sensory inputs are all biologically realistic and they access the CPG network and entrain its output instead of directly modifying activities of motoneurons. In the two-level architecture the movement of hip joints are coordinated under a higher level rhythm generator instead of directly coupling hip pattern generators of two legs.

Our developed controller and simulation not only demonstrates that stable bipedal walking in the sagittal plane can be achieved through a biologically based neural controller, but can also be used as a platform to help researchers test hypotheses related to the neural control of human locomotion.

References

1. Guertin, P.A.: Central pattern generator for locomotion: anatomical, physiological, and pathophysiological considerations. Front. Neurol. **3**, 4–15 (2013)
2. Mackay-Lyons, M.: Central pattern generation of locomotion: a review of the evidence. Phys. Ther. **82**, 69–83 (2002)
3. Ijspeert, A.J.: Central pattern generators for locomotion control in animals and robots: a review. Neural Netw. **21**, 642–653 (2008)
4. Grillner, S., Zangger, P.: On the central generation of locomotion in the low spinal cat. Exp. Brain Res. **34**, 241–261 (1979)
5. Pearson, K.G., Rossignol, S.: Fictive motor patterns in chronic spinal cats. J. Neurophysiol. **66**, 1874–1887 (1991)
6. Dimitrijevic, M.R., Gerasimenko, Y., Pinter, M.M.: Evidence for a spinal central pattern generator in humans. Ann. N. Y. Acad. Sci. **860**, 360–376 (1998)
7. Dickinson, M.H., et al.: How animals move: an integrative view. Science **288**, 100–106 (2000)
8. Pearson, K.G.: Common principles of motor control in vertebrates and invertebrates. Ann. Rev. Neurosci. **16**, 265–297 (1993)
9. Rossignol, S., Dubuc, R., Gossard, J.P.: Dynamic sensorimotor interactions in locomotion. Physiol. Rev. **86**, 89–154 (2006)
10. McCrea, D.A.: Spinal circuitry of sensorimotor control of locomotion. J. Physiol. **533**, 41–50 (2001)
11. Pearson, K.G.: Generating the walking gait: role of sensory feedback. Prog. Brain Res. **143**, 123–129 (2004)
12. Taga, G., Yamaguchi, Y., Shimizu, H.: Self-organized control of bipedal locomotion by neural oscillators in unpredictable environment. Biol. Cybern. **65**, 147–159 (1991)
13. Aoi, S., Tsuchiya, K.: Locomotion control of a biped robot using nonlinear oscillators. Auton. Robots **19**, 219–232 (2005)
14. Ogihara, N., Yamazaki, N.: Generation of human bipedal locomotion by a bio-mimetic neuro-musculo-skeletal model. Biol. Cybern. **84**, 1–11 (2001)
15. Paul, C., Bellotti, M., Jezernik, S., Curt, A.: Development of a human neuro-musculo-skeletal model for investigation of spinal cord injury. Biol. Cybern. **93**, 153–170 (2005)
16. Geyer, H., Herr, H.: A muscle-reflex model that encodes principles of legged mechanics produces human walking dynamics and muscle activities. IEEE Trans. Neural Syst. Rehabil. Eng. **18**(3), 263–273 (2010)
17. Van der Noot, N., Ijspeert, A.J., Ronsse, R.: Biped gait controller for large speed variations, combining reflexes and a central pattern generator in a neuromuscular model. In: IEEE International Conference on ICRA 2015, pp. 6267–6274 (2015)
18. Morimoto, J., Endo, G., Nakanishi, J., Cheng, G.: A biologically inspired biped locomotion strategy for humanoid robots: modulation of sinusoidal patterns by a coupled oscillator model. IEEE Trans. Robot. **24**(1), 185–191 (2008)

19. Righetti, L., Ijspeert, A.J.: Programmable central pattern generators: an application to biped locomotion control. In: Proceedings of 2006 IEEE International Conference on ICRA 2006, pp. 1585–1590 (2006)

20. Klein, T.J., Lewis, M.A.: A physical model of sensorimotor interactions during locomotion. J. Neural Eng. **9**(4), 1–14 (2012)

21. Cofer, D.W., Cymbalyuk, G., Reid, J., Zhu, Y., Heitler, W.J., Edwards, D.H.: AnimatLab: a 3D graphics environment for neuromechanical simulation. J. Neurosci. Methods **187**(2), 280–288 (2010)

22. Fischer, M., Witte, H.: Legs evolved only at the end. Philos. Trans. R. Soc. A **365**, 185–198 (2006)

23. Plagenhoef, S., Evans, F.G., Abdelnour, T.: Anatomical data for analyzing human motion. Res. Q. Exerc. Sport **54**, 169–178 (1983)

24. Zucker, R.S., Regehr, W.G.: Short-term synaptic plasticity. Annu. Rev. Physiol. **64**(1), 355–405 (2002)

25. Rybak, I.A., Stecina, K., Shevtsova, N.A., Lafreniere-Roula, M., McCrea, D.A.: Modeling spinal circuitry involved in locomotor pattern generation: insights from deletions during fictive locomotion. J. Physiol. **577**, 617–639 (2006)

26. Dietz, V., Duysens, J.: Significance of load receptor input during locomotion: a review. Gait and Posture **11**, 102–110 (2000)

27. Pang, M.Y.C., Yang, J.F.: The initiation of the swing phase in human infant stepping: importance of hip position and leg load. J. Physiol. **528**(2), 389–404 (2010)

28. Kadaba, M.P., Ramakrishnan, H.K., Wootten, M.E.: Measurement of lower extremity kinematics during level walking. J. Orthop. Res. **8**, 383–392 (1990)

Towards Unsupervised Canine Posture Classification via Depth Shadow Detection and Infrared Reconstruction for Improved Image Segmentation Accuracy

Sean Mealin$^{(\boxtimes)}$, Steven Howell, and David L. Roberts

North Carolina State University, Raleigh, USA
{spmealin,robertsd}@csc.ncsu.edu, schowel2@ncsu.edu

Abstract. Hardware capable of 3D sensing, such as the Microsoft Kinect, has opened up new possibilities for low-cost computer vision applications. In this paper, we take the first steps towards unsupervised canine posture classification by presenting an algorithm to perform canine-background segmentation, using depth shadows and infrared data for increased accuracy. We report on two experiments to show that the algorithm can operate at various distances and heights, and examine how that effects its accuracy. We also perform a third experiment to show that the output of the algorithm can be used for k-means clustering, resulting in accurate clusters 83 % of the time without any preprocessing and when the segmentation algorithm is at least 90 % accurate.

1 Introduction

Recently there has been a growing interest in the use of technology to interpret dogs' postures and behaviors. Existing work in this space has fallen into two main categories: on-body systems that rely primarily on inertial measurement units (*c.f.*, [1]) and off-body systems that use computer vision techniques (*c.f.*, [7]). On-body techniques have proven effective for classifying postures, but aren't useful for identifying dogs' position or orientation in space.

Previous work has used a Microsoft Kinect and supervised learning algorithms [7] for canine posture detection; however, they identified several challenges. Specifically, changes in illumination negatively impacted the segmentation of the dog from the background, and the accuracy of the classification algorithm. Further, identifying the dog's paws proved difficult due to noise in the data produced by the Kinect and the proximity of the paws to the floor. Another practical challenge with their system is that supervised algorithms require the collection of sizable corpora of labeled data to allow a model to be built and successfully validated. Depending on the algorithm and dog, these hand-labeled data may be required for every dog, which does not scale to larger applications and may prove difficult for non-technical experts (*i.e.*, pet owners) to provide.

We propose that an unsupervised approach for canine posture detection would avoid several of those problems including the collection and labeling of a

© Springer International Publishing Switzerland 2016
N.F. Lepora et al. (Eds.): Living Machines 2016, LNAI 9793, pp. 155–166, 2016.
DOI: 10.1007/978-3-319-42417-0_15

training dataset. Our approach uses both depth and infrared data to perform canine-background segmentation more accurately than just relying on depth data alone. As a heuristic for locating the dog in the scene, we use depth shadows, while infrared data helps to differentiate the dog's paws from the ground. We also present empirical data from two experiments: the first showed that the Kinect must be positioned within 2 m of the canine for best performance, while the second experiment indicated that the algorithm is capable of detecting the canine at most angles. We then conclude with some preliminary results showing that the output of the algorithm is suitable for k-means clustering for body part differentiation, as a step towards unsupervised classification. Since the focus of this paper is on the performance of the algorithm in different environments, and to ensure that our results are as reproducible and rigorous as possible, we exclusively used two life-sized anatomically correct canine models for the experiments. In future work, we will expand our experiments to include a wide variety of canine breeds.

2   Related Work

For more than three decades the problem of image segmentation has been studied in the field of computer vision (*c.f.*, [4,6,11]). With the prevalence of hardware like the Microsoft Kinect that is capable of detecting depth as well as color, techniques for segmentation, object tracking, and gesture recognition have changed to take advantage of the additional data [3]. One method for segmentation uses depth shadows as a feature for increased accuracy. A depth shadow is a region in the image that is set to an invalid value, due to a number of possible reasons; identifying that reason can assist with segmentation of objects in the scene [2].

In addition to allowing 3D interaction for humans, the Kinect has proven to be a useful tool for a variety of animal related tasks. Previous studies have examined the suitability of the Kinect for monitoring cows [9], and estimating the activity and weight of pigs [5,12]. The use of the infrared data returned by the Kinect has also enabled the automatic recognition of rat posture and behavior [10] using supervised machine learning algorithms. For entertainment purposes, the Kinect has tracked cats for human-feline games [8].

To our knowledge, the only study focusing specifically on the use of the Kinect with dogs resulted in a system for canine wellness [7]. The study used depth data provided by the Kinect as input to a Structural Support Vector Machine (SSVM), which then classified seven body parts of the dog. The SSVM is a supervised machine-learning algorithm, which required the manual labeling of a training dataset based on data collected during their trials. Our approach may aid this approach by improving SSVM accuracy through the use of infrared data to more accurately capture the location and contours of the canine's body parts; however, our main goal is to lay the foundation for an unsupervised posture classification system, which removes the burden of manually annotating data, and training a classifier for each dog.

3 Approach

We used a Microsoft Kinect 2.0 to capture depth and infrared data, and a laptop with an Intel Core i7-3720QM processor and 16 gigabytes of memory for the analysis. The Kinect is capable of capturing color frames at 1080p resolution at 30 frames per second, and depth and infrared frames at a resolution of 512×424 pixels at 30 frames per second. According to Microsoft, for optimal results, the Kinect should be positioned between 0.5 m and 4 m from the subject, and should not be in direct sunlight.

For the purposes of this work, we separated the data capture and analysis processes. We used the Kinect SDK 2.0 to save each frame to a file without any preprocessing. We implemented the segmentation algorithm in Python 3.5, relying on the Python bindings of OpenCV 3.1, Numpy 1.9.3, and Scipy 0.16.1.

Our approach has two steps: first, we isolate a region of interest (ROI), which (ideally) contains the canine in the scene; and second, we refine the depth data associated with the canine. For this discussion, an image consists of both depth frame and the corresponding infrared frame. It is necessary to take an image of the background without the canine present to do background subtraction. Let IMG denote an image with the canine present, while BG denotes an image of the background without the canine. A subscript D or I refers to the depth or infrared portion of an image respectively; for example, IMG_D refers to the depth data of an image which contains a canine.

To find the ROI, we convert a copy of IMG_D and BG_D to binary images using an inverse binary threshold filter. This effectively singles out the depth shadows in the image. As explained by Deng et al. [2], a depth shadow results when the infrared light used by the Kinect for distance measurements is occluded by an object in the scene. We then apply the morphological operation dilate to the binary image of BG_D; dilate is a standard image manipulation technique provided by OpenCV which causes thin lines in an image to become thicker. That will cause the depth shadows in the background of the scene to become more prominent. We then use background subtraction to remove the depth shadows that naturally occur in the scene due to fixed objects, while also reducing the noise that sometimes leads to false shadows when an object does not reflect infrared light back to the Kinect. To close any holes in the depth shadow, which could interfere with the next step, we then re-run the morphological operation on the subtracted image. At this point, the most prominent depth shadow is the outline of the dog, so contour detection returns the bounding box of the canine.

We next convert the portion of IMG_I that corresponds to the ROI identified in the IMG_I to a threshold image. The value for the threshold function can be taken from a part of the image that contains mainly background, such as the edge of the bounding box. On the generated threshold image, we run contour detection, giving preference to the contour closest to the center of the image if there are multiple detected. That contour serves as a bit mask, removing all depth readings not covered by the contour. This process results in all depth readings being set to 0, with the exception of the canine in the scene.

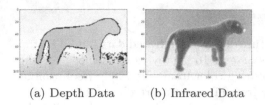

(a) Depth Data   (b) Infrared Data

Fig. 1. Image showing the depth data of a dog (a), where the paws are indistinct, and image of the infrared data (b), where the paws are well-defined.

Noise in the Kinect depth readings causes details of objects to get lost when near flat surfaces, such as the floor. This problem is exacerbated by uneven services, such as carpeting. Figure 1(a) is an example where the dog's paws are almost completely lost due to noise. Our approach uses infrared data to recover the information from the noise. Where the depth data is noisy, the infrared data maintains a clean contour in most circumstances. Figure 1(b) shows the same view, only in infrared, where the paws are clearly defined against the floor.

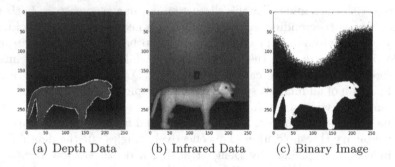

(a) Depth Data   (b) Infrared Data   (c) Binary Image

Fig. 2. Image showing the effects of infrared reflections on depth data (a), infrared data (b), and a binary image to highlight the dog and reflection (c).

Vertical flat objects, such as walls, cause the infrared data from the Kinect to be especially noisy. As an example, Fig. 2(a) is the depth data, Fig. 2(b) is the infrared data of the same scene, and Fig. 2(c) is the result of applying a binary threshold filter. The Figure shows that the infrared reflection causes an imaginary object at the top of the image. To reduce this noise look for a nearby depth shadow; if there is not one, the algorithm ignores the object. If there are multiple objects with depth shadows, the algorithm selects the one that is most central.

4   Experiment One

To understand the physical constraints of the Kinect's positioning on the algorithm, we captured a total of 14 images of a canine in two locations at multiple distances and heights. A human then assigned a binary score based on whether the canine was visible in the output of the algorithm, which determined whether the algorithm was successful at that distance-height combination. We used this data to determine the best position of the Kinect for future experiments.

4.1   Experiment One Methodology

Fig. 3. A color image of the dog in location 1, side facing the camera.

In order to get clean and repeatable data, our experiment used a life-sized anatomically correct stuffed yellow Labrador Retriever, measuring 95 cm long from nose to tail, 29 cm wide, and 55 cm tall. The stuffed dog was in a standing posture, with head turned slightly to the left of center. We positioned the canine sideways relative to the Kinect in every image for this experiment to provide an ideal angle for segmentation. Figure 3 shows a sample of the stuffed dog's positioning.

The Kinect does not work best in direct sunlight, so both locations were indoors. Location 1 was in a room that had vinyl flooring, with the dog located in the corner of the room. We selected this to examine if the algorithm had trouble with the slant of the two walls creating the corner. Due to space restrictions, we were only able to capture images of the dog at distances of 1 m, 1.5 m, and 2 m from the dog, and at heights of 1 m and 1.5 m above ground level, for a total of 6 positions. Location 2 contained carpeted flooring, doorways to open spaces, and various pieces of furniture within the image. We chose this location to see how carpet and clutter effected the performance of the algorithm. The Kinect was positioned at distances of 1 m, 2 m, 3 m, and 4 m away from the dog, and at heights of 1 m and 1.5 m above ground level, resulting in 8 positions.

4.2   Experiment One Results

To be successful, we determined that the output of the algorithm must contain the canine either completely, or partially, no matter how small. We were as lenient on the criteria as possible, to only filter out the positions that the algorithm completely failed to find the canine in the scene. In location 1, the algorithm was able to identify the canine in 4 out of the 6 images. At distances 1 m and 1.5 m, and at heights 1 m and 1.5 m, the canine was visible in the output of the algorithm. In the two remaining locations, at distance 2 m and heights 1 m and 1.5 m, the infrared reflections on the two walls proved to be too much for the algorithm, which resulted in output containing the wall instead of the canine.

In location 2, the Kinect was able to locate the dog in 6 of the 8 images, at distances 1 m, 2 m, and 3 m, at both heights 1 m and 1.5 m. When positioned 4 m away from the canine at either height, the area taken up by the canine was too small for the algorithm, causing it to focus on other objects in the scene.

We found that when infrared light reflects off a surface directly at the Kinect, such as the wall directly behind the canine in location 2, the algorithm is more successful at filtering it out, compared to the angled surfaces that create the corner in location 1. We conclude that when possible, the Kinect should not be directly facing a corner created by two walls. We also note that the algorithm was able to find the dog when it was either small or large compared to the total area of the image, indicating the algorithm is resilient to dog size.

5   Experiment Two

To evaluate the effect of the canine's orientation on the algorithm, we captured images of three canines at different angles in the same two locations as before, at the best distances and heights identified by the first experiment.

5.1   Experiment Two Methodology

The two canines used in this experiment consisted of two life-sized anatomically correct stuffed dogs. The first was the stuffed yellow Labrador Retriever previously described, and the second an identical black one. We posed the two dogs in five positions: directly facing the Kinect, directly away from the Kinect, sideways relative to the Kinect, and at 45° angles between those three positions; those angles are designated as front, back, left, front-left, and back-left respectively below.

The two locations were the same as the first experiment. Since the Kinect failed to isolate the canine at a distance of 2 m in location 1, we limited the distances to 1 m and 1.5 m. Likewise, 4 m failed to result in useful output in location 2, so we limited the Kinect to distances 1 m, 2 m, and 3 m, at both heights as before.

5.2   Experiment Two Results

A human scored the output of the algorithm based on three criteria:

1. Located canine: at least some part of the canine must be visible in the output for this criteria to be considered successful.
2. Isolated canine: The canine must be the only object in the output for this criteria to be considered successful.
3. Present body parts: A point is awarded for each of the head, torso, tail, and all four paws if they are visible in the output of the algorithm. We did not penalize the algorithm if a body part was visually occluded, as determined by examining the color image. The final score is reported as the ratio of the number of detected body parts to the number of visible body parts.

Results for Criteria 1 and 2: For the first two criteria, in every image, the algorithm either exclusively identified the canine, or completely failed. There were no images where the canine was not completely isolated in the output, so we simultaneously report on the first two criteria below.

Table 1 contains the results for location 1 with the stuffed dogs. The algorithm was able to locate both stuffed dogs at distance 1 m at both heights, with the exception of the black lab in the front-left position at $h = 1.5$ m. At $D = 1.5$ m, the algorithm had trouble finding the dogs. At the lower height, it missed the black lab in every position, while at the upper height it was able to find it in three out of five positions.

Table 2 contains the results for location 2 with the stuffed dogs. Encouragingly, both the yellow and black labs were identified in every image through distance 2 m. At 3 m neither dog was detected reliably by the algorithm, suggesting that a distance of 2 m or less is ideal.

Table 1. Whether the stuffed yellow lab or stuffed black lab was detected in location 1 at a given distance and height, and at a given position. Each row is a position, while each column is a distance-height measured in meters. If a Y appears, it means that the stuffed yellow lab was detected, while a B means the stuffed black lab was detected.

Posture	1D1H	1D1.5H	1.5D1H	1.5D1.5H
Front	Y,B	Y,B	Y	Y,B
Front-Left	Y,B	Y	Y	
Left	Y,B	Y,B	Y	Y
Back-Left	Y,B	Y,B		Y,B
Back	Y,B	Y,B	Y	Y,B

Table 2. Whether the stuffed yellow lab or stuffed black lab was detected in location 2 at a given distance and height, and at a given position. Each row is a position, while each column is a distance-height measured in meters. If a Y appears, it means that the stuffed yellow lab was detected, while a B means the stuffed black lab was detected.

Posture	1D1H	1D1.5H	2D1H	2D1.5H	3D1H	3D1.5H
Front	Y,B	Y,B	Y,B	Y,B	B	
Front-Left	Y,B	Y,B	Y,B	Y,B		B
Left	Y,B	Y,B	Y,B	Y,B	Y,B	Y,B
Back-Left	Y,B	Y,B	Y,B	Y,B	Y,B	Y,B
Back	Y,B	Y,B	Y,B	Y,B		

Results for Criteria 3: The complete results for the third criteria are in Tables 3 and 4 for location 1 and location 2 respectively. The score is for the algorithms performance for both the yellow and black labs combined.

In location 1, the algorithm had trouble finding all parts of the canines. At distance 1 m, it consistently found both dogs, however it only scored 100 % when the canine was in the front and back positions, depending on the height.

Table 3. The score of the stuffed yellow lab and stuffed black lab in location 1. Each row is a position, while each column is a distance-height measured in meters. N/A means that neither dog was detected.

Posture	1D1H	1D1.5H	1.5D1H	1.5D1.5H
Front	100.00 %	75.00 %	100.00 %	87.50 %
Front-Left	83.33 %	60.00 %	60.00 %	N/A
Left	90.00 %	70.00 %	80.00 %	90.91 %
Back-Left	60.00 %	75.00 %	N/A	100.00 %
Back	87.50 %	100.00 %	100.00 %	100.00 %

Table 4. The score of the stuffed yellow lab and stuffed black lab in location 2. Each row is a position, while each column is a distance-height measured in meters. N/A means that neither dog was detected.

Posture	1D1H	1D1.5H	2D1H	2D1.5H	3D1H	3D1.5H
Front	100.00 %	100.00 %	100.00 %	87.50 %	100.00 %	N/A
Front-Left	90.00 %	100.00 %	80.00 %	70.00 %	N/A	100.00 %
Left	100.00 %	100.00 %	100.00 %	100.00 %	90.00 %	100.00 %
Back-Left:	100.00 %	100.00 %	100.00 %	100.00 %	91.67 %	91.67 %
Back	100.00 %	100.00 %	100.00 %	100.00 %	N/A	N/A

For distance 1.5 m, the algorithm sometimes failed to find a canine at all, but interestingly, scored the highest at h = 1.5 m with the lowest score being 87.5 %.

Table 4 shows that the algorithm did very well through the 2 m distance mark, where the lowest score was 70 %. Even at distance 3 m, when the algorithm started to struggle to find the dogs, the lowest score was 90 %, indicating that it only missed at most one body part.

6   Experiment Three

As a step towards the unsupervised recognition of canine posture, we present some preliminary results of k-means clustering on the output of the algorithm. Whereas the third criteria for the second experiment indicated which body parts were visible after segmentation as judged by a human, this experiment examines how successful an unsupervised learning technique, k-means clustering, is at identifying those body parts.

6.1   Experiment Three Methodology

We converted the depth data provided by our algorithm to a point cloud, using spatial transforms provided by the Kinect. For the below examples, we did no further processing of the data before running k-means clustering.

For this experiment, we ran clustering on all of the images from the second experiment, a total of 100 images. For each image, we ran k-means with $k = 7$ to account for the background and up to six distinct body areas we hoped to identify. A human then scored each image based on whether the clusters covered (depending on visibility in the raw image) up to six distinct body parts: the head, torso, rump, tail, and two sets of paws. We counted the algorithm correct if it grouped the two front and two back paws together.

6.2   Experiment Three Results

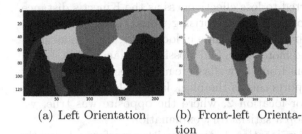

(a) Left Orientation   (b) Front-left Orientation

Fig. 4. An image showing the dog in the left orientation (a). There are different clusters for its head, torso, rump, front legs, back legs, and tail. In the front-left orientation (b) there are different clusters on its head, torso, rump, front legs, and back legs (the tail wasn't visible in the raw image).

Table 5 contains the results of the clustering algorithm, scored by human graders. Note, for these results, we do not report a percentage of clusters that were correct, but an all-or-nothing classification. Since clustering is an important step for unsupervised classification, we were as critical of the criteria for success as possible. The first column represents a threshold for the criteria from the second study. Out of the 50 images that scored a 100%, k-means resulted in sensible clusters for 41 images. Interestingly, as the accuracy of visible body parts decreased to 90%, the clustering algorithm was still able to maintain a high success rate.

Figure 4(a) is an example of very good results, where the background is cleanly distinguished from the dog and all six body parts are clustered neatly. Figure 4(b) illustrates another example of a successful result where the head, torso, rump, front paws, and back paws are all clustered as well. Because the tail wasn't visible and k-means was set to find seven clusters, an extra cluster appears on the chest.

Table 5. Results of the clustering algorithm for images of varying quality as determined by criteria three of the second experiment. The left column indicates a range of accuracy as defined in the second study. The next columns are the number of images for that accuracy range, the number of images that were successfully clustered, and a success rate given by dividing those two columns.

Percent of un-occluded body parts visible	Number of images	Number of successfully clustered images	Success rate
100 %	50	41	82.00 %
90 % - 99.99 %	12	10	83.33 %
80 % - 89.99 %	12	6	50.00 %
70 % - 79.99 %	8	5	62.50 %
60 % - 69.99 %	6	2	33.33 %
0 % (Dog Not Found)	12	0	0.00 %
Total	100	64	64.00 %

7   Discussion

As expected, the algorithm started to lose effectiveness as the Kinect's distance and height increased in the two locations. Surprising to us, the algorithm was able to find the stuffed yellow lab more of the time in location 1, while in location 2, the stuffed black lab was found more often. One possible explanation is that in location 1, The infrared reflections from the walls were more effective at drowning out the black lab. The data supports this as the black lab was lost as the Kinect could see more of the walls. For location 2, where the opposite was true, we require additional data to develop a satisfactory explanation.

Another observation is that in location 1, the algorithm performed slightly better at the taller height when $D = 1.5$ m. An explanation is that the increased height, which forced the Kinect's angle to be more sharply pointed towards the ground actually helped reduce the impact of the infrared reflections. This could indicate that when working in environments with unavoidable reflections, the Kinect should be positioned higher for optimal results.

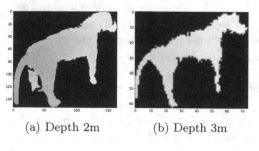

(a) Depth 2m   (b) Depth 3m

Fig. 5. Images showing the dog in the back-left position at 2 m (a) and 3 m (b). Its head, torso, tail, and *three* legs are clear at 2 m, while its head, torso, tail, and *two* legs are clear at 3 m.

In location 2, the algorithm gives satisfactory results at distances 1 m and 2 m. In all of the captured images, at both heights, the output was either perfect or missed at most three body parts for both dogs. After eliminating $D = 2$ m and $H = 1.5$ m, the algorithm only missed two body parts in the worst case, a tail and a paw. Based on these data, we suggest that the Kinect be positioned a maximum of 2 m from the dog for optimal results, which Fig. 5(a) demonstrates. Interestingly, even when that is exceeded, the algorithm can still produce acceptable results, as demonstrated by Fig. 5(b).

8   Future Work

In the future, we need to collect more data to better understand how the color of the dog and coat length effects the algorithm's accuracy. We were unable to explain why the algorithm was able to find the stuffed yellow lab and not the stuffed black lab in location 1, while being able to find the opposite in location 2 at extreme distances. We also will expand testing to real dogs, with a focus on dogs of various sizes, colors, and coat lengths. In conjunction with this, we are going to introduce additional postures such as when the canine is sitting or laying down.

As another way to increase the total accuracy of the algorithm, we plan to introduce multiple Kinect units. We believe that it would help when much of the canine's body parts are visually occluded, such as in the front or back positions. Having a second Kinect at another angle would also enable us to create larger point clouds, and capture the dog in more dimensions, which could potentially help with the k-means clustering algorithm.

9   Conclusion

In this paper, we presented an algorithm capable of canine-background segmentation in images captured by the Microsoft Kinect. The algorithm used depth shadows to identify the location of the canine, and infrared data to accurately find parts of the canine located near other surfaces, such as the floor.

We reported on two experiments that evaluate the positioning of the Kinect and the algorithm's performance as the position of the dog changes. The first experiment showed that the algorithm's performance is degraded by infrared reflections, however, the negative effects can be somewhat minimized by raising the height of the Kinect to angle it down towards the ground. The latter indicated that the algorithm can identify the canine in a variety of angles at various distances. We plan to run additional studies with a variety of dog breeds to more-fully evaluate the algorithm.

We finally presented the preliminary results of k-means clustering on the output of the algorithm. The clusters were sensibly formed in 83 % of images when the segmentation algorithm is at least 90 % accurate. It is our belief that adding preprocessing steps such as smoothing, and using multiple Kinect units could possibly increase the accuracy of the clusters. In the future, we will further refine the algorithm sensor placement with data gathered by using real, rather than stuffed, dogs.

Successfully identifying the position and orientation of a dog in an environment with little or no expensive-to-obtain hand-labeled data is a critical first step towards enabling novel computer-mediated interactions for dogs. The potential applications in this space are enormously-broad, ranging from computer assisted training, to welfare monitoring, and live evaluation of working dogs. The work presented in this paper represents an important first step towards realizing unsupervised localization and posture identification using commodity sensing hardware.

Acknowledgments. This material is based upon work supported by the National Science Foundation, under both Graduate Research Fellowship Grant No. DGE-1252376 and NSF grant 1329738. Any opinion, findings, and conclusions or recommendations expressed in this material are those of the authors and do not necessarily reflect the views of the National Science Foundation.

References

1. Brugarolas, R., Roberts, D., Sherman, B., Bozkurt, A.: Machine learning based posture estimation for a wireless canine machine interface. In: 2013 IEEE Topical Conference on Biomedical Wireless Technologies, Networks, and Sensing Systems (BioWireleSS), pp. 10–12. IEEE (2013)
2. Deng, T., Li, H., Cai, J., Cham, T.J., Fuchs, H.: Kinect shadow detection and classification. In: Proceedings of the IEEE International Conference on Computer Vision Workshops, pp. 708–713 (2013)
3. Han, J., Shao, L., Xu, D., Shotton, J.: Enhanced computer vision with microsoft kinect sensor: a review. IEEE Trans. Cybern. **43**(5), 1318–1334 (2013)
4. Haralick, R.M., Shapiro, L.G.: Image segmentation techniques. Comput. Vis. Graph. Image Process. **29**(1), 100–132 (1985)
5. Kongsro, J.: Estimation of pig weight using a microsoft kinect prototype imaging system. Comput. Electron. Agric. **109**, 32–35 (2014)
6. Peng, B., Zhang, L., Zhang, D.: A survey of graph theoretical approaches to image segmentation. Pattern Recogn. **46**(3), 1020–1038 (2013)
7. Pistocchi, S., Calderara, S., Barnard, S., Ferri, N., Cucchiara, R.: Kernelized structural classification for 3d dogs body parts detection. In: 2014 22nd International Conference on Pattern Recognition (ICPR), pp. 1993–1998. IEEE (2014)
8. Pons, P., Jan, J., Catal, A.: Developing a depth-based tracking systems for interactive playful environments with animals. In: 12th International Conference on Advances in Computer Entertainment Technologies, Second International Congress on Animal Human Computer Interaction (2015)
9. Salau, J., Haas, J., Thaller, G., Leisen, M., Junge, W.: Developing a multi-kinect-system for monitoring in dairy cows: object recognition and surface analysis using wavelets. Anim. Int. J. Anim. Biosci., 1–12 (2016)
10. Wang, Z., Mirbozorgi, S.A., Ghovanloo, M.: Towards a kinect-based behavior recognition and analysis system for small animals. In: 2015 IEEE Biomedical Circuits and Systems Conference (BioCAS), pp. 1–4. IEEE (2015)
11. Zhang, Y.J.: A survey on evaluation methods for image segmentation. Pattern Recogn. **29**(8), 1335–1346 (1996)
12. Zhu, Q., Ren, J., Barclay, D., McCormack, S., Thomson, W.: Automatic animal detection from kinect sensed images for livestock monitoring and assessment. In: 2015 IEEE International Conference on Computer and Information Technology; Ubiquitous Computing and Communications; Dependable, Autonomic and Secure Computing; Pervasive Intelligence and Computing (CIT/IUCC/DASC/PICOM), pp. 1154–1157. IEEE (2015)

A Bio-Inspired Model for Visual Collision Avoidance on a Hexapod Walking Robot

Hanno Gerd Meyer[1]($\boxtimes$), Olivier J.N. Bertrand[2], Jan Paskarbeit[1],
Jens Peter Lindemann[2], Axel Schneider[1,3], and Martin Egelhaaf[2]

[1] Biomechatronics, Center of Excellence 'Cognitive Interaction
Technology' (CITEC), University of Bielefeld, Bielefeld, Germany
hanno.meyer@uni-bielefeld.de
[2] Department of Neurobiology and Center of Excellence 'Cognitive Interaction
Technology' (CITEC), University of Bielefeld, Bielefeld, Germany
[3] Embedded Systems and Biomechatronics Group, Faculty of Engineering
and Mathematics, University of Applied Sciences, Bielefeld, Germany

Abstract. While navigating their environments it is essential for
autonomous mobile robots to actively avoid collisions with obstacles.
Flying insects perform this behavioural task with ease relying mainly
on information the visual system provides. Here we implement a bio-
inspired collision avoidance algorithm based on the extraction of nearness
information from visual motion on the hexapod walking robot platform
HECTOR. The algorithm allows HECTOR to navigate cluttered environments while actively avoiding obstacles.

Keywords: Biorobotics · Bio-inspired vision · Collision avoidance ·
Optic flow · Elementary motion detector

1 Introduction

Compared to man-made machines, insects show in many respects a remarkable
behavioural performance despite having only relatively small nervous systems.
Such behaviours include *complex flight or walking manoeuvres, avoiding collisions, approaching targets* or *navigating in cluttered environments* [11]. Sensing
and processing of environmental information is a prerequisite for behavioural
control in biological as well as in technical systems. An important source of
information is *visual motion*, because it provides information about self-motion,
moving objects, and also about the 3D-layout of the environment [7].

When an agent moves through a static environment, the resulting visual
image displacements (*optic flow*) depend on the speed and direction of ego-
motion, but may also be affected by the nearness to objects in the environment.
During *translational* movements, the optic flow amplitude is high if the agent
moves fast and/or if objects in the environment are close. However, during *rotational* movements, the optic flow amplitude depends solely on the velocity of ego-
motion and, thus, is independent of the nearness to objects. Hence, information

© Springer International Publishing Switzerland 2016
N.F. Lepora et al. (Eds.): Living Machines 2016, LNAI 9793, pp. 167–178, 2016.
DOI: 10.1007/978-3-319-42417-0_16

about the depth structure of an environment can be extracted parsimoniously during translational self-motion as distance information is immediately reflected in the optic flow [7]. To solve behavioural tasks, such as *avoiding collisions* or *approaching targets*, information about the depth structure of an environment is necessary. Behavioural studies suggest, that flying insects employ flight strategies which facilitate the neuronal extraction of depth information from optic flow by segregating flight trajectories into translational and usually much shorter rotational phases (*active-gaze-strategy*, [4]). Furthermore, the neuronal processing of optic flow has been shown to play a crucial role in the control of *flight stabilisation*, *object detection*, *visual odometry* and *spatial navigation* [3].

In flying insects, optic flow is estimated by a mechanism that can be modelled by *correlation-type elementary motion detectors* (EMDs, [2]). A characteristic property of EMDs is that the output does not exclusively depend on velocity, but also on the pattern properties of a moving stimulus, such as its contrast and spatial frequency content. Hence, nearness information can not be extracted unambiguously from EMD responses [5]. Rather, the responses of EMDs to pure translational optic flow have been concluded to resemble a representation of the *relative contrast-weighted nearness* to objects in the environment, or, in other words, of the contours of nearby objects [18].

Recently, a simple model for collision avoidance based on EMDs was proposed [1]. The model is based on three successive processing steps: (a) the extraction of (contrast-weighted) nearness information from optic flow by EMDs, (b) the determination of a collision avoidance direction from the map of nearness estimates and (c) the determination of a collision avoidance necessity, i.e. whether to follow (i.e. potential obstacles are close) or not to follow the collision avoidance direction (i.e. potential obstacles are still rather distant). When coupled with a goal direction, the algorithm is able to successfully guide an agent to a goal in cluttered environments without collisions.

In this study, this collision avoidance model was implemented on the insect-inspired hexapod walking robot HECTOR [15]. In contrast to flight, walking imposes specific constraints on the processing of optic flow information. Due to the mechanical coupling of the agent to the ground the perceived image flow is superimposed by continuous rotational components about all axes correlated to the stride-cycle [9]. Therefore, nearness estimation from optic flow during translational walking might be obfuscated, potentially reducing the reliability of the collision avoidance algorithm. Further, in contrast to the 360° panoramic vision used in [1], a fisheye lens was mounted on the front segment of the robot with its main optical axis pointing forward, limiting the field of view for retrieving optic flow information.

In the following, the implementation of the collision avoidance and vision-based direction control on the robotic platform will be described and the performance assessed in artificial and natural cluttered environments in a simulation framework of HECTOR. After optimisation of parameters, HECTOR will be able to successfully navigate to predefined goals in cluttered environments while avoiding collisions.

2 The Simulation Framework

The performance of the model of collision avoidance and direction control was assessed in a dynamics simulation of HECTOR which is coupled with a rendering module. The simulation allows parameter optimisation and tests in different virtual environments. The simulation framework is depicted in Fig. 1 and can be separated into four processing modules: (a) *walking controller*, (b) *robot simulation*, (c) *renderer* and (d) *vision-based directional controller*.

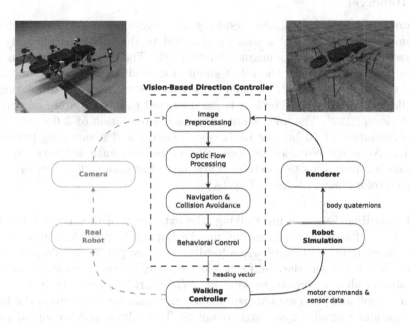

Fig. 1. The simulation framework used for testing the implementation of the visual collision avoidance model on HECTOR. The dashed box (*Vision-Based Direction Controller*) indicates the algorithm used for controlling the robot's behaviour based on nearness estimation from optic flow.

2.1 Robot Simulation and Walking Controller

The hexapod robot HECTOR is inspired by the stick insect *Carausius morosus*. For its design, the relative positions of the legs as well as the orientation of the legs' joint axes have been adopted. The size of the robot was scaled up by a factor of 20 as compared to the biological example which results in an overall length of roughly 0.9 m. This size allows the robot to be used as an integration platform for several hard- and software modules. All 18 drives for the joints of the six legs are serial elastic actuators. The mechanical compliance of the drives is achieved by an integrated, sensorised elastomer coupling [14] and is the foundation for the robot's ability to passively adapt to the structure of the substrate during walking. The bio-inspired walking controller is a conversion of the WALKNET approach [17] and allows the robot to negotiate rough terrain [15]. Abstracting the complex task of leg coordination, only the heading vector must be provided externally, e.g. by a vision-based direction controller as proposed here.

To simulate a multitude of controller parameters, a dynamics simulation has been set up based on ODE (Open Dynamics Engine) which also simulates the elastic joint actuation. The HECTOR simulator is controlled by the same controller framework as the physical robot.

2.2   Renderer

To obtain optic flow information resulting from ego-motion in different virtual environments the images of a camera attached to the robot's main body are rendered using the graphics engine Panda3D [6]. The robot's orientation and position are obtained from the robot simulation module. To emulate the wide field of view of insect eyes [21], virtual camera images are rendered simulating an equisolid fisheye lens [12]. The lens is parametrised to a horizontal and vertical field of view of 192°. The images obtained have a resolution of 400×400 pixels with a resolution of 10 bit per RGB color channel and a sampling frequency of 20 Hz. Although blowflies possess color vision, evidence suggests that the pathways involved in motion detection are monochromatic [20]. Therefore, only the green color channel is used (Fig. 2A).

Head Stabilisation. During walking, the extraction of distance information on the basis of optic flow processing may be impaired by stride-coupled image shifts. For example, walking blowflies hardly ever show purely translational loco-motion phases. Rather, they perform relatively large periodic rotations of their body around all axes due to walking [9]. While stride-induced body rotations around the roll and pitch axes are compensated by counter-rotations of the head, body rotations around the yaw axis are not [8]. To minimise stride-coupled image displacements, movements of the camera around the roll and pitch axis are compensated in simulation. This is achieved by setting the roll and pitch angles of the camera to fixed values independent of the movement of the robot's main body, effectively keeping the center of the optical axis of the camera parallel to the ground plane.

2.3   Vision-Based Direction Controller

The sequences of camera images obtained from the renderer are processed by the vision-based direction controller, which can be subdivided into four processing steps:

(a) *preprocessing of images*, in order to emulate the characteristics of the visual input of flying insects,
(b) estimation of a relative nearness map by *processing of optic flow* via EMDs,
(c) computation of a *collision avoidance direction* based on the relative nearness of objects and a goal direction, and
(d) *controlling the walking direction* of the robot.

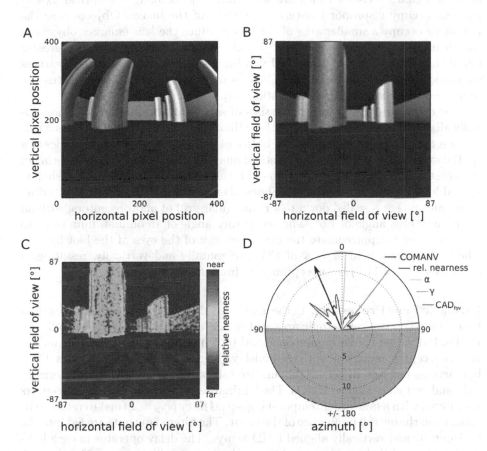

Fig. 2. (A) *Camera image* of a virtual environment rendered with an equisolid fish-eye lens. Image has been resized and reduced to 8-bit dynamic range for reproduction. (B) Camera image *remapped to a rectilinear representation, spatially filtered and scaled to an array of photoreceptors*. Each pixel position represents a luminance value as perceived by the according photoreceptor. Image has been resized and reduced to 8-bit dynamic range for reproduction. (C) *Relative contrast-weighted nearness map* μ_r obtained from optic flow estimation via horizontally and vertically aligned EMDs. The color-code depicts near (red) and far (blue) estimated relative nearnesses. (D) *Polar representation of the relative nearness averaged over the azimuth* (blue). The arrow (black) depicts the sum of nearness vectors ($COMANV$) used for determining the *collision avoidance direction* CAD_{fov} (red). Based on the weighted sum of the direction to a goal α (yellow) and the CAD_{fov}, a heading direction γ (green) is computed. The greyed out area indicates the surrounding of the robot *not* covered by the field of view. (Color figure online)

(a) **Image Preprocessing.** The non-linear fisheye lens, used here, possesses distortion characteristics which are such that objects along the optical axis of the lens occupy disproportionately large areas of the image. Objects near the periphery occupy a smaller area of the image. Since the lens enlarges objects in the vicinity of the optical axis, those objects are transmitted with much greater detail than objects in the peripheral viewing region, thus, obfuscating nearness estimation from optic flow. Hence, images obtained from the fisheye lens are remapped to a rectilinear representation [12].

The compound eye of insects consists of a two-dimensional array of hexagonally aligned ommatidia comprising the retina. Each ommatidium contains a lens and a set of photoreceptor cells. The lattice of ommatidia has characteristics of a spatial low-pass filter and blurs the retinal image. To mimic the spatial filtering of the eye, the remapped images are filtered by a two-dimensional Gaussian-shaped spatial low-pass filter according to Shoemaker et al. [19]. After spatial filtering, each input image is scaled down to a rectangular grid of photoreceptors, with an interommatidial angle of $1.5°$. The acceptance angle of an ommatidium is set to $1.64°$ in order to approximate the characteristics of the eyes of the blowfly [16]. The grid covers a field of view of $174°$ horizontally and vertically, resulting in an array of 116×116 photoreceptors (i.e. luminance values) (Fig. 2B).

(b) **Optic Flow Processing.** In the model used here (see [18]), optic flow estimation is based on two retinotopic arrays of either *horizontally* or *vertically* aligned EMDs. Individual EMDs are implemented by a multiplication of the delayed signal of a receptive input unit with the undelayed signal of a neighbouring unit. Only interactions between direct neighbours are taken into account, for both horizontally and vertically aligned EMDs. The luminance values from the photoreceptors are filtered with a first-order temporal high-pass filter ($\tau_{hp} = 20$ ms) to remove the mean from the overall luminance of the input. The filtered outputs are fed into the horizontally and vertically aligned EMD arrays. The delay operator in each half-detector is modelled by a temporal first-order low-pass filter ($\tau_{lp} = 35$ ms). Each EMD consists of two mirror-symmetric subunits with opposite preferred directions. Their outputs are subtracted from each other. For each retinotopic unit the motion energy is computed by taking the length of the motion vector given by the combination of the responses of a pair of the horizontal h_{EMD} and the vertical v_{EMD} at a given location (x,y) of the visual field:

$$\mu_{r(x,y)} = \sqrt{v_{\text{EMD}}^2(x,y) + h_{\text{EMD}}^2(x,y)} \tag{1}$$

The array of the absolute values of these local motion vectors μ_r resembles a map of *contrast-weighted relative nearness* to objects in the environment [18], providing information about the contours of nearby objects (Fig. 2C).

(c) **Navigation and Collision Avoidance.** Once the relative nearness map μ_r is known, collision avoidance is achieved by moving away from the maximum nearness value (e.g. objects that are close) (see [1]). However, the contrast-weighted nearness map also depends on the textural properties of the environment. To reduce the texture dependence, the nearness map is averaged along the elevation ϵ, giving the average nearness for a given azimuth ϕ. Each of these averaged nearness values can be represented by a vector in polar coordinates, where the norm of the vector is the averaged nearness, and its angle corresponds to the azimuth. The sum of these vectors points towards the average direction of close objects (Fig. 2D). This vector is denoted *center-of-mass-average-nearness-vector* ($COMANV$; [1])

$$COMANV = \sum \left(\begin{pmatrix} \cos(\phi) \\ \sin(\phi) \end{pmatrix} \frac{1}{n} \sum \mu_r(\epsilon, \phi) \right), \tag{2}$$

where n is the number of elements in the azimuth. The inverse of the $COMANV$ vector, scaled to the horizontal field of view θ of the photoreceptor array, points away from the closest object and, thus, can be used as the direction of the robot to avoid collisions (*collision avoidance direction*, CAD_{fov}; Fig. 2D; [1]):

$$CAD_{fov} = \frac{-\arctan(COMANV_y, COMANV_x)}{\frac{2\pi}{\theta}} \tag{3}$$

The length of the $COMANV$ vector increases with nearness and apparent size of objects. Its length is a measure of the *collision avoidance necessity* (CAN; [1]):

$$CAN = \| COMANV \|. \tag{4}$$

The CAN measure is used to control the heading direction γ of the robot between *avoiding collisions* and *following the direction to a goal* (α; Fig. 2D) [1]:

$$\gamma = W(CAN) \cdot CAD_{fov} + (1 - W(CAN)) \cdot \alpha \tag{5}$$

W is a sigmoid weighting function based on the CAN:

$$W(CAN) = \frac{1}{1 + \left(\frac{CAN}{n_0}\right)^{-g}}, \tag{6}$$

and driven by a gain g and a threshold n_0 [1].

(d) Behavioural Control of the Walking Robot. The walking direction of the robot is controlled based on the heading direction γ obtained from estimating relative nearness values to objects from optic flow and the goal direction. Information about the spatial structure of an environment can only be extracted from optic flow during translational movements, as rotational flow components do not provide distance information [7]. Inspired by the *active-gaze-strategy* employed by flying insects [4], the control of walking direction is implemented by segregating the motion trajectory into *translational* and *rotational phases*. During translation, the robot moves forward with a constant velocity of 0.2 m/s for 4 s (corresponding to 80 camera images), while averaging the heading direction γ.

After that, the robot switches to a rotational state and turns towards the averaged heading direction γ until the vector is centred in the field of view. When the optic flow field is estimated by EMDs, the nearness estimations also depend on the motion history due to the temporal filters. Hence, the optic flow obtained during the rotational phase interferes with the optic flow measurements during the translational phase. This effect decreases over time. Therefore, the heading direction γ is only averaged for the last 3 s of the translational phase.

Due to the restricted horizontal field of view, no information about the nearness of objects outside of the camera's field of view can be obtained (grey area in Fig. 2D). However, as the camera is pointing forward along the direction of walking during translation, information about the nearness of objects sidewards or behind the robot is not essential. In situations where the goal direction does *not* reside within the field of view, the CAN is set to zero, effectively inducing a turn of the robot in the rotational phase until the goal direction is centered in the field of view.

3   Visual Collision Avoidance in Cluttered Environments

Parameter Optimisation. The implementation of the collision avoidance model in the dynamics simulation of HECTOR was tested in several cluttered environments. In a first step, the threshold n_0 and gain g of the weighting function W [see Eq. (6)] were optimised in an artificial environment. The environment consisted of a cubic box with a cylindrical object placed in the center (see Fig. 3B–D). Both, the box and the object were covered with a Perlin noise texture. The robot was placed at a starting position (S) in front of the object, facing a goal position (G) behind the object. The distance between starting position and goal was set to 10 m. For each of the possible parameter combinations of the gain $g = [1.0, 2.0, ..., 10.0]$ and threshold $n_0 = [0.0, 1.0, ..., 20.0]$ the trajectory length for reaching the goal (G) without colliding with the object was taken as a benchmark of the performance of the collision avoidance model (see Fig. 3A). A collision was assumed if the position of the camera crossed a radius of 1.5 m around the center of the object (*black dashed circle*, Fig. 3B–D) and the respective combination of parameters was discarded. For each combination of parameters 3 trials were performed.

If the threshold n_0 is set to *low* values, the computation of the heading direction γ [see Eq. (5)] mainly depends on the collision avoidance direction CAD_{fov}, whereas the goal direction α is only taken into account to a small extent. Hence, the robot will more likely avoid collisions than navigate to the goal (G). Further, a steeper slope of the sigmoid weighting function W, set by the gain g, leads to higher temporal fluctuation of the heading direction γ. As a consequence, when setting the threshold to $n_0 = 0.0$ and the gain to $g = 10.0$, the resulting trajectories were relatively long (Fig. 3A) and showed erratic movement patterns. However, all trajectories reached the goal position for the given parameter combination (Fig. 3B). Due to the robot following the collision avoidance direction CAD_{fov}, in several cases the goal direction did not reside within the field of view,

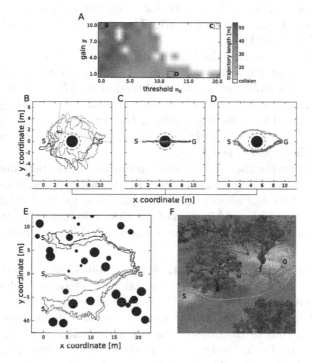

Fig. 3. (A) *Length of simulated trajectories* (color-coded) in a cubic box with a single object (see B–D) for different combinations of the weighting function parameters gain $g = [1.0, 2.0, ..., 10.0]$ and threshold $n_0 = [0.0, 1.0, ..., 20.0]$ [see Eq. (6)]. The size of the box was $14\,\mathrm{m} \times 14\,\mathrm{m} \times 10\,\mathrm{m}$ (length × width × height) and the radius of the object $r = 1\,\mathrm{m}$ (height $h = 10\,\mathrm{m}$). The walls of the box and the object were uniformly covered with a Perlin noise texture (scale $= 0.05$). When the trajectory crossed a circle of a radius of $1.5\,\mathrm{m}$ around the center of the object (dashed line in B–D) a collision was assumed (white areas). (B–D) *Simulated trajectories (n = 10) in a cubic box with a single object* (filled circle). Starting positions are given as S and goal positions as G. Weighting function parameters were set to (B) $g = 10.0$ and $n_0 = 0.0$, (C) $g = 10.0$ and $n_0 = 20.0$ and (D) $g = 1.0$ and $n_0 = 12.0$. The *grey dotted lines* in B indicate the main optical axis before and after recentering the goal direction in the visual field. (E) *Simulated trajectories in a cubic box with randomly placed objects* (filled circles) for different starting positions $(S_1–S_3)$. The size of the box was $25\,\mathrm{m} \times 25\,\mathrm{m} \times 10\,\mathrm{m}$ (length × width × height). The radius of each object $(n = 30$; height: $10\,\mathrm{m})$ was set randomly in a range from $0.25\,\mathrm{m}$ to $1.0\,\mathrm{m}$. The walls of the box and the objects were uniformly covered with a Perlin noise texture (scale $= 0.05$). Weighting function parameters were set to $g = 1.0$ and $n_0 = 12.0$ (see A and D). For each starting position 3 trajectories are shown. It is notable, that the variability for trajectories with the same starting positions arises due to the initialization of the robot with differing body postures, effectively influencing the initial perceived image flow. (F) *Simulated trajectories* $(n = 5)$ *in a reconstructed natural environment*. Weighting function parameters were set to $g = 1.0$ and $n_0 = 12.0$ (see A and D). The distance between starting position (S) and goal position (G) was 48.83 m.

resulting in a recentering of the goal vector along the main optical axis (as indicated by the *grey dashed lines* in Fig. 3B). This strategy led to reaching the goal position for all trajectories.

In contrast, when setting the threshold n_0 to *high* values, the computation of the heading vector γ mainly takes the goal direction α into account, whereas the influence of the collision avoidance direction (CAD_{fov}) is reduced. As a consequence, the robot will more likely follow the direction to the goal without avoiding obstacles. Therefore, when setting the threshold to $n_0 = 20.0$ and the gain to $g = 10.0$, the robot directly approached the goal position, consequently, colliding with the object (Fig. 3A and C).

Figure 3D shows the trajectories for a combination of the parameters gain and threshold which resulted in short trajectory lengths without collisions ($n_0 = 12.0$, $g = 1.0$). Here, the robot almost directly approached the goal, while effectively avoiding the object. This combination of the threshold and gain parameters was used in subsequent simulations in more complex cluttered environments.

Artificial Cluttered Environment. After optimisation of the threshold and gain parameters the performance of the collision avoidance model was tested in an Artificial cluttered environment (Fig. 3E) which was set up in a cubic box. Several cylindrical objects ($n = 30$) were placed at random positions in the x,y-plane. The radius of the objects was set individually to a random number of $r = [0.25, 1.0]$ m. Both, the box and the objects were covered with a texture generated from Perlin noise. The robot was placed at different starting positions (S_1–S_3), with the main optical axis oriented in parallel with the x-axis. The distance to the goal position was $d_{1,3} = 21.36$ m for the starting positions S_1 and S_3 and $d_2 = 20$ m for the starting position S_2. The parameters of the sigmoid weighting function W were set to $n_0 = 12.0$ and $g = 1.0$ according to Fig. 3A. For each starting position 3 trajectories were simulated. In all cases the robot successfully reached the goal position without collisions and without encountering local minima (see however [1] for a more detailed analysis).

Natural Cluttered Environment. We further tested the performance of the collision avoidance model in a reconstructed natural environment (Fig. 3F). The environment consisted of a dataset obtained from several laser scans [22]. The starting (S) and goal position (G) were set so that the robot had to avoid collisions with trees to reach the location of the goal. Also in the natural environment – which substantially differs in the textural pattern properties from the tested artificial environment – for all trials ($n = 5$) the combination of weighting function parameters $n_0 = 12.0$ and $g = 1.0$ resulted in trajectories successfully leading to the goal, without colliding with objects.

4   Conclusion

A prerequisite for autonomous mobile robots is to navigate their environments while actively avoiding collisions with obstacles. Whereas, to perform collision avoidance, autonomous robots nowadays normally rely on active sensors (e.g. laser range finders [13]) or extensive computations (e.g. Lucas-Kanade optic flow computation [10]), insects are able to do so with minimal energetic and computational expenditure by relying mainly on visual information. We implemented a bio-inspired model of collision avoidance in simulations of the hexapod walking robot HECTOR solely based on the processing of optic flow by correlation-type elementary motion detectors (EMDs). EMDs have previously been accounted for playing a key role in the processing of visual motion information in insects [3]. As could be shown, although the responses of EMDs to visual motion are entangled with the textural properties of the environment [5], the relative nearness information obtained from optic flow estimation via EMDs is sufficient to direct HECTOR to a goal location in cluttered environments without colliding with obstacles. This holds true either for *artificially* generated environments as well as for a *reconstructed natural* environment, which substantially differ in their textural pattern properties. Moreover, by employing behavioural strategies such as (a) an *active-gaze strategy* and (b) *active head stabilisation* – both also found in insects – the influence of rotational optic flow components which potentially obfuscate the estimation of relative nearness information from optic flow is reduced. Hence, on the physical robot a prototype for mechanical gaze-stabilisation has been implemented and is currently compared to a software implementation.

The simulation results shown here will serve as a basis for the implementation of more complex bio-inspired models for visually-guided navigation in hardware which is currently under development. These models will comprise strategies for navigation and search behaviour based on the insect-inspired processing of optic flow.

Acknowledgments. This work has been supported by the DFG Center of Excellence Cognitive Interaction TEChnology (CITEC, EXC 277) within the EICCI-project. We thank Dr. Wolfgang Stürzl for kindly providing us with a dataset of a laser scanned outdoor environment.

References

1. Bertrand, O.J., Lindemann, J.P., Egelhaaf, M.: A bio-inspired collision avoidance model based on spatial information derived from motion detectors leads to common routes. PLoS Comput. Biol. **11**(11), e1004339 (2015)
2. Borst, A.: Modelling fly motion vision. In: Feng, J. (ed.) Computational Neuroscience: A Comprehensive Approach, pp. 397–429. Chapman and Hall/CTC, Boca Raton, London, New York (2004)
3. Borst, A.: Fly visual course control: behaviour, algorithms and circuits. Nat. Rev. Neurosci. **15**(9), 590–599 (2014)
4. Egelhaaf, M., Boeddeker, N., Kern, R., Kurtz, R., Lindemann, J.P.: Spatial vision in insects is facilitated by shaping the dynamics of visual input through behavioral action. Front. Neural Circuits **6**(108), 1–23 (2012)

5. Egelhaaf, M., Kern, R., Lindemann, J.P.: Motion as a source of environmental information: a fresh view on biological motion computation by insect brains. Front. Neural Circuits 8(127), 1–15 (2014)

6. Goslin, M., Mine, M.R.: The panda3d graphics engine. Computer 37(10), 112–114 (2004)

7. Koenderink, J.J.: Optic flow. Vis. Res. 26(1), 161–179 (1986)

8. Kress, D., Egelhaaf, M.: Head and body stabilization in blowflies walking on differently structured substrates. J. Exp. Biol. 215(9), 1523–1532 (2012)

9. Kress, D., Egelhaaf, M.: Impact of stride-coupled gaze shifts of walking blowflies on the neuronal representation of visual targets. Front. Behav. Neurosci. 8(307), 1–13 (2014)

10. Lucas, B.D., Kanade, T., et al.: An iterative image registration technique with an application to stereo vision. In: IJCAI, vol. 81, pp. 674–679 (1981)

11. Matthews, R.W., Matthews, J.R.: Insect Behavior. Springer, Netherlands (2009)

12. Miyamoto, K.: Fish eye lens. JOSA 54(8), 1060–1061 (1964)

13. Montano, L., Asensio, J.R.: Real-time robot navigation in unstructured environments using a 3d laser rangefinder. In: Proceedings of the 1997 IEEE/RSJ International Conference on Intelligent Robots and Systems, IROS 1997, vol. 2, pp. 526–532. IEEE (1997)

14. Paskarbeit, J., Annunziata, S., Basa, D., Schneider, A.: A self-contained, elastic joint drive for robotics applications based on a sensorized elastomer coupling - design and identification. Sens. Actuators A Phys. 199, 56–66 (2013)

15. Paskarbeit, J., Schilling, M., Schmitz, J., Schneider, A.: Obstacle crossing of a real, compliant robot based on local evasion movements and averaging of stance heights using singular value decomposition. In: 2015 IEEE International Conference on Robotics and Automation (ICRA), pp. 3140–3145. IEEE (2015)

16. Petrowitz, R., Dahmen, H., Egelhaaf, M., Krapp, H.G.: Arrangement of optical axes and spatial resolution in the compound eye of the female blowfly calliphora. J. Comp. Physiol. A 186(7–8), 737–746 (2000)

17. Schilling, M., Hoinville, T., Schmitz, J., Cruse, H.: Walknet, a bio-inspired controller for hexapod walking. Biol. Cybern. 107(4), 397–419 (2013)

18. Schwegmann, A., Lindemann, J.P., Egelhaaf, M.: Depth information in natural environments derived from optic flow by insect motion detection system: a model analysis. Front. Comput. Neurosci. 8(83), 1–15 (2014)

19. Shoemaker, P.A., Ocarroll, D.C., Straw, A.D.: Velocity constancy and models for wide-field visual motion detection in insects. Biol. Cybern. 93(4), 275–287 (2005)

20. Srinivasan, M., Guy, R.: Spectral properties of movement perception in the dronefly eristalis. J. Comp. Physiol. A 166(3), 287–295 (1990)

21. Stürzl, W., Böddeker, N., Dittmar, L., Egelhaaf, M.: Mimicking honeybee eyes with a 280 field of view catadioptric imaging system. Bioinspir. Biomim. 5(3), 036002 (2010)

22. Stürzl, W., Grixa, I., Mair, E., Narendra, A., Zeil, J.: Three-dimensional models of natural environments and the mapping of navigational information. J. Comp. Physiol. A 201(6), 563–584 (2015)

MIRO: A Robot "Mammal" with a Biomimetic Brain-Based Control System

Ben Mitchinson and Tony J. Prescott[⊠]

Department of Psychology and Sheffield Robotics, University of Sheffield,
Western Bank, Sheffield, S10 2TN, UK
{b.mitchinson,t.j.prescott}@sheffield.ac.uk

Abstract. We describe the design of a novel commercial biomimetic brain-based robot, MIRO, developed as a prototype robot companion. The MIRO robot is animal-like in several aspects of its appearance, however, it is also biomimetic in a more significant way, in that its control architecture mimics some of the key principles underlying the design of the mammalian brain as revealed by neuroscience. Specifically, MIRO builds on decades of previous work in developing robots with brain-based control systems using a layered control architecture alongside centralized mechanisms for integration and action selection. MIRO's control system operates across three core processors, P1-P3, that mimic aspects of spinal cord, brainstem, and forebrain functionality respectively. Whilst designed as a versatile prototype for next generation companion robots, MIRO also provides developers and researchers with a new platform for investigating the potential advantages of brain-based control.

1 Introduction

Many robots have been developed that are animal-like in appearance; a much smaller number have been designed to implement biological principles in their control systems [1, 2]. Of these, even fewer have given rise to commercial platforms that demonstrate the potential for brain-based, or *neuromimetic*, control in real-world systems. Building on more than two decades of research on robots designed to emulate animal behavior and neural control [3–7] —that has developed key competences such as sensorimotor interaction, orienting, decision-making, navigation, and tracking—we teamed with an industrial designer, experts in control electronics, and a manufacturer, to create an affordable animal-like robot companion. The resulting platform, MIRO, was originally designed to be assembled in stages, where each stage constitutes a fully-operational robot that demonstrates functionality similar to that seen in animals. These stages loosely recapitulate brain development, as well as, to some extent, brain/phylogenetic evolution. with the finalized robot emulating some of the core functionality of a generalized mammal. Previous publications have reported on the potential of the MIRO robot as a biomimetic social companion [8] and as a platform for education and entertainment [9]. In this article we (i) describe the principles of brain-based control that have inspired MIRO, (ii) outline the morphology and hardware design, and (iii) detail the three key levels of the MIRO control architecture and the functionality to which they give rise.

© Springer International Publishing Switzerland 2016
N.F. Lepora et al. (Eds.): Living Machines 2016, LNAI 9793, pp. 179–191, 2016.
DOI: 10.1007/978-3-319-42417-0_17

We end our article by briefly discussing some of the trade-offs we have made in developing a functioning brain-based robot as a commercial product.

2 Principles of Mammal-Like Brain-Based Control

Living, behaving systems display patterns of behavior that are integrated over space and time such that the animal controls its effector systems in a coordinated way, generating sequences of actions that maintain homeostatic equilibrium, satisfy drives, or meet goals. How animals achieve behavioral integration is, in general, an unsolved problem in anything other than some of the simplest invertebrates. This has also been called the problem of architecture, and it is equally as problematic for robots as it is for animals [10]. Today's robots are notoriously "brittle" in that their behavior—which may appear integrated and coordinated with respect to a well-defined task—can rapidly break down and become disintegrated when task parameters go outside those anticipated by the robot's programmers. Animals can also go into states of indecision and disintegration when challenged by difficult situations [11], but generally show a robustness and capacity to quickly adapt that is the envy of roboticists [12].

We believe that neuroscience and neuroethology have important lessons for robotics concerning the problem of architecture. Specifically, theoretical and computational analyses of animal nervous systems point to the presence of "hybrid" control architectures that combine elements of reactive control with integrative mechanisms that operate both in space, coordinating different parts of the body, and in time, organizing behavior over multiple time-scales (for discussion, see, [1, 13–15]).

One key principle, whose history dates at least to the 19th century neurologist John Hughlings Jackson [16], is that of layered architecture. A layered control system is one in which there are multiple levels of control at which the sensing apparatus is interfaced with the motor system [17]. It is distinguished from hierarchical control by the constraint that the architecture should exhibit *dissociations*, such that the lower levels still operate, and exhibit some sort of behavioral competence, in the absence (through damage or removal) of the higher layers but *not* vice versa. A substantial body of the neuroscience literature can be interpreted as demonstrating layered control systems in the vertebrate brain; layering has also been an important theme in the design of artificial control systems, for instance, for autonomous robots [18]. The notion of a layered architecture has been mapped out in some detail in the context of specific types of behavior. For example, in [15], we described how the vertebrate defense system—the control system that protects the body from physical harm—can be viewed as being instantiated in multiple layers from the spinal cord (reflexes), through the hindbrain (potentiated reflexes), midbrain (coordinated responses to species-specific stimuli), forebrain (coordinated responses to conditioned stimuli), and cortex (modification of responses according to context). In this system the higher layers generally operate by modulating (suppressing, potentiating, or modifying) responses generated by the lower layers.

Whilst the brain shows clear evidence of layered control there are other important governing principles in its organization. Indeed, a system that worked by the principles of layered control alone would be too rigid to exhibit the intelligent, flexible behavior that mammals are clearly capable of. One proposal, stemming from the research of the neurologist Wilder Penfield, is of a centralized, or *centrencephalic*, organizing principle whereby a group of central, sub-cortical brain structures serves to coordinate and integrate the activity of both higher- and lower-level neural systems [19]. Candidate structures include the midbrain reticular formation—which may be important in integrating behavior within the brainstem, and in regulating behavior during early development—and the basal ganglia, a group of mid- and forebrain structures that we have argued play a critical role in action selection. We have previously developed several embodied models of these brain systems (see Fig. 1) and have demonstrated their sufficiency to generate appropriate behavioral sequences for mobile robots engaged in activities such as simulated foraging [4, 20].

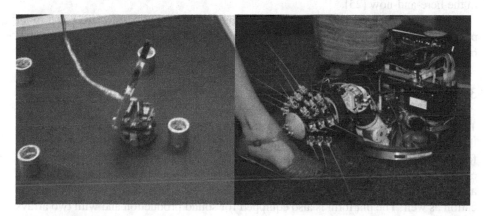

Fig. 1. Neurorobotic models of control architectures. Left: [4] embedded a model of the vertebrate basal ganglia in a table-top robot and showed its ability to control action selection and behavioural sequencing for a simulated foraging task. Right: Shrewbot [5] is one of series of whiskered robots developed to explore the effectiveness of brain-based control architectures in generating life-like behaviour.

Our research on biomimetic robot control architectures is predicated on the notion that the principles of both centrencephalic organization and layered control are at work in mammalian brains and can be co-opted to generate coordinated and robust behavior for robots. Over recent years we have developed a number of neurorobotic models to further test this proposition [5, 21], of which the MIRO robot is the first commercial instantiation.

A further question with regard to the problem of control architecture concerns the fundamental units of selection. The neuroethology literature suggests a decomposition of control into behavioral sub-systems that then compete to control the animal (see [15, 22], an approach that has been enthusiastically adopted by researchers in behavior-based robotics (see, e.g. [23]). An alternative hypothesis emerges from the literature on spatial attention, particularly that on visual attention in primates including humans [24]. This approach suggests that actions, such as eye movements and reaches towards targets, are generated by first computing a 'salience map' that integrates information about the relevance (salience) to the animal of particular locations in space into a single topographic representation. Some maximization algorithm is then used to select the most salient position in space towards which action is then directed. Of course, the approaches of behavioral competition and salience map competition are, again, not mutually exclusive and it is possible to imagine various hierarchical schemes, whereby, for instance, a behavior is selected first and then a point in space to which the behavior will be directed. In the mammalian brain, sensorimotor loops involving the cortex, superior colliculus, basal ganglia, and midbrain areas such as the periaqueductal gray, interact to control how the animal orients towards or away from different targets and what actions and behaviors are then selected with respect to these targets [15]. Other structures provide contextual information based on past experience—the hippocampal system, for instance, contributes to the animal's sense of time and place—thereby promoting better decisions in the here-and-now [25].

In the following we briefly describe the physical instantiation of MIRO as a robot platform and then return to the question of how MIRO has been designed to support a brain-based control architecture.

3   The MIRO Platform

The MIRO platform (see Fig. 2) is built around a core of a differential drive base and a three degree-of-freedom (DOF) neck (lift, pitch, yaw). Additional DOFs include two for each ear (curl, rotate), two for the tail (droop, wag), and one for the eyelids (open/close). Whilst these latter DOFs target only communication, the movements of the neck and body that serve locomotion and active sensing play a significant role in communication as well. The platform is also equipped for sound production and with two arrays of colored lights, one on each side, both elements serving communication and/or emotional expression.

All DOFs in MIRO are equipped with proprioceptive sensors (potentiometers for absolute positions and optical shaft encoders for wheel speed). Four light level sensors are placed at the corners of the base, two task-specific 'cliff sensors' point down from its front face, and four capacitive sensors are arrayed along the inside of the body shell providing sensing of direct human contact. In the head, stereo microphones (in the base of the ears) and stereo cameras (in the eyes) are complemented by a sonar ranger in the nose and an additional four capacitive sensors over the top and back of the head (behind the ears). Accelerometers are present in both head and body.

MIRO has a three-level processing stack (see below). Peripheral components are reached on an I2C bus from the 'spinal processor' (ARM Cortex M0), which communicates via SPI with the 'brainstem processor' (ARM Cortex M0/M4 dual core), which in turn communicates via USB with the 'forebrain processor' (ARM Cortex A8). All peripherals and some aspects of processing are accessible from off-board through WiFi connectivity (with MIRO optionally configured as a ROS—Robot Operating System—node), and the forebrain processor can be reprogrammed if lower-level access is required (lower processors can be re-programmed if desired, though with more onerous requirements to respect the specifics of the platform).

Fig. 2. The MIRO prototype companion robot. Some example MIRO behavior can be seen at https://youtu.bc/x4tya6Oj5sU

4   Control Architecture of the MIRO Robot

As explained above, a fundamental feature of the MIRO control architecture is its layered form as further illustrated in Fig. 3 (which is not exhaustive but includes the key architectural elements). Processing loops are present at many different levels, generating actuator control signals based on sensory signals and current state. In addition, higher systems are able to modulate the operation of loops lower down (a few examples are shown in the figure) thus implementing a form of subsumption [18]. Each layer builds upon the function of those below, so that the architecture is best understood from the bottom-up. We place the loops into three groups, each loosely associated with a broad region of the mammalian central nervous system, as follows.

Spinal Cord
The first layer, which we denote "spinal cord", provides two types of processing. The first is signal conditioning—non-state-related transformations that can be applied unconditionally to incoming signals. This includes robot-specific operations such as removing register roll-overs from the shaft encoder signals, but also operations with biological correlates.. An example of the latter is automatic acquisition of the zero-point of accelerometer signals (accounting for variability of manufacture) which is functionally comparable to sensory habituation in spinal cord neurons. The resulting "cleaned" or "normalized" signals provide the input to the second type of processing in this layer—reflex loops. A bilateral "cliff reflex" inhibits forward motion of each wheel at the lowest level if the corresponding cliff sensor does not detect a floor surface. A parallel "freeze reflex" watches for signals that might indicate the presence of another agent (a tilting acceleration-due-to-gravity vector, or touch on any of the touch sensors) and inhibits all motions, sounds, and lighting effects when

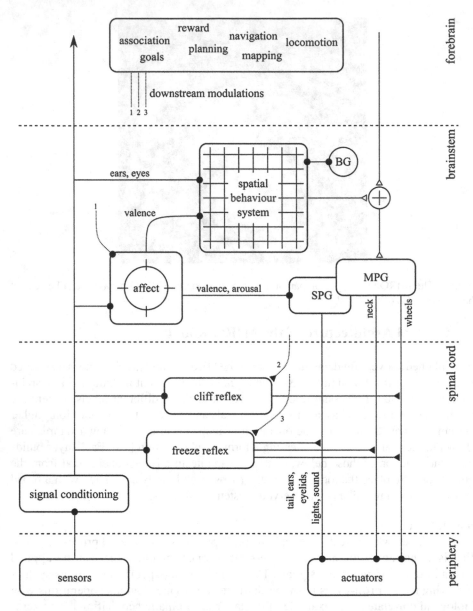

Fig. 3. Control architecture of MIRO loosely mapped onto brain regions (spinal cord, brainstem, forebrain). Signal pathways are excitatory (open triangles), inhibitory (closed triangles), or complex (closed circles). See text for description of components.

triggered. If left alone, MIRO will slowly recover and begin to move and vocalize once more.

All of the reflexes can be inhibited by higher systems, allowing them to be "switched off" if a higher-level understanding of MIRO's context demands it. Overall, this layer can be characterized as implementing "reactive control".

Brainstem

We group some of the most central elements of MIRO's biological control system into the second layer, denoted "brainstem". This layer is concerned with simple action selection, the computation and maintenance of affective state, simple spatial behaviors and the generation of motor patterns to drive the actuators.

Affect is represented using a circumplex model derived from affective neuroscience [26], that comprises a two-dimensional state representing valence (unpleasantness, pleasantness) and arousal. Fixed transforms map events arising in MIRO's sensorium into changes in affective state: for example, stroking MIRO drives valence upwards, whilst striking him on the head drives valence down. Baseline arousal is computed from a number of sources including the real-time clock. That is, MIRO has a circadian rhythm, being more active during daylight hours. General sound and light levels also affect baseline arousal, whilst discrete events cause acute changes of affective state (very loud sound events raise arousal and decrease valence, for example).

MIRO expresses affect in a number of ways. Most directly, a set of "social" pattern generators (SPG) drive the light displays, as well as movement of the ears, tail, and eyelids, so as to indicate affective state [8]. Meanwhile, MIRO's vocalization model, a complete generative mechano-acoustic model of the mammalian vocal system [27], is modulated by affect, so that MIRO's voice can range from morose to manic, angry to relaxed. More indirectly, MIRO's movements are modulated also by affect: low/high arousal slows/speeds movement, and very low arousal leads to a less upright posture of the neck.

The other major system in the brainstem layer is a spatial behavior system modeled on the management of spatial attention and behavior in superior colliculus and related nuclei in mammals [28]. This system comprises a topographic salience map of the space around MIRO's head which is driven by aspects of both visual and aural inputs. One filter generates positive salience from changes in brightness in camera images, so that movement is typically a key generator of salience. Another, alongside, uses a Jeffress model [29] to localize the source of loud sound events so that a representation of their intensity can be added to the salience map at the appropriate location. Other aspects of these sensory streams, as well as signals from other sensory modalities, can be configured to contribute to this global salience map in a straightforward way [30].

Simple hard-coded filters generate behavioral plans from this map: "where" is computed as the maximum of the map; "what" is computed by combining MIRO's current affective state with the nature of the stimulus (for example its size, location, or temporal nature). The system generates behavioral plans including "orient" (turn to visually "foveate" the stimulus), "avert" (turn away from the stimulus), "approach" and "flee" (related behaviors with locomotion components), and assigns a priority (a scalar value) to each plan. A model of the basal ganglia (BG) [4, 22] is then used to select, with persistence and pre-emption, one of these plans for execution by the motor plant at any one time. Overall, this system corresponds closely to similar, hard-wired, behavior systems that have been identified in several animal species, including rodents [31] and amphibians [14].

This loop is closed through a motor pattern generator (MPG) that takes as input behavioral plans and generates time series signals for the actuators. Any behavioral plan is encoded as an open-loop trajectory for a point in the frame of reference of one of the robot's kinematic links. In all current plans, the point chosen corresponds to a "generalized sensory fovea" [30, 31] just in front of the nose; thus, MIRO is "led by the nose" as a behavioral plan executes. The MPG comprises a kinematic self-model that is computed by moving the guided point and then identifying the remaining parameters (undriven) of the model through a principle of "least necessary movement", starting with the most distal DOFs. This computation is performed using a non-iterated coordinate descent procedure. The lack of iteration limits the quality of the approximate solution, but is very cheap to compute, biologically plausible, and performs reasonably well. In previous work, we have used an adaptive filter model as a pre-processing stage to this MPG, greatly improving accuracy, and suggested that this may be a role played by mammalian cerebellum [31].

A second, distinct, kinematic self-model is used to estimate MIRO's configuration for the interpretation of sensory signals. The model combines motor efferent signals with sensory afferent (proprioceptive) signals through a complementary filter [32] to derive a timely estimate of MIRO's instantaneous configuration. This configuration is available as an input to the analysis of data with a spatial component; for example, it determines the optical axis of the cameras when a video frame was captured.

The brainstem layer contains several other sub-systems that have biological correlates. For instance, sleep dynamics are implemented as a relaxation oscillator, with wakefulness and exhaustion the two oscillator states. Thus, MIRO spends around five in every twenty minutes "asleep", expressed by closed eyes and a lowered head. Motor reafferent noise is present in MIRO's sensory streams in several forms—particular sources include obstruction of the cameras by blinking of the eyelids, corruption of video frames through self-motion (blurring), and the presence of audio noise whilst motors are active. All of these forms of noise are eliminated from the incoming data streams by gating, based on efferent and afferent cues of their presence. Thus, for example, MIRO will not attempt to detect motion when it is, itself, in motion. Selective suppression of sensory streams during some forms of motion is also a feature of biological vision [33].

Forebrain
MIRO's forebrain control systems are under present development. Figure 2 gives an indication of the character of components that are anticipated for this layer. The nature of the control architecture allows that these "higher" systems can be built on top of the existing layers, taking advantage of already-implemented functionality. For example, a higher system intended to perform task-specific orienting would not need to replicate the orienting system that is already present. Rather, a suitable modulation can be applied to the existing spatial salience filters, or additional filters added, and the orienting behavioral plan can be "primed" [30]. The result is a tendency to perform orienting towards the primed region of signal space (or physical space). In MIRO, all lower systems are amenable to modulation; highlighted in the diagram are implemented

modulation routes allowing affect to be driven by influences from the forebrain layer or reflexes to be inhibited completely allowing the recovery of direct control. Current research is directed at implementing a spatial cognition module modeled on the mammalian hippocampus that will support inhibition-of-return during exploratory behavior and will allow the robot to learn about, and navigate to, important sites such as a home 'bed'.

The centrality of the basal ganglia model to any extension of the motor repertoire is notable. Since there is only one motor plant, only one motor pattern should be selected at any one time (simultaneous activation of multiple motor plans through the same output space constituting a motor error). Therefore, some selection mechanism is required so that only one plan is disinhibited at any one time. There is substantial support for the hypothesis that the vertebrate basal ganglia is such a centralized selection mechanism, that may implement a form of optimal decision-making between competing actions, that operates across the different layers of the neuraxis, and that has contributed to the flexibility and scalability of the vertebrate brain architecture [4, 10, 22, 31].

Processing Stack of MIRO Robot

The layers of MIRO's biomimetic control architecture are mirrored in their implementation distributed across three on-board processors as shown in Fig. 4. A fourth level of processing, denoted "P4", is available by inclusion of off-board systems into the control stack. The rationale for this arrangement has a pedagogic aspect (ease of understanding) but the key benefits are functional.

One important feature is that the control latency of loops through the lowest reprogrammable processor, P1, can be as low as a few milliseconds. This contrasts very favorably with the control latency through an off-board processor, P4, which—even under favorable conditions—can be hundreds of milliseconds. The inherent unreliability of wireless communications means that off-board latency can, on occasion, be longer still. Thus, safety critical aspects of the control policy, such as that implemented by the cliff reflex, will display superior performance if implemented in P1 versus, say, P4. There is, unsurprisingly, a continuum of latencies from P1 (~10 ms) through P2 (~30 ms), P3 (~50-200 ms), and P4 (100 ms or more).

Conversely, computational power (as well as energy consumption) increases as we move upwards through the processing stack (see Figure). This means that there is also a continuum of "competence", or control sophistication. P1 can respond fast, but lacks the power to make sophisticated decisions. P2 is able to perform spatial processing, and respond quickly to the spatial nature of events, but lacks the power to perform pattern discriminations or image segmentation, say. P3 is more capable still, but as an on-board processor on a battery-powered mobile robot still has tight computational constraints. The characteristic of increasing latency and control sophistication as we move up through the different levels of layered architecture is shared by vertebrate brains [15]. Latencies of escape reflexes implemented in spinal cord can be ~10 ms, but reflex responses are relatively unsophisticated and involve minimal signal processing; meanwhile, midbrain responses to visual events begin after 50 ms [34], whilst classification of objects by human visual cortex begins to emerge at around 100 ms [35] but allows a much more sophisticated response.

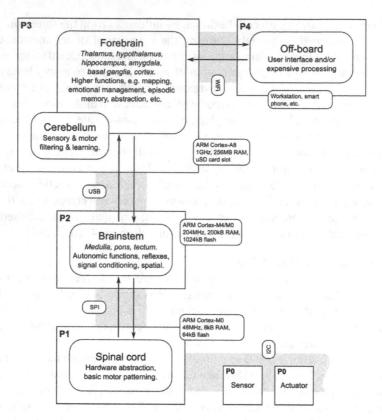

Fig. 4. The biomimetic MIRO control architecture is implemented across three on-board processors, P1-3, each loosely associated with a broad region of the mammalian brain. The design physically displays dissociation, since P2 (and, to a more limited extent, P1) is able to control the robot independently of higher processing layers. "P4" denotes off-board processing, and "P0" non-reprogrammable peripheral-specific processors.

This distribution of substrates from "fast and simple" through to "slow and sophisticated" may be a potentially useful design element for many robots. One aspect of "simple", that can easily be forgotten, is "less likely to fail". Especially during development, sophisticated systems such as P3 are highly prone to transient failures; having lower-level systems that protect the robot from possible damage will be beneficial. At the same time, higher processors can be put to sleep when they are not required, saving power and leaving lower processors to watch for events that may turn out to be behaviorally-relevant. The downside to this tiered processing stack is design complexity. Cost, however, may not be a serious concern, since the simpler processors are rather cheap parts.

A concern specific to robotics is increasing accessibility as we move up the stack. Running new control code on P4 can be as simple as pressing a key or clicking a mouse; on P3, at least a network file transfer will be required, and perhaps also a cross-compile step; changing the control code in P2 requires reprogramming the majority of the sectors of the on-board FLASH, an operation that takes a few seconds, and requires the harnessing of P3 as a mediator; reprogramming P1 requires that the robot be powered down and undergo a minor wiring change first, before being updated, reconfigured and powered back up. Changing the code in any P0 (non-reprogrammable) processor requires installing a new part. Thus, the development cycle tends to favour placing code that is changing often (typically sophisticated) higher up, and code that is more stable (typically simple) lower down. Whilst brain evolution is fundamentally different from this style of robot design, there is a similar tendency in nature towards conservation of structure and function towards the lower end of the neuraxis (spinal cord/brainstem), and increase in flexibility and adaptability at the upper end (cortex) [15].

5 Conclusion

We began with the goal of creating an affordable animal-like robot in which we could embed a biomimetic control architecture that we had previously developed on expensive bespoke robotic platforms. Important constraints in the design process, that were later relaxed but still strongly influenced the outcome, were the need to have a platform that could operate in an integrated way at multiple stages of construction, and that no single component should cost more than $10. We have found that a brain-based design is actually well-suited to these challenges of incremental construction and use of cheap, off-the-shelf parts. During the course of evolution, the mammalian brain has adapted and scaled to many different body types and ecological niches; the MIRO robot shows that future living machines, built of non-biological components such as plastic and silicon, can also make use of layered control architectures inspired by, and abstracted from, those we find in animals.

Acknowledgments. The development of the MIRO robot was funded by Eaglemoss Publishing and Consequential Robotics with contributions from Sebastian Conran, Tom Pearce, Victor Chen, Dave Keating, Jim Wyatt, Maggie Calmels, and Emily Collins. Our research on layered architectures was also supported by the FP7 WYSIWYD project (ICT-612139), and the EPSRC BELLA project (EP/I032533/1).

References

1. Prescott, T.J., et al.: Embodied Models and Neurorobotics. In: Arbib, M.A., Bonaiuto, J.J. (eds.) From Neuron to Cognition via Computational Neuroscience. MIT Press, Cambridge (in press)
2. Floreano, D., Auke, J., Ijspeert, J., Schaal, S.: Robotics and neuroscience. Curr. Biol. **24**(18), R910–R920 (2014)
3. Prescott, T.J., Ibbotson, C.: A robot trace-maker: modeling the fossil evidence of early invertebrate behavior. Artif Life **3**, 289–306 (1997)
4. Prescott, T.J., et al.: A robot model of the basal ganglia: behaviour and intrinsic processing. Neural Netw. **19**(1), 31–61 (2006)
5. Pearson, M.J., et al.: Biomimetic vibrissal sensing for robots. Philos. Trans. R. Soc. Lond. B Biol. Sci. **366**(1581), 3085–3096 (2011)

6. Lepora, N.F., et al.: Optimal decision-making in mammals: insights from a robot study of rodent texture discrimination. J. R. Soc. Interface **9**(72), 1517–1528 (2012)
7. Mitchinson, B., et al.: Biomimetic tactile target acquisition, tracking and capture. Robot. Auton. Syst. **62**(3), 366–375 (2014)
8. Collins, E.C., Prescott, T.J., Mitchinson, B.: Saying it with light: a pilot study of affective communication using the MIRO robot. In: Wilson, S.P., Verschure, P.F.M.J., Mura, A., Prescott, T.J. (eds.) Living Machines 2015. LNCS, vol. 9222, pp. 243–255. Springer, Heidelberg (2015)
9. Collins, E.C., et al.: MIRO: a versatile biomimetic edutainment robot. In: 12th Conference on Advances in Computer Entertainment, Iskandar, Malaysia (2015)
10. Prescott, T.J.: Forced moves or good tricks in design space? landmarks in the evolution of neural mechanisms for action selection. Adapt. Behav. **15**(1), 9–31 (2007)
11. Hinde, R.A.: Animal Behaviour: a Synthesis of Ethology and Comparative Psychology. McGraw-Hill, London (1966)
12. McFarland, D., Bosser, T.: Intelligent Behaviour in Animals and Robots. MIT Press, Cambridge (1993)
13. Verschure, P.F.M.J., Krose, B., Pfeifer, R.: Distributed adaptive control: the self-organization of structured behavior. Robot. Auton. Syst. **9**, 181–196 (1992)
14. Arbib, M.A., Liaw, J.S.: Sensorimotor transformations in the worlds of frogs and robots. Artif. Intell. **72**(1–2), 53–79 (1995)
15. Prescott, T.J., Redgrave, P., Gurney, K.N.: Layered control architectures in robots and vertebrates. Adapt. Behav. **7**(1), 99–127 (1999)
16. Jackson, J.H.: Evolution and dissolution of the nervous system. In: Taylor, J. (ed.) Selected Writings of John Hughlings Jackson. Staples Press, London (1884/1958)
17. Prescott, T.J.: Layered control architectures. In: Pashler, H. (ed.) Encyclopedia of Mind, pp. 464–467. Sage, London (2013)
18. Brooks, R.A.: A robust layered control system for a mobile robot. IEEE J. Robot. Autom. **RA-2**, 14–23 (1986)
19. Penfield, W.: Centrencephalic integrating system. Brain **81**, 231–234 (1958)
20. Humphries, M.D., Gurney, K., Prescott, T.J.: Is there a brainstem substrate for action selection? Philos. Trans. R. Soc. Lond. B Biol. Sci. **362**(1485), 1627–1639 (2007)
21. Prescott, T.J., et al.: Whisking with robots: From rat vibrissae to biomimetic technology for active touch. IEEE Robot. Autom. Mag. **16**(3), 42–50 (2009)
22. Redgrave, P., Prescott, T., Gurney, K.N.: The basal ganglia: a vertebrate solution to the selection problem? Neuroscience **89**, 1009–1023 (1999)
23. Brooks, R.A.: New approaches to robotics. Science **253**, 1227–1232 (1991)
24. Gandhi, N.J., Katnani, H.A.: Motor functions of the superior colliculus. Annu. Rev. Neurosci. **34**, 205–231 (2011)
25. Fox, C., et al.: Technical integration of hippocampus, basal ganglia and physical models for spatial navigation. Front. Neuroinformatics, **3**(6) (2009)
26. Posner, J., Russell, J.A., Peterson, B.S.: The circumplex model of affect: an integrative approach to affective neuroscience, cognitive development, and psychopathology. Dev. Psychopathol. **17**(3), 715–734 (2005)
27. Hofe, R., Moore, R.K.: Towards an investigation of speech energetics using 'AnTon': an animatronic model of a human tongue and vocal tract. Connection Sci. **20**(4), 319–336 (2008)
28. Mitchinson, B., Prescott, T.J.: Whisker movements reveal spatial attention: a unified computational model of active sensing control in the rat. PLoS Comput. Biol. **9**(9), e1003236 (2013)

29. Jeffress, L.A.: A place theory of sound localization. J Comp Physiol Psychol. **41**(1), 35–39 (1948)
30. Mitchinson, B.: Attention and orienting. In: Lepora, P.T.J.N., Verschure, P.F.M.J. (eds.) Living Machines: A Handbook of Research in Biomimetic and Biohybrid Systems OUP: Oxford (in press)
31. Prescott, T.J., et al.: The robot vibrissal system: understanding mammalian sensorimotor co-ordination through biomimetics. In: Krieger, P., Groh, A. (eds.) Sensorimotor Integration in the Whisker System, pp. 213–240. Springer, New York (2015)
32. Higgins, W.T.: A comparison of complementary and Kalman filtering. IEEE Trans. Aerosp. Electron. Syst. **AES-11**(3), 321–325 (1975)
33. Burr, D.C., Morrone, M.C., Ross, J.: Selective suppression of the magnocellular visual pathway during saccadic eye movements. Nature **371**(6497), 511–513 (1994)
34. Redgrave, P., Prescott, T.J., Gurney, K.: Is the short-latency dopamine response too short to signal reward error?. Trends Neurosci. **22**(4), 146–151 (1999)
35. Thorpe, S.J.: The speed of categorization in the human visual system. Neuron **62**(2), 168–170 (2009)

A Hydraulic Hybrid Neuroprosthesis for Gait Restoration in People with Spinal Cord Injuries

Mark J. Nandor[1,2(✉)], Sarah R. Chang[1,3], Rudi Kobetic[1], Ronald J. Triolo[1,4], and Roger Quinn[2]

[1] Louis Stokes Cleveland Department of Medical Affairs Medical Center, Cleveland, USA
mnandor@aptcenter.org, rkobetic@fescenter.org
[2] Department of Mechanical Engineering, Case Western Reserve University, Cleveland, USA
rdq@case.edu
[3] Department of Biomedical Engineering, Case Western Reserve University, Cleveland, USA
src74@case.edu
[4] Department of Orthopedics, Case Western Reserve University, Cleveland, USA
rxt24@case.edu

Abstract. The Hybrid Neuroprosthesis (HNP) is a hydraulically actuated exoskeleton and implanted Functional Electrical Stimulation (FES) system that has been designed and fabricated to restore gait to people with spinal cord injuries. The exoskeleton itself does not supply any active power, instead relying on an implanted FES system for all active motor torques. The exoskeleton instead provides support during quiet standing and stance phases of gait as well as sensory feedback to the stimulation system. Three individuals with implanted functional electrical stimulation systems have used the system to successfully walk short distances, but were limited in the flexion torques the stimulation system could provide.

1 Introduction

1.1 Functional Electrical Stimulation Enabled Gait

Restoring gait in people with spinal cord injuries is a long sought after and desired goal in rehabilitation research. This has been accomplished through orthotic bracing/exoskeletons (both passive and active), functional electrical stimulation (FES), and a combination or bracing and FES.

A spinal cord injury is a breakdown of communication – muscle activation signals cannot be sent through the damaged spinal column, however in many cases the subject's muscles are still receptive to commands. FES utilizes an external command unit to issue commands to the user's own muscles.

Electrodes placed on the skin have been utilized for their relative ease of setup, focusing on eliciting hip and knee extensor muscles for standing, and eliciting a withdrawal reflex for stepping. A typical surface stimulation consists of 4–6 channels [10, 11]. Implanted systems have been able to provide a greater number of channels (upwards of 48) [9], with that additional bandwidth providing a greater degree of control [12]. FES gait is typically run open loop as a timed pattern of muscle triggerings.

© Springer International Publishing Switzerland 2016
N.F. Lepora et al. (Eds.): Living Machines 2016, LNAI 9793, pp. 192–202, 2016.
DOI: 10.1007/978-3-319-42417-0_18

The primary disadvantage of FES enabled is the quick onset of muscle fatigue. Combined with the open loop nature of the system, walking distances are typically short.

1.2 Orthotics for People with Spinal Injuries

Many different orthoses have been designed for use by people with spinal cord injuries. Typically used is a Reciprocating Gait Orthosis (RGO), a passive device that when standing will lock the user's knees and ankles in a neutral position, and mechanically couple contralateral hip motion – hip flexion on one side causes hip extension in the other. By prevention bilateral hip flexion, this arrangement is statically stable as well as facilitates a reciprocating gait. However, the locked knees force the user to exert a large amount of upper extremity forces to achieve sufficient foot/floor clearance during swing.

1.3 Hybrid FES/Orthotic Devices

Combining electrical stimulation with an external orthotic has been accomplished previously in a variety of forms. Early efforts [2–4] focused on combining stimulation with passive orthotics, such as a reciprocating gait orthosis. However, joint kinematics was still restricted by the orthoses used.

Later efforts sought to create a more natural looking gait by enabling the joints of the orthotic to lock and unlock based on sensor feedback [5, 6]. These devices relied on stimulation to provide all active motor power. More recent developments include efforts to combine stimulation with a fully active, robotic exoskeleton [7, 8].

Discussed here is an exoskeleton that utilizes passive electrohydraulic joints and FES that combines the strengths while negating the weaknesses of each individual approach. It utilizes implanted stimulation (capable of providing relatively high torques - 60 Nm in hip flexion, 63 Nm in hip extension, 15 Nm in knee flexion, 80 Nm in knee extension [9]) for all of the propulsive power. Like an RGO, the exoskeleton is capable of locking the knee joints and reciprocally coupling the hip joints, providing stability during quiet standing and stance phases of gait. Unlike an RGO, the exoskeleton is able to automatically change its joint constraints as needed, such as unlocking the knee of the swing leg, allowing for knee flexion and toe clearance. This reduces the need for compensatory mechanisms.

2 Design Overview

2.1 System Architecture Overview

The complete HNP exoskeleton (as shown in Fig. 1) consists a mechanical frame (a torso and two leg uprights) carrying 3 hydraulic subsystems – each knee contains a Dual State Knee Mechanism (DSKM), and the hip joints are controlled by a hydraulic Variable Constraint Hip Mechanism (VCHM). Figure 2 shows the hydraulic circuitry contained within each device. Additionally, the exoskeleton carries a variety of sensors and electronics capable of providing wireless, independent operation outside of a lab environment, including joint angle encoders, force sensitive resistors in the soles of the shoe, and an inertial measurement unit carried on the torso of the exoskeleton.

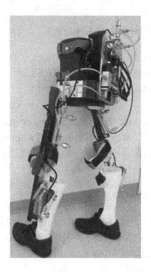

Fig. 1. The Hybrid Neuroprosthesis exoskeleton

VCHM and DSKM

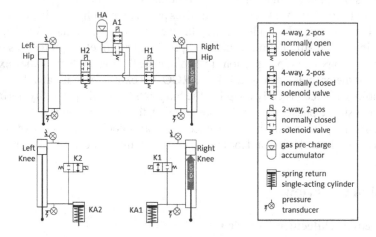

Fig. 2. The HNP hydraulic circuitry diagram

2.2   Dual State Knee Mechanism

The purpose of the DSKM is to lock the knee and provide support during quiet standing and the stance phases of gait, while unlocking and providing free, unimpeded motion during the swing phase of gait, with knee flexion and extension torques being provided by the FES system. Determination between stance and swing phases of gait is performed by the onboard sensors and microcontrollers.

To accomplish this, each DSKM consists of a 7/8 in. bore, 3 in. stroke hydraulic cylinder (Clippard Minimatic), with the rod and blind side of the cylinder connected to each other. To control the flow of fluid, a two-way/two-position solenoid valve (Allenair Corporation) is placed in between. In its default unpowered state, the solenoid valve is closed, preventing fluid flow from occurring and consequently the cylinder from extending and retracting. A small accumulator (Bimba Corporation) is also placed in the circuit to accommodate the fluid volume differential between blind side and rod side. Maximum operation pressure is limited to 600 psi, the maximum rating of the hydraulic hose utilized.

To translate the linear motion of the cylinder to rotary motion of the joint, a three bar linkage is utilized. Other transmission concepts such as a rack and pinion gear set were considered but dismissed as being too large and heavy. The linkage was designed and optimized according to a few specifications:

- Minimum range of motion: 10° of hyperextension to 110° of flexion.
- Locking torque at neutral: minimum 70 Nm.
- Minimum posterior protrusion of the hydraulic cylinder.

Figure 3 shows the final locking torque capability of the DSKM as a function of knee angle.

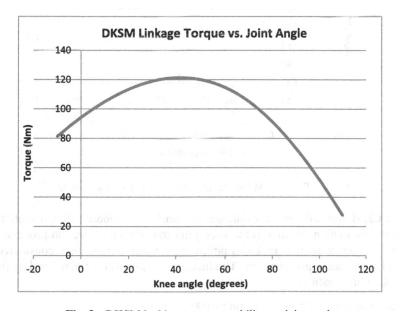

Fig. 3. DKSM locking torque capability vs. joint angle

2.3   Variable Constraint Hip Mechanism

Unlike the knee joint hydraulics, which is two independent units, both hip joints are controlled by a single mechanism. The hydraulics of the VCHM consist of two 7/8 in. bore, 3 in. stroke hydraulic cylinders (one for each hip joint), three four way/two position hydraulic valves (Hydraforce), and a small accumulator. Like the DSKM, the VCHM is designed to let each hip joint operate independently locked an unlocked. Additionally, it is capable of reciprocally coupling hip motion. By directly connecting the rod and blind sides of the two cylinders, hip flexion on one side with cause hip extension on the contralateral side and vice versa.

In its unpowered state, the VCHM defaults to the reciprocally coupled hip state. Combined with the locked knee default/unpowered state of the DSKM, in the case of complete battery discharge or unintended power failure, the HNP exoskeleton will simply act as a conventional reciprocating gait orthosis (RGO), a traditional orthotic device prescribed to people with spinal cord injuries. This is an important safety feature of the HNP exoskeleton – in the case of malfunction or power loss, the user is never stranded, but can still utilize the exoskeleton as a familiar unpowered orthotic (Fig. 4).

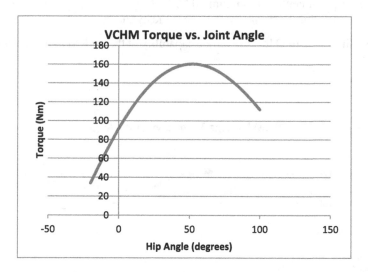

Fig. 4. The VCHM locking torque capability vs. joint angle

The VCHM also utilizes 3 bar linkages to translate the linear cylinder motion into rotary hip joint motion. Similar to the knee joint transmission selection process, other transmission options such as rack and pinion gear transmissions were considered and dismissed as too heavy and complex. The linkage was again constrained and optimized by the following conditions;

– Range of motion – 20° of extension to 100° of flexion.
– Locking Torque at neutral – minimum 70 Nm.
– Minimum posterior protrusion of the hydraulic cylinder.

2.4   Mechanical Elements of the Exoskeleton

The mechanical frame of the exoskeleton is designed to resist maximum force from the cylinders. At 2482 kPa (600 psi), each cylinder is capable of exerting over 1596 N (360 lbs.) of force. Finite element analysis was performed to ensure structural integrity under maximum load. By placing all cylinder mounts in double shear, the bending moment due to cylinder force on the exoskeleton frame is minimized, and the exoskeletal frame is only moderately stressed. Peak Von Neumann stresses from the simulation are under 29 MPa (4.2 ksi.). The frame is constructed of 6061-T6 aluminum, with a yield strength of 276 Mpa (40 ksi).

The shank portions of the exoskeleton uprights are designed to accept standard thermoformed ankle foot orthoses (AFOs). These are prescribed and fabricated by a certified orthotist and are custom fit to each individual user for best possible fit and prevention of pressure sores during use.

The exoskeleton is designed to accommodate users from 1.52 m (60 in.) to 1.93 m (76 in.) with a 100 kg. (220 lb.) maximum weight limit. To facilitate this, each leg segment can adjust its length, and the torso unit is able to adjust both width and anterior/posterior position of the hip centers. Total device weight is 17 kg (38 lbs.). VCHM valves, accumulators, and electronic microprocessor, stimulation boards, battery, and voltage regulator are all carried on the posterior portion of the torso, as shown in Fig. 5.

Fig. 5. Posterior mounted VCHM hydraulics and exoskeleton electronics

2.5   Electronics

Part of the exoskeleton's purpose is to provide enough sensory feedback to determine what phase of gait the user is in, closing the loop with the actions of the stimulation system. To accomplish that, the following sensors are placed onto the exoskeleton:

– Encoders at the hip and knee joints, reporting joint angle.
– Force sensitive resistors within the soles of the shoes, capable to detecting heel strike and toe off events
– A nine axis inertial measurement unit (IMU) placed on the torso component of the exoskeleton. This device reports absolute orientation of the torso in space.

These sensor readings are all read by a custom designed PCB with the appropriate signal conditioning circuitry and an 8-bit microprocessor (Atmega Corp.). The microcontroller is responsible for running the gait event detector (GED), and issuing commands to the hydraulic valves and stimulation system. The user issues commands via a wireless finger switch held in the hand with four buttons.

The microcontroller communicates via Bluetooth to a nearby computer, running a graphical user interface. This interface provides real time sensor data, displays the current phase of gait determined by the GED, and records the sensor data for later analysis.

3 Device Evaluation

3.1 Locking Compliance

The exoskeleton must be able to lock and support the user's body weight when necessary, such as during single/double stance phases of gait. To test this functionality, the exoskeleton joints were individually connected to a robotic dynamometer (Biodex Medical Systems, Shirley, NY), capable of applying a measured torque while the joint is hydraulically locked. The resultant angular displacement is shown in Figs. 6 and 7.

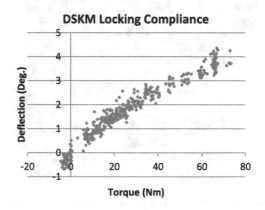

Fig. 6. DSKM locking compliance

3.2 Passive Resistance

Because all of the joint torque is provided by the stimulation, it is important to minimize the passive resistance of system. To measure this passive resistance, the exoskeleton joints were again attached to the dynamometer. While the joints were commanded to be

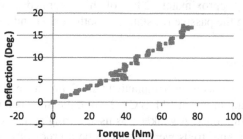

Fig. 7. VCHM locking compliance

free, the dynamometer was commanded to rotate at constant angular velocity; recording the torque it was exerting to maintain that speed. Figures 8 and 9 display the data collected for the DSKM and VCHM.

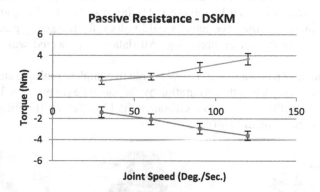

Fig. 8. DSKM passive resistance torque vs. speed

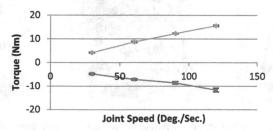

Fig. 9. VCHM passive resistance vs. joint speed

At the knee joint, approximately 20 % of the torque generated by stimulation in knee flexion was needed to overcome the hydraulic passive resistance at 120°/s. 4 % of the maximum generated extension torque is necessary to overcome the resistance in extension. At the hip joint, approximately 23 % of the generated stimulation torque was necessary to overcome the passive resistance in both flexion and extension.

3.3   Experimental Use

To date, three different subjects with implanted stimulation have walked with the HNP, with levels of injury ranging from T11 to C7. Two of the three had previous experience walking with stimulation only. A walker was used during all trials to provide stability and support. Early walking trials were meant to be a proof of concept and allow for debugging and fine tuning of the controller/stimulation patterns.

All subjects transferred from their wheelchair onto a chair containing the exoskeleton for donning. Standing was initiated by button press, at which time the exoskeleton joints would unlock and after a short delay; stimulation of the extensor muscles was ramped up for standing. Once standing, the exoskeleton joints were locked and the encoders zeroed. Initiation of walking was triggered with another button press, with each subsequent step triggered by another button press. During walking, the hip joints remained coupled at all times, and the knee joints would unlock for the swing phase leg while remaining locked for the stance phase leg. All data was collected wirelessly with a computer.

While all subjects (shown in Fig. 10) were able to complete multiple steps, but were ultimately limited in walk length and duration by fatigue of flexor muscles, limiting knee flexion (causing insufficient foot floor clearance) and hip flexion (limiting step length and making step initiation difficult).

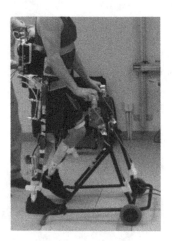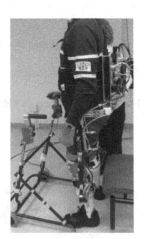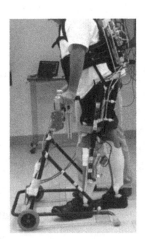

Fig. 10. The HNP in use, on 3 separate volunteers with implanted stimulation

Figure 11 shows data collected from 4 steps from a single trial. The hip angle plot shows the effect of the hip-hip reciprocal coupling – driving the swing hip into flexion causes the stance hip to extend. The data also shows the lack of knee flexion – stimulation was unable to provide more than 30° of knee flexion.

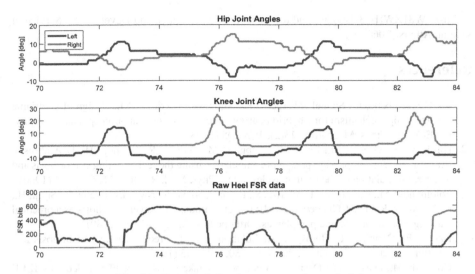

Fig. 11. Recorded data from an experimental trial

3.4   Conclusion and Future Work

Described here is an electromechanical exoskeleton that is meant to expand on previous FES + RGO gait research. Like previous hybrid FES + RGO experiments, this HNP derives all of its active torques from stimulation. Unlike a purely passive RGO, the exoskeleton described here utilizes an onboard sensors, microcontroller and hydraulic actuators to detect the user's phase of gait, and apply joint constraints accordingly.

Ultimately, the limitation of the HNP system is muscle fatigue. While this exoskeleton described here has the capability to unlock the knee joints for the swing phase of gait, fatigue of the knee flexors means the user is unable to take full advantage of that design feature. Similarly, fatigue of hip flexor muscles limits overall step length and walk duration.

Future controller work will take advantage of the onboard telemetry and use that to modulate the stimulation patterns. Typical FES enabled gait (including the preliminary work shown here) is simply a timed pattern run open loop, without feedback. With this exoskeleton, the sensors should be able to detect the onset of fatigue and increase the muscle activation levels to compensate.

Additionally, hardware revisions that aim to reduce the passive resistance of the joints, as well as provide a small amount of assistive power can further delay the onset of muscle fatigue, and increase walking speed and distance. Concepts that place small knee flexion assist springs in the DKSM, or small assistive motors (sized to allow stimulation to still provide a majority of the necessary walking torques) at each hip in the VCHM have been discussed.

Acknowledgements. The author would like to thank all APT Center and FES Center investigators and engineers that contributed to this work, including Dr. Musa Audu, Kevin Foglyano, and John Schnellenberger. This work was supported by grants from the Department of

Defense (W81XWH-13-1-0099), and the Rehabilitation Research and Development Service of Veteran Affairs (B0608-R).

References

1. To, C.S., Kobetic, R., Schnellenberger, J.R., Audu, M.L., Triolo, R.J.: Design of a variable constraint hip mechanism for a hybrid neuroprosthesis to restore gait after spinal cord injury. IEEE/ASME Trans. Mechatron. **13**(2), 197–205 (2008)
2. Hirokawa, S., Grimm, M., Le, T., Solomonow, M., Baratta, R.V., Shoji, H., D'Ambrosia, R.D.: Energy consumption in paraplegic ambulation using the reciprocating gait orthosis and electrical stimulation of the thigh muscles. Arch. Phys. Med. Rehabil. **71**, 687–694 (1990)
3. Solomonow, M., Baratta, R., Hirokawa, S., Rightor, N., Walker, W., Beaudette, P., Shoji, H., D'Ambrosia, R.: The RGO generation II: muscle stimulation powered orthosis as a practical walking system for thoracic paraplegics. Orthopedics **12**, 1309–1315 (1989)
4. Isakov, E., Douglas, R., Berns, P.: Ambulation using the reciprocating gait orthosis and functional electrical stimulation. Paraplegia **30**, 239–245 (1992)
5. Goldfarb, M., Durfee, W.: Design of a controlled-brake orthosis for FES-added gait. IEEE Trans. Rehabil. Eng. **4**(1), 13–24 (1996)
6. Goldfarb, M., Korkowski, K., Harrold, B., Durfee, W.: Preliminary evaluation of a controlled-brake orthosis for FES-aided gait. IEEE Trans. Neur. Syst. Rehabil. Eng. **11**(3), 241–248 (2003)
7. Ha, K.H., Murray, S.A., Goldfarb, M.: An approach for the cooperative control of FES with a powered exoskeleton during level walking for persons with paraplegia. IEEE Trans. Neur. Syst. Rehabil. Eng. (99), 1
8. del-Ama, A., Gil-Agudo, Á., Pons, J., Moreno, J.: Hybrid FES-robot cooperative control of ambulatory gait rehabilitation exoskeleton. J. NeuroEng. Rehabil. **11**(1), 27 (2014)
9. Kobetic, R., Marsolais, E.B.: Synthesis of paraplegic gait with multichannel functional neuromuscular stimulation. IEEE Trans. Rehabil. Eng. 2(2), 66–79 (1994)
10. Marsolais, E.B., Kobetic, R.: Development of a practical electrical stimulation system for restoring gait in the paralyzed patient. Clin. Orthop. (233), 64–74 (1988)
11. Alojz, K., et al.: Gait restoration in paraplegic patients: a feasibility demonstration using multichannel surface electrode FES. J. Rehabil. R&D/Veterans Adm. Dept. Med. Surg. Rehabil. R&D Serv. **20**(1), 3–20 (1983)
12. Kobetic, R., Triolo, R.J., Marsolais, E.B.: Muscle selection and walking performance of multichannel FES systems for ambulation in paraplegia. IEEE Trans. Rehabil. Eng. **5**(1), 23–29 (1997)

Principal Component Analysis of Two-Dimensional Flow Vector Fields on Human Facial Skin for Efficient Robot Face Design

Nobuyuki Ota, Hisashi Ishihara$^{(\boxtimes)}$, and Minoru Asada

Graduate School of Engineering, Osaka University, Suita, Japan
ishihara@ams.eng.osaka-u.ac.jp

Abstract. In this study, deformation patterns of an adult male lower face are measured and analyzed for efficient face design for android robots. We measured flow vectors for 96 points on the right half of the lower face for 16 deformation patterns, which are selected from Ekman's action units. Namely, we measured 16 flow vector fields of facial skin flow. The flow vectors were created by placing ink markers on the front of the face and then video filming various facial motions. A superimposed image of vector fields shows that each point moves in various directions. Principle component analysis was conducted on the superimposed vectors and the contribution ratio of the first principal component was found to be 86 %. This result suggests that each facial point moves almost only in one direction and different deformation patterns are created by different combinations of moving lengths. Based on this observation, replicating various kinds of facial expressions on a robot face might be easy because an actuation mechanism that moves a single facial surface point in one direction can be simple and compact.

1 Introduction

Facial expression is one of the important communication channels for humans; therefore, designers of communication robots have tried to replicate human facial expressions on robot faces [1,2,4–14]. When humans change facial expressions, entire skin surfaces deform in complicated patterns, which is difficult for robot designers to replicate. To replicate several facial expressions on a single robot face, the designers should know how each point on a human face moves for several facial deformation patterns.

Traditionally, robot face designers [2,4,6,7,10,11,13,14] have decided the locations of facial moving points and their directions based on Ekman's action units (AUs), which are summarized and introduced in the Facial Action Coding System (FACS) [3]. AUs are facial deformation patterns created when one or more facial muscles are activated, and FACS explains that different facial expressions are realized by different combinations of one or several action units.

© Springer International Publishing Switzerland 2016
N.F. Lepora et al. (Eds.): Living Machines 2016, LNAI 9793, pp. 203–213, 2016.
DOI: 10.1007/978-3-319-42417-0_19

However, AUs are not enough to design effective actuation mechanisms for a robot face because only verbal qualitative descriptions of deformations are introduced in FACS. Instead, quantitative information such as flow vector fields must be used to effectively design robot faces. Cheng et al. [2] have pointed out this issue and measured facial deformations with a 3D scanner. However, their observation was limited to only four typical facial expressions, which were smile, anger, sadness, and shock.

Therefore, in this study, we measured every conceivable deformation pattern especially in the lower face of a Japanese adult male. Namely, we measured flow vector fields of facial skin. Then, two kinds of compensation processing for the measured vector fields, such as head movement compensation and neutral face matching, are executed. Finally, Principal Component Analysis (PCA) was conducted to determine variations in movements and flow field trends of each facial point.

2   Method

2.1   Motion Patterns

FACS introduces sixteen deformation patterns in the lower face and they are selected for this experiment. Table 1 summarizes the selected deformation patterns. Each AU has its own AU number and name defined in FACS. Basically,

Table 1. Deformation patterns (Ekman's action units) measured in this experiment

AU number	Name
9	Nose Wrinkler
10	Upper Lip Raiser
11	Nasolabial Furrow Deepener
12	Lip Corner Puller
13	Sharp Lip Puller
14	Dimpler
15	Lip Corner Depressor
16	Lower Lip Depressor
17	Chin Raiser
18	Lip Pucker
20	Lip Stretcher
22	Lip Funneler
23	Lip Tightener
24	Lip Presser
25	Lips Part
28	Lips Suck

the name contains information about the responsible part of the face (e.g., lip corner) and its deformation pattern (e.g., puller).

FACS describes how to activate each AU and an example of its associated facial image. A Japanese adult male (hereafter demonstrator) activated each AU and his facial deformations were measured in this experiment.

2.2 Measurement

Ninety six measurement points were identified on the demonstrator's skin, and ink markers were placed on them, as shown in Fig. 1. Each marker was approximately 2–5 mm in size, and their locations were decided so that they were placed not only on anatomically distinctive points, such as corners of the mouth and the top of the nose, but also on other plane surfaces such as the cheek.

The movements (or flows) of the markers for each AU were recorded by video filming (SONY HANDYCAM HDR-CX420) at the front of the demonstrator. While being filmed, the demonstrator began with a neutral face, activated one of the selected AUs, and kept its deformation for several seconds. Sixteen video recordings were created, each of which contains both the demonstrator's neutral face and his deformed one.

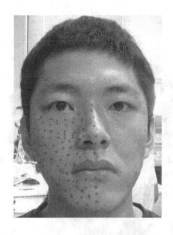

Fig. 1. Locations of ink markers on the right half of a Japanese adult male's lower face

2.3 Data Processing

To measure flow vectors of the markers, two video frames were selected from each video recording of each AU. The first one contains the demonstrator's neutral face and the second one contains his deformed one.

By comparing pixel coordinates of each marker in these two images, we can roughly calculate each marker's flow vector. However, two kinds of image corrections are necessary to calculate more precise flow vectors and to superimpose several flow vectors of the same marker. The first one is the Head Movement Compensation. While activating each AU, the demonstrator's head can move; therefore, the markers' pixel coordinates can move even if the marker does not move on face. The second one is Neutral Face Matching between different AUs to superimpose several vectors for the same marker. Even though the demonstrator tried to show the same neutral face with the same head position in each video recording session, these neutral faces can not be exactly the same. Therefore, starting points for flow vectors of the same marker are not in the same position between different AUs.

Head Movement Compensation. For head movement compensation, five reference points on the facial images shown in Fig. 2 were used. Reference points were placed on the top of the nose, on the mid-point between the eyes, and on the mid-point between the corner of the eye and the ear. These points are known to be relatively static on humans faces [2]. In addition to these three points, two additional reference points were placed on the top of the head and the chin.

The sizes and positions of the two images for each AU were adjusted so that each reference point matched. This adjustment was manual because matching five points was relatively simple.

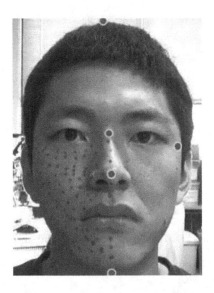

Fig. 2. Reference points on a facial image for head movement compensation

Neutral Face Matching. For neutral face matching between different AUs, first, the neutral face image for AU9 was randomly chosen as a reference image. Affine transformations were applied to 15 sets of 97 marker coordinates (96 ink markers and the reference point on the top of the chin) in neutral face images so that these marker sets would match with a marker set of the reference image.

Two-dimensional coordinate vector $X_{i,j}$ of a marker i in an image j is converted to a coordinate vector $X'_{i,j}$ by the equation

$$X'_{i,j} = A_j X_{i,j} + b_j,$$

where A_j is a linear transformation matrix and b_j is a translation vector for an image j. The purpose is to find the parameters of A_j and b_j so that the coordinate vectors $X'_{i,j}(i = 1, \ldots, 97)$ match with vectors $X_{i,9}$ for an neutral face image of AU9.

To find the appropriate parameters, Steepest Descent Method was implemented with an error function defined as

$$F(A_j) = \sqrt{\Sigma_{i=1}^{97} \|X_{i,9} - X'_{i,j}\|^2} .$$

3   Result

3.1   Flow Vector Fields

Figures 3 and 4 show the measured flow vector fields for AU12 and AU16, respectively. Green arrows represent flow vectors, and the facial images depict deformation for each AU. These flow vectors were obtained after the Head Movement Compensation. After the compensation, the maximum average error of five reference points between two images was 1.9 pixels. Considering the actual head size was 275 mm and the head image size was 600 pixels, the estimated error was 0.9 mm.

These figures show how facial deformations are extensive and complicated and how the flow fields are different between AUs. For example, in AU12, which is Lip Corner Puller, we can see entire surfaces around the cheek, lips, and jaw move. The flow vectors near the lip corner have longer lengths, and their

Fig. 3. Measured skin flow vector field for AU12 (Lip Corner Puller) (Color figure online)

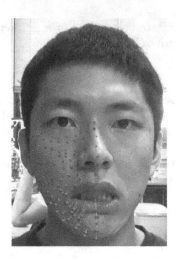

Fig. 4. Measured skin flow vector field for AU16 (Lower Lip Depressor) (Color figure online)

directions are aligned toward the top side of the face. On the other hand, in AU16, which is Lower Lip Depressor, we can see the entire surface below the mouth movement, and these directions are not aligned toward the side bottom.

3.2 Superimposed Field

Figure 5 represents the superimposed vector field before Natural Face Matching, which is the compensation processing to match vector starting points for the same marker. Vector colors indicate the vector angle. We can find that the starting points are not matched.

On the other hand, the starting points are matched well after Neutral Face Matching, as shown in Fig. 6. From this figure, we can determine how far and in which directions each point on the face moved. The point near the face midline moved only vertically. However, other points moved in several directions. The points near the mouth moved farther than points near the nose or the eye. The maximum vector length was 22 mm and it appeared at the corner of the lower lip for AU13.

Thus, the flow vectors seem to be too complex to replicate on a robot face. However, there seems to be a trend in the flow lines that connect the chin, the corner of the eye, and the corner of the lips. We will analyze this trend in the next section.

3.3 Principal Component Analysis

Figure 7 shows the first and second principal components of each marker's flow vectors whose starting points are matched by Neutral Face Matching. The directions of two double-headed arrows for each marker represent the directions of

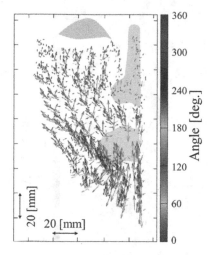

Fig. 5. Superimposed vector field before Neutral Face Matching. Starting points are not matched. The positions of the eye, nose, and mouth are indicated by gray areas as a reference. (Color figure online)

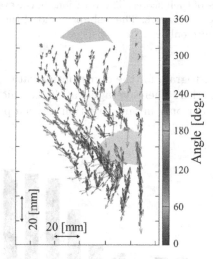

Fig. 6. Superimposed vector field after Neutral Face Matching. Starting points are matched. The positions of the eye, nose, and mouth are indicated by gray areas as a reference. (Color figure online)

the first and second principal components, respectively, while the lengths of the two arrows represent the variances of their scores. Namely, the longer the arrows, the farther the marker moved for different AUs. We found that the arrows closer to the mouth are longer.

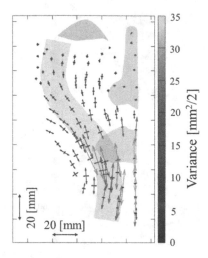

Fig. 7. The first and second principal components of flow vectors for each marker. The directions of two double-headed arrows for each marker represent the directions of the first and second principal components, respectively, while the lengths of the two arrows represent variances of their scores. The positions of the eye, nose, and mouth are indicated by gray areas as a reference. The s-shape trend of the flow lines is indicated by a green line. (Color figure online)

Figure 8 shows the histogram for the contribution ratios of the first principal components for every marker. Their average was 86 %, which means that almost all marker movement occur only in one direction at each point.

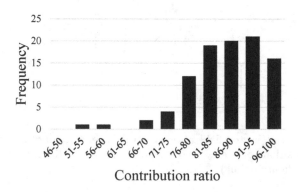

Fig. 8. Histogram for the contribution ratios of the first principal components for every marker

Figure 9 shows the histogram for the minimum reproduction errors of every flow vectors when every markers move only in the directions of each first principal component. The reproduction error of each marker in each AU is its second

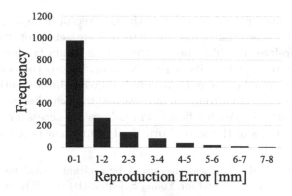

Fig. 9. Histogram for the reproduction errors of flow vectors when every markers move only in the directions of each first principal component

principal component score. Over eighty percent of the reproduction errors was under 2 [mm] and over 95 percent of them was under 4 [mm].

Furthermore, we can find an s-shape trend of flow lines that connect the side of the chin, the corner of the lips, the side of the cheek, and the corner of the eye. This trend is represented in Fig. 7 as a green line.

4   Conclusion

The main findings of the measurement and analysis of 16 patterns of facial flow vector fields are as follows:

– Although facial surfaces seem to move in a complicated manner when several deformation patterns are expressed, almost all movements occur only in one direction in each measured point on the face.
– Such principal directions for all of the points are continuous through the entire facial surface; such continuous flows form an s-shape trend.
– Moving lengths for each point for several deformation patterns vary more significantly around the mouth.

These findings suggest that various kinds of facial expressions are created by combinations of different movement lengths of facial surface points, each of which moves almost only in one direction.

These observations can facilitate replicating various kinds of facial expressions on a robot. This is because an actuation mechanism that moves a facial surface point in one direction can be simple and compact while a mechanism that moves the point in several directions would be complex.

However, several problems remain to be solved to obtain further effective design policies for face robots. First, the flow vectors lack depth information since we obtained these vectors from two-dimensional images. Therefore, three-dimensional flow vectors should be obtained for more precise analysis. Second, the number of points required to replicate the flow vector fields on a robot should be less than 96 because that is too many to prepare actuation mechanisms for each point. Therefore, neighbor points that move similarly should be treated as a single point representing its peripheral area. Third, the second principal component should not be ignored. Although its contribution ratio is low, skin flows of the second principal component exist and they can be dominant in some of deformation patterns. Further flow field analyses are necessary on each AU to understand when flows of the second principal component occur.

Acknowledgment. This work was supported by a Grant-in-Aid for Specially Promoted Research No. 24000012 and for Young Scientist (B) No. 15K18006.

References

1. Bickel, B., Kaufmann, P., Skouras, M., Thomaszewski, B., Bradley, D., Beeler, T., Jackson, P., Marschner, S., Matusik, W., Gross, M.: Physical face cloning. ACM Trans. Graph. **31**(4), 118:1–118:10 (2012)
2. Cheng, L.C., Lin, C.Y., Huang, C.C.: Visualization of facial expression deformation applied to the mechanism improvement of face robot. Int. J. Soc. Robot. **5**(4), 423–439 (2013)
3. Ekman, P., Friesen, W.V., Hager, J.C.: Facial action coding system (FACS): manual. In: A Human Face (2002)
4. Hanson, D., Andrew, O., Pereira, I.A., Zielke, M.: Upending the uncannyvalley. In: Proceedings of the International Conference on Artificial Intelligence (2005)
5. Hashimoto, M., Yokogawa, C.: Development and control of a face robot imitating human muscular structures. In: Proceedings of the International Conference on Intelligent Robots and Systems, pp. 1855–1860 (2006)
6. Hashimoto, T., Hiramatsu, S., Kobayashi, H.: Development of face robot for emotional communication between human and robot. In: Proceedings of the International Conference on Mechatronics and Automaton, pp. 25–30 (2006)
7. Hirth, J., Schmitz, N., Berns, K.: Emotional architecture for the humanoid robot head ROMAN. In: Proceedings of the International Conference on Robotics and Automation, pp. 2150–2155 (2007)
8. Ishiguro, H.: Android science: conscious and subconscious recognition. Connection Sci. **18**(4), 319–332 (2006)
9. Ishihara, H., Yoshikawa, Y., Asada, M.: Realistic child robot Affetto for understanding the caregiver-child attachment relationship that guides the child development. In: IEEE Proceedings of the International Conference on Development and Learning, pp. 1–5 (2011)
10. Kobayashi, K., Akasawa, H., Hara, F.: Study on new face robot platform for robot-human communication. In: 8th International Workshop on Robot and Human Interaction, pp. 242–247 (1999)
11. Lin, C.Y., Cheng, L.C., Tseng, C.K., Gu, H.Y., Chung, K.L., Fahn, C.S., Lu, K.J., Chang, C.C.: A face robot for autonomous simplified musical notation reading and singing. Robot. Auton. Syst. **59**(11), 943–953 (2011)

12. Minato, T., Yoshikawa, Y., Noda, T., Ikemoto, S., Ishiguro, H., Asada, M.: CB2: a child robot with biomimetic body for cognitive developmental robotics. In: Proceedings of the 7th IEEE-RAS International Conference on Humanoid Robots, pp. 557–562 (2007)
13. Tadesse, Y., Priya, S.: Graphical facial expression analysis and design method: an approach to determine humanoid skin deformation. J. Mech. Robot. 4(2), 1–16 (2012)
14. Yu, Z., Ma, G., Huang, Q.: Modeling and design of a humanoid robotic face based on an active drive points model. Adv. Robot. 28(6), 379–388 (2014)

Learning to Balance While Reaching:
A Cerebellar-Based Control Architecture
for a Self-balancing Robot

Maximilian Ruck[1,2], Ivan Herreros[1(✉)], Giovanni Maffei[1],
Martí Sánchez-Fibla[1], and Paul Verschure[1,3]

[1] SPECS, Technology Department, Universitat Pompeu Fabra, Carrer de Roc
Boronat 138, 08018 Barcelona, Spain
maximilianruckl@gmail.com, {ivan.herreros,
giovanni.maffei,marti.sanchez,paul.verschure}@upf.edu
[2] Department of Mechanical Engineering, Westfälische Hochschule, University
of Applied Sciences, Bocholt, Gelsenkirchen, Germany
[3] ICREA, Institució Catalana de Recerca i Estudis Avançats, Passeig Lluís
Companys 23, 08010 Barcelona, Spain

Abstract. In nature, Anticipatory Postural Adjustments (APAs) are actions that
precede predictable disturbances with the goal of maintaining a stable body
posture. Neither the structure of the computations that enable APAs are known
nor adaptive APAs have been exploited in robot control. Here we propose a
computational architecture for the acquisition of adaptive APAs based on current theories about the involvement of the cerebellum in predictive motor
control. The architecture is applied to a simulated self-balancing robot
(SBR) mounting a moveable arm, whose actuation induces a perturbation of the
robot balance that can be counteracted by an APA. The architecture comprises
both reactive (feedback) and anticipatory-adaptive (feed-forward) layers. The
reactive layer consists of a cascade-PID controller and the adaptive one includes
cerebellar-based modules that supply the feedback layer with predictive signals.
We show that such architecture succeeds in acquiring functional APAs, thus
demonstrating in a simulated robot an adaptive control strategy for the cancel-
lation of a self-induced disturbance grounded in animal motor control. These
results also provide a hypothesis for the implementation of APAs in nature that
could inform further experimental research.

Keywords: Cerebellar control · Anticipatory postural adjustments ·
Self-balancing robot · Adaptive control

1 Introduction

Nowadays self-balancing robots are becoming pervasive in everyday situations and
there exist numerous prototypes that explore new configurations and application
domains. However, most of the commercial self-balancing robots (SBRs) use solely

Research supported by socSMC-641321—H2020-FETPROACT-2014.

N.F. Lepora et al. (Eds.): Living Machines 2016, LNAI 9793, pp. 214–226, 2016.
DOI: 10.1007/978-3-319-42417-0_20

pre-programmed feedback control [1]. In this paper we explore the use of adaptive control techniques in order to deal with a problem that will arise the moment SBRs are to include additional actuators. That is, how to adapt and prevent the momentary loss of balance that occurs once a SBR changes its body configuration by extending a manipulator to reach for objects, thereby modifying the position of its center of mass?

Our approach to solve this problem is grounded in nature, namely in experimental psychology and neuroscience. On the one hand, humans and other bipedal primates are faced with the same difficulty when they reach for objects. They have to adapt their stance in order to preserve balance. Indeed, bipedal animals and SBRs are both different instances of inverted pendulum systems. From a control theory perspective, one can interpret the extension of one's arm as self-induced disturbances. In humans, changes in body posture or in muscular activity that precede these types of disturbances are known as anticipatory postural adjustments (APAs) [2]. Experimental manipulations have shown that healthy subjects rapidly acquire APAs, adjusting them both in timing and amplitude to cancel the effect of predictable disturbances. Such APAs usually comprise the predictive activation of the same muscles that before learning were recruited reactively [3]. Therefore, learning can be characterized as a (partial) transference of the responses from feedback to feed-forward control [4]. Additionally, patients with damages in the cerebellum show impaired and non-adaptive APAs. This evidence, together with neural recording in monkeys, has pointed to the cerebellum as the critical substrate for the acquisition of APAs [3].

To simulate the acquisition of APAs by the cerebellum we apply a model of cerebellar learning. The classical theory of cerebellar learning originated in the late 60 s [5, 6]. According to that theory, the cerebellum works as a supervised learning device where inputs from climbing fiber pathway, one of its two input pathways, instruct the processing of the information that comes via the mossy fiber pathway, the second input pathway. As a result, the cerebellum adaptively adjusts a mapping of mossy fiber inputs into Purkinje cell output activity. From an information processing perspective, the inferior olive, via its climbing fiber afferents, informs the cerebellar cortex about an error while plasticity at the cerebellar cortex aims at changing the cerebellar output to reduce that error. Cerebellar anatomy and physiology will not be treated in detail in this paper, but the reader can find a computationally oriented review in [7]. The basic intuition is that the cerebellum receives simultaneously error signals (through the climbing fiber pathway) and context information through the mossy fiber pathway. That context information can relate both to states of the world or the agent (as in our case, where it codes the agent's state of being next to move the arm). Finally, the coincidence between context and error information allows for the acquisition of adaptive outputs that can be issued once contexts that preceded errors are re-encountered.

Cerebellar-based controllers are commonly used in the robotic community, especially following the cerebellar model for articulatory control (CMAC) [8]. Indeed, CMAC and other cerebellar-based models have already been applied to the task of controlling a SBR [9–12]. However, the research described here is novel in the following ways:

Firstly, it addresses the problem of adjusting the reactions of the two-wheeled robot to changes in its body configuration. This problem will be relevant for SBR with additional actuators, such as wheeled *humanoids*. Secondly, the mixed feed-forward/feedback architecture is built according to the hypothesis that the cerebellum facilitates fine motor control by acquiring sensory predictions. This is in contrast with the still prevailing view that interprets the cerebellum as an inverse model issuing motor

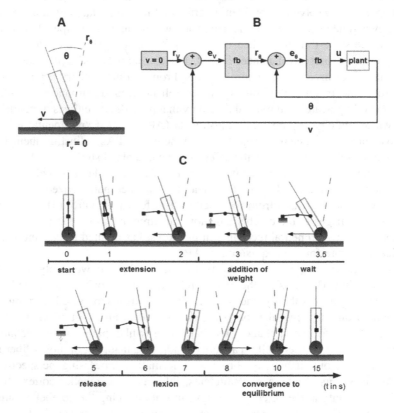

Fig. 1. Scheme of SBR and task. **A.** Display of the SBR from the lateral side balancing on a surface. The state variables θ and v are indicated in relation to their reference states r_θ and r_v respectively. **B.** The feedback motor control system consists of two overlapping feedback controllers (*fb*). The outer feedback controller reacts to the mismatch e_v of reference velocity r_v and velocity (v). Such controller is set with low gains, thus it is slowly reacting and changing the reference angle r_θ. Whereas the inner feedback controller is highly reactive to sudden changes and issues motor commands (u) according to the error e_θ. The motor command computed from the feedback control system (grey) is conveyed to the plant (blue), that models the robot's dynamical behavior. **C.** Trajectory of SBR with feedback control. The plant (blue) starts the trial at $t_0 = 0$ s in an equated position (timeline grey). The extension of the gripper (black) is initiated at $t_1 = 1$ s and ends at $t_2 = 2$ s. At $t_3 = 3$ s extra weight is attached to the gripper and released at $t_4 = 5$ s. The flexion of the gripper is initiated at $t_5 = 6$ s and finishes at $t_6 = 7$ s. Afterwards the robot returns steadily into a balanced state. The arrow at the wheel indicates the current velocity and sets the angular reference state r_θ. The black line displays the current angular orientation of the robot, whereas the dashed black line as well as the shape represents the desired angular orientation of the robot (all values are approximated simply to indicate the relations between the different variables at different stages of the task). (Color figure online)

commands [3, 13, 14]. Whereas Kawato and collaborators used sensory errors to acquire motor commands [15], our model acquires sensory error-predictions based in sensory errors.

Operationally, we have defined the task of balancing the SBR as that of maintaining the linear velocity of its wheel axis at zero. Maintaining balance while reaching forward becomes a disturbance-rejection task with a self-generated, hence predictable, disturbance. At the basic and non-adaptive level, this task is managed by cascade PID controller that uses both the linear velocity and the angular position of the robot body to compute the control signal. The reliance on two plant outputs implies that two separate error signals (velocity and angle errors) drive each layer of the PID controller (Fig. 1A −B). The outer layer of the PID, based on the velocity error computes the target angle that would eventually keep the robot still and balanced. The inner layer aims at reducing the error between the current and the previous target angle layer. Note that in terms of performance, only the error in velocity matters. As long as the error in velocity is minimized, the error in angular position is irrelevant.

That duplicity in the feedback controller allows interfacing an adaptive feed-forward module with each level of the reactive layer. In one case, supplying the outer PID controller with predicted errors in the velocity (p_v), or the second case driving the inner PID with predicted errors in the angle (p_θ). That made also possible to test the effect of providing both PID controllers with predictive inputs simultaneously. Thus, a third novelty of our adaptive approach is that even if there is only one degree of freedom to control, we propose to use a hierarchy of adaptive modules that mirrors the hierarchy of the feedback controller.

To summarize, we propose (1) that cerebellar-based control strategies can be applied to endow SBRs with anticipatory behaviors that resemble those observed in animals, especially as they have been measured in the APA paradigm; (2) that the adaptive control strategy can be successful even if, instead of directly accessing the plant, the adaptive component or components only affect behavior by recruiting the feedback control with sensory predictions; and (3) that even if we faced a single degree-of-freedom problem, the use of two feed-forward adaptive components that map directly onto different stages of the feedback control can entail an increase in performance compared to strategies that use a single predictive component.

2   Methods

A. Robot Model and Feedback Controller. A two-wheeled SBR is a vehicle able to keep its center of mass above a pivot (the wheels' axis) (Fig. 1-A). In the standard configuration the robot makes use of a feedback control loop to balance and withstand disturbances.

We model the robot with a standard non-linear state-space equations of a balancing system [16]. The rigid body equations model two masses attached by rigid link, where the mass at the base is actuated by a force (u) parallel to the ground. The model has three state variables: angular position θ, linear velocity v and angular acceleration $\dot{\theta}$. The model equations are given below.

$$\frac{d}{dt}\begin{bmatrix} \theta \\ v \\ \dot\theta \end{bmatrix} = \begin{bmatrix} \dot\theta \\ \dfrac{-ml\sin(\theta)\dot\theta^2 + mg\left(\frac{ml^2}{J_t}\right)\sin(\theta)\cos(\theta) - cv - \gamma lm\cos(\theta)\dot\theta + u}{M_t - m\left(\frac{ml^2}{J_t}\right)\cos(\theta)^2} \\ \dfrac{-ml^2\sin(\theta)\cos(\theta)\dot\theta^2 + M_t gl\sin(\theta) - cl\cos(\theta)v - \gamma\left(\frac{M_t}{m}\right)\dot\theta + l\cos(\theta)u}{J_t\left(\frac{M_t}{m}\right) - m(l\cos(\theta))^2} \end{bmatrix} \tag{1}$$

The parameters of the model are set to match the approximate values of a physical prototype currently under development (see Table 1).

Using this model of the plant, reaching the gripper and loading a weight are simulated as changes in the relative position (p_{gr}) and total weight of a mass situated at the end of the gripper (m_{gr}) that affect the moment of inertia (J_t), the total mass (M_t) and the distance from the wheel axis to the center of mass (1). Specifically, during the reaching we simulated a forward movement of the mass, perpendicular to the robot body. After extending it, the gripper is loaded (e.g. by another agent) with an extra weight. To complete the task, the robot must maintain balance with the additional load for a few seconds, release the load and retract the gripper. At each time the new center of masses (c.o.m.) and moments of intertia are computed, and the relative change in the c.o.m. angle with respect to the previous time step is added to the state variable θ.

The robot stabilizes by using a cascade *PID* controller (Fig. 1B) that is defined as follows:

$$r_\theta(t) = -k_p^v v(t) - k_i^v \int_0^t v(\tau)d\tau \tag{2}$$

$$u(t) = k_p^\theta (r_\theta(t) - \theta(t)) + k_d^\theta \left(\dot r_\theta(t) - \dot\theta(t)\right) \tag{3}$$

A first (outer) proportional-integral (*PI*) controller receives as input the linear error in velocity (e_v), which by definition of the problem is equal to minus the velocity (v), and outputs a target angle (r_θ). Such target angle becomes the reference for a second (inner) proportional-derivative (*PD*) controller. The inner controller then receives a signal coding the current error in the angular position (e_θ), computed comparing the actual angle (θ) with r_θ. Indeed, this cascade PID controller can be interpreted as a two step strategy to solve the balancing problem: first computing a desired *stance* (r_θ), in this case, an inclination of the robot's body, and secondly, computing the motor command (u) that brings the robot to that target stance. We set the parameters of the feedback controller such robot is stable near the equilibirum point. Note that reactive control maintained stability of the SBR during all stages of the reaching task, at the expenses of introducing large errors in the linear velocity.

B. Basic Cerebellar Model. Our adaptive feed-forward modules desing is based on the cerebellar microcircuit. In the cerebellum, each microcircuit integrates diverse inputs coming from the mossy/parallel fiber into a single output relayed by Purkinje cells/cerebellar deep-nuclei. According to the cerebellar learning theory, the integration of the mossy-fiber information is determined (or instructed) by a highly specific input coming from the inferior olive via the climbing fibers. In the same line of the adaptive filter theory of the cerebellum, we have chosen to model a cerebellar microcircuit as an adaptive linear element, implementing a decorrelation learning or Widrow-Hoff learning rule. More precisely, the model uses an adaptive linear element that combines a set of basis to produce an output signal.

Each basis signals ($p(t)$) results from a double exponential convolution of an input signal. In our case the inputs to each set of bases are triggers that, through different labeled lines, anticipate each of the six stages of the task (e.g., reaching onset, reaching offset, load weight, unload weight and onset and offset of the retraction). In our implementation we generated 30 bases for each input, with gives a total of approximately 180 bases. To have a diverse set of signals to combine, each basis implements a filter with different time constants (the details of the basis generation have already been introduced in [17]). The bases are mixed according to a set of weights $w(t)$. Using the convention that both $w(t)$ and $p(t)$ are column vectors, the output of the cerebellum ($y(t)$) can be generated as follows:

$$y(t) = w(t)^T p(t) \tag{4}$$

The weights are updated according to the decorrelation learning rule [18] (also known as Widrow-Hoff rule [19]). In this particular experiment the following version of the update-rule is used:

$$\frac{d}{dt} w_c = -\eta e(t) p(t - \delta) \tag{5}$$

where w_c, intialized at 0 at each trial onset, stores the change in weights to be applied at the end of the trial; η is a learning rate; $e(t)$, the error signal; and δ, the anticipatory delay. The anticipatory delay is a crucial parameter, distinctive of our modelling approach to cerebellar function (see [4, 17, 20]). It operationalizes the idea that to avoid an error occuring at time t the cerebellum should have acted at time $t - \delta$. The rule updates the weights by associating errors at a given time with the information encoded in the filter bases (p) δ seconds earlier. Setting δ requires approximate knowledge of the plant (or closed-loop) dynamics. That is, it requires knowing the lag in the response dynamics of the system, in order to anticipate responses earlier enough. Additionally, this parameter also implies that cerebellar responses in this model are anticipatory, which is not the standard in computational models of the cerebellum (e.g., see [21]).

C. Structure of the Adaptive Layer. In our modelling of a cerebellar microcircuit, each adaptive element receives layer through its climbing fiber pathway an error signal computed in the reactive control layer and sends output to the associated error feedback controller. The mossy/parallel fiber information that is provided to each element consists of a series of time varying signals (alpha-like functions) that are triggered by a motor intention. The wiring of the cerebellar components to implement an adaptive layer is displayed in Fig. 2. The two cerebellar controllers (V_CRB and A_CRB, for velocity and angle, respectively) receive the error signals that drive each individual PID controller, namely e_v and e_θ. Based on these teaching signals, they acquire the predictive outputs p_v and p_θ, which are added to the actual measured errors, driving the feedback controllers. Hence, the feed-forward predictive commands issued by both cerebellar components affect behavior only after passing through the feedback controllers.

Regarding the inputs that enter the cerebellar models through the mossy fiber pathway, the different basis signals generated in each cerebellar model represent the response to an input signal encoding motor intentions (MI in Fig. 2). This signals can also be thought of as efference copies arriving from the system controlling the robot's arm which allow the generation of a feed-forward sensory error signal that, driving the feedback controllers, cancels the error introduced by the self-induced disturbance.

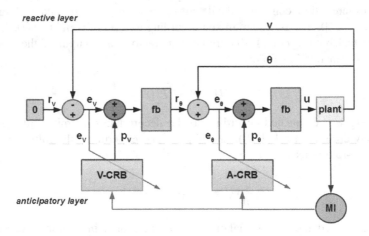

Fig. 2. Two-layered control architecture of the SBR. The reactive motor control system is extended with a second anticipatory layer. The anticipatory layer learns to associate the motor intention of the plant with the error signal e_x and outputs the counterfactual sensory error prediction p_x. The prediction is incorporated into the control loop and processed from the feedback controller (fb). The cerebellar controllers V-CRB and A-CRB (refer to veleocity and angle respectively) are implemented in cooperation.

D. Simulations. The two-wheeled SBR and the motor control architectures are implemented in *MATLAB 2015a Simulink* and run in a laptop computer. The parameters are detailed in Table 1. The learning experiments were conducted in 15 s duration trials. The structure of each trial is detailed in Fig. 1C. The simulations, deterministic and noiseless, were performed with a fixed size ODE solver, with a fixed time-step of 5 ms.

In the simulations we used three different versions of the architecture, depending upon whether both or just one of the adaptive cerebellar-like components were instantiated. That way we could first asses the individual contribution of each feed-forward module to performance and the effect of having both adaptive modules working in synergy. We refer to the architecture that only includes V_CRB as the velocity-only adaptive (VOA) architecture, to the one including only A_CRB as the angle-only adaptive (AOA) architecture and the one instantiating both modules as the dual adaptive (DA) architecture.

3   Results

We examine first the performance of the full control architecture (DA) that comprises two adaptive feed-forward controllers. The DA architecture succeeds in reducing the errors significantly. Indeed, the primary measure of performance, the error in the velocity e_v signal, decreases from a range spanning from −0.7 to 1.1 m/s to a range spanning from −0.4 to 0.2 m/s (Fig. 3A). Hence, after learning, the disturbance caused by the actuation of the robot arm was greatly reduced in amplitude. The errors in the angular position (Fig. 3B) were also reduced after learning, albeit in a lesser extent.

Regarding the output of the predictive controllers (Fig. 3C and D), they resemble smoothed (and in the case of the p_v, advanced in time) versions of the errors measured in the first trial.

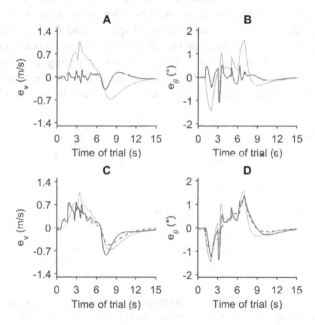

Fig. 3. A, B. First error of naive (*gray*) and last (*color*) error of trained agent for velocity (**A**) and angle (**B**). **C, D.** Complete sensory error signal of a trained robot for velocity (**C**) and angle (**D**). Dashed line indicates the adaptive signal (*p_v and p_θ, respectively*) and the colored solid lines, the sum of adaptive and reactive error signals.

In a second set of trials we separately assess the impact of the delay parameters on each module's performance after. We use only one adaptive the VOA and AOA architectures (see Methods). In each case only one of the feedback controllers receives predictive signals. The most relevant results are obtained with the VOA architecture. First, this architecture, which only includes one adaptive component, yields a reduction of the relative root mean square error (rRMSE) for the velocity of approximately 95 % (Fig. 4A). This means that, this architecture succeeds almost completely in decreasing displacement while executing the reaching actions. Moreover, a large range of values of the anticipatory delay parameter δ_v, from 150 ms to 600 ms achieves an almost equal performance.

The results for the AOAarchitecture are markedly different. Performance, measured in term of does not improve for any of the values tested for the anticipatory delay parameter (δ_θ) (Fig. 4A). Indeed, in terms of avoiding displacements induced by the reaching actions, the AOA architecture always performs worse than an architecture relying only in the PID cascade (without anticipation). Secondly, the error in the angular position is actually reduced, even beyond the level attained by the VOA architecture (Fig. 4B). The final difference between the AOA and the VOA architecture according to the different values of their delay parameters is, that in the VOA architecture there is clear and wide range of values of preferred values for δ_v, whereas performance in the AOA architecture performance increases as the δ_θ decreases.

These results can be summarized as follows: First, in both cases the adaptive components succeed in predicting their corresponding errors signals, as evidenced by the marked decreases in their measured errors (Figs. 4A and B). Secondly, predicting errors associated with the reference of the closed-loop (e_v) is relevant increasing performance in the disturbance-rejection task, whereas acquiring predicting the error signal of the inner feedback controller (e_θ) does not decrease the performance error (e_θ). Finally, performance of the VOA architecture was very tolerant to changes in the key parameter (δ_v).

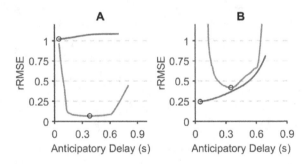

Fig. 4. Relative root mean square errors (rRMSE) as function of the anticipatory delay. **A, B.** Effect of the anticipatory delay of A-CRB (blue) and V-CRB (red) in the velocity error (**A**) and the angular error (**B**). For both figures normalization is performed using the cumulative sum of the errors in the first trial compared to the errors in the last five trials. The black circle indicates the delay with the least error in the respective space. (Color figure online)

Finally, we compare the results of the three different architectures; the two including only one adaptive component (VOA and AOA), and the one with two (DA). Indeed, after the previous results it is unclear to what extent the adaptive component for the error in the angle has any positive impact in performance. However, the outcome of this last set of simulations suggests that working in synergy with the adaptive component for e_v (V_CRB) the adaptive component for e_θ (A_CRB) ameliorates performance as well (Fig. 5A and C). First, the DA architecture reduces slightly the error in velocity (Fig. 5A). Secondly, the DA architecture shows an improvement relative to the VOA in terms of control effort. Indeed, taking pure feedback control as the baseline and

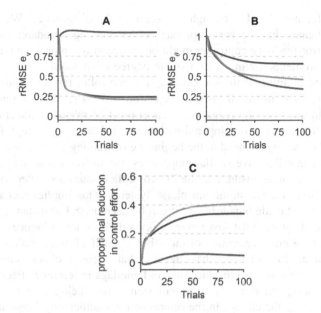

Fig. 5. Performance over session trials. **A, B.** rRMSE of the velocity (**A**) and the angle (**B**) for three different architectures, VOA (red), AOA (blue), DA (purple) during the trials. **C.** Motor effort reduction over trials for three different architectures. Motor effort reduction is calculated as the ratio between the accumulated squared motor output at each trial over the one on the first trial. (Color figure online)

computing the proportional reduction in the integral of the absolute control, the VOA architecture minimizes the control effort by 33 % and the DA architecture by 41 % (Fig. 5C). Hence, the DA architecture not only improves the balance of the robot while reaching, it also decreases the control effort.

4 Conclusions

We have introduced a cerebellar-based control architecture that allows a simulated wheeled-SBR to avoid displacements and maintain balance while actuating an arm, loading and unloading weight.

Even if the simulated robot was already able to maintain balance by means of the feedback control provided by a cascade of PID controller, the introduction of supplementary feed-forward predictive signals improved the performance reducing the RMSE by a 95 % over the pure feedback strategy (Fig. 4A).

Notably, even if the feedback controller is two-staged, almost all of the performance gain in terms of error reduction comes from providing the outer feedback controller with predictive signals. Indeed, even though the cascade PID includes two error-correction steps, in control theory terms, there is just one external reference signal – i.e., behavior has only one goal that is to keep the robot still. Controlling the position (inclination) is merely an instrumental step that enables the robot to remain still. Therefore, our results seem to imply that acquiring a predictive signal to complement the external reference signal is enough for rejecting the disturbance. We believe that such result indicates that the basic approach of connecting an adaptive-anticipatory module to an error-feedback controller might be sufficient for improving performance a wide range of problems dealing with a single degree of freedom.

However, the addition of the second adaptive controller further improves the performance of the SBR in the disturbance-rejection task in terms of reduction of the control effort. In this case, the reduction in effort must be related to the smoothness of the predicted error signal (p_θ) compared with the actual error signal (e_θ). Even if both error signals, the one experienced at the beginning of training and the one entering the inner feedback controller have similar amplitudes, the slower transients present in the input signal after learning ensure that the PD-controller issues a smaller control signal.

Wheeled SBRs, are non-minimum phase systems. This implies that any error in velocity must be first made bigger before it can be decreased. For instance, a still robot that has to move forward must first move backwards in order to induce a forward-tilt that, causing a transient acceleration of the robot's center of mass, makes the forward translation stable. In practice, that means that error-feedback controllers for non-minimum phase systems introduce unavoidable lags in reference tracking. Hence, the benefit for anticipation in those system stems from being able to bypass such latency by predicting the change in the reference by a sufficiently large interval. This property explains the results for the anticipatory delay parameter (δ_v) of the adaptive velocity signal controller. Indeed, in order to achieve a significant gain of performance, this parameter must be set to at least 150 ms.

From a pure neuroscience or neuro-robotic perspective these results illustrate how the cerebellum, which is considered to be crucial for accurate motor control, can enable precise predictive motor control operating as a purely sensory-prediction device. Additionally, the results show that the same cerebellar model can be wired to error-feedback controllers with different properties: a memory-less PD controller and a controller with an internal state. In nature, this result may illustrate how the same cerebellar algorithm could contribute to feedback computations located in brain structures with very different dynamics, such as the motor cortex or the spinal cord.

In the context of APAs, our implementation of a control architecture embodies a theory of how anticipatory postural adjustments could occur in nature. Of course, there are substantial differences between controlling a self-balancing robot and maintaining a bipedal stance, e.g., in terms of body structure, means of actuation, etc. However, our solution does suggest, that as a general principle, APAs could be acquired using sensory-error predictions that are sent to a reactive controller.

Due to the non-linearity and intrinsic difficulty of controlling SBRs, acquiring an anticipatory motor command that precisely controls the robot stance while it changes its body configuration and loads objects is not a trivial task. With our results we show, in simulation, that a very general-purpose architecture can succeed in this task using motor learning principles from neuroscience literature. As further work, the robustness of the control approach could be assessed either by adding noise in the sensors and/or actuators in simulations or by mounting it in a real robot. And finally, the context information could be enriched, coding also for the intended velocity of the arm movement and/or expected weight, thereby allowing the generalization of the APA to a set of disturbances.

Acknowledgements. The research leading to these results has received funding from the European Commission's Horizon 2020 socSMC project (under agreement number: socSMC-641321H2020-FETPROACT-2014) and by the European Research Council's CDAC project: (ERC-2013-ADG 341196).

Appendix

See Table 1.

Table 1. Simulation parameters.

feedback control	P	I	D	feedforward control	V-CRB	A-CRB
Outer fb	-0.070	-0.020	—	number of basis	150	210
Inner fb	35	—	0.01	learning rate η	0.00258	0.00258
				anticipatory delay δ	0.380	0.050
				time constants for bases (see [18])		
				τ_{slow}		1.000
				τ_{fast}		3.000
				f_{inh}		3.000

physical properties of the robot				
	M_t	1.580	kg	mass of SBR
	m	0.040	kg	mass of wheels
	m_{Gr}	0.030	kg	mass of gripper
	m_{extra}	0.005	kg	extra weight
	g	9.800	m/s^2	gravity
	l	0.0331	m	distance of center of mass to wheels
	J_t	$1.9828 * 10^{-3}$	m^{2}kg	moment of inertia
	Υ	0.010	Nms	rotational friction
	c	0.100	Ns/m	rolling friction
of the arm movement				
	h	0.150	m	height of gripper
	p_{gr}	0 to 0.100	m	extension of gripper
	t_{accel}	0.100	s	duration of gripper acceleration
	J_{gr}	$6.7500 * 10^{-4}$	m^2 kg	moment of inertia of gripper
	T	0.0045	Nm	torque of recoil
	$\ddot{\theta}_{recoil}$	6.670	rad/s2	applied angular acceleration (T/J_{gr})

References

1. Chan, R.P.M., Stol, K.A., Halkyard, C.R.: Review of modelling and control of two-wheeled robots. Annu. Rev. Control **37**(1), 89–103 (2013)
2. Massion, J.: Movement, posture and equilibrium: Interaction and coordination. Prog. Neurobiol. **38**(1), 35–56 (1992)
3. Horak, F.B., Diener, H.C.: Cerebellar control of postural scaling and central set in stance. J. Neurophysiol. **72**(2), 479–493 (1994)
4. Maffei, G., Herreros, I., Sánchez-Fibla, M., Verschure, P.F.: Acquisition of anticipatory postural adjustment through cerebellar learning in a mobile robot. In: Lepora, N.F., Mura, A., Krapp, H.G., Verschure, P.F., Prescott, T.J. (eds.) Living Machines 2013. LNCS, vol. 8064, pp. 399–401. Springer, Heidelberg (2013)
5. Marr, D.: A theory of cerebellar cortex. J. Physiol. **202**(2), 437–470 (1969)
6. Albus, J.S.: A theory of cerebellar function. Math. Biosci. **10**(1–2), 25–61 (1971)
7. Dean, P., Porrill, J., Ekerot, C.-F., Jörntell, H.: The cerebellar microcircuit as an adaptive filter: experimental and computational evidence. Nat. Rev. Neurosci. **11**(1), 30–43 (2010)
8. Albus, J.: A new approach to manipulator control: The cerebellar model articulation controller (CMAC). J. Dyn. Syst. Meas. Control **97**(3), 220–227 (1975)
9. Li, C., Li, F., Wang, S., Dai, F., Bai, Y., Gao, X., Kejie, L.: Dynamic adaptive equilibrium control for a self-stabilizing robot. In: 2010 IEEE International Conference on Robotics and Biomimetics, ROBIO 2010, pp. 609–614 (2010)
10. Chiu, C.H., Peng, Y.F.: Design and implement of the self-dynamic controller for two-wheel transporter. In: IEEE International Conference on Fuzzy Systems, pp. 480–483 (2006)
11. Ruan, X., Chen, J.: On-line NNAC for two-wheeled self-balancing robot based on feedback-error-learning. In: Proceedings - 2010 2nd International Workshop on Intelligent Systems and Applications, ISA 2010 (2010)
12. Tanaka, Y., Ohata, Y., Kawamoto, T., Hirata, Y.: Adaptive control of 2-wheeled balancing robot by cerebellar neuronal network model. In: 2010 Annual International Conference of the IEEE Engineering in Medicine and Biology Society, EMBC 2010, pp. 1589–1592 (2010)
13. Gao, J.H., Parsons, L.M., Bower, J.M., Xiong, J., Li, J., Fox, P.T.: Cerebellum implicated in sensory acquisition and discrimination rather than motor control. Science **272**, 545–547 (1996)
14. Miall, R.C., Wolpert, D.M.: Forward models for physiological motor control. Neural Netw. **9**(8), 1265–1279 (1996)
15. Kawato, M., Furukawa, K., Suzuki, R.: A hierarchical neural-network model for control and learning of voluntary movement. Biol. Cybern. **57**(3), 169–185 (1987)
16. Astrom, K.J., Murray, R.M.: Feedback Systems: An Introduction for Scientists and Engineers (2012)
17. Herreros, I., Maffei, G., Brandi, S., Sanchez-Fibla, M., Verschure, P.F.M.J.: Speed generalization capabilities of a cerebellar model on a rapid navigation task. In: 2013 IEEE/RSJ International Conference on Intelligent Robots and Systems, pp. 363–368 (2013)
18. Fujita, M.: Adaptive filter model of the cerebellum. Biol. Cybern. **45**(3), 195–206 (1982)
19. Widrow, B., Lehr, M.A., Beaufays, F., Wan, E., Bilello, M.: Adaptive signal processing. In: Proceedings of the World Conference on Neural Networks, p. 11 (1993)
20. Herreros, I., Verschure, P.F.M.J.: Nucleo-olivary inhibition balances the interaction between the reactive and adaptive layers in motor control. Neural Netw. **47**, 64–71 (2013)
21. Wolpert, D.M., Kawato, M.: Multiple paired forward and inverse models for motor control. Neural Netw. **11**, 1317–1329 (1998)

Optimizing Morphology and Locomotion on a Corpus of Parametric Legged Robots

Grégoire Passault[✉], Quentin Rouxel, Remi Fabre, Steve N'Guyen, and Olivier Ly

LaBRI, University of Bordeaux, Bordeaux, France
{gregoire.passault,quentin.rouxel,remi.fabre,
steve.nguyen,olivier.ly}@labri.fr

Abstract. In this paper, we describe an optimization approach to the legged locomotion problem. We designed a software environment to manipulate parametrized robot models. This environment is a platform developed for future experiments and for educational robotics purpose. It allows to generate dynamic models and simulate them using a physics engine. Experiments can then be made with both morphological and controller optimization. Here we describe the environment, propose a simple open loop generic controller for legged robots and discuss experiments that were made on a robot corpus using a black-box optimization.

1 Introduction

Eadweard Muybridge did an early work in studying animals locomotion, analyzing photographs of animals during different phases of strides [1]. What is noticeable is that there is a lot of similarities in how different animals move at corresponding paces.

These similarities are explained by studies on Central Pattern Generators (CPGs), that pointed out that the nervous system is able to produce rhythmic motor patterns even in vitro [2]. This is called fictive motor patterns, and this points out that animals locomotion may be based on hardware encoded *a prioris*.

Simple physical models such as the SLIP (spring-loaded inverted pendulum), can also describe running animals. It is interesting to mention that animals with two to eight legs can fit this model when running, where groups of legs acts in concert so that the runner is an effective biped [3].

One of our goals is to explore how some classical gaits like the trot, and in lesser extent the walk, could appear naturally by optimizing trajectory-based locomotion controllers. We explore how the locomotion solutions of animals, provided by hundreds of millions years of evolution, corresponds to solutions issued from statistical optimization on pure physics based criteria, and in this way appear to be canonical.

What we propose here is to optimize the locomotion trajectories on a parametric robot corpus as well as morphology/anatomy. In this context, we propose a controller with *a priori* knowledge (such as symmetry) allowing to simplify the search space of the optimization process.

© Springer International Publishing Switzerland 2016
N.F. Lepora et al. (Eds.): Living Machines 2016, LNAI 9793, pp. 227–238, 2016.
DOI: 10.1007/978-3-319-42417-0_21

This work is based on a technology developed by authors allowing to simulate a parametric family of legged robots. It is part of the Metabot [4] project, which consist in developing an open-source low cost robot for education and research.

We will first describe the architecture of the system that was developed, and then the generic controller that can make all the robots of the corpus walk. Then, we will discuss the optimization experiments that were made and the obtained results.

2   Related Works

The GOLEM project [5] proposed to evolve both the robot's morphology and its controller, using neural networks and quasi-static simulation. They generated manufacturable walking robots without any *a priori*.

More recently, Disney Research [6] proposed a system allowing casual users to design and create 3D-printable creatures. They proposed a method to generate motion using statically stable moves on a robot model given by the user. They also proposed a way to generate printable geometry from the custom parts.

The presented work is not restricted to statically stable motions. Instead, dynamic locomotion patterns are explored - which mean that it is not assumed that the robot have to be statically stable at each time step. The ability to print the robots is also proposed.

[7] co-evolved pure-virtual robots in dynamics simulator and proposed a script language for robot structures.

[8] produced real-world robots whose morphology and controller were optimized by simulation. The study underlines an important gap between simulation the physical world. Our simulation tries to reduce this gap by including an estimation of the shocks and the backlash.

[9] studied the legged locomotion space, trying to adapt the locomotion when there is damaged parts using pre-computed behaviour-performance maps.

There are also several works on fixed architecture robot's locomotion tuning, like the teams that worked on the Aibo robots [10] or Bostom Dynamic's Little Dog [11], or the locomotion learning from scratch [12].

In our system, not only the robot locomotion but also its morphology is optimized.

3   System Architecture

3.1   Robots Modelling

To model the robots, we developed a custom tool that uses the OpenSCAD [13] language as backend. This language can be used to describe 2D or 3D models using code, which makes extensive parametrization possible. This code is actually based on constructive solid geometry (CSG), where boolean operations are applied to basic objects to combine them. OpenSCAD can produce meshes and CSG trees, which contains the basic objects, boolean operations and transform matrices.

We customized OpenSCAD to add metadata in the code. We extended the parts description language to add markers in the CSG tree so that we can automatically retrieve information such as reference frames such as anchor or tips. Thus, we are able to specify in the generated part some reference frames that we can retrieve once the part is compiled and to tag specific parts of the CSG tree.

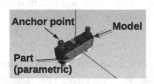

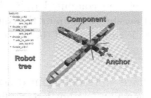

Fig. 1. A component contains models, parts and anchor points.

Fig. 2. An overview of the robot editor, using parametric components library.

Moreover, we also added the notion of *component*, which contains *models* of "real-life" existing objects (such as motors), *parts* that should be manufactured to make the robot and *anchors* that are points where other compatible anchors can be attached (see Fig. 1). Components can have parameters, that can modify some features (for instance mechanical lengths) of the parts that it contains.

We then created a components library and an editor that can be used to instantiate and attach components together. The edited robot is actually represented as a tree with components as nodes and anchors as child relation (see Fig. 2)[1]. We also added global robot parameters that can impact multiple components. This results in parametric robots that can be re-generated with new parameters to change some morphological features.

All the parts of a robot can be exported, generating meshes that can be used for example for 3D printing.

3.2 Kinematic and Dynamics

Since we know the transformation matrices for each part of the robot tree, we can compute the kinematic model for a given robot topology and its parameters. Leg tips are also tagged in the components.

It is also possible to compute the dynamics model. To achieve this, parts and models are turned into small voxels (cubes) of typically 1 mm^3. This can be done using even-odd rule [14] on each point of the bounding box. With these cubes, we can deduce the volume of the part and the mass distribution (see Fig. 3). The center of mass can then be known using a weighted average and the inertia with an integration of the cubes.

---

[1] Videos showing the editor is available at https://www.youtube.com/watch?v=smHctwi05Ic.

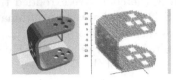

Fig. 3. Voxels are computed from the actual part meshes to compute the complete dynamic properties of the model.

Fig. 4. Robot parts and models and its pure shape collisions approximation for simplified collision computations.

Parts have a pre-determined homogeneous density, and models (for example a well-known motor) can have a fixed overall mass, in this case, the density is adjusted to fit the mass.

All the parts have a corresponding (parametric) pure shape approximation. This is made easy thanks to the constructive geometry, as it is already based on basic objects. Most of the part collision approximations simply consist in removing items from the tree such as screw holes. Then, we parse the CSG tree and directly retrieve the pure shapes (see Fig. 4). To avoid auto-collision glitches, these shapes can be automatically "retracted" (each pure shape is slightly downsized, so that near shapes such as a motor horn don't collide because of numerical approximations when rotating).

We can then inject all these information into a dynamics engine.

3.3   Corpus

We designed a corpus containing 22 robots, with 4 to 8 legs and 8 to 24 motors. The parts are designed to be manufacturable (3D printable). The modeled motor is similar to an on-shelf low-cost motor (Robotis XL-320). Robots vary in shape and thus have different morphological parameters (see Fig. 5).

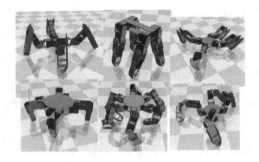

Fig. 5. An overview of six robots of the corpus.

All the robots are facing the X-axis, which is the arbitrary "front" of the robot. Some degrees of freedom range were limited artificially to disambiguate the kinematic (see below).

4   Generic Controller

Here we introduce a generic controller for locomotion that takes a robot model and generates motor trajectories. This controller allows to steer the robot according to a given speed vector.

4.1   Principle

The locomotion task to be solved consists in following a given speed and rotational velocity described by the dynamic controls $(\dot{x}, \dot{y}, \dot{\theta})$. Since the robots have a significant number of degrees of freedoms, it is clear that there are many different ways to achieve this. For more convenience, the legs are ordered using their anti-clockwise position around the z-axis. Thus, the front left leg is always the first one and the front right the last.

In order to reduce the parameter's space, we consider that the robot's height during the walk remains constant and that the body yaw and roll are null.

If the robot is indeed following the dynamic controls, the leg trajectory on the ground during the support phase will then be determined. We then use iterative inverse kinematics using simple stochastic method (we apply random variations on the angles and check if one of these variation lead to a result closest to the target and iterate again until it doesn't). Since we are dealing with small number of degrees of freedom per leg, this is efficient enough (This consumes far less computational power than the dynamics engine itself). Thus, we are able to control the tip position in the Cartesian (x, y, z) coordinates.

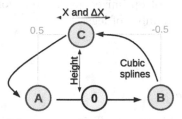

Fig. 6. Model of the leg trajectory that uses cubic splines and three locus, the "0" point is the position of the tip when the dynamic controls are null.

The leg trajectory is described using splines. In order to ensure smooth motions (and finite speeds) we have chosen cubic splines position for trajectories representation. We define three couples point/velocity (or locus) for this spline. The first two are defining the segment followed by the leg on the floor and the last one being the point reached when the leg rises (C).

Here, the two loci on the floor have both a fixed position and speed: the position is determined by the steps length and the speed is null during support phase. The third locus height is a parameter that defines how high the robot will rise its legs. Its position and speed belong to the optimized parameters of the controller that change the leg trajectory when raised (see Fig. 6).

We assume that the support duration is the same for each leg and is a parameter of the controller. Leg phases (i.e. synchronization) are also parameters. The first leg (front left) is the reference phase that can be changed along with all the other leg phases accordingly. The frequency of the whole stride is another parameter.

Finally, the robot posture (i.e. the position of the legs when dynamic controls are null and the robot is static) can be parametrized with x, y and z, where z is the height of the robot and x, y are multiple of the leg position on the original robot model (if x = 1 and y = 1 the position of the legs are exactly the same as in the model, if x = 0.5 and y = 2, the legs are near the center of the robot along the sagittal plane and farther along the frontal plane, the symmetry is preserved).

4.2   Parameters Summary

- **locus x, speed and height:** are locus parameters (see above and Fig. 6)
- **support:** the duration of the support of each leg
- p_2, p_3, ... p_n (where n is the number of legs): the leg phases (p_1 is the "reference" leg, so there is only a set of parameters for a given gait)
- **frequency:** the number of stride per second
- **x, y and z:** are posture parameters that define the robot posture when the dynamic controls are null

Thus, the total number of optimized parameters for the walk controller is $(n - 1) + 8$ where n is the number of legs.

5   Experiments

5.1   Dynamics Engine

We use the Bullet physics library, an efficient open-source dynamics engine [15]. In this case, it can simulate rigid bodies, torque controllable hinges and collisions.

To simulate the motors, we applied torques directly on the components. Joints control is made by applying a target velocity from the position error (proportional), and then applying a target torque from the velocity error (proportional). To do so we use gains manually tuned. The maximum motor speed and torque curve is considered, the maximum torque that can be applied depends linearly on the current speed, reducing to zero when the max speed is reached.

In order to get more realistic results, we added a backlash simulation on the degrees of freedom, using a cone/twist constraint.

The collisions information can be retrieved from the dynamic simulation. They were used to detect both auto-collisions between the robot's parts and collisions between other parts than legs with the floor to add a penalty on the score (to avoid crawling behaviors for instance).

Simulations were made using a time accuracy (Δt) of 1 ms, and all the parts had a friction coefficient of 0.5.

In order to make experiments possible, morphological parameters were rounded to millimeters and a cache system was added to avoid compiling and voxelizing parts that were already met before (see Fig. 7).

5.2 Optimization

We use the black-box evolutionary algorithm CMA-ES [16,17] as optimizing algorithm, with the following meta-parameters (all the problem parameters were normalized between 0 and 1): algorithm: BIPOP CMA-ES, restarts: 3, elitism: 2, f-tolerance 10^{-6}, x-tolerance 10^{-3}.

Notice that the *f tolerance* is maybe the more critical meta-parameter, because it determines how accurately the score should be tuned. The optimization indeed stops when the score is not optimized more than this tolerance during a certain number of iterations. In our case, since each fitness evaluation typically takes a few seconds, a bad value for this can lead to hours of useless computation trying to tune meaningless small values.

The dimension of experiments vary with morphological parameters and number of legs (because of the phases) from 15 to 20.

5.3 Score

The goal of the first experiment consists in walking forward, i.e. going as far as possible on the X axis. Two scores were tested, the first is simply the inverse of the distance walked (minimization), and the second is the inverse of the distance walked multiplied by the energy cost, which is the sum of all the impulses ($N.m.s$) sent to the motors.

A second experiment consists in reaching checkpoints. Four near points have to be attained successively by walking forward and turning to get the body aligned with the target. This introduces the trajectory control which adds a few more parameters to be optimized. Here, giant steps were used on the score with each checkpoint passed. In order to help the convergence of the optimization, if the last checkpoint is not reached, the score is related to the distance to the next missed checkpoint. If the last checkpoint is reached, two scores were tested, the first is the duration of the experience (trying to minimize it), and the second is the duration multiplied by the energy cost.

Finally, during this two experiments, we optimize the walk using a small fixed frequency to compare the resulting gait.

5.4 Workflow

The overall architecture is shown on Fig. 7. CMA-ES runs parallel simulations with parameters set, each simulation first compile the dynamics of the robot, trying to hit the (filesystem) cache, speeding up the process. Simulations are headless, but can also stream their state and are viewed using a custom 3D client GUI.

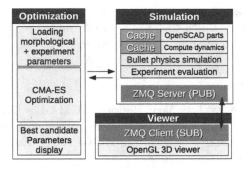

Fig. 7. Software architecture.

6   Results

While there is no guarantee that CMA-ES finds the global optimum, the optimization still provides very satisfactory sets of parameters for each robot.

6.1 Leg Phases

A first result can be viewed on the leg phases of the optimized robots. The (p_2, p_3, p_4) plot of all optimized quadruped robots (all experiments mixed) is shown on Fig. 8, where all the points appear to be along a long quasi ellipsoid.

The trot gait is on the middle of the ellipsoid, the walk on one side and the "cross" walk on the other side (see Fig. 10). Doing a Principal Component Analysis (PCA) reveals that the first axis explains 65 % of the variance and the second one 32 % (see Fig. 9).

6.2 Locus

We plotted the histogram of locus for all experiments, resulting in Fig. 11.

We can notice that the locus X values are mostly in the $[0, 0.5]$ interval, which means that the leg trajectory will go slightly forward when rising.

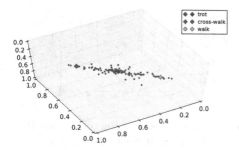
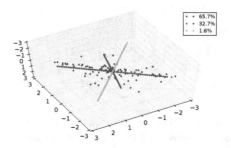

Fig. 8. p_2, p_3 and p_4 for all optimized quadruped are along an ellipsoid (p_3 was shifted by 0.5 to center it) (Color figure online)

Fig. 9. PCA analysis of the points from Fig. 8, data are standardized and the legend corresponds to the proportion of variance explained by each of the principal components. (Color figure online)

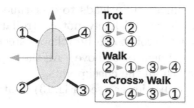

Fig. 10. Different quadruped gaits mentioned in this paper. The diagram on the right is the order of the rising legs (which correspond to leg phases).

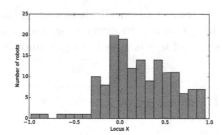

Fig. 11. Histogram of locus X for all experiments (see Fig. 6).

6.3 Walking with Low Frequency

We also tried optimizations with fixed frequencies that were not tuned.

We observe that this results in higher robot postures (see Fig. 12). Which indeed reduce energy cost but lose stability. More robots were able to exhibit a walk (or cross walk) during this experiment.

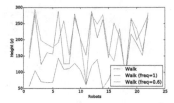
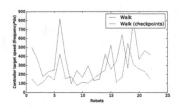

Fig. 12. The posture height (z) of the optimized robots from the corpus for optimized frequency and constrained frequency. (Color figure online)

Fig. 13. The speed of the robots that the controller tries to reach ($frequency * dx$) for the walking and checkpoints experiments (with energy cost in the score). (Color figure online)

6.4 Checkpoints

The "walk forward" experiment often leads to over-tuned running robots, that sometime seem very unrealistic and likely not controllable. Moreover, walking forward doesn't prove that the robot can be reliably steered. This is why introduced the checkpoints experiment (see above). All the experiments converged to robots that successfully reach all the checkpoints.

The resulting robots walk slower (see Fig. 13) that those from the walk experiment.

6.5 Fast Walk

The experiment that consists in a fast walk disregarding the energy cost unsurprisingly converges to bigger robots in every cases (see Fig. 14). This is likely because the reachable space of legs is simply bigger, allowing bigger steps.

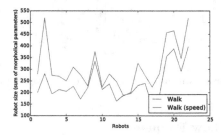

Fig. 14. The size of robots (which is the sum of morphological parameters) are compared in the walking experiment based on energy cost score and walking experiment based on pure distance score. (Color figure online)

7   Conclusion and Future Work

We presented a framework allowing to generate and simulate parametric multi-legged robots. We also proposed a simple generic controller and used it through simulations experiments on a corpus of parametric robots. This controller was able to produce locomotion for all the robots of the corpus, but also to control them to reach arbitrary checkpoints[2].

Analysis of the simulations revealed that the search space could be reduced, like the leg phases for quadruped robots and the locus position that is almost always on the front. We also noticed that reducing the robot's energy cost imply a smaller morphology in every simulations, while making it steerable (with checkpoints experiments) reduces the speed. Moreover, the obtained leg phases for optimized quadruped robots seems very similar to animals trot and walk.

In a future work, the phases ellipsoid result will be checked on real robots with an appropriate experimental setup.

Some other optimizable parameters could also be added to the controller, like the orientation of the body.

The controller could also be improved in order to be able to handle robots with spine and legs kinematic chains that share degrees of freedom, thus greatly extending the capabilities of the robots.

It would also be interesting to compare CMA-ES with other optimizing algorithm for this specific experiment. For example a multi-objective optimization method could allow to find not only "optimal" candidates but also optimal populations of candidates.

Finally we could also produce automatically robot morphologies from scratch without a pre-defined corpus. This raises some problems like kinematic singularities, upstream filtering of bad candidates (for example all the motors on the same plane) and particularly adapting the optimization algorithm.

References

1. Muybridge, E.: Animals in Motion. Courier Corporation (2012)
2. Marder, E., Bucher, D.: Central pattern generators and the control of rhythmic movements. Curr. Biol. **11**(23), R986–R996 (2001)
3. Holmes, P., Full, R.J., Koditschek, D., Guckenheimer, J.: The dynamics of legged locomotion: models, analyses, and challenges. SIAM Rev. **48**(2), 207–304 (2006)
4. Grégoire Passault, F.P., Rouxel, Q., Ly, O.: Metabot: a low-cost legged robotics platform for education (Submitted)
5. Pollack, J.B., Lipson, H.: The GOLEM project: evolving hardware bodies and brains. In: 2000 Proceedings of the Second NASA/DoD Workshop on Evolvable Hardware, pp. 37–42. IEEE (2000)
6. Megaro, V., Thomaszewski, B., Nitti, M., Hilliges, O., Gross, M., Coros, S.: Interactive design of 3D-printable robotic creatures. ACM Trans. Graph. (TOG) **34**(6), 216 (2015)

---

[2] A video of the obtained behaviors is available: https://www.youtube.com/watch?v=GF1KM7JrmC0.

7. Marbach, D., Ijspeert, A.J.: Co-evolution of configuration and control for homogenous modular robots. In: Proceedings of the Eighth Conference on Intelligent Autonomous Systems (IAS8), BIOROB-CONF-2004-004, pp. 712–719. IOS Press (2004)
8. Samuelsen, E., Glette, K.: Real-world reproduction of evolved robot morphologies: automated categorization and evaluation. In: Mora, A.M., Squillero, G. (eds.) Applications of Evolutionary Computation. LNCS, vol. 9028, pp. 771–782. Springer, Heidelberg (2015)
9. Cully, A., Clune, J., Tarapore, D., Mouret, J.-B.: Robots that can adapt like animals. Nature **521**(7553), 503–507 (2015)
10. Hengst, B., Ibbotson, D., Pham, S.B., Sammut, C.: Omnidirectional locomotion for quadruped robots. In: Birk, A., Coradeschi, S., Tadokoro, S. (eds.) RoboCup 2001. LNCS (LNAI), vol. 2377, pp. 368–373. Springer, Heidelberg (2002)
11. Neuhaus, P.D., Pratt, J.E., Johnson, M.J.: Comprehensive summary of the institute for human and machine cognition's experience with little dog. Int. J. Robot. Res. **30**(2), 216–235 (2011)
12. Maes, P., Brooks, R.A.: Learning to coordinate behaviors. In: AAAI, pp. 796–802 (1990)
13. The programmers solid 3D CAD modeller. http://www.openscad.org/
14. Hormann, K., Agathos, A.: The point in polygon problem for arbitrary polygons. Comput. Geom. **20**(3), 131–144 (2001)
15. Boeing, A., Bräunl, T.: Evaluation of real-time physics simulation systems. In: Proceedings of the 5th International Conference on Computer Graphics and Interactive Techniques in Australia and Southeast Asia, pp. 281–288. ACM (2007)
16. Hansen, N., Ostermeier, A.: Completely derandomized self-adaptation in evolution strategies. Evol. Comput. **9**(2), 159–195 (2001)
17. Multithreaded C++11 implementation of CMA-ES family for optimization of nonlinear non-convex blackbox functions. https://github.com/beniz/libcmaes/

Stick(y) Insects — Evaluation of Static Stability for Bio-inspired Leg Coordination in Robotics

Jan Paskarbeit[1(✉)], Marc Otto[2], Malte Schilling[3], and Axel Schneider[1,4]

[1] Biomechatronics Group, Center of Excellence 'Cognitive Interaction Technology' (CITEC), University of Bielefeld, Bielefeld, Germany
jpaskarbeit@uni-bielefeld.de
[2] Robotics Research Group, Faculty of Mathematics and Computer Science, University of Bremen, Bremen, Germany
[3] Neuroinformatics Group, Center of Excellence 'Cognitive Interaction Technology' (CITEC), University of Bielefeld, Bielefeld, Germany
[4] Embedded Systems and Biomechatronics Group, Faculty of Engineering and Mathematics, University of Applied Sciences, Bielefeld, Germany

Abstract. As opposed to insects, todays walking robots are typically not constructed to withstand crashes. Whereas insects use a multitude of sensor information and have self-healing abilities in addition, robots usually rely on few specialized sensors that are essential for operation. If one of the sensors fails due to a crash, the robot is unusable. Therefore, most technical systems require static stability at all times to avoid damages and to guarantee utilizability, whereas insects can afford occasional failures. Despite the failure tolerance, insects also possess adhesive, "sticky" pads and claws at their feet that allow them to cling to the substrate, thus reducing the need for static stability. Nevertheless, insects, in particular stick insects, have been studied intensively to understand the underlying mechanisms of their leg coordination in order to adapt it for the control of robots. This work exemplarily evaluates the static stability of a single stick insect during walking and the stability of a technical system that is controlled by stick insect - inspired coordination rules.

1 Introduction

Man-made, technological systems designed after biological examples are usually built of different materials as compared to the natural systems. This leads to additional constraints for the technical design which, in some cases, contradict the transfer of the original, bio-inspired idea. In contrast, biological systems often have to conform to side conditions that are of no importance to the technical system. To resolve this antagonism, a careful abstraction is important when only subsystems or substructures are transferred from the biological example to the technological system, in particular if mass, inertia or dimensions differ. This work highlights this perspective for the concept of static stability in the six-legged walking robot HECTOR, depicted in Fig. 1(b), modeled after the stick insect *Carausius morosus* as shown in Fig. 1(a). Unlike most robots, insects can

© Springer International Publishing Switzerland 2016
N.F. Lepora et al. (Eds.): Living Machines 2016, LNAI 9793, pp. 239–250, 2016.
DOI: 10.1007/978-3-319-42417-0_22

afford to disregard stability. If they lose static stability, they either fall with negligible negative consequences or they cling to the ground with their tarsi. Their inviolability is e.g. due to their small size and mass combined with the compound materials they are made from. HECTOR is neither designed to fall over nor are there any special ground attachment (anchoring) mechanisms at its leg tips. This work starts with an exemplary evaluation of a walking stick insect and then focuses on the question under which conditions the leg coordination framework in bio-inspired WALKNET-type controllers [12] provides stable walking. It finishes with conclusions on how to add stability as a side-condition to the walking controller without negative interferences with the bio-inspired leg coordination.

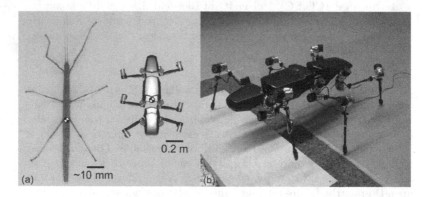

Fig. 1. (a) Biological model, *Carausius morosus*, and rendering of HECTOR in top view. The CoMs are marked in both cases. (b) Image of HECTOR while climbing over a small obstacle.

2   Evaluation of Static Stability for Walking Stick Insects

The static stability of stick insects has been discussed in literature with different outcomes. According to Jander, stick insects maintain static stability during tripod walking [6], whereas Kindermann found that stick insects topple backwards if they stand on loose ground and one of the hind legs is lifted [7]. This view is supported when looking at the detailed kinematics data of walking sticks as published recently by Theunissen et al. [15,16]. As an example, in one trial ("Animal12_110415_00_22", visualized in Fig. 2) the insect would be deemed multiple times statically unstable (shown in red in Fig. 2(c) and (d)). Even though three legs are on the ground the centre of mass (CoM) leaves the support polygon (CoM was assumed to be located between the hind coxae, see [2]). The footfall pattern for this run is shown in Fig. 2(b). Although at any point in time at least three feet are on the ground (black bars indicate stance phases), the stability margin varies. The distance between the projection of the Center of Mass (CoM) onto the walking surface and the closest line of the support polygon is shown in Fig. 2(a). Negative distances stand for stable, positive distances for unstable walking phases. As can be seen in conjunction with the footfall pattern in Fig. 2(b), the distance increases every time one of the hind legs (L3, R3) is lifted. Even though the exact contact positions might vary depending on the unknown orientation of the tarsi, this would not change the result as the CoM leaves the support polygon by a considerable amount.

As the accompanying video of the trial shows, the stick insect does to not topple even if the CoM is outside the support polygon since the insect is able to cling to the ground with its tarsi. This points in the same direction as the findings of Kindermann.

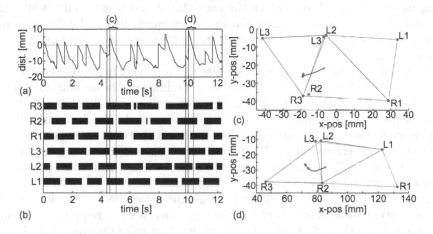

Fig. 2. Experimental data from walking stick insect (based on data from [16]). (a) and (b) show the distance of the CoM to the support polygon and the corresponding footfall pattern over time. A negative distance in (a) corresponds to a statically stable posture, a positive distance denotes an unstable posture. In (c) and (d), exemplary foot positions are plotted for three points in time that correspond to the colored, vertical lines in (a) and (b). The connecting lines between the footpoints represent the support polygons for the respective postures. The positions of the CoM that correspond to the times for which the support polygons are shown, are marked by colored circles. The movement of the CoM is plotted as a black line.

3   Bioinspired Leg Coordination

For bioinspired control of walking in hexapod and quadrupedal robots there are two distinct general approaches which both do not necessarily inflict stability on the robot. On the one hand, open-loop control applies precomputed or planned control signals to the actuators [5]. As an advantage, such control approaches are not affected by sensory noise and the communication does not deal with latencies from sensory signals. On the other hand, in many cases animals have to deal with highly unpredictable environments, e.g. for a stick insect when walking in uneven terrain or when climbing on twigs. Such approaches require close consideration of feedback from the environment which is mediated through sensory input. Closed-loop controllers allow for adaptive behavior by taking sensory feedback into account.

As mentioned, in locomotion those different approaches might fulfill different roles. This work focuses on slow walking as an adaptive behavior which takes the sensed environment into account. An example for such a controller is WALKNET which is inspired by experiments on the walking behavior of stick insects [12]. Walking in a hexapod robot poses a difficult problem as it requires the coordinated movement of six legs. In walking, mainly two different behaviors are distinguished on the leg level: On the one hand, as depicted in Fig. 3(b), the swing movement [14] during which the leg is lifted and protracted towards the Anterior Extreme Position (AEP). On the other hand, during the stance movement a leg supports the body as it is retracted towards its Posterior Extreme Position (PEP) [11].

In WALKNET, the leg coordination is distributed in six individual controllers, one for each leg. This is inspired by biological findings on the organization of the control system in stick insects [12]. Each of these leg controllers has to decide which of the two behaviors (swing or stance) is executed. The decision is driven by sensory signals and in particular the transitions between the two behaviors are initiated by sensory input: A leg continues with a swing movement until it touches the ground. During the stance phase, the leg is pushed backwards in order to propel the body forwards. The stance movement continues until the leg reaches its PEP. As each leg cycles through those two behaviors individually, a coordination mechanisms between the leg controllers is required. A simple solution is given through a set of local coordination rules inspired by biological insights on stick insect behavior [12]. Section 4 will deal with the question of static stability in the light of this coordination scheme. The first three coordination rules act directly on the transition from stance to swing to prolong or shorten the stance phase (for effective directions for all rules see Fig. 3(a)):

Rule 1 prolongs the stance phase of the receiver leg if the sender leg is in swing phase. After the sender leg starts its stance phase, the influence persists for a short time at a reduced strength. The duration is speed-dependent. **Rule 2** facilitates the start of the receiver leg's swing phase after the sender leg established ground contact at the end of its own swing phase. Between the detection of ground contact and the activation of the influence, a short delay is provoked. **Rule 3** facilitates the start of the receiver leg's swing phase whenever the sender leg crosses a threshold position on its way towards the PEP. The influence lasts only for a short time. The location of the threshold position is speed-dependent and differs for ipsilaterally and contralaterally neighbouring legs.

Rules 4–6 act in different ways on the coordination. These rules have seldomly been implemented and are missing in this implementation as well as in the reference implementation of [12]. Early, theoretical analyses of some of the rules, though not in combination, are given in [1]. For a detailed description of the coordination rules, see [12]. WALKNET has successfully been tested in dynamics simulations and on robots [4], most recently on the robot HECTOR [10]. It has proven to produce stable walking patterns at different velocities and leading to different types of insect gaits. In addition, the WALKNET approach has been used for a simple strategy on curve walking [11]. However, in tight curves the working ranges of the individual legs are unequal. Legs situated on the inner side of the curve are barely moving while the outer legs' working ranges increase (a detailed analysis of curve walking in stick insects is provided in [3]). These unequal stance distances represent a challenging situation for WALKNET. This work analyzes systematically how this affects stability and how the parameters of the coordination rules have to be adapted to produce stable walking.

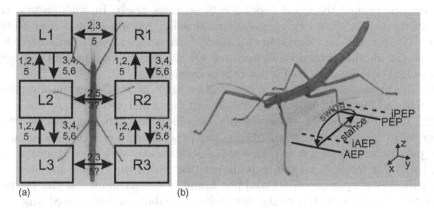

(a)   (b)

Fig. 3. (a) Directions of influence for the coordination rules within WALKNET. (b) Schematic depiction of a stance-swing-cycle for the left middle leg. The leg stances from AEP to PEP. Both positions are shifted relative to the constant intrinsic positions (iAEP, iPEP) by the coordination rules.

4 Static Stability in Forward and Curve Walking for a Technical System

To test the suitability of the described bio-inspired leg coordination without modifications for the control of a hexapod robot, a physics simulation of HECTOR was used. To assess stability, the virtual robot was commanded to walk on flat terrain for a given time. At the end of each trial, the fraction of iterations was determined, during which the robot was statically stable. For this evaluation, the first 5 s of each trial were excluded as this work is not concerned with the transient behaviour at the beginning of walking. Therefore, a fraction of 1 signifies that the robot has been statically stable after the first seconds. A reduced stability fraction indicates unstable situations during the experiment.

The bio-inspired leg coordination is based mostly on the shifting of the PEP of each leg (based on the constant, intrinsic PEP (iPEP), triggered by its neighbouring legs). Therefore, the unit of rule strengths will be given in meters.

For the first experiments, the walking controller as reported by Schilling et al. was used for slow and fast walking [12]. To evaluate the influence of the individual coordination rule strengths, the intensities of the coordination rules were varied within intervals of $[-0.5\,m,\ 0.0\,m]$ for rule 1 and $[0\,m,\ 0.25\,m]$ for rules 2 and 3. The ipsilateral and contralateral influences of rule 3 were adjusted individually, as their start thresholds are computed differently (see [12]). As the PEP-shifts of rule 3 in the front legs are found to be stronger than in the hind legs [12], a fixed ratio of 3:1 is assumed and only the values for the front legs will be given in the results. Since the leg coordination is influenced by the initial posture, each run was conducted from three different, predefined postures to rule out peaks in stability due to a incidentally chosen perfect starting posture. All data that is shown represents the worst result for these three trials. In stick insects, the load of the legs is sensed by campaniform sensilla that are located close to the joints [17]. For the data shown in this work, force sensors were simulated at the leg tips. To obtain an immediate feedback for the detection of ground contacts, a threshold of $0.001\,N$ is used if not otherwise specified.

In the first experiment, the robot walked with two different speeds (slow: $0.15\,m/s$ and fast: $0.25\,m/s$) using the default controller settings that are given in [12]. Only the strengths of the coordination influences (the rule-specific PEP-shifts) were varied systematically during the experiment. For slow walking, multiple parameter sets were found that create permanently stable coordination (see Fig. 4(a)). For the fast walking experiment, for which the results are shown in Fig. 4(b), no such parameter set could be found. The maximum stability fraction was 0.94. Therefore, in 6 % of the iterations the robot was considered unstable. In practice, a short duration of instability will not necessarily result in a crash. Due to the inertia of the main body it takes some time for the robot to tilt. Also, if a leg is in swing phase during the tilting movement, it may strut the robot as soon as it touches ground and switches to stance phase. Therefore, the virtual as well as the real robot have successfully performed multiple walks – even on rough terrain [10]. Nevertheless, for the general application of the bio-inspired controller on real hexapod robots, the stability should be maintained in all situations and independent of the initial posture.

To achieve this, the complete set of controller parameters has been optimized using Simulated Annealing [8], a global optimization technique which involves a probabilistic search function. In addition to the already mentioned variation of the PEP-shifts, the deactivation delay of rule 1 and the activation delay of rule 2 were optimized. Also, the contact force threshold between the leg and the ground, at which the leg switches to stance at the end of a swing phase, was optimized.

With the parameter set listed in Table 1, a stability fraction of 0.998 could be achieved for fast walking. Although the optimization focused only on fast walking, with these optimized parameters, slow walking was still found to be statically stable. Although this result misses the goal of permanent, static stability, the controller will likely preserve the robot from falling in almost all situations during forward walking.

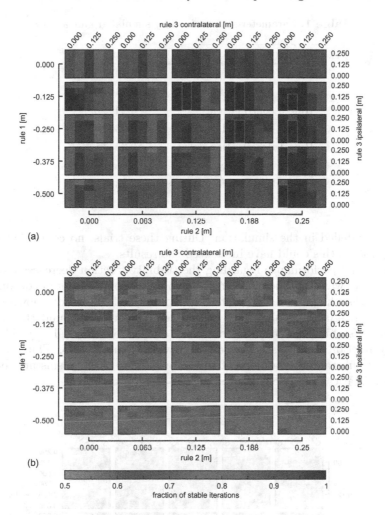

Fig. 4. Fraction of stable iterations per total iterations for different strengths of the coordination rules. (a) shows the results for slow walking (0.15 m/s), (b) for fast walking (0.25 m/s). The fields marked by red frames represent the parameter sets that achieve maximum stability fractions. (Color figure online)

As has been shown in [12], after some initial steps, the WALKNET-based controller usually generates a regular gait. In the data collected in this work, this is observable as well. However, during straight forward movement, the stance distances of the legs are equal in average. Therefore, the question remains how the controller handles situations, in which the stance distances of the legs vary, e.g. in curve walking. To examine this question, the robot was commanded to walk curves with a constant radius as shown in Fig. 6(a). The deviations from a perfect circular trajectory are due to the inherent compliance of HECTOR that

Table 1. Parameter set obtained by simulated annealing

Parameter	Value
Rule 1	$-0.0765\,\mathrm{m}$
Rule 2	$0.2464\,\mathrm{m}$
Rule 3 contralateral	$0.2499\,\mathrm{m}$
Rule 3 ipsilateral	$0.0364\,\mathrm{m}$
Rule 1 deactivation delay	$0.3547\,\mathrm{s}$
Rule 2 start delay	$0.0591\,\mathrm{s}$
Ground contact force threshold	$3.7385\,\mathrm{N}$

was also modelled in the simulation. During these trials, no course correction was applied, as this could have influenced the results.

The overall performance regarding the stability fraction decreases considerably for curve walking. Therefore, in order to find an optimal set of coordination rule strengths, the parameter variation was performed as already shown for forward walking. As Fig. 5 shows, no parameter set could be found that creates comparable stability fractions as obtained during forward walking. The best stability fraction that was achieved for curve walking is 0.85. Thus, it seems likely that substantial adaptions of the co-ordination mechanisms are required to achieve statically stable curve walking.

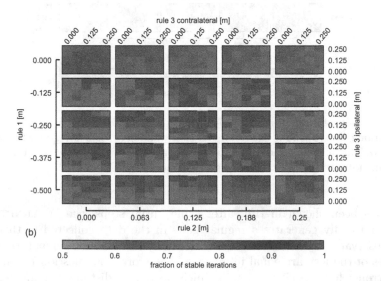

Fig. 5. Fraction of stable iterations per total iterations for different strengths of the coordination rules during fast curve walking ($0.25\,\mathrm{m/s}$). The fields marked by red frames represent the maximum stability fractions. (Color figure online)

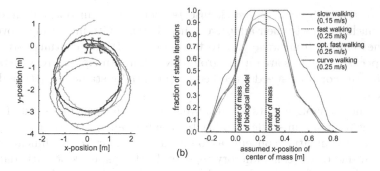

Fig. 6. (a) Exemplary trajectories for curve walking. (b) Maximum fraction of stable iterations over all iterations for different virtually shifted CoMs. The position of the CoM is given relative to the onset of the hind legs. (Color figure online)

Regarding stability, one of the most relevant differences between the insect and the robot is the shifted CoM. Whereas the CoM lies roughly between the hind leg onsets in the insect, in the robot, it is located slightly in front of the middle legs (see Fig. 1(a)). To analyze the influence of this aspect, the stability fraction was re-evaluated by *virtually* shifting the location of the CoM along the longitudinal axis of the robot. In practice, a real shift of the CoM would probably result in different outcomes, especially for the extreme positions that would make coordinated locomotion impossible.

Figure 6 shows the highest stability fractions that could be achieved for any of the coordination rule parameter sets for slow and fast walking (first experiment), optimized fast and curve walking for different CoMs (relative to the position of the hind legs). For forward walking, the peak stability can be obtained for a CoM close to that of the robot. For curve walking, the optimal CoM lies closer to the back of the robot. Although these results do not reflect the actual outcome if the CoM would have been actually shifted, it is both notable and plausible that the stability fraction would be reduced for an insect-like positioned CoM.

5   Conclusion

Unlike most robots, insects typically have the ability to cling to the ground. Depending on their habitat they might still require static stability to keep from toppling over. An example are grass-cutting ants that usually walk on loose ground [9]. On the other hand, stick insects like *Carausius morosus* live on branches, often hanging upside-down. For them, the substrate is rigid enough to hold on to it. Thus, static stability is not a key requirement for them in their natural habitat. However, for walking on plain ground, different results have been reported regarding their ability to maintain static stability. As Jander reports, the lift-off of hind legs is triggered only when the CoM can be supported by the remaining legs on the ground [6]. Kindermann observed backward toppling if the insects could not cling to the ground and one of the hind legs was lifted [7].

Despite the stability issue, stick insects seem to be an optimal model for walking machines due to their adaptive gait patterns. With increasing movement speed, they change from wave gait to tetrapod and tripod continuously. Thus, for low speeds, many legs are on the ground which in principle increases stability and resilience. For high speeds, the fastest potentially stable gait (tripod gait) is adopted.

While these adaptive gaits are desirable for walking machines, due to their requirement to maintain static stability to prohibit falling, the direct adaptability of the leg coordination concepts as found in stick insects for technical systems was unclear due to the contradictory reports. In this paper, kinematic data of a walking stick insect is analyzed regarding the position of its CoM relative to the support polygon. It was found that the stability margin decreases whenever one of the hind legs switches from stance to swing phase. For two exemplary situations, it was shown that the CoM left the support polygon, thus indicating an unstable posture. As a consequence, the suitability of the coordination concepts for walking machines, in the light of their need for unconditioned stability, is unclear.

Based on the setup of the bio-inspired robot HECTOR, the coordination rules as reported by [12] have been evaluated for their ability to maintain static stability in different scenarios. The most relevant differences between the robot and its biological model, *Carausius morosus*, are the inability of the robot to cling to the ground and a shift of its CoM towards the onsets of the middle legs as opposed to the hind legs in the insect. The inability to cling to the ground makes the robot prone to falling if the stability is not maintained by safe arrangement of the support polygon at any time. In a physics simulation, the robot, controlled by the bio-inspired coordination concept, was tested in three different scenarios and a multitude of control parameters. For slow forward walking at speeds of 0.15 m/s, multiple sets of coordination parameters were found that maintained static stability throughout the test runs. For fast forward walking at 0.25 m/s, none of the parameter sets could stabilize the robot at all times. With an optimized parameter set, however, a stability fraction of 0.998 could be achieved. Since the coordination rules have been derived from observation of forward walking stick insects, the concept of WALKNET and the implemented coordination rules are not explicitly designed for curve walking. During tests, at which the robot was commanded to walk curves, the stability was significantly decreased to stability fractions of about 0.85.

As the evaluation of the kinematic data showed, stick insects do not rely on static stability for locomotion. In comparison, technical systems such as the hexapod robot HECTOR must maintain stability at all times. HECTOR has been modeled explicitly after the model of the stick insect. Thus, for example, the relative positions of the legs have been preserved (scaled up by a factor of 20). The results suggest that the existing, bio-inspired coordination rules alone do not guarantee unconditioned stability in a robot like HECTOR or other robot setups without ground attachment mechanisms or specifically adapted leg distances.

However, to benefit from the adaptability of bio-inspired leg coordination, provided by WALKNET, in arbitrary multi-legged robot setups, additional control layers must be introduced. One way to guarantee stability would be the observation of the distance between the center of gravity and the closest line of the support polygon. If the body movement would lead the CoM too close to the border of the support polygon, the movement could be modified to counteract the danger of toppling. Also, if a leg is supposed to switch to swing phase, it could be checked prior to its lift-off whether this stance-swing-switch would endanger the stability and, if required, the stance phase could be extended. This can easily be formulated as an additional, technological rule that amends the otherwise purely bio-inspired rule set of WALKNET.

Another approach would be a cognitive layer that detects whenever a situation occurs that does not comply with the requirements of the technical system [13]. In these cases, an internal model of the robot could be used to search for adequate solutions of the problem. In the cases depicted in Fig. 2, in which the lift-off of a hind leg compromised the stability of the insect, a solution could be the repositioning of the opposite hind leg such that it can support the center of gravity.

For robust locomotion on complex terrain, combinations of the mentioned approaches could combine the elegance of insect locomotion with the requirements of technical systems.

Acknowledgments. This work has been supported by the DFG Center of Excellence 'Cognitive Interaction TEChnology' (CITEC, EXC 277) within the EICCI-project.

References

1. Calvitti, A., Beer, R.D.: Analysis of a distributed model of leg coordination. I. Individual coordination mechanisms. Biol. Cybern. **82**(3), 197–206 (2000)
2. Cruse, H.: The function of the legs in the free walking stick insect, Carausius morosus. J. Comp. Physiol. B **112**(2), 235–262 (1976)
3. Dürr, V., Ebeling, W.: The behavioural transition from straight to curve walking: kinetics of leg movement parameters and the initiation of turning. J. Exp. Biol. **208**(12), 2237–2252 (2005)
4. Espenschied, K.S., Quinn, R.D., Beer, R.D., Chiel, H.J.: Biologically based distributed control and local reflexes improve rough terrain locomotion in a hexapod robot. Robot. Auton. Syst. **18**(1–2), 59–64 (1996)
5. Ijspeert, A.J.: Central pattern generators for locomotion control in animals and robots: a review. Neural Netw. **21**(4), 642–653 (2008)
6. Jander, J.: Mechanical stability in stick insects when walking straight and around curves. In: Gewecke, M., Wendler, G. (eds.) Insect Locomotion, pp. 33–42. Paul Parey, Berlin, Hamburg (1985)
7. Kindermann, T.: Positive Rückkopplung zur Kontrolle komplexer Kinematiken am Beispiel des hexapoden Laufens: Experimente und Simulationen. Ph.D. thesis, Universität Bielefeld (2003)
8. Kirkpatrick, S., Gelatt, C.D., Vecchi, M.P.: Optimization by simulated annealing. Science **220**, 671–680 (1983)

9. Moll, K., Roces, F., Federle, W.: How load-carrying ants avoid falling over: mechanical stability during foraging in *Atta vollenweideri* grass-cutting ants. PLoS ONE **8**(1), e52816 (2013)

10. Paskarbeit, J., Schilling, M., Schmitz, J., Schneider, A.: Obstacle crossing of a real, compliant robot based on local evasion movements and averaging of stance heights using singular value decomposition. In: IEEE International Conference on Robotics and Automation, ICRA 2015, Seattle, WA, USA, 26–30 May 2015, pp. 3140–3145 (2015)

11. Schilling, M., Paskarbeit, J., Schmitz, J., Schneider, A., Cruse, H.: Grounding an internal body model of a hexapod walker - control of curve walking in a biological inspired robot–control of curve walking in a biological inspired robot. In: Proceedings of IEEE/RSJ International Conference on Intelligent Robots and Systems, IROS 2012, pp. 2762–2768 (2012)

12. Schilling, M., Hoinville, T., Schmitz, J., Cruse, H.: Walknet, a bio-inspired controller for hexapod walking. Biol. Cybern. **107**(4), 397–419 (2013)

13. Schilling, M., Paskarbeit, J., Hoinville, T., Hüffmeier, A., Schneider, A., Schmitz, J., Cruse, H.: A hexapod walker using a heterarchical architecture for action selection. Front. Comput. Neurosci. **7** (2013)

14. Schumm, M., Cruse, H.: Control of swing movement: influences of differently shaped substrate. J. Comp. Physiol. A **192**(10), 1147–1164 (2006)

15. Theunissen, L., Bekemeier, H., Dürr, V.: Stick insect locomotion (2014). toolkit.cit-ec.uni-bielefeld.de/datasets/stick-insect-locomotion-data

16. Theunissen, L.M., et al.: A natural movement database for management, documentation, visualization, mining and modeling of locomotion experiments. In: Duff, A., Lepora, N.F., Mura, A., Prescott, T.J., Verschure, P.F.M.J. (eds.) Living Machines 2014. LNCS, vol. 8608, pp. 308–319. Springer, Heidelberg (2014)

17. Zill, S.N., Schmitz, J., Büschges, A.: Load sensing and control of posture and locomotion. Arthropod Struct. Dev. **33**(3), 273–286 (2004)

Navigate the Unknown: Implications of Grid-Cells "Mental Travel" in Vicarious Trial and Error

Diogo Santos-Pata[1(✉)], Riccardo Zucca[1], and Paul F.M.J. Verschure[1,2]

[1] Universitat Pompeu Fabra, SPECS group, N-RAS,
Roc Boronat, 138, 08018 Barcelona, Spain
{diogo.pata,riccardo.zucca,paul.verschure}@upf.edu
[2] ICREA, Barcelona, Spain
http://www.specs.upf.edu

Abstract. Rodents are able to navigate within dynamic environments by constantly adapting to their surroundings. Hippocampal place-cells encode the animals current location and fire in sequences during path planning events. Place-cells receive excitatory inputs from grid-cells whose metric system constitute a powerful mechanism for vector based navigation for both known and unexplored locations. However, neither the purpose or the behavioral consequences of such mechanism are fully understood. During early exploration of a maze with multiple discrimination points, rodents typically manifest a conflict-like behavior consisting of alternating head movements from one arm of the maze to the other be- fore making a choice, a behavior which is called vicarious trial and error (VTE). Here, we suggest that VTE is modulated by the learning process between spatial- and reward-tuned neuronal populations. We present a hippocampal model of place- and grid-cells for both space representation and mental travel that we used to control a robot solving a foraging task. We show that place-cells are able to represent the agents current location, whereas grid-cells encode the robots movement in space and project their activity over unexplored paths. Our results suggest a tight interaction between spatial and reward related neuronal activity in defining VTE behavior.

Keywords: Biomimetics · Navigation · Grid-cells · Mental travel · VTE

1 Introduction

Representing ones' location in space and decode future trajectories are essential features for surviving in the world. Rodents are exceptional foragers capable of extensively explore large and complex environments, find their way back home and adapt to dynamic environments. Their robotic counterparts, however, still find difficulties in integrating their path over space, decide routes to take and adapt to novel environmental configurations. Studies on the rodent hippocampus have reported multiple cell types encoding for spatial properties, such as animal

© Springer International Publishing Switzerland 2016
N.F. Lepora et al. (Eds.): Living Machines 2016, LNAI 9793, pp. 251–262, 2016.
DOI: 10.1007/978-3-319-42417-0_23

position and orientation. Place-cells, a type of neuron encoding for specific locations of the explored environment have been found in the dentate gyrus, CA3 and CA1 regions of the rodent hippocampus [15]. Since the discovery of place-cells, the hippocampus has been considered to play a fundamental role in spatial navigation and representation.

Place-cells receive their excitatory projections from the entorhinal cortex where grid-cells are found [1]. The firing activity of grid-cells in the medial entorhinal cortex is typically arranged in an hexagonal pattern tessellation covering the explored environment. Grid-cells encode the environment through multiple scales that progressively increase along the dorsal-ventral axis of the medial entorhinal cortex, allowing to represent space at multiple levels of resolution. Grid-cells have been suggested to serve as a metric component necessary for place-cells in the proper hippocampus to tune their activity to specific locations of the explored environment. Indeed, computational models of place-cell formation have relied on the input-output transformation of grid-cells signals through inhibitory network competition processes [14].

An universal metric system combined with place-specific firing activity allowing to decode spatial positions is a crucial feature for optimal spatial navigation. However, in order to reach goal-locations, one needs to be able to plan future trajectories and make sense of previously encoded spatial memories. Hippocampal place-cells are not only active at their specific locations, but they have been also found to fire in spatial sequences during both locomotion, a phenomenon called theta-sequences [5], and stationary periods, the so-called sharp-wave ripples [17]. Place-cells fire in spatial sequences when the animal is at decision points [5] of a multiple t-maze configuration. Such mechanism of spatial representation encoding future trajectories, allowing the animal to decide which action to take, is a major advantage of foraging animals when compared to the current capabilities of their robotic counterparts.

Despite a myriad of studies regarding grid-cells formation and functionality, the grid-cells metric system is still far from being exploited in robot navigation, path-planning and the so called "mental traveling". Because grid-cells activate based on attractor dynamics, show periodicity, and are modulated by the animal movements, their primary role in path integration is widely accepted and such type of signaling is a powerful candidate to build spatial representations of the explored environments. However, it has been recently proposed that a second function of grid-cells is to perform "mental traveling" processes [10,11]. Because their activity is independent of environmental sensory cues, they constitute a potentially robust context-independent metric system for dynamic environments. Furthermore, a mechanism has been proposed in which grid-cells potentially drive vector-based navigation to both known and unexplored locations [11].

However, the mechanisms generating such sequences of future locations are still unclear. "Mental travel" implies a neural signal of spatial navigation without actual motor movement [10,11]. Thus, "mental travel" refers to an imagery process of translating ones' body in space while remaining stationary. Because grid-cells activity is independent of environmental sensory cues or the spatial context, they constitute a potentially robust context independent metric system to 'virtually' explore dynamic environments. Given their underlying low-dimensional continuous attractor dynamics in synaptic connectivity [16], as well as their ramp-driven spiking behavior [4], grid-cells do not need intrinsic learning in order to project their activity into imaginary locations. Thus, at the computational level, motion related signals such as orientation and direction are sufficient to generate grid-like activity [6]. Hippocampal theta oscillations have been suggested to drive the hippocampal representational mode [10]. That is, during the first half of a theta cycle, the animal computes its current location, while during the second half, the animal computes future trajectories, through a 'virtual speed' input arriving to this representational system. However, the origin of this virtual speed signal the role of hippocampal theta oscillations in driving the hippocampal mode remains unclear.

Here, we argue that septal/hippocampal theta generated signals function as a representation mode switching mechanism and that the *virtual* speed signal for performing "mental travel" is encoded in the actual behavior of the animal. Thus, there is no need of extra anatomical projections in order to perform simulated vector based navigation. Previously, we have presented an hippocampal based robotic system for spatial representation where populations of head-direction-cells encoding for robot's orientation and grid-cells performing path-integration were sufficient to drive place-cells encoding the robot's current location [2,3]. In this study, we present an improved implementation of the previously presented hippocampal system. Here, grid-cells not only perform path-integration used for the place-cells location encoding, but also allow the robot to anticipate future locations through "mental travel" processes. Being able to simulate a trajectory without physically navigating within the environment allows to constantly update planned trajectories as well as predict where do novel routes will lead to. Thus, it seems reasonable to develop such mechanisms to be employed on real world foraging robots.

2 Methods

2.1 Neural Populations

In order to simulate future trajectories, a foraging robot has to be able to understand the properties of an environment and being capable of situating itself in space. To do so, we first implemented rate-based grid-cell neurons equally distributed among 6 modules (N = 6000), each with different grid-scale properties (see [2,3,6] for a detailed explanation of the model's implementation). Grid-cells activity was initialized with random activity between 0 and 1/N (number of neurons in each grid-cell module) and were modulated by the speed-vector of the robot movements at each time step. As in [6], the activity of each cell at time t was defined using a linear transfer function given by:

$$B_i(t+1) = A_i(t) + \sum_{j=1}^{N} A_j(t)W_{ij} \qquad (1)$$

where, A_i is an average normalization mechanism to maintain stability of the network (see [6] for details), i and j are indexes of cells in the network and W_{ij} is the synaptic weight between cell i and j. Because grid-cells are periodic and modulated by the robot's velocity, their synaptic weights were set based on low-continuous attractor mechanisms. The attractor dynamics were defined through the synaptic distribution within each grid-cell modules as:

$$W_{ij} = I exp \left(-\frac{\|c_i - c_j\|^2_{tri}}{\sigma^2} \right) - T \tag{2}$$

where, $I(= 0.3)$ is the synaptic strength intensity parameter, $\sigma(= 0.24)$ modulates the synaptic distribution Gaussian size, and $T(= 0.05)$ is a parameters defining inhibitory and excitatory connections in the synaptic distribution, tri regulates the bump of activity to be moved along a toroidal network accordingly with the robot's movements (see [6] for more details).

A second neural population was set to reproduce place-cells like activity over the environment. That is, each cell should increase its firing rate at specific locations of the explored environment. If so, the robot will be able to develop a spatial representation and awareness of its current position at each time.

Similarly to [14], we implemented 10000 rate-based place-cells and their activity was modulated by the excitatory inputs arriving from multiple grid-cells along the dorsal-ventral axis with progressive scales:

$$F_{place} = I_{grid} \cdot H(I_{grid} - (1 - K) \cdot grid^{max}(r)) \tag{3}$$

where, H is a Heaviside function, $grid^{max}(r)$ is the maximum activity projected from a grid- to a place-cell at position r, and $K(= 0.9)$ determines the threshold for every place-cell to compete through the winner-take-some process:

$$F^i_{place} = \begin{cases} F^i_{place}, & \text{if } F^i_{place} \geq place^{max} \cdot (1 - K) \\ 0, & \text{otherwise} \end{cases} \tag{4}$$

where F^i_{place} is the activity of place-cell i and $place^{max}$ is the activity of the place-cell with higher firing rate at each time. Because we did not impose a specific synaptic weights distribution from grid- to place-cells, place-cells were susceptible of having multiple and spread firing fields. In order to refine place-cells receptive fields, during the exploration phase we set a potentiation rule determining that when the amount of active place-cells would exceed 25 % of total number of place-cells, a Hebbian learning mechanism [13] would be applied to the 5 % strongest activated cells, while the grid- to place-cells synaptic weights of the remaining 95 % of neurons would be suppressed.

We have described the emergence of VTE events during deliberative stages of task contingencies learning [8,9]. Because such behavior is involved in goal-directed navigation, usually towards a reward location, the hippocampal formation alone would not be sufficient to trigger VTE. We thus implemented an approximation of striatal cells encoding for reward locations [12]. Reward-cells were modeled through rate-cells activity with bidirectional connectivity between hippocampal place-cells. Learning association between the place-cells population vector and reward-cells was performed through a single layer perceptron. Thus, two sources of input arrive to reward-cells. On one hand, every time the simulated agent crosses a reward location, reward-cells are set to high firing rate. On the other hand, place-cells activity is also projected into reward-cells, allowing to decode reward locations during "mental traveling" towards the rewarded arm of the maze (see Fig. 1C). A single hippocampal-reward perceptron circuit was implemented and output activity of its reward-cell was considered as a mean-field approximation of striatal output. Synaptic weights between reward- and place-cells were initially set to random values ranging from 0 to 1, with an additional bias unit. The learning rate (γ) of hippocampal-reward associations was set to 0.7. During learning, reward cells activity was set as:

$$Reward_r = \sum_{j=1}^{N} PC_j \cdot W_{rj} \qquad (5)$$

where PC_j corresponds to the activity of every place-cell connected to reward-cell (r), and W_{rj} is the synaptic weight between these two cells. At every update of the reward-cell activity, the error between the expected activity provided from the reward signaling when the agent crossed a reward location and the actual reward-cell activity was given by:

$$Reward_{Error} = Reward_{Expected} - f(Reward_r) \qquad (6)$$

where, $f(Reward_r)$ is a step-function given by:

$$f(Reward_r) = \begin{cases} 1, & \text{if } Reward_r \geq 0, \\ 0, & \text{otherwise.} \end{cases} \qquad (7)$$

The synaptic weights between place- and reward-cells was finally updated given by:

$$W_{rj} = \gamma \cdot Reward_{Error} \cdot PC_j \qquad (8)$$

where γ is the learning rate for the synaptic weights update.

2.2 Robotic Setup

Our setup consisted of a virtual agent performing a navigation task towards previously explored locations. First, in order to test whether our implementation was properly set, we allowed the agent to randomly explore an open field arena. Thus, we could assess grid- and place-cells rate maps (Fig. 2). To validate our system we have configured a virtual environment resembling a T-maze alternate task. At every time the virtual agent arrived to the T-maze decision point, it could orient towards either left or right sides of the maze until eventually decide to move on to the orientation correspondent alley. Trajectories from decision-point until arriving back to that same location were programmatically defined.

Given the structure of our neural network implementation resembling the hippocampal formation and striatal reward mechanisms, the simulated agent could modulate its decisions based on the reward-cells activity every time it would "mental travel" towards a specific alley of the maze. Thus, after learning place-reward contingencies, the agent's *go signal* was triggered every time reward-cells would exceed an activity rate of 0.75. Because at the decision point the agent was set to randomly orient to the maze alleys and perform grid-cells "mental travel" simulation, it would affect both place- and reward-cells activity, by means of virtually traveling towards a reward-location. Thus, the agent could learn the best action to take in order to get its reward. A maximum of 8 "mental travel" simulations (4 towards the right arm) was set so that the agent would not get stuck at the decision-point when no *go signal* was triggered. In those trials, the agent randomly picked a direction to move on. Moreover, in this study, we considered as VTE events, those trials at which the agent could not take a decision based on the reward-cell activity.

3   Results

In order to quantify the effects of grid-cells "mental travel" in VTE behavior, we implemented the described hippocampal model in a simulated agent. Six populations of grid-cells with distinct scales were generated (Fig. 2-bottom). A total of 6000 cells were made available to project excitatory input to the place-cells population where individual cells place-fields showed to be specific to unique locations of the explored environment (Fig. 2-top).

Open Field Arena. In the first quantification session, we set the agent to randomly explore the open field arena where no target was assigned and contact with environmental borders was avoided. Figure 2 illustrates rate-maps examples of both grid- and place-cells obtained from this exploratory session. Thus, we could assess that a grid-to-place spatial representation from our hippocampal implementation was properly functioning. Rate-maps of four grid-cells belonging to different populations depict changes in grid-scale (Fig. 2), mimicking the dorsal-ventral axis of medial entorhinal cortex layer 2 [6,7].

T-maze Test. We then tested our simulations in an alternate task mimicking a T-maze configuration (Fig. 3). Rate-maps of grid-cells during exploration of the T-maze were obtained. As expected, firing-fields from single cells covered multiple regions of the maze, implicating that position could not be assessed from single cells activity.

We have described our hippocampal implementation for a simulated agent performing navigational tasks. Nonetheless, hippocampal specific activity *per se* is not sufficient to trigger VTE. Additionally to our previous implementation [2,3], we added a population of rate-based neurons mimicking the role of striatal cells. Thus, we expected that by learning associations between place-cells population vector and reward-cells activity when the agent is at the reward-location,

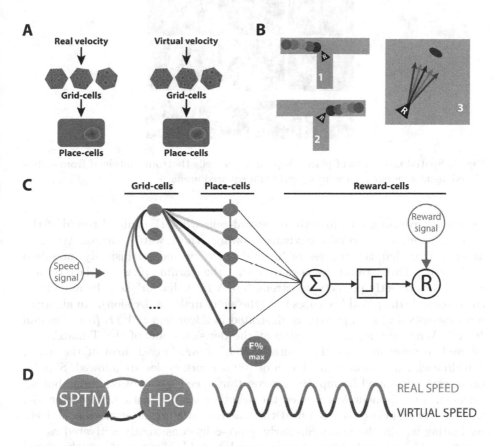

Fig. 1. Spatial representation and 'virtual' exploration. (**A**) Grid-cells receive either real velocity inputs from the robot movements or 'virtual' velocity inputs for decoding future positions. (**B**) Situations of 'mental' traveling. At decision points, the robot simulates exploration towards left (1) and right (2) possible trajectories. In an open field arena (3), the agent probes optimal orientations to achieve a target location. (**C**) Artificial neural network implementation. Speed signal input grid-cells. Each grid-cell is synaptically connected to every cell in its network forming an all-to-all connectivity matrix per population. The synaptic weights are modulated by the Cartesian distance of each cell pair in the population matrix, as illustrated in the cell at the top (color code ranges from inhibitory to excitatory weights). Grid-cells activity is projected to the population of place-cells where competition through inhibitory interneurons tune individual cells to specific spatial locations. Place-cells population vector is than projected into reward-cells such that a reward signal is obtained when place-cells activity match with activity at reward location. (**D**) Illustration of hippocampal/septal theta generation and its correspondent real/virtual speed alternation mode. (Color figure online)

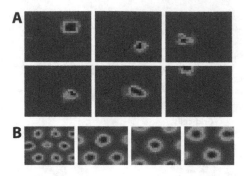

Fig. 2. Spatial rate maps of place- (top) and grid-cells (bottom) obtained from a simulated agent randomly navigating withing a squared arena.

we would have been able to activate reward-cells while the animal rested at the decision-point, but virtually navigated towards the reward-location. We have hypothesized that the process of 'mental travel' would intrinsically modulate VTE events. Furthermore, we suggested that the learning phase of place-reward associations directly predicts the strength of VTE behavior (i.e., the amount of turns towards the possible choices until the agent makes a decision). An algorithmic description of such process is illustrated in *Algorithm 1: VTE from "mental travel"*. When the agent is situated at the decision-point of the T-maze, it is allowed to perform a set of "mental travels" towards each arm of the maze. Within each simulation, a maximum of 100 network cycles are allowed. Simulations were performed by inputting a constant speed scalar of one unit, but its orientation component was defined the agent's head orientation. At each cycle, the activity of grid-cells resembled their activity when the agent was actively navigating within the maze. Similarly, place-cells consistently activated as the agent would be crossing their respective place-field. Figure 4 shows the correlation of place-cells population activity during 'mental travel' events for the rewarded location, suggesting a proper decoding of future outcomes. Because reward-cells activity has been associated with specific population vector activity of hippocampal place-cells, by virtually crossing the rewarded location, also reward-cells would become active, and thus triggering the *go signal* of the decision process. In case the amount of simulation events reached the maximum number of allowed choices, the agent's orientation was set to towards the correct arm and navigation started.

Changed Task Contingencies. We have hypothesized that the learning process between place- and reward-cells would modulate VTE events when an animal is at the decision-point of a T-maze task where the reward is placed in one arm of the maze. Moreover, as observed in Tolman's studies [8], VTE events increase after changing of task contingencies, that is, varying the starting position or the location of the reward. In order to assess that our implementation of hippocampal and striatal networks in combination with mechanisms of "mental travel"

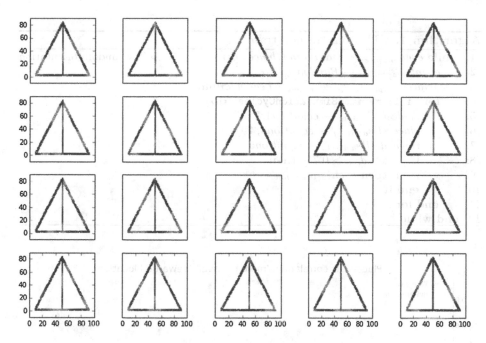

Fig. 3. Rate maps of grid-cells obtained from the agent performing an alternate task within a t-maze configuration. (Color figure online)

would reproduce similar effects in a simulated navigational task, we have tested a virtual agent in situations where task contingency would be manipulated. A total of 14 sessions were performed, VTE events quantified and averaged across sessions. Because our agent began navigation with randomly assigned weights between place- and reward-cells, we would expect that VTE events would be high at the initial stage of exploration. Indeed, that was confirmed at the very first trial where 'mental traveling' was performed towards both possible decision orientations and no activation of reward-cells above 0.7 of its maximum rate (1.0) was obtained (Fig. 5). However, because the learning rate (γ) of our learning mechanism between place- and reward-cells was set to 0.75, the simulated agent was able to perform a decision toward the correct arm at the third trial. Again, studies on task contingency changes have shown that after altering the reward location to the opposite site of the maze makes the animal engaging VTE behavior, a period where rodents look back and forth towards possible routes to take. We have tested increases in look-ahead events after removing the reward from the learned location and place it at the opposite arm of the maze. As expected, the amount of VTE events increased after reconfiguration of the task, an effect from dissociating the population vector activity of place-cells and the absence of reward-related activity (Fig. 5).

Algorithm 1. VTE from "mental travel"

```
 1: while event_number < maximumChoices do          ▷ Try random orientation
 2:     event_orientation ← Random_Left||Right
 3:     gridcells_input ← speed_virtual + headorientation
 4:     for i 1: maximumSimulationCycles do
 5:         gridcells_activity ← equation1, 2
 6:         placecells_activity ← equation3, 4
 7:         rewardcells_activity ← equation5
 8:         if reward_cell > 0.75 then
 9:             return event_orientation
10:         end if
11:     end for
12: end while
```

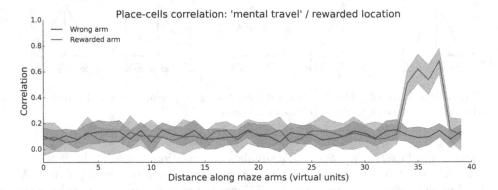

Fig. 4. Correlation of place-cells population vector along both left and right arm of the maze at decision point. (Color figure online)

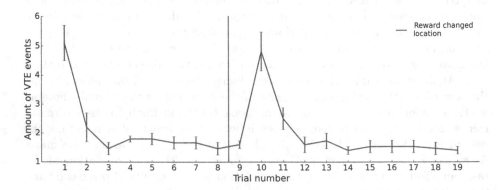

Fig. 5. Modulation of VTE events during learning (early trials) and after reward replacement (trial 9). (Color figure online)

4   Discussion

Situating oneself in space and being able to reach target locations is a major survival feature that every organism must accomplish. The rodent hippocampus is an example of a great deal of resources to be deployed on real world foraging robots. Electrophysiological studies have revealed a myriad of cell type to be encoding spatial properties. Head-direction cells in both thalamic anterodorsal and hippocampal postsubicular and early medial entorhinal cortex layers, provide the proper hippocampus with a *sense* of global orientation [18]. From a computational perspective, orientation signals help to maintain a relationship between encoded locations and build a trajectory scheme to reach target locations [3]. Similarly, grid-cells found in the medial entorhinal cortex allow the animal to cover explored environments with tessellating patterns and obtain a metric system with global coordinates facilitating the encoding of specific locations [1]. Finally, place-cells are the ultimate positional encoding mechanism, allowing the animal not only to situate itself in space, but also, to decode future trajectories to reach a target [5,17]. The implementation of grid- and place-cells in navigating machines has been accomplished. However, there is more that the hippocampal region can offer to robotics. In this study, we have brought recently developed theories of the grid-cells functionality into a simulated robotic platform. Specifically, we have presented a mechanism in which by 'mentally traveling' towards future trajectories with minimal computational effort and zero energetic cost, a simulated agent is able to: (1) predict future locations through specific trajectories; (2) constantly monitor the value of a chosen route and; (3) 'mentally' explore never taken paths. We have shown that a simulated agent could decode what was about to encounter by mentally probing its options on a T-maze configuration scenario. Such mechanism might be of great advantage for planning routes in highly dynamic environments, where unknown routes must be considered. In this study we have combined a previously described implementation of the hippocampal formation for space representation [2] and path-planning [3]. Here, we have extended our previous work by integrating a neural population mimicking the striatal rewarding circuitry motivating the animal towards specific spatial locations. Our results suggest a mechanism where neural activity in populations encoding spatial locations directly modulate animal decision-making process and consequent VTE behavior. Our implementation of grid-cells "mental travel" resembles the activation of theta-sequences in hippocampal formation found in [5]. Further investigation would reveal neural dynamics in depicting future trajectories during quiescent and ongoing navigation.

Acknowledgments. The research leading to these results has received funding from the European Research Council under the European Union's Seventh Framework Programme (FP7/2007-2013)/ERC grant agreement 341196; CDAC.

References

1. Hafting, T., et al.: Microstructure of a spatial map in the entorhinal cortex. Nature **436**(7052), 801–806 (2005)
2. Pata, D.S., Escuredo, A., Lallée, S., Verschure, P.F.M.J.: Hippocampal based model reveals the distinct roles of dentate gyrus and CA3 during robotic spatial navigation. In: Duff, A., Lepora, N.F., Mura, A., Prescott, T.J., Verschure, P.F.M.J. (eds.) Living Machines 2014. LNCS, vol. 8608, pp. 273–283. Springer, Heidelberg (2014)
3. Maffei, G., Santos-Pata, D., Marcos, E., Sanchez-Fibla, M., Verschure, P.F.: An embodied biologically constrained model of foraging: from classical and operant conditioning to adaptive real-world behavior in DAC-X. Neural Netw. **72**, 88–108 (2015)
4. Domnisoru, C., Kinkhabwala, A.A., Tank, D.W.: Membrane potential dynamics of grid cells. Nature **495**(7440), 199–204 (2013)
5. Johnson, A., David Redish, A.: Neural ensembles in CA3 transiently encode paths forward of the animal at a decision point. J. Neurosci. **27**(45), 12176–12189 (2007)
6. Guanella, A., Kiper, D., Verschure, P.: A model of grid cells based on a twisted torus topology. Int. J. Neural Syst. **17**(04), 231–240 (2007)
7. Brun, V.H., et al.: Progressive increase in grid scale from dorsal to ventral medial entorhinal cortex. Hippocampus **18**(12), 1200–1212 (2008)
8. Tolman, E.C.: Cognitive maps in rats and men. Psychol. Rev. **55**(4), 189 (1948)
9. Redish, A.D.: Vicarious trial and error. Nat. Rev. Neurosci. **17**(3), 147–159 (2016)
10. Sanders, H., et al.: Grid cells and place cells: an integrated view of their navigational and memory function. Trends Neurosci. **38**(12), 763–775 (2015)
11. Bush, D., et al.: Using grid cells for navigation. Neuron **87**(3), 507–520 (2015)
12. van der Meer, M.A., et al.: Triple dissociation of information processing in dorsal striatum, ventral striatum, and hippocampus on a learned spatial decision task. Neuron **67**(1), 25–32 (2010)
13. Hebb, D.O.: The Organization of Behavior: A Neuropsychological Theory. Psychology Press, New York (2005)
14. de Almeida, L., Idiart, M., Lisman, J.E.: The input output transformation of the hippocampal granule cells: from grid cells to place fields. J. Neurosci. **29**(23), 7504–7512 (2009)
15. O'Keefe, J., Dostrovsky, J.: The hippocampus as a spatial map. Preliminary evidence from unit activity in the freely-moving rat. Brain Res. **34**(1), 171–175 (1971)
16. Yoon, K.J., et al.: Specific evidence of low-dimensional continuous attractor dynamics in grid cells. Nat. Neurosci. **16**(8), 1077–1084 (2013)
17. Pfeiffer, B.E., Foster, D.J.: Hippocampal place-cell sequences depict future paths to remembered goals. Nature **497**(7447), 74–79 (2013)
18. Taube, J.S., Muller, R.U., Ranck, J.B.: Head-direction cells recorded from the postsubiculum in freely moving rats. I. Description and quantitative analysis. J. Neurosci. **10**(2), 420–435 (1990)

Insect-Inspired Visual Navigation for Flying Robots

Andrew Philippides[1]([⊠]), Nathan Steadman[1], Alex Dewar[2], Christopher Walker[1], and Paul Graham[2]

[1] Department of Informatics, Centre for Computational Neuroscience and Robotics, University of Sussex, Brighton, UK
{andrewop,N.Steadman,Chris.Walker}@sussex.ac.uk
[2] School of Life Sciences, Centre for Computational Neuroscience and Robotics, University of Sussex, Brighton, UK
{A.Dewar,paulgr}@sussex.ac.uk

Abstract. This paper discusses the implementation of insect-inspired visual navigation strategies in flying robots, in particular focusing on the impact of changing height. We start by assessing the information available at different heights for visual homing in natural environments, comparing results from an open environment against one where trees and bushes are closer to the camera. We then test a route following algorithm using a gantry robot and show that a robot would be able to successfully navigate a route at a variety of heights using images saved at a different height.

Keywords: Visual navigation · Insect-inspired robotics · Visual homing · UAVs · Flying robots

1 Introduction

Navigation is a vital ability for animals and robots. In the latter, where GPS is unavailable or unreliable, visual homing methods can be used. For flying robots, the limitations on payload mean that it is important that any algorithms are efficient in terms of computation as this reduces power consumption. It is therefore natural that engineers turn for inspiration to insects such as ants and bees [1], who use vision to navigate long distances through complex natural habitats despite limited neural and sensory resources [2–5]. In this spirit, we have previously developed a view-based homing algorithm based on the behavior of desert ants. While this algorithm has been tested for ground-based robots in simulation, here we test the ability of this algorithm to generalize to a flying robot.

The first generation of biomimetic algorithms for insect-like navigation were inspired by the fact that an insect's use of vision for navigation is often a retinotopic matching process (Ants: [2, 4, 5]; Bees: [3]; Hoverflies: [6]; Waterstriders: [7]; Review: [8]) where remembered views are compared with the currently experienced visual scene in order to set a direction or drive the search for a goal. Insect-inspired robotic models of visual navigation have thus been dominated by snapshot-type models where a single view of the world, as memorized from the goal location, is compared to the current view in order to drive a search for the goal ([3]; for review see [9]). Importantly, Zeil and colleagues [10, 11] used simple metrics based on the sum-square differences in pixel

© Springer International Publishing Switzerland 2016
N.F. Lepora et al. (Eds.): Living Machines 2016, LNAI 9793, pp. 263–274, 2016.
DOI: 10.1007/978-3-319-42417-0_24

values between images, and thus agnostic of the details of the model used, to analyse the range over which a single image can be used for visual homing.

This work and others showed that, while snapshot models work in a variety of environments, these approaches are limited in that they generally allow for navigation only in the immediate vicinity of the goal [10–13]. We therefore developed a familiarity-based model of route navigation in which individual views are used as a visual compass to recall the direction the agent was facing at that point, rather than the direction to the location as in most snapshot models. This allowed us to develop a holistic route memory which allowed a ground-based agent to navigate through simulated natural environments with route showing many characteristics of ant routes [14]. Following on from this work, similar models have been shown to work with a biologically plausible neural network in simulation [15] and on a robot in a natural environment [16].

In all these previous works, however, all images were taken from the ground level. To assess the whether this model can also be used for a flying robot, we thus need to test the model with images gathered from different heights. To do this we follow the methods of [10, 11, 17] to first assess the extent over which single images gathered from different heights through two natural environments can provide information for visual homing. We show that in line with Zeil [10], a snapshot stored at one height can successfully be used as either an attractor snapshot or visual compass for an agent travelling at a different height. We then test our route navigation algorithm with similar data gathered using a high precision gantry robot in an indoor environment. The success of the algorithm provides proof of principle that an aerial robot could use our route navigation algorithm despite being at a different height to that at which the original route was travelled.

2 Measuring the Informational Content of Natural Images from Different Heights

To get an understanding of the information that is present in images for homing, we follow the procedures of [10, 11, 17] and use simple metrics to estimate the range over which a single image can be used either as an attractor type snapshot, to recover a direction towards a goal, or as a visual compass, to recall the heading at which the agent was facing when the goal image was stored.

2.1 Data Collection and Image Processing

The process starts by capturing sets of images using a panoramic imaging device within two natural environments at different heights (Fig. 1). Data was collected using a Kodak PixPro SP360 camera to take panoramic images at regular intervals at a range of heights from straight-line routes in two locations. The heights investigated were 0 cm, 40 cm, 70 cm, 100 cm, 150 cm and 200 cm. With the exception of 0 cm, when the camera was placed directly upon the ground, images were taken from a tripod with an in-built spirit gauge, allowing each image to be taken at a level setting. In addition, it allowed us to examine the impact of pitch by including an image taken with a 45° downward tilt at each of the heights and locations investigated. The tripod was aligned to a common heading to ensure all images from each location were taken facing in the same direction.

Fig. 1. Sample images from two locations. A: Location A – open transect through Queen's Park, Brighton. B: Location B – transect through wooded copse in East Brighton Park. Images unwrapped from fish-eye lens via PixPro software.

Location A: The initial dataset was collected from a field in Queen's Park, Brighton. These transects covered a distance of 36 m taken in 2 m increments through an open area. This location was selected as it is an open environment without any nearby trees or foliage and in which the trees which constitute the skyline were at least 50 m from the camera location with the majority being several hundred metres away (Fig. 1A). This means the visual data changed slowly relative to movement along the route.

Location B: The second dataset was collected from a small copse in East Brighton Park, Brighton. Transects of 12 m were covered in 1 m intervals. The treeline was at no point further than 5 m from the camera in any of the images taken. The decreased interval between images was used to compensate for the small distance of the route. Due to the forested nature of this site (Fig. 1B), the route passed noticeably between shade and light with distinct lens flare on certain images taken. The images were not manipulated to compensate for this.

The images were processed in three ways: they were first unwrapped from a fish-eye image to a panoramic image using the software provided with the PixPro camera. Images were then scaled-down to 408 × 1632 pixels, converted to grey-scale and the sky homogenized to a uniform value of 255 (white). The first two stages were achieved with Matlab functions imresize and rgb2gray, respectively. Sky homogenization was also performed in Matlab automatically and quite roughly by thresholding the blue-channel at a value of 170 (determined by trial and error) and setting above 170 values to 255. Occasionally, parts of clouds were below the 170 threshold, and so we also converting any isolated 'not-sky' pixels which were not connected to the ground to 255. Sky homogenization is necessary as the light gradient can provide a very strong, but spurious cue, by which homing algorithms can gain information.

2.2 Assessing the Region in Which an Image Can Be Used for Visual Homing

By measuring the image difference between a reference image and images from surrounding points, we can build an image difference function (IDF) that shows how images change with distance from a goal view [10]. The image difference between two images X and Y is defined as:

$$IDF(X, Y) = \frac{1}{P} \sum_i \sum_j |X(i,j) - Y(i,j)| \tag{1}$$

where X(i,j) is the pixel in the i'th row and j'th column of image X and P is the number of pixels. Notice that this value is dependent on the alignment of image X and Y to a common heading. The IDF for an image from Location A is shown in Fig. 2.

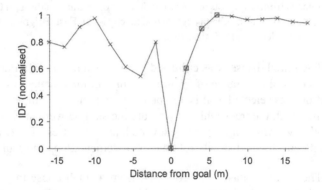

Fig. 2. An example IDF from Location A at a height of 100 cm. IDF values are shown as a proportion of the maximum and plotted against distance from the goal. x's mark points where images were taken from. Points within the catchment area are marked with red squares. Noise in the IDF limits the catchment area to the left of the goal while to the right it extends 6 m.

To assess the range over which a single image can be used for navigation, we quantify the region around a goal image in which the difference between images increases with increasing distance from the reference image. The presence of an increasing IDF is significant as it shows that the information needed for view-based homing is available [10, 17]. We therefore define the catchment area (CA) of the IDF as the region where the IDF is generally increasing. Note that as we are applying this to a straight line transect this is not strictly an 'area' but an indication of the distance over which a single image can be used. Nevertheless we use the term 'catchment area' to be consistent with the generally used terminology. To estimate this region, we take the number of consecutive locations spreading out from the goal position where the IDF is strictly increasing relative to the direction of movement from the goal (Fig. 2). This is a lower bound on the region within which an image could be used for homing as a single uneven/bad image can introduce a spurious minimum in this discrete data (e.g. point at distance -2 in Fig. 2). While this could be overcome by taking more images and smoothing the data in some way, as we are using the same metric to compare two data sets, we use the raw data as a lower bound. Likewise, the fact that the distribution of catchment areas is asymmetric (goal positions at the centre have CAs that can extend in both directions unlike those at the edge) is not so important as we are comparing the same positions for different heights.

Figure 3A-B shows that images gathered at different heights do carry sufficient information for visual homing. While there is quite a spread of results at each height from different goal positions, a good proportion of the route can be traversed using a single image. As expected, a greater distance could be traversed with a single image in Location A than Location B. Interestingly, in the more cluttered Location B, the performance increased with increasing height. The cause of this can be seen in Fig. 4. When there are nearby objects, a limited field of view in the elevation access means that close objects dominate the skyline more for lower heights, leading to inaccuracy in the matching process. We also tested routes taken with a level camera against goal images where the camera was angled at 45° to the ground plane (Fig. 3 C-D). The idea of this was to see whether the effect of pitch changed with changing height. For this data set, the impact of pitch was detrimental to the route coverage as has been seen in [18] and did not vary with height, but at least in Location A, there remains some information for homing.

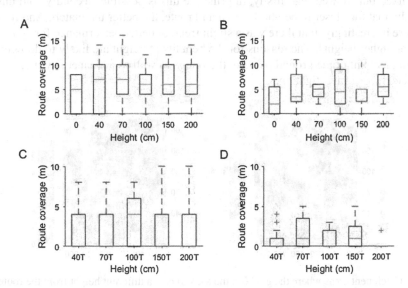

Fig. 3. Catchment areas in locations A (A and C) and B (B and D). X-labels identify the height of the route. In the top row (A and B), data is shown for goal images taken from the route tested. In contrast, height labels followed by a 'T' (bottom row, C-D) indicate that the goal images only were captured at a 45° tilt. Red lines indicate the median catchment area, whole boxes cover the 25th and 75th percentile and whiskers extend to 1.5 times the inter-quartile range. (Color figure online)

We next tested whether images that were taken at a particular goal height, could be successfully used to home for a robot travelling at a different height (Fig. 5). Results

Fig. 4. The impact of increased height at Location B. **A** - Location B taken at a height of 40 cm. **B** - Location B at a height of 150 cm. The increased elevation in B results in a more distinct outline of the treeline to the left of the image. This in turn leads to better IDF comparisons as there is now more visual information.

were mixed but showed that: firstly, in principle this is possible; secondly, and unsurprisingly, that the closer in height the goal and route, the better the match; And finally, and more interestingly, that there was a slight trend of better performance for goals and routes at higher heights. The reasons for the benefits of height are likely to do with the lessening amount of the ground plane in the image as well as a cleaner skyline, though this needs further investigation.

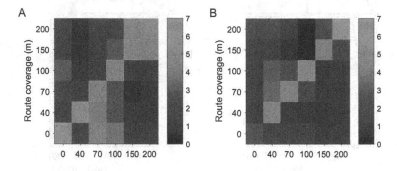

Fig. 5. Catchment areas where the goal (on the x-axis) is at a different height from the route (y-axis). Colours show the mean catchment area across all goal positions in Location A (A) and B (B). (Color figure online)

2.3   Assessing the Region Over Which Views Can Be Used to Recall a Heading

While the analysis of the catchment area of natural images demonstrates that the information required for visual navigation is present, to more directly link it to our familiarity based algorithm we need to repeat the analysis using views as a visual compass. We therefore next determine what we termed rotational catchment areas (RCAs) [17], which indicate the region in which an agent would be able to use a goal image to recall the heading it was facing when the goal images was stored.

To determine the RCA, one first calculates the rotational IDF (RIDF) by evaluating the IDF between a reference image and the current image rotated (in silica) in steps of 1° of azimuth resulting in a 1 × 360 RIDF (Fig. 6). The minimum value in the RIDF defines an orientation of the current image which gives the closest match with the reference image. In the vicinity of the reference image these orientations will be similar to the reference image orientation [10] and so can be used to recall a heading or a movement direction for a forward facing agent [17]. We define the rotational catchment area as the region spreading out from the location of the reference image where the minimum in the RIDF is less than 45° from the true orientation of the reference image (Fig. 6).

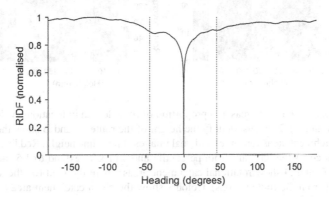

Fig. 6. Sample RIDF using an image from Location A, 100 cm height, as both goal and current image. IDF is normalised to the maximum value and plotted against the degree of azimuthal rotation. Dashed lines mark the 45° threshold region. In this case the best match is at 0° meaning the agent is facing in the same direction as when the image was stored.

Analysis of RCAs for routes and goals at the same height shows that there are strong visual compass cues across the routes (Fig. 7 A-B) with much increased catchment areas compared to the IDF analysis. In Location A, performance increases with height of the route. In Location B the highest routes seem to slightly underperform but this is likely a ceiling effect (the maximum possible RCS is 11 m for Location B. Comparing goals at different heights to routes again there are large catchment areas for the visual compass with much more matching across different heights than was evident in the IDF data. In Location A, there seem to be two broad blocks of matching with all routes from 100 cm down broadly matching while the two highest routes match each other. The picture for Location B is similar but with a more broad matching across heights. While the rather crude measure of mean RCA and the ceiling effect for Location B means more investigation is needed, these results bode well for our route following algorithm to function at different heights as it is based on the visual compass method.

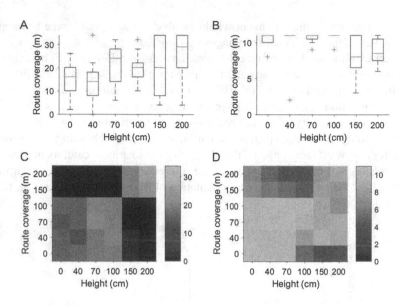

Fig. 7. Rotational catchment areas as a proportion of route length in locations A (left: A and C) and B (right: B and D). X-labels identify the height of the route. A and B show the distribution of rotational catchment areas for routes and goal images at the same height. Red lines indicate the medians, boxes cover the 25th and 75th percentile and whiskers extend to 1.5 times the inter-quartile range. C and D show rotational catchment areas where the goal (on the x-axis) is at a different height from the route (y-axis). Colours show the mean catchment area across all goal positions in Location A (C) and B (D). (Color figure online)

3   Route Navigation

We next test route navigation for images gathered at different heights using the famili-arity-based algorithms described in [14]. However, as we need to test the algorithm over a region, we need to use a more controlled robotic platform and move to an indoor gantry robot.

The first version of the algorithm, dubbed 'Perfect Memory' as it stores all training views, proceeds as follows. The agent first traverses a (here pre-defined) route storing greyscale, panoramic images at set distances (represented by red crosses in Fig. 8). Crucially, the stored views are oriented in the direction in which the agent was "facing" at that point. In order to recapitulate the route, the agent then obtains a best-guess of the correct heading by comparing the current and stored views across rotations, with the rotation yielding the smallest image difference (from among all stored views) taken as the agent's goal bearing. The second method involves the use of an InfoMax familiarity-based neural network [14] and is referred to as Infomax. In this variant, instead of remembering all the views and comparing the current and stored views on an individual basis, a single layer neural network is trained with the stored views to learn the familiarity of the training views using an Infomax training rule [19]. The network is then presented with the current view at a range of rotations and the orientation which yields the smallest activation (and thus the greatest familiarity) is taken as the best-guess heading.

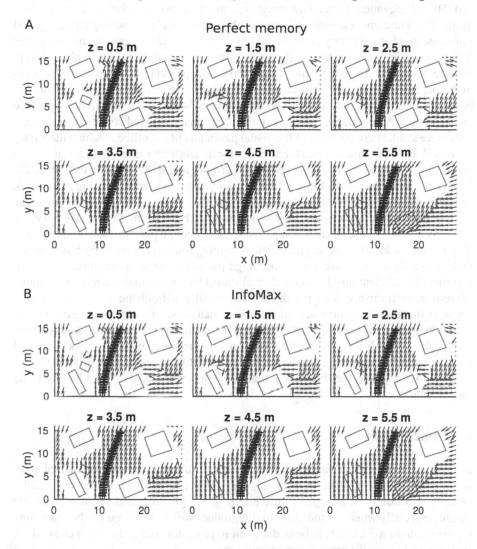

Fig. 8. Estimated headings for different homing algorithms at a range of heights using the robot gantry. **A** – Best-guess headings using the Perfect Memory algorithm. Reliable headings are obtained over a range of locations and the algorithm is robust to changes in height. Blue arrows indicate the headings, red crosses the locations for different stored views and green lines the positions of the cardboard boxes. The scale, as described in the text, is in reality a tenth of that shown in the figure. **B** – Best-guess headings using the InfoMax algorithm. Performance is virtually equivalent to that obtained with the Perfect Memory algorithm, despite requiring far fewer computational resources. Colour scheme is as for A. (Color figure online)

To test these two algorithms, we used an indoor gantry robot, which is comprised of a panoramic camera which outputs a 720 × 60 pixel image covering 360° of azimuth and 50° of elevation (extending roughly equally above and below the camera's 'horizon'). The camera is mounted onto a robot arm capable of moving along x, y and z dimensions to any arbitrary point within our 2.7 × 1.8 × 1.2 m arena. As this was rather small for our purposes, we treated the arena as though it (and the objects within it) were 10 times as large and scaled the agent's movements appropriately. Henceforth distances in this paper refer to the new artificial scale. For this test we placed a number of cardboard boxes of different heights (min = 0.98 m; max = 6.04 m) within the arena to provide visual stimuli. Additionally, the walls of the arena were removed so more distal visual cues were available in the form of the visual panorama of the office in which the gantry is housed. First we acquired the stored views for an arbitrary route within the arena at a separation of 25 cm ($n = 60$; indicated by red crosses in Fig. 8). We then calculated the best-guess headings for the Perfect Memory and InfoMax algorithms (Fig. 8A and B, respectively) for images collected from the gantry at a range of x, y and z coordinates.

Figure 8 shows that robust performance is given by both the Perfect Memory and InfoMax flavours of the visual homing algorithm. Although performance is understandably poorer for locations where the original training route is obscured by boxes, nonetheless, the algorithms mostly yield headings parallel to the training route, indicating that there is sufficient distal visual information (in the office) to drive reliable homing. Moreover, performance does not decay substantially with changes in height, which suggests that even quite dramatic amounts of visual noise in a real-world environment would be unlikely to lead the agent astray. The quality of performance is particularly remarkable for the InfoMax case, where the only data required in memory is a 644 × 644 matrix of weights, as opposed to the entire cache of stored views (58 × 720 × 60 pixels in total). Accordingly, the time taken to compute a heading is also considerably faster for InfoMax.

4   Conclusions

Here we have shown that parsimonious ant-inspired homing strategies are suitable for aerial robots. Training data collected at one height can be used to robustly recall a heading direction from a range of different heights, indicating that variations in flight height, especially when not too close to the ground, will be tolerated by the algorithm. However, more work needs to be undertaken to prove this performance in closed loop systems, and to test the route algorithm outdoors.

The robustness of performance of visual compass based methods across the range of perspectives also gives hope that route memories could be transferred between robots. The use of a holistic route memory, which is easy to transfer between robots, will aid this endeavor and is the subject of on-going work. In particular, we are interested to see if for instance a UAV could be used to follow a path specified by a ground-based robot, or vice versa. This has implications in, for instance, search and rescue operations where different robots could be used for exploration and retrieval.

Finally, it would be interesting to observe how the flight height of bees vary during foraging. However, currently positional data on bees is generally taken from radar where information on height is unavailable, though efforts are on-going to improve these radar systems which could prove very insightful.

Acknowledgements. Thanks to the anonymous reviewers for helpful suggestions. This work was funded by the Newton Agri-Tech Program RICE PADDY project (no. STDA00732). AP also received funding from the European Union's Seventh Framework Programme for research, technological development and demonstration under grant agreement no. 308943.

References

1. Graham, P., Philippides, A.: Insect-inspired vision and visually guided behavior. In: Bhushan, B., Winbigler, H.D. (eds.) Encyclopedia of Nanotechnology, pp. 1122–1127. Springer, Netherlands (2015)
2. Wehner, R., Räber, F.: Visual spatial memory in desert ants. Cataglyphis Bicolor. Experientia **35**, 1569–1571 (1979)
3. Cartwright, B.A., Collett, T.S.: Landmark learning in bees - experiments and models. J. Comp. Physiol. **151**, 521–543 (1979)
4. Wehner, R.: Desert ant navigation: how miniature brains solve complex tasks. J. Comp. Physiol. A. **189**, 579–588 (2003). Karl von Frisch lecture
5. Wehner, R.: The architecture of the desert ant's navigational toolkit (Hymenoptera: Formicidae). Myrmecol News **12**, 85–96 (2009)
6. Collett, T.S., Land, M.F.: Visual spatial memory in a hoverfly. J. Comp. Physiol. A. **100**, 59–84 (1975)
7. Junger, W.: Waterstriders (Gerris-Paludum F) compensate for drift with a discontinuously working visual position servo. J. Comp. Physiol. A. **169**, 633–639 (1991)
8. Collett, T.S., Graham, P., Harris, R.A., Hempel-De-Ibarra, N.: Navigational memories in ants and bees: memory retrieval when selecting and following routes. Adv. Study Behav. **36**, 123–172 (2006)
9. Möller, R., Vardy, A.: Local visual homing by matched-filter descent in image distances. Biol. Cybern. **95**, 413–430 (2006)
10. Zeil, J., Hofmann, M., Chahl, J.: Catchment areas of panoramic snapshots in outdoor scenes. J. Opt. Soc. Am. A: **20**, 450–469 (2003)
11. Stürzl, W., Zeil, J.: Depth, contrast and view-based homing in outdoor scenes. Biol. Cybern. **96**, 519–531 (2007)
12. Smith, L., Philippides, A., Graham, P., Baddeley, B., Husbands, P.: Linked local navigation for visual route guidance. Adapt. Behav. **15**, 257–271 (2007)
13. Smith, L., Philippides, A., Graham, P., Husbands, P.: Linked local visual navigation and robustness to motor noise and route displacement. In: Asada, M., Hallam, J.C.T., Meyer, J.-A., Tani, J. (eds.) SAB 2008. LNCS (LNAI), vol. 5040, pp. 179–188. Springer, Heidelberg (2008)
14. Baddeley, B., Graham, P., Husbands, P., Philippides, A.: A model of ant route navigation driven by scene familiarity. PLoS Comput. Biol. **8**(1), e1002336 (2012)
15. Ardin, P., Peng, F., Mangan, M., Lagogiannis, K., Webb, B.: Using an insect mushroom body circuit to encode route memory in complex natural environments. PLoS Comput. Biol. **12**(2), e1004683 (2016)

16. Kodzhabashev, A., Mangan, M.: Route following without scanning. In: Wilson, S.P., Verschure, P.F.M.J., Mura, A., Prescott, T.J. (eds.) Living Machines 2015. LNCS, vol. 9222, pp. 199–210. Springer, Heidelberg (2015)
17. Philippides, A., Baddeley, B., Cheng, K., Graham, P.: How might ants use panoramic views for route navigation? J. Exp. Biol. **214**, 445–451 (2011)
18. Ardin, P., Mangan, M., Wystrach, A., Webb, B.: How variation in head pitch could affect image matching algorithms for ant navigation. J. Comp. Physiol. A. **201**(6), 585–597 (2015)
19. Lulham, A., Bogacz, R., Vogt, S., Brown, M.W.: An infomax algorithm can perform both familiarity discrimination and feature extraction in a single network. Neural Comput. **23**, 909–926 (2011)

Perceptive Invariance and Associative Memory Between Perception and Semantic Representation USER a Universal SEmantic Representation Implemented in a System on Chip (SoC)

Patrick Pirim[✉]

Brain Vision Systems, Paris, France
patrick.pirim@bvs-tech.com
http://www.bvs-tech.com

Abstract. USER (Universal SEmantic Representation) is a bio-inspired module implemented in a system on a chip (SoC), which builds a link between multichannel perception and semantic representation. The input data are projected into a generic bioinspired higher dimensional non-linear semantic space with high sparsity. A pooling of these semantic representations (global, dynamic and structural) is done automatically by a set of dynamic attractors embedding spatio-temporal histograms, being drastically more efficient than back-propagation. A supervised learning is used to build the association between the invariant multimodal semantic representations (histogram results) and the labels ('words'). The invariant recognition is achieve thanks to multichannel multiscale dynamic attractors and bilinear representations - imitating brain attentional processes. USER modules can be cascaded, allowing to work at different levels of abstraction (or complexity). Due to its low consumption, small size and minimal price, USER targets deep learning, robotics, and Internet of Things (IoT) applications.

Keywords: USER · Invariant representation · Dynamic attractor · Pattern recognition · Semantic representation · Associative memory · Multi-sensory · Deep learning · Sparsity · Embedded system · SoC · IoT

Introduction. Semantic representations allowed by the invariance of perceptive situations are of tremendous importance for humans. Here we propose a generic framework ascribing a semantic representation to a combination of pluralities of senses, with the following properties: 1. Filtering of the input sensor cues towards three feature maps: Global map (Gm), Dynamic map (Dm) and Structural map (Sm), 2. Extraction of semantic elementary representations through dynamic attractors (DA) on each of the feature maps, 3. Dynamical control of the sensor range in order to obtain perceptive invariance during the first processing, perceptive invariance despite the environment's variability being a prerequisite to

© Springer International Publishing Switzerland 2016
N.F. Lepora et al. (Eds.): Living Machines 2016, LNAI 9793, pp. 275–287, 2016.
DOI: 10.1007/978-3-319-42417-0_25

build a universal single sensory semantic representation, 4. Association of elementary semantic representations into a unique global representation (*Learning* mode). 5. Ability to associate a 'word' to a given perception (*Request* mode) and ability to retrieve the word associated to a given situation (*Get* mode), and 6. Silicon implementation allowing scalability and low power consumption.

1 Electronic Model of Neural Population Activities

Figure 1 shows (from top to bottom):

Biologically-Inspired Implementation. The visual cortex exhibits three remarkable properties: blobs perception, local motion, and line orientation [5]. These three generic functions exist all over the entire cortex [1]. Each of these features is 2D mapped (cortical column organization), with data in w bits as input, into $2w$ bits Global map (Gm), Dynamic map (Dm) and Structural map (Sm). In order to account for the topographical organization of the cortical columns, a Position map (Pm) is added.

Features Extractor. The main difference between biological and electronic implementations lays between the parallel (biology) or serial processing (hardware) However, the time to process the information are equivalent since the fast electronic processing compensates its lack of parallelism. The feature extractors are hard-wired embedded as a set of four filters, i.e. the four maps (with outputs in $2w$ bits):

- Global map (Gm): the filter capitalizes the general instantaneous characteristics of sensory data.
- Dynamic map (Dm): the filter capitalizes their time-change characteristics.
- Structural map (Sm): the filter capitalizes their geometrical characteristics.
- Position map (Pm) is the position of the current data, at each elementary step time, within the actual temporal sample.

By contrast with biological processing that is event-driven (spikes), the electronic processing is time-driven with temporal sequence of frames involving the sub-sequences of spatial informations. It is a global process implementing a temporo-spatial transform [11].

Dynamic Attractor. The $2w$ bits data from the four maps are processed in a bihistogram module. Simultaneously, two classifier modules compare these data, each with two bounds. In each of them, a binary output is validated whenever the data are inside these bounds. The bounds of the first classifier module are updated internally at each end of cycle. Those of the second one are updated externally. The later module has the function of an elementary request (R). A logical AND ("spike") on all eight binary outputs is sent to the four bihistogram modules, as validation signals for their computations. This loop is essential for automatic convergence on the optimal information. In vision for example, the two attributes (for bihistogram computation) are hue and saturation for Gm,

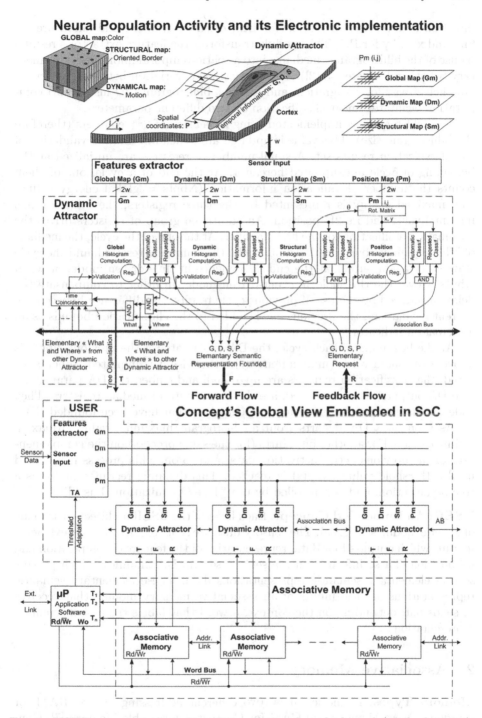

Fig. 1. USER presentation. *Top*: Transcription from biological (cerebral cortex) function to electronic implementation. *Middle*: Scheme of the dynamic attractor (DA). *Bottom*: Concept's global view embedded in SoC.

direction and velocity of motion for Dm, oriented gradient and curvature for Sm, and x and y for Pm. This module transforms the feature extractor's results, by use of the bihistograms' medians, into a spatio-temporal information (elementary 'what' and 'where' [4, 12]). The geometrical covariation between patterns is an efficient way for recognition, and for that bihistogram (2D histogram) computation facilitates linear classification on two different parameters.

The bihistogram is implemented using a memory of 2^{2w} counters (therefore the bihistogram size). This value is updated after each information validation of the classification results set. A small number of registers are initialized at the beginning of cycle and updated hereafter. At the end of a cycle, one of them counts the number of validation information (Nbpts = global energy value). The maximal energy is represented by the three registers, the bihistogram's maximal value and its 2D position. An additional group of registers drives the automatic classification for the first classifier. At the end of the cycle, the medians of bihistogram are outputted. Additionally, the counter is resetted, and the cycle starts over again, either at the end of sequence in time-driven mode, or when Nbpts is over the threshold in event-driven mode. In order to compensate the delay introduced by the use of previous cycle results, the difference between the current median and the previous one is added to the classification bounds as an anticipation indicator.

At the beginning of each cycle, the DA is activated if the previous Nbpts is above a threshold value (minimal energy), otherwise it is inhibited. When activated, the classification registers are upgraded, and embed the DA parameters. The DA outputs, when activated, are the four groups of median registers. They code for the elementary semantic representations that have been founded (F).

As soon as a DA departs from an histogram maximum value (Rmax, a "winner-take-all"), another one starts (for the same process) on the complementary of the previous active activation data, and so on. This process is repeated until a threshold value (Nbpts) is reached. This dynamic recruitment forms a tree-organization, and is controlled by the "Tree Organization" bus (T).

USER: An Embedded Concept. Each layer of the USER architecture is made of generic modules (Fig. 6c). The sensory data are sent to a first feature extractor module which outputs four data, representative of the feature decomposition and its position. These information are sent to several DA modules, which process several elementary semantic representations data. These elementary semantic representations are the inputs of an associative memory module that receives or sends out, depending on the context, a word that labels the global semantic representation.

2   Associative Memory

Memory Types. Memories allow two different addressing modes, RAM for Random Access Memory and CAM for Content-Addressable Memory [7]. Generally only one function is integrated, such as RAM in conventional PC or CAM for Internet use. In our case, the semantic elementary representations are associated with words, the addresses of those data being transparent. We need two memory modules, each one with selectable CAM or RAM functions, for these two kinds of data, a first one linked to word representation, and the second one to elementary semantic representations. Both types of data are mapped by internal addresses.

Figure 2 shows the three features of functioning with the required memory types:

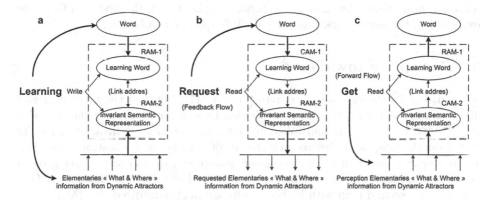

Fig. 2. Associative memory modes. In *Learning* mode, the multi-modes perception is associated with a labelling word, and memorized in the associative memory. In *Request* mode, from a given word, the associative memory delivers the elementary semantic representations for searching the result in the multi-modes perception. In *Get* mode, from the multi-modes perception, the word is extracted.

Learning Mode in order to memorize the association between the elementary semantic representation from the perception and a dedicated word (Fig. 2 *left*). The two data are stored in two memory modules addressed by a common address. The two memory modules are set in RAM.

Request Mode allows to find -for a given word- the corresponding pattern in the perceptual environment (Fig. 2 *middle*). In this case, the first memory module is set in CAM with word as the content and delivers an address to the second memory module, set in RAM function. The output result is the corresponding elementary semantic representations.

Get Mode in order to find the dedicated word from the elementary semantic representations of the perceptual environment (Fig. 2 *right*). A CAM is used in the second memory module with semantic elementary representation as the content, and delivers an address to the first memory module set in RAM. The output result is the corresponding word.

Electronic implementation of the associative memory with two non-volatile memories (either RAM or CAM). Its conception is a delicate process, difficult to implement, involving high energy consumption, as also high price. To account for all theses constraints, a Magneto-resistive Random Access Memory (MRAM, non-volatile) has been selected. The CAM function is then emulated with two memory blocs, one (addressed by the data) stores a virtual address that is used to address the second memory bloc in order to store this data. The complete associative memory is obtained with four memory blocks, controlled by a small microprocessor. The best compromise is obtained when data of W bits are stored in 2^W addresses. With today's market, W=16 or 32 is a good choice (available on shelf), and allows an implementation on one chip of 4 Mb with a 64 K words learning limit. Since the three mode functions (Learning, Request and Get) are executed in the same clock time, the clock frequency can be low.

3   Perceptive Invariance

Vision. Fig. 3a shows a standard acuity curve [17]. Between the fovea axis and 20° retinal eccentricity, the acuity is divided by a factor of 10; and thus the size represented by a pixel is multiplied by 100. The perceptive invariance results here from the adaptation between two scales of sensory representations (Fig. 3b). From sensor data, a first low resolution mapping (MAP1) with DAs gives a contextual pattern representation, that is detected as a region of interest (ROI), and drives a second map with higher resolution in standardized size (MAP2). It is necessary for MAP1 that its resolution is adjusted to obtain only a single DA. Then, the loop adjusts the resolution of MAP2 so that the two bihistogram's medians (elementary semantic representations) are inside useful ranges.

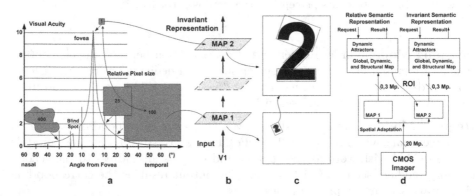

Fig. 3. Perceptive invariance in vision. (a) Standard acuity curve [17]. (b) MAPs representations. (c) MAPs example detailed in Fig. 4. (d) Electronic implementation. Note that the ratio between MAP1 and MAP2 definitions (20 Mp and VGA 0.3 Mp) is similar to the ratio of visual acuities at fovea and 20°.

Figure 3c illustrates this concept with the invariant representation of the word 'digit 2'. In MAP1 the DA perceives the element as a blob. Then this DA drives

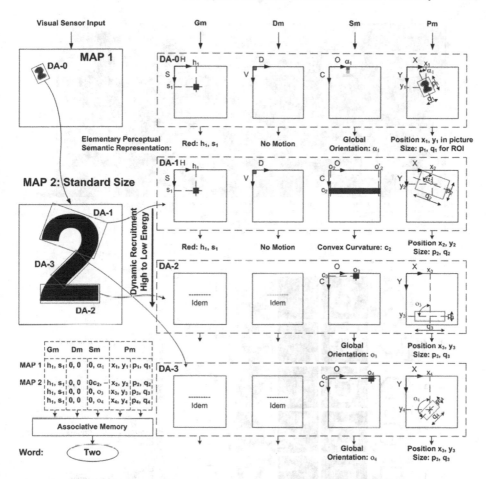

Fig. 4. Detailed functioning of the dynamic attractors. MAP1 is the contextual representation in environment. *Top line:* a single dynamic attractor (DA) delivers the elementary semantic representation. MAP2 is the representation in standard size (perceptive invariance). *Bottom left:* associative memory results. (Color figure online)

the computation towards this ROI and other DAs detect parts of this more detailed pattern in MAP2. Figure 3d shows how high-definition sensors (20 Mp or more) may be connected to a spatial adaptation function that carries two signals, one low-resolution of the global view and the second with high-resolution inside the ROI. The latter signal is driven by the DA of MAP1 and each signal is connected to a USER process.

Audition. In the mammalian cochlea, the basilar membrane's (BM) mechanical responses are amplified, and frequency tuning is sharpened through active feedback from the electromotile outer hair cells (OHCs). To be effective, OHC feedback must be delivered to the correct region of the BM and introduced at the appropriate time in each cycle of BM displacement [6]. Frequency analysis,

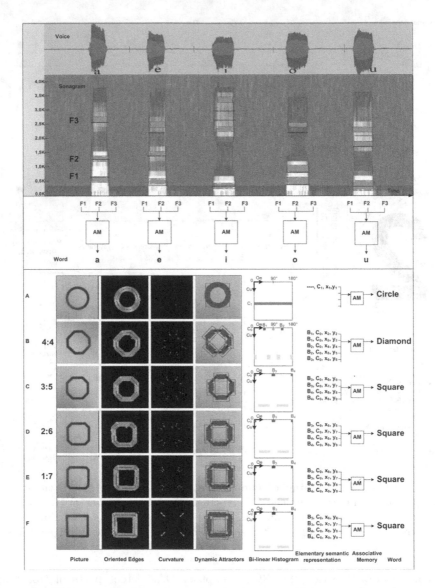

Fig. 5. Audition recognition tests (*Top*) *Upper top*: Voice recording when vowels are successively pronounced *Middle*: Corresponding sonagrams with the first three formants (F1, F2, F3) founded by the dynamic attractors. *Schemes below*: Words (a, e, i, o, u) labelling the vowels are obtained from these formants thanks to the associative memories (AM). **Visual recognition tests** (*Bottom*) *Column 1*: Picture. Same six drawings as used in [3,14]. *Columns 2 and 3*: Oriented Edges and Curvature maps. *Column 4*: Dynamic attractors. The classification map is shown in blue and the positions of the dynamic attractor in red. *Column 5*: Bihistograms with oriented edges in abscissae and curvature in ordinate. *Column 6*: Elementary semantic representations founded by the dynamic attractors. *Last column*: Words obtained by AM in Get mode. (Color figure online)

compression of dynamic-range and amplification are three major signal process-ing functions of the cochlea [13]. After that, the auditory signal is converted into a normalized sonagram. On a low time-scale, MAP1 selects a main blob (patch) with a single DA, that becomes the ROI for MAP2. Inside MAP2, the different formants are found with additional DAs in order to obtain the sound's semantic representations.

4   Results from Dynamic Attractors

Visual Invariance: Figure 4 details the functioning initiated in Fig. 3c. MAP1 is the context representation in environment. A single DA delivers the elementary semantic representations: (h1, s1) for Gm, (0, 0) for Dm, (0, α1) for Sm, and (x1, y1, p1, q1) for Pm. Values translate into: *"A red element at x1, y1, size p1, q1, with orientation α1, not moving"*. MAP2 is the representation in standard size (perceptive invariance). Each DA is recruited by decreasing energy (rank order coding [18]). The first DA found in MAP2 detects a curvature (with its direction). Values then translate into: *"A red curved and (bottom oriented) convex element, at x2, y2, with size p2, q2, not moving"*. The second DA detects *"A red horizontal line at bottom of MAP2"* and the third one *"A red oblique line in the central part of MAP2"*. Altogether these three sentences uniquely characterize the word 'digit 2'. The bottom left part shows the associative memory function with, on one side these elementary semantic representations and, on the other side the equivalent word 'digit 2'.

Visual and Audition Applications. Figure 5 (top) shows how the USER concept deals with a simple auditory perception processed as a visual input. From the sonagram, the perception of the three first formants represents a specific vowel. Figure 5 (bottom) shows the set of six drawings used in [3,14], ranging from exact circle to exact square. The result shows a clear-cut separation between the semantic representations. The drawing A is clearly recognized as a circle, the drawing B as a diamond, and other drawings, C to F, as squares.

5   Toward Multi-sensory Integration

A first multi-integration application has been performed on the Psikharpax rat [2] with a 220 nm chip technology names BIPS. Since, the new chip USER, in 28 nm, is an ongoing process for first quarter 2017 delivery.

Figure 6 shows the integration of the different modules explained above: A. Detailed scheme between two different perceptions. B. Extension to multi-perceptions in parallel. C. Serial extension for hierarchical processing.

Each elementary semantic representation is the result of bihistogram com-putation with $2^{2.w}$ bits. A minimal configuration of a single dynamic attractor typically uses the three results of barycenter position (x, y), of structural char-acteristics (curvature and oriented gradient), and of the dynamic attractor's size (p, q). Thus, the input data of memory is $2^{3.2.w}$ bits, and thus 2^{30} bits for

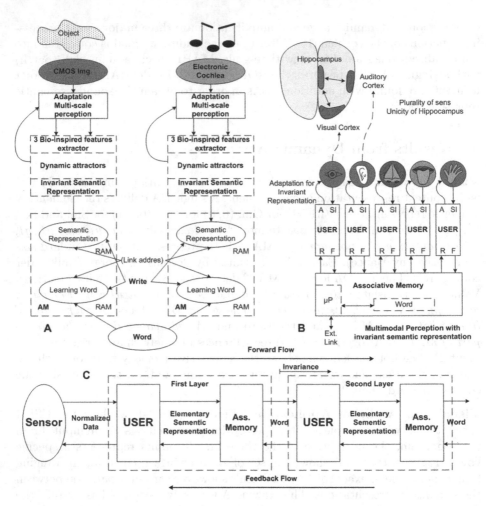

Fig. 6. Scalability in multi-sensory integration. In A, a first integrative level shows that the same module can be used independently for two different sensory inputs, vision and audition, with only one simple link. This link consists in the common internal addresses between the two associative memories. **In B**, the same process is generalized to other sensory inputs in parallel. Three main generic properties are shared: 1. The same module is used for each sensory input. 2. Each time a feedback is sent back to the sensor that allows the adaptation for invariant representation. 3. The link of the common internal addresses between associative memories described above is extended to all sensory processes. **In C**, for each sensory input, the process is extended to a hierarchical serial processing along which the duration of temporal sequences increases. The grouping of USER modules with one associative memory forms one layer. The layers are serially connected bidirectionally [15]. In forward flow, the sensor data is sent to the first layer. Its output 'word' is then sent as the input to a second layer whose output 'word' will represent a set of previous words. In feedback flow, each layer may anticipate the previous layer processing. For example, in forward flow a first layer may extract phonemes, a second one syllables, and a third one spoken words. When the beginning of a spoken word is heard, the feedback flow allows to anticipate the end the spoken word.

w=5 that is a range of only 32 values. An optimized memory size [16] of 2^{30} is already a large size when embedded. With w=5 and a map of 0.3 Mp, the USER module have inputs in 5 bits and includes 8 dynamic attractors, each of them integrating 4 bihistograms, each of 1024×19 bits memory sizes. A 100 Hz sampling of the world is enough for human representation, leading to 100 MHz silicon chip computation. With inputs in only 5 bits (32 levels), a USER module allows huge semantic structural elementary representations. Then an associative memory with a size of $256 K \times 16$ bits (standard package) is enough for a quite large brain-like memory.

6 Discussion

The USER concept is the result of the knowledge accumulation gained from perceptive industrial applications using bio-inspired ASIC conceptions. An initial combination of histogram chips allowed to make a DA [8] that became a generic tool, initially used to detect color blobs (global perception). Then for video sensors, the dynamic perception was added [9]. Later the structural perception was included in a new chip for Advanced Driver Assistance Systems (ADAS) [10]. These three features [5] and the DA [14] are supported by neurobiological data, confirming the fact that the nature uses it on a daily basis.

The architecture generalizes real-time processes at different levels. First, the USER module transforms incoming data into elementary semantic representations. Secondly, a combination of USER modules and associative memories forms a layer with perceptive invariance ability (as soon as the first layer), followed by cascading processes mixing sensory data. Thirdly, the USER module continuously performs forward and feedback flows. This latter feedback flow allows the previous layer to anticipate its perceptive process. As a consequence, the architecture allows fast single-trial learning.

The same module processes different perceptive functions at successive time steps producing invariant representations. It is a compact and efficient solution since it allows the use of associative memories with reasonable size for real-time complex applications with low energy consumption (200 mW), small physical size (50 cm3) and cheap price with reasonable process (28 nm). IoT, ADAS, and DeepMind are clearly good targets for such applications. The semantic representation allows to build a complete application using processing elements that match words in the human language, allowing an easy and efficient communication between developers.

7 Conclusion

The concept of electronic USER modules is a new way of performing invariant perceptive processing and producing semantic representation with any type of sensory input. We can now test and implement a low-cost and low-energy electronic device that labels any multimodal sensory input in a set of words defining basic useful categories, thanks to automatic extraction of invariant properties.

The generic aspect of the USER's layers will open new researches and applications in robotics and in IoT, allowing to communicate through these universal semantic representations. The USER devices being more aware of the world around them, will be more capable, more autonomous, safer and easier to use.

References

1. Bach-y-Rita, P.: Theoretical aspects of sensory substitution and of neurotransmission-related reorganization in spinal cord injury. Spinal Cord **37**(7), 465–474 (1999)
2. Bernard, M., N'Guyen, S., Pirim, P., Guillot, A., Meyer, J.-A., Gas, B.: A supramodal vibrissa tactile and auditory model for texture recognition. In: Doncieux, S., Girard, B., Guillot, A., Hallam, J., Meyer, J.-A., Mouret, J.-B. (eds.) SAB 2010. LNCS, vol. 6226, pp. 188–198. Springer, Heidelberg (2010)
3. Colgin, L.L., Leutgeb, S., Jezek, K., Leutgeb, J.K., Moser, E.I., McNaughton, B.L., Moser, M.B.: Attractor-map versus autoassociation based attractor dynamics in the hippocampal network. J. Neurophysiol. **104**, 35–50 (2010)
4. Goodale, M.A., Milner, A.D.: Separate visual pathways for perception and action. Trends Neurosci. **15**(1), 20–25 (1992)
5. Hubel, D.H., Wiesel, T.N.: Receptive fields of single neurons in the cat's striate cortex. J. Physiol. (London) **148**, 574–591 (1959)
6. Nilsen, K.E., Russell, I.J.: The spatial and temporal representation of a tone on the guinea pig basilar membrane. Proc. Nat. Acad. Sci. **97**(22), 11751–11758 (2000)
7. Pagiamtzis, K., Sheikholeslami, A.: Content-addressable memory (CAM) circuits and architectures: a tutorial and survey. IEEE J. Solid-State Circuits **41**(3), 712–727 (2006)
8. Pirim, P.: Method and device for real-time processing of a sequenced data flow, and application to the processing and digital video signal representing a video image, Patent FR2611063 (1987)
9. Pirim, P.: Method and device for real-time detection, localization and determination of the speed and direction of movement of an area of relative movement in a scene, Patent FR2751772 (1996)
10. Pirim, P.: Automated method and device for perception associated with determination and characterisation of borders and boundaries of an object of a space, contouring and applications, Patent WO2005010820 (2003)
11. Pirim, P.: Generic bio-inspired chip model-based on spatio-temporal histogram computation: application to car driving by gaze-like control. In: Lepora, N.F., Mura, A., Krapp, H.G., Verschure, P.F.M.J., Prescott, T.J. (eds.) Living Machines 2013. LNCS, vol. 8064, pp. 228–239. Springer, Heidelberg (2013)
12. Pirim, P.: Processeur de perception bio-inspiré: une approche neuromorphique. Techn. Ingénieur. IN-220 (2015). (in French)
13. Rasetshwane, D.M., Gorga, M.P., Neely, S.T.: Signal-processing strategy for restoration of cross-channel suppression in hearing-impaired listeners. IEEE Trans. Biomed. Eng. **61**(1), 64–75 (2014)
14. Rennó-Costa, C., Lisman, J.E., Verschure, P.F.: A signature of attractor dynamics in the CA3 region of the hippocampus. PLoS Comput. Biol. **10**(5), e1003641 (2014)
15. Touzet, C.F.: The theory of neural cognition applied to robotics. Int. J. Adv. Robot. Syst. **12**, 74 (2015)

16. Somasundaram, M.: Circuit to generate a sequential index for an input number in a pre-defined list of numbers, Patent US7155563 B1, 26 Dec 2006
17. Strasburger, H., Rentschler, I., Jüttner, M.: Peripheral vision and pattern recognition: a review. J. Vis. **11**(5), 1–82 (2011)
18. Van Rullen, R., Gautrais, J., Delorme, A., Thorpe, S.: Face processing using one spike per neurone. Biosystems **48**(1–3), 229–239 (1998)

Thrust-Assisted Perching and Climbing for a Bioinspired UAV

Morgan T. Pope[✉] and Mark R. Cutkosky

Stanford University, Stanford, CA 94305, USA
mpope@stanford.edu
http://bdml.stanford.edu

Abstract. We present a multi-modal robot that flies, perches and climbs on outdoor surfaces such as concrete or stucco walls. Although the combination of flying and climbing mechanisms in a single platform extracts a weight penalty, it also provides synergies. In particular, a small amount of aerodynamic thrust can substantially improve the reliability of perching and climbing, allowing the platform to maneuver on otherwise risky surfaces. The approach is inspired by thrust-assisted perching and climbing observed in various animals including flightless birds.

1 Introduction

In the aftermath of an earthquake or other disaster, small unmanned air vehicles such as quadrotors are ideal for initial response. They are unaffected by debris on the ground and can fly rapidly to remote and cluttered sites. Unfortunately, they are typically hampered by a short mission life – often 20 min or fewer for platforms weighing less than 0.5 kg. However, if these vehicles can perch and crawl on vertical surfaces, they can extend their missions to hours while performing surveillance, inspection or communication functions. Once perched, they are also relatively unaffected by poor weather; when conditions improve, they can take off and return home or fly elsewhere. Not surprisingly, many small flying animals also perch frequently, resting and taking shelter between flights.

Although multi-modal flying, perching and crawling operation confers advantages, it also extracts a penalty. For small air vehicles, where every gram counts, the mechanisms used for clinging and crawling can easily double the weight of the vehicle, with a corresponding deterioration in flying performance. Fortunately, there are also synergies to be realized by combining climbing and flying capabilities. These synergies are explored in SCAMP (Stanford Climbing and Aerial Maneuvering Platform).

SCAMP is a small robot that combines a commercial Crazyflie 2.0™ quadrotor with insect- and bird-inspired perching and climbing mechanisms. It can fly outdoors and land on unprepared vertical surfaces such as concrete or stucco walls. Like various flying and gliding animals, it also uses aerodynamic effects to enhance perching and climbing.[1]

---

[1] A video of SCAMP in operation is available at https://www.youtube.com/watch?v=bAhLW1eq8eM.

© Springer International Publishing Switzerland 2016
N.F. Lepora et al. (Eds.): Living Machines 2016, LNAI 9793, pp. 288–296, 2016.
DOI: 10.1007/978-3-319-42417-0_26

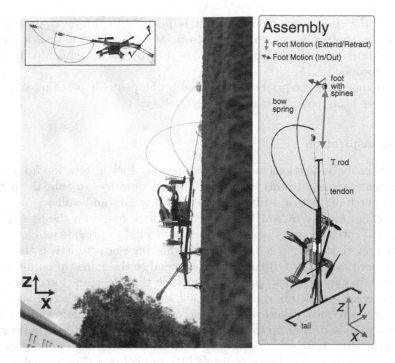

Fig. 1. (left) SCAMP climbing (inset: SCAMP in flight). (right) Climbing mechanism with two servo motors, one to extend/retract feet along the wall, and one to pull feet in/out from wall surface, attaching or detaching spines.

2   Related Work

Steadily shrinking microprocessors, sensors and actuators have lead to an explosion in the popularity of micro air vehicles in general, and quadrotors in particular. However, limits on battery life and small-scale aerodynamic efficiency lead to an acknowledged deficiency in mission life, which tends to be short – 30 min or less [1]. A solution to this problem is to fly to a target area and then perch, with the rotors powered down to extend battery life. Motivated by this realization, the last decade has seen steady advances in climbing and perching robots. One particular impetus was DARPA's *Perch and Stare* competition, which, although it did not produce a drone that could complete all of the the target objectives, advanced the state of the art for future efforts to build on [2].

Other perching efforts have used darts [3], spines [4,5] or adhesives or suction [6–10] to perch on vertical and even overhanging surfaces. SCAMP also draws particularly on robots that use arrays of miniature spines to climb rough vertical walls with two, four, or six legs [11–13]. Another related subject concerns multimodal robots that combine airborne and vertical locomotion abilities [14–16].

Most of the previous work on maneuvers for perching has utilized off-board visual tracking and computation (e.g., [17,18]). However, for the smallest and lightest of the cited perching vehicles, it is possible to simply fly into a wall, utilizing the contact forces on impact to achieve a grip that resists rebound forces. As described in Sect. 3, SCAMP takes this approach, augmented by thrust from the rotors.

To our knowledge, no prior robot is capable of repeatedly perching and climbing outdoors using passive attachment technology.

2.1   Bioinspiration

The mechanisms and actions used for perching also draw inspiration from gliding and perching animals. In this section, we restrict ourselves to animals that, like SCAMP, perch on vertical surfaces, such as tree trunks and walls.

The spines used in SCAMP are similar to those found on the legs of many insects, that catch on small asperities (bumps or pits) to provide purchase on all but very smooth surfaces [19, 20]. The distinguishing characteristic is that spines engage asperities that are (i) larger than or equal to the spine tip radius and (ii) have a surface angle and friction coefficient such that the spine force vector falls inside the resulting friction cone. This point is considered further in Sect. 3.1.

In contrast to most robotic approaches to flying and perching, animals maintain aerodynamic control up to the instant of contact. Thrust-assisted perching draws inspiration from aerial strategies used by a wide range of species, from ants [21, 22] to frogs [23], flying squirrels [24] and colugos [25], as well as many birds that perch on tree trunks, and similar vertical surfaces. In particular, birds that climb tree trunks often use their tails to provide a positive contact force, reducing the "pull-in" force required at the front feet [26], a strategy similar to that adopted in SCAMP. Other birds, including flightless birds, use their wings to enable them to run up otherwise insurmountable surfaces [27,28], a strategy analogous to thrust-assisted climbing in SCAMP.

From an energetic standpoint, the strategy of climbing up vertical surfaces and then gliding down to a nearby tree or other target can be favorable [29], which may explain its popularity across many species.

3   SCAMP Operations

When flying, SCAMP carries its perching and climbing apparatus above the rotors (Fig. 1 inset), where they add 11 g to the 28 g Crazyflie platform. With this overhead addition, the platform's flight time reduces from 5 min to 3.5 min. Hence, SCAMP relies on perching to achieve useful mission durations.

The behaviors programmed for SCAMP are depicted in Fig. 2, with labels matching those in the following text. SCAMP perches by flying, tail-first, into a wall at speeds of $\approx$1–2 m/s. This brisk approach reduces the effects of wind that could make it difficult to hover next to a wall. As the tail impacts the wall (a. Contact) the internal accelerometer measures a large signal, which triggers the pitch-up maneuver (2. Pitch up). At its simplest, this consists of disabling the internal stabilization control and allowing the rotors to pull the climbing

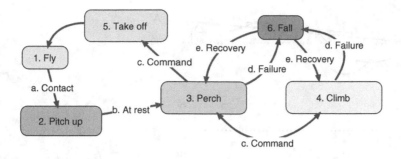

Fig. 2. State-event diagram for SCAMP – labels match numbers in text.

mechanism parallel to the wall. SCAMP then falls slightly until the feet engage asperities on the surface and the robot comes to rest (3. Perch). SCAMP detects a failure to attach (d. Failure) by monitoring whether the vertical acceleration falls below 0.8 g during the first few hundred milliseconds after impact.

Once attached, SCAMP can climb up or down (4. Climb), using servos that move the feet in the extend/retract (z) and in/out (x) directions (Fig. 1, right). The extend/retract servo rotates a carbon fiber rod back and forth in the (y, z) plane, pulling on tendons connected to the two feet. Carbon fiber bow springs, with roughly constant force over the range of travel (up to 95 mm), keep the tendons taut. A second servo rotates a T-shaped rod with wire guide loops at its tips, which pull the tendons and feet into or away from the wall surface. The feet use small spines to catch on asperities and provide forces primarily parallel and slightly into the wall, supporting the robot's weight and counteracting the pitch-back moment produced by the robot's center of mass, which is approximately 40 mm out from the wall surface. The spines are adapted from those used in [4].

Based on wireless commands (c. Command), SCAMP switches between perching and climbing up or down. At present, the external wireless commands are issued by a human operator. However, the details of operation while perching or climbing are computed and controlled using the onboard microprocessor. If the accelerometer detects a fall, SCAMP attempts to arrest the fall (e. Recovery) by briefly turning on the rotors to generate thrust.

On command, SCAMP can position its servos, normally used for climbing, into a special configuration that causes its spring-loaded take-off arm, located at the rear of the platform, to swing into the wall. The tip of the arm has a spine that catches on the wall so that, as SCAMP loses its grip, it pitches backward into the horizontal flight orientation and resumes flight (1. Fly). When the servos return to their normal climbing configuration, a string attached to one of the servos pulls the take-off arm back into its cocked position, visible in Fig. 1.

3.1 Thrust-Assisted Perching and Climbing

Using thrust simplifies perching and increases climbing reliability in several ways.

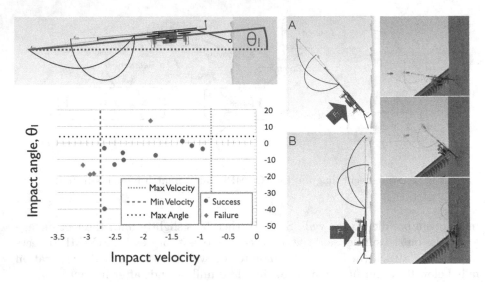

Fig. 3. (top left) Diagram of impact angle θ_I. (bottom left) Observed range of impact angles and velocities leading to successful perching. (A) Rotors and kinetic energy tilt SCAMP upward. (B) Rotors spin down as perch stabilizes.

Pitch-up Maneuver: Whereas previous fixed-wing airplanes and quadrotors (e.g. [4,17,18]) initiated a pitch-up maneuver immediately prior to perching – to reduce speed and orient their spines correctly for attachment – SCAMP is light enough that it can fly directly into a wall and use aerodynamic thrust to tilt its body upward, aligning itself with the wall. As SCAMP tilts upward, its rotor thrust is enhanced by ground effects, which produce an inward force, normal to the wall. This force increases rapidly as SCAMP reaches vertical, suppressing the rebound that previous perching vehicles had to absorb with springs and damping foam. When SCAMP is aligned vertically, the rotors need only 10 % of their maximum thrust to hold the platform against the wall.

This approach is robust within some angle and velocity constraints. At impact angles $\theta_I > 5°$, the point of impact is above the center of mass (Fig. 3) and torque is initially applied in the wrong direction, increasing the chance of failure. The impact velocity must be also large enough to create an acceleration larger than that experienced during normal flight conditions. Approximately 1 g is a safe acceleration threshold, corresponding to ≈ 0.8 m/s. Initial velocities above ≈ 2.8 m/s produce large vibrations and a large initial rebound, which can lead to SCAMP aligning incorrectly. The perching process and empirical limits are summarized in Fig. 3.

Climbing: Thrust is equally effective at increasing the robustness of climbing. Figure 4 shows how the application of a small amount of rotor thrust can dramatically change the effective loading angle for a microspine engaging a surface. In the profile image, θ_{Ln} represents the angle of the load force at the front feet,

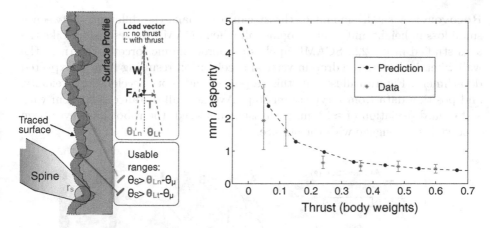

Fig. 4. (left) Traced concrete surface corresponding to 10μm spine tip. Adding thrust shifts load vector from θ_{Ln} to θ_{Lt}, increasing asperity density. (right) Stochastic asperity model with thrust compared to empirical data for concrete. (Color figure online)

which depends on the location of the center of mass and the length from the feet to the tail. Adding a small amount of thrust shifts the load vector from θ_{Ln} to θ_{Lt}. Only a small amount of thrust (10 % of the maximum) is needed to shift the load force from pointing slightly outward to slightly inward.

The surface shown in Fig. 4 is taken from [12] and represents a concrete cinderblock profiled with a stylus. The spine tip (10μm radius) sliding over this surface creates a traced surface [30], shown in yellow. On this traced surface, the "perchable regions" are those for which the surface orientation is such that

$$\theta_s > \theta_{Lt} - \theta_\mu \tag{1}$$

where θ_s is the angle of the local inward surface normal with respect to vertical, $\theta_\mu = \arctan \mu$ and μ is the coefficient of friction.

Without thrust, there are just four regions (shown in magenta) that satisfy Eq. 1; with a small amount of thrust there are eight regions (shown in green). More generally, the distribution of asperities on a surface can be modeled approximately as a spatial Poisson process. For a spine dragging along a region of the surface, the distribution of lengths between asperities is approximately described by an exponential random variable, with probability density function

$$f_x(x; \lambda) = \lambda e^{-\lambda x} \qquad\qquad x \geq 0 \tag{2}$$

where x is the distance traveled and λ is the number of asperities per unit length; λ in turn is a function of the traced surface, θ_{Lt}, and μ.

The plot on the right side of Fig. 4 compares the predicted average distance between asperities with data obtained by dragging a loaded spine along a concrete block and varying the load angle to correspond to different amounts of thrust. The error bars represent the 95 % confidence intervals for our estimation of the average distance between asperities based on an exponential distribution.

Recovery: A small amount of thrust can also make the difference between a small loss in height and a catastrophic fall when SCAMP misses a step. Like the ants studied in [21,22], SCAMP applies an aerodynamic force to return to the wall when it detects the drop in vertical acceleration resulting from unexpected detachment. Figure 5 illustrates this response, and plots acceleration, velocity, and position data from a typical misstep. Average fall distance is ≈14 cm with a standard deviation of ≈12 cm. If thrust is not used, the robot has never been observed to re-engage with the surface.

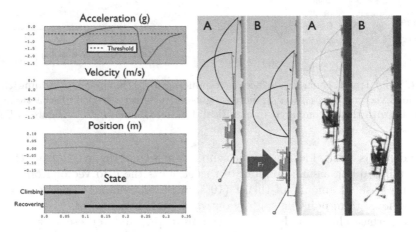

Fig. 5. Data, illustration, and images for climbing recovery. When vertical acceleration drops below 0.5 g (A), SCAMP turns its rotors on to return its feet to the wall, where they engage with asperities and arrest downward motion (B).

4   Conclusions

A robot that functions in two distinct locomotory modes must make compromises in weight and complexity to accommodate the different interaction mechanics, control approaches, and power requirements for the modes of interest. While this usually means that the resulting design is heavier than a conventional single-purpose robot, the additional locomotory modes can act to complement each other. In the case of SCAMP, the perching mechanism extends the effective time aloft for a flying machine that is interested in observing a fixed area; the rotorcraft itself, on the other hand, aids climbing through both its sensitive onboard accelerometers and its ability to create aerodynamic forces, allowing the robot to recover from failures and to increase the asperity density.

Acknowledgements. Support for this work was provided by NSF IIS-1161679 and ARL MAST MCE 15-4. We gratefully acknowledge the help of H. Jiang, C. Kimes, W. Roderick and C. Kerst in conducting experiments.

References

1. Keennon, M.T., Grasmeyer, J.M.: Development of the black widow and microbat mavs and a vision of the future of mav design. In: AIAA International Air and Space Symposium and Exposition: The Next 100 Years, pp. 14–17 (2003)
2. Prior, S.D., Shen, S.-T., Erbil, M.A., Brazinskas, M., Mielniczek, W.: HALO the winning entry to the DARPA UAVForge challenge 2012. In: Marcus, A. (ed.) DUXU 2013, Part III. LNCS, vol. 8014, pp. 179–188. Springer, Heidelberg (2013)
3. Kovač, M., Germann, J., Hürzeler, C., Siegwart, R.Y., Floreano, D.: A perching mechanism for micro aerial vehicles. J. Micro-Nano Mechatron. **5**, 77–91 (2010)
4. Lussier Desbiens, A., Asbeck, A.T., Cutkosky, M.R.: Landing, perching and taking off from vertical surfaces. Int. J. Rob. Res. **30**(3), 355–370 (2011)
5. Kovac, M., Germann, J.M., Hurzeler, C., Siegwart, R., Floreano, D.: A perching mechanism for micro aerial vehicles. J. Micro-Nano Mechatron. **5**, 77–91 (2009)
6. Anderson, M.L., Perry, C.J., Hua, B.M., Olsen, D.S., Parcus, J.R., Pederson, K.M., Jensen, D.D.: The sticky-pad plane and other innovative concepts for perching UAVs. In: 47th AIAA Aerospace Sciences Meeting (2009)
7. Daler, L., Klaptocz, A., Briod, A., Sitti, M., Floreano, D.: A perching mechanism for flying robots using a fibre-based adhesive. In: IEEE International Conference on Robotics and Automation (2013)
8. Kalantari, A., Mahajan, K., Ruffatto III., D., Spenko, M.: Autonomous perching and take-off on vertical walls for a quadrotor micro air vehicle. In: 2015 IEEE International Conference on Robotics and Automation (ICRA), pp. 4669–4674. IEEE (2015)
9 Tsukagoshi, H., Watanabe, M., Hamada, T., Ashlih, D., Iizuka, R.: Aerial manipulator with perching and door-opening capability. In: 2015 IEEE International Conference on Robotics and Automation (ICRA), pp. 4663–4668. IEEE (2015)
10. Liu, Y., Sun, G., Chen, H.: Impedance control of a bio-inspired flying and adhesion robot. In: 2014 IEEE International Conference on Robotics and Automation (ICRA), pp. 3564–3569. IEEE (2014)
11. Clark, J., Goldman, D., Lin, P.C., Lynch, G., Chen, T., Komsuoglu, H., Full, R.J., Koditschek, D.E.: Design of a bio-inspired dynamical vertical climbing robot. In: Robotics: Science and Systems (2007)
12. Asbeck, A., Kim, S., Cutkosky, M.R., Provancher, W.R., Lanzetta, M.: Scaling hard vertical surfaces with compliant microspine arrays. Int. J. Rob. Res. **25**, 14 (2006)
13. Spenko, M.J., Haynes, G.C., Saunders, J.A., Cutkosky, M.R., Rizzi, A.A., Full, R.J.: Biologically inspired climbing with a hexapedal robot. J. Field Rob. **25**, 223–242 (2008)
14. Estrada, M., Hawkes, E.W., Christensen, D.L., Cutkosky, M.R., et al.: Perching and vertical climbing: Design of a multimodal robot. In: 2014 IEEE International Conference on Robotics and Automation (ICRA), pp. 4215–4221. IEEE (2014)
15. Myeong, W., Jung, K., Jung, S., Jung, Y., Myung, H.: Development of a drone-type wall-sticking and climbing robot. In: 2015 12th International Conference on Ubiquitous Robots and Ambient Intelligence (URAI), pp. 386–389. IEEE (2015)
16. Dickson, J.D., Clark, J.E.: Design of a multimodal climbing and gliding robotic platform. IEEE/ASME Trans. Mechatron. **18**(2), 494–505 (2013)
17. Mellinger, D., Shomin, M., Kumar, V.: Control of quadrotors for robust perching and landing. In: Proceedings of the International Powered Lift Conference, pp. 119–126 (2010)

18. Thomas, J., Pope, M., Loianno, G., Hawkes, E.W., Estrada, M.A., Jiang, H., Cutkosky, M.R., Kumar, V.: Aggressive flight for perching on inclined surfaces. J. Mech. Rob. **8**(5), Article ID. 051007 (2016). doi:10.1115/1.4032250

19. Dai, Z., Gorb, S.N., Schwarz, U.: Roughness-dependent friction force of the tarsal claw system in the beetle pachnoda marginata (coleoptera, scarabaeidae). J. Exp. Biol. **205**(16), 2479–2488 (2002)

20. Gorb, S.N.: Biological attachment devices: exploring nature's diversity for biomimetics. Philos. Trans. R. Soc. Lond. A: Math. Phys. Eng. Sci. **366**(1870), 1557–1574 (2008)

21. Yanoviak, S.P., Dudley, R., Kaspari, M.: Directed aerial descent in canopy ants. Nature **433**(7026), 624–626 (2005)

22. Yanoviak, S.P., Munk, Y., Kaspari, M., Dudley, R.: Aerial manoeuvrability in wingless gliding ants (cephalotes atratus). Proc. Roy. Soc. London B: Biol. Sci. **277**, 2199–2204 (2010). doi:10.1098/rspb.2010.0170

23. McCAY, M.G.: Aerodynamic stability and maneuverability of the gliding frog polypedates dennysi. J. Exp. Biol. **204**(16), 2817–2826 (2001)

24. Paskins, K.E., Bowyer, A., Megill, W.M., Scheibe, J.S.: Take-off and landing forces and the evolution of controlled gliding in northern flying squirrels Glaucomys sabrinus. J. Exp. Biol. **210**(Pt. 8), 1413–1423 (2007)

25. Byrnes, G., Lim, N.T.L., Spence, A.J.: Take-off and landing kinetics of a free-ranging gliding mammal, the Malayan colugo (Galeopterus variegatus). Proc. Roy. Soc. B: Biol. Sci. **275**, 1007–1013 (2008)

26. Fujita, M., Kawakami, K., Higuchi, H.: Hopping and climbing gait of japanese pygmy woodpeckers (picoides kizuki). Comp. Biochem. Physiol. Part A: Mol. Integr. Physiol. **148**(4), 802–810 (2007)

27. Dial, K.P.: Wing-assisted incline running and the evolution of flight. Science **299**(5605), 402–404 (2003)

28. Bundle, M.W., Dial, K.P.: Mechanics of wing-assisted incline running (WAIR). J. Exp. Biol. **206**, 4553–4564 (2003)

29. Norberg, R.Å.: Why foraging birds in trees should climb and hop upwards rather than downwards. Ibis **123**(3), 281–288 (1981)

30. Okamura, A.M., Cutkosky, M.R.: Feature detection for haptic exploration with robotic fingers. Int. J. Rob. Res. **20**(12), 925–938 (2001)

The EASEL Project: Towards Educational Human-Robot Symbiotic Interaction

Dennis Reidsma[1]([✉]), Vicky Charisi[1], Daniel Davison[1], Frances Wijnen[1], Jan van der Meij[1], Vanessa Evers[1], David Cameron[2], Samuel Fernando[2], Roger Moore[2], Tony Prescott[2], Daniele Mazzei[3], Michael Pieroni[3], Lorenzo Cominelli[3], Roberto Garofalo[3], Danilo De Rossi[3], Vasiliki Vouloutsi[4], Riccardo Zucca[4], Klaudia Grechuta[4], Maria Blancas[4], and Paul Verschure[4,5]

[1] Human Media Interaction/ELAN,
University of Twente, Enschede, The Netherlands
{d.reidsma,v.charisi,d.p.davison,f.m.wijnen,
j.vandermeij,v.evers}@utwente.nl
[2] Sheffield Robotics, University of Sheffield, Sheffield, UK
{d.s.cameron,s.fernando,r.k.moore,t.j.prescott}@sheffield.ac.uk
[3] E. Piaggio Research Center, University of Pisa, Pisa, Italy
{daniele.mazzei,michael.pieroni,lorenzo.cominelli,
roberto.garofalo,d.derossi}@centropiaggio.unipi.it
[4] SPECS, DTIC, N-RAS, Universitat Pompeu Fabra, Barcelona, Spain
{vicky.vouloutsi,riccardo.zucca,klaudia.grechuta,
maria.blancas,paul.verschure}@upf.edu
[5] Catalan Institute of Advanced Studies (ICREA), Barcelona, Spain

Abstract. This paper presents the EU EASEL project, which explores the potential impact and relevance of a robot in educational settings. We present the project objectives and the theorectical background on which the project builds, briefly introduce the EASEL technological developments, and end with a summary of what we have learned from the evaluation studies carried out in the project so far.

Keywords: Synthetic Tutoring Assistant · DAC · Education · Child robot interaction · Architecture · Evaluation

1 Introduction

This paper presents the EU EASEL project ("Expressive Agents for Symbiotic Education and Learning"), which explores the potential impact and relevance of a robot in educational settings. EASEL targets Human Robot Symbiotic Interaction (HRSI) in the domain of education and learning. Symbiosis is taken here as the capacity of the robot and the person to mutually influence each other,

D. Reidsma—This project has received funding from the European Union Seventh Framework Programme (FP7-ICT-2013-10) as part of EASEL under grant agreement no. 611971. Coauthors are grouped by institute.

© Springer International Publishing Switzerland 2016
N.F. Lepora et al. (Eds.): Living Machines 2016, LNAI 9793, pp. 297–306, 2016.
DOI: 10.1007/978-3-319-42417-0_27

and alter each other's behaviour over different time-scales (within encounters and across encounters). Based on perception of the social, communicative and educational context, the robot responds to the student in order to influence their learning progress.

The impact of EASEL developments crucially depends on the combination of two domains. The field of human robot interaction concerns conversational and social interaction between humans and robots. This involves short term interactions as well as the development of a relation over longer time through repeated interactions. The field of learning and education concerns principles and practices of how a student learns in interaction with other people and learning materials (Charisi et al. 2015). The theoretical work carried out in EASEL concerns the integration of insights from these two fields.

Clearly, the resulting tutoring assistant(s) need to be evaluated. EASEL achieves this through a combination of lab studies and in-the-wild studies in schools, museums, and daycare centers. The studies carried out in EASEL range across the combination of the two above domains, from studies focusing on the development of a longer term relation between human and robot to studies focusing on the exact effect of certain tasks on the learning process and outcome.

The paper is structured as follows. We introduce the EASEL objectives in Sect. 2. Section 3 briefly discusses the interplay between the two above mentioned domains of learning and social interaction. Section 4 focuses on the technology developed in EASEL. Section 5 summarizes the EASEL evaluation studies carried out so far, placing them in the framework between HRI and education. Finally, we tie the results together in Sect. 6, looking at the future directions we need to go to achieve the ultimate aim of EASEL: social robots that are a transformative contribution to the classroom of the future.

2   EASEL Objectives

EASEL aims to deliver a new set of Robotic Based Tutoring Solutions: a Synthetic Tutoring Assistant that incorporates key features of human tutors and other proven approaches capable of instructing a human user and learns from their interactions during large time scales. To this end, new approaches are developed for acquiring social context from sensor data, modeling the student's learning process, and determining the appropriate and most effective strategies for delivering the learning material. The results are incorporated in a social dialog carried out between student and robot, which supports the student in the learning task. The end result of EASEL is a unique and beyond the state of the art social robot based tutoring system that comprises a learning model of the user, a synthetic agent control system for symbiotic interaction establishment, a computational framework of social affordances and a multi modal analysis system for subject's social and affective state analysis. In addition, the project yields guidelines for the design and development of the appropriate robot behaviour toward children in various possible robot roles in the contexts of school and museum environments.

3   Learning, Education, and Robots

The theoretical viewpoint from which we approach learning and education is described by Charisi et al. (2015). Below, we briefly describe this view and show the potential areas of contribution for social robots.

Vygotsky (1978) describes learning as a socio-cultural process. In this process, the student's learning is mediated in two ways. Firstly, by "physical tools": the learning materials such as books, computers, and other tools. Secondly, learning is mediated by "social tools": other people who participate in the learning situation. Figure 1 shows how learning happens in a triadic interaction between student, materials, and other persons.

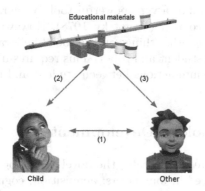

Fig. 1. Learning as a triadic interaction between student, learning materials, and another person (in this case: a robot)

When we look at the interaction between student and learning materials, there is the student interacting directly with the learning materials. We can observe the actions the student takes in the learning task, the utterances and expressions, and the performance in the task. Furthermore, there are the things going on in the mind of the student. These include attitude towards task (is it interesting? is it hard? is it relevant?), self-efficacy with respect to the task (can I do this task? am I doing well right now?), mind set in learning (Dweck 2012) (e.g., directed at risk avoidance / aversive to failure? directed at growth and learning? willingness to make mistakes? curiosity?), and other factors.

When we look at the role that an "other person" can play in the student's learning process, we see three possibilities. The other can be *more knowledgeable* (e.g., teacher or more advanced fellow student), *differently knowledgeable* (a fellow student doing the task together with, or alongside of, the student, see Dillenbourg (1999)), or even *less knowledgeable*. Things the other person could do in their various roles are, for example, explain to student, be explained to by student, encourage student, praise student in various ways, give good or bad example as fellow student, and many other things. In this way, the other person can influence both the observable behavior and the mental state of the student. The effectiveness of these actions depends to a large extent on the relation between the student and the other: factors such as trust or likability will influence to what extent the student is willing to modify his or her actions or perceptions in response to the other's suggestions and contributions.

Given this context, a robot in class can serve a unique mixed role. It is a computer, and as such it can present learning materials to the student in a smart and adaptive way based on the student's skills and progress: the robot as smart learning material. At the same time, a robot is a social entity, more so than a computer. As such, it can fulfill the role of a more or differently knowledgeable "other person" across the entire spectrum from teacher or more advanced fellow student to less knowledgeable peer being taught to by the student. It is this social role of the robot that we are concerned with here. Like many others, we believe this makes the robot a very powerful tool for learning because learning takes place in a social context (Vygotsky 1978). However, the effectiveness of the robot depends among other things on the social believability and the quality of the relation between student and robot. This requires us to also look at short and long term affective interactions between student and robot (Cameron et al. submitted).

4 EASEL Technology Developments

The EASEL architecture incorporates the novel technology developments of the project, and is structured in four layers: acquisition, cognitive modules, dialog and behavior planning, and robot behavior realisation. Vouloutsi et al. (2016) describe how this architecture is a practical incarnation of the conceptual architecture of the Distributed Adaptive Control (DAC) theory (Verschure 2012) of the design principles underlying perception, cognition and action. This section briefly introduces the main components developed as part of EASEL.

4.1 Acquisition

The acquisition layer of EASEL consists of various modules integrating audio-visual analysis and acquisition of physiological signals from the user. The speech recognition is based on the open source Kaldi speech recognizer (Povey et al. 2011), with an EASEL specific vocabulary, language model and recognition grammar. Its specific speaker adaptation solution makes it very robust with respect to interfering speech from the robot itself. Audiovisual scene analysis is done using the SceneAnalyzer (Zaraki et al. 2014), which builds upon several other libraries to deliver quick and robust integrated recognition of multimodal features of the users and their behaviour. Physiological signals can be acquired from the user using non-obtrusive sensor patches that can be integrated in the robot or the learning materials, which allows signal analysis to take place without sensors worn strapped to arm/head/body of user.

4.2   Cognitive Modules

The cognition of the tutoring assistant consists of the memory and decision modules. The memory module stores the current state of the user and the learning task, and has the potential to store longer term memory in repeated interaction, which allows the tutor to build up a user model of a learner from all information in the network. The allostatic control module (Vouloutsi et al. 2013) regulates the tutoring process, learning optimal strategies for delivering the learning content in the right order and at the right difficulty levels, adapted to the student's characteristics and capabilities. The exercise generator, finally, delivers learning exercises of exactly the right nature and difficulty level that is requested by the tutoring models.

4.3   Dialog and Behavior Planning

The cognitive modules deliver the interaction goals of the tutoring agent. These are then translated by the dialog and behavior planning modules into actual utterances and expressions to be realised by the robot. We use the Flipper dialog manager of ter Maat and Heylen (2011) for managing the dialog, and the BML realizer ASAPRealizer by van Welbergen et al. (2012). Flipper offers flexible dialog specification via information state and rule based templates to trigger information state changes as well as behaviour requests. ASAPRealizer offers easy, configurable, control of multimodal choreographed behaviours across several robots using the BML language, which abstracts away from specific motor control by exposing more general behaviour specifications to the dialog manager. ASAPRealizer realizes the requested behaviours on robotic embodiments by directly accessing the motion primitives of the embodiments (see below).

4.4   Content Presentation and Behavior Generation

The learning content is presented in two ways: through the EASEL Scope tablets, and through utterances and expressions of the robot. The EASEL Scope offers a mixed reality interface that allows the student to interact with the learning scenario materials. It can be used to present additional information to the child about the learning materials. This allows the system to vary between different ways of scaffolding the learning of the user. The two main robots used in EASEL are the Robokind Zeno R25[1] and the FACE robot (Lazzeri et al. 2013; Mazzei et al. 2014). For both robots, controllers have been developed that offer access to the robot's motion control primitives in a comparable way, allowing the system to present the same content through different robots.

The first versions of the complete EASEL architecture have been deployed already in various settings and used in a number of studies; the final year of the project will see several deployments of the system in real life contexts over longer periods of time.

---

[1] http://www.robokindrobots.com/.

5   EASEL Studies on Robots in Learning

In order to evaluate the integrated solutions developed in EASEL, an ongoing series of experiments is conducted throughout the project. In this section we discuss results we have achieved so far. We looked at the impact of robot behavior and characteristics on the way students *perceive the robot*, the potential for impact on the *longer term (social) relationship* between student and robot, the impact of the robot on the *learning outcome*, and the impact of the robot on the *learning process* that the student goes through. In most experiments these students were children of primary school age since these are the main target group for EASEL. Most studies were conducted with a Robokind Zeno R25 robot; a few were carried out with an iCub.

5.1   Impact of Robot Behavior and Characteristics on Perception of Robot by Student

Regarding direct perception of the robot by the children interacting with it, EASEL studies have looked at the childs affective and social responses related to gender (Cameron et al. 2015b) and age (Cameron et al. 2015c). Children interacted with a robot during a Simon Says game in two studies. Results of both studies show a clear gender difference: boys respond more positively, showing more smiling and liking, while girls display less smiling with an expressive robot. Furthermore, the second study revealed an age difference: on average older children considered the robot to be significantly more like a machine than the younger children did. Follow-up studies will look at the influence of age on perception of animacy and gender of the robot Zeno by boys and girls from different age groups.

Regarding responses of users to the robot during a task, another EASEL study found that children seem to respond differently to the robot depending on whether it seems to be autonomously responsive to speech from the children or its speech understanding is visibly mediated by the experiment controller. In the former case children seem to display more anticipatory gaze awaiting the robot's responses, in contrast to more reactive gaze after the robot started responding (Cameron et al. 2016a). In several of the studies we saw that children attempted to engage the robot in social interactions, displaying behavior such as conversational turn-taking and socially oriented gaze and spontaneous utterances towards the robot, and they would sometimes modify their speech utterances (following robot errors, participant corrects robot then repeats question more slowly and clearly, emphasizing key words).

5.2   User Behavior Towards the Robot: Potential for Longer Term Relation Between Robot and Student

In the previous section we saw that children seem to be willing to treat the EASEL robot as a social partner, and respond socially to it, depending on its behavior. We are also interested in finding out to what extent the robot's behavior could influence the forming of a longer term relationship.

One EASEL study focused on the impact of robot-stated phrases, relating to its limitations or its intentions, on individuals liking, perceptions of robot competence, and willingness to assist the robot (Cameron et al. 2015a, 2016b). Results of this study showed that robot behavior can influence the user's willingness to use the robot in the future. This study provides new evidence that strategies used by individuals in interpersonal relationship development can be extended to apply to social robotics HRI.

We also looked at the impact of the *activities* that robot and child share on the potential for longer term relationship. Davison et al. (2016) looked at children who engage in both an educational task and a physical exercise with a peer-like robot, or only in educational tasks. Results, although not significant enough, suggest that sharing an additional physical and extra-curricular activity might promote social perception of the robot.

For future EASEL studies we are planning to focus on the details of relationship development over a longer period of time and in repeated interaction with the system.

5.3   Impact of Robot on Learning Outcome

We looked at the effect on learning performance related to different behavior types of the robot: tutor-like behavior vs peer-like behavior, and various ways of implementing the robot's gaze behavior. The impact was measured on the perception and subjective experience of the user and on task performance and learning outcome (Blancas et al. 2015; Vouloutsi et al. 2015). Regarding learning outcome, experiments so far indicate that although we can measure improvement in performance successfully with a post-test, the improvement was not yet significantly different between conditions of different robot behavior. We will address this aspect further in longer term studies, since we expect that this effect will depend on longer term interactions.

5.4   Impact of Robot on Learning Process

The final aspect that the EASEL project targets is impact of the robot (behavior) on the learning *process*. In the final EASEL study discussed here, children had to do an inquiry learning task in one of two conditions: with a social robot, or with an interactive tablet. In both conditions the content of the task was the same, including the spoken instructions issued by the robot or tablet. In both conditions, the child was invited to verbally explain their thoughts. Important steps in an inquiry learning task are related to generating explanations. It is well known that explaining learning content to someone else is a powerful source of learning (Bargh and Schul 1980); children often gain a deeper understanding of the material when they are asked to verbalize their thoughts and reasoning to others (Coleman et al. 1997; Holmes 2007).

The main hypothesis for this study was that the social nature of the robot, compared to the tablet, would trigger the child to verbalize their thoughts more easily (faster and/or longer responses). To measure this we looked at both the verbalization by the child, and their perception of the tablet or robot as a social entity.

Results tentatively indicated that children tended to verbalize more and respond faster to questions when working with the robot. It seems that the robot provided a more intuitive context for verbalization than the tablet. The results of the exit interview suggested that the robot was indeed seen as a more social entity compared to a tablet. For example, statements like: "I taught the robot", or "the robot was curious" were given by the children in the exit interview, and children spontaneously addressed the robot as 'robot', asking it its opinion, etcetera.

6   Discussion

We presented the EASEL project, which aims to deliver a new state of the art in Synthetic Tutoring Assistants. The conceptual and technological developments so far have resulted in an integrated system capable of deriving contextual information from audiovisual sensors, modeling students and learning, learn strategies for effective teaching and deliver the teaching material through social dialog.

An important observation in the evaluation studies is the fact that child-robot interaction, be it in learning or in other domains, is challenging to evaluate. So far we have applied a number of novel and traditional methods that gave us sensible results and significant differences between conditions, but not all methods worked equally well (or at all) (Charisi et al. submitted). The evaluation of child-robot interaction is a topic that we will address in future work. Clearly, we are not alone in this position, as shown by the growing number of workshops, symposia, and panels dedicated to this topic.

Nevertheless, we feel that we managed to start exploring the potential impact of robots in education across the whole spectrum: the perception of the robot by the student, the student's responses to the behavior and characteristics of the robot; the potential for longer term relationship; the possibility for robot to influence the learning process, and the learning outcome. We see increasing evidence toward the positive effects of the EASEL robots social behavior on how children approach learning tasks as well as on the learning outcome. We are now preparing the final longer term evaluations in which we combine all these aspects in one setup, with the aim of showing how the EASEL robots can be a transformative contribution to the classroom of the future.

References

Bargh, J.A., Schul, Y.: On the cognitive benefits of teaching. J. Educ. Psychol. **72**(5), 593–604 (1980)

Blancas, M., Vouloutsi, V., Grechuta, K., Verschure, P.F.M.J.: Effects of the robot's role on human-robot interaction in an educational scenario. In: Wilson, S.P., Verschure, P.F.M.J., Mura, A., Prescott, T.J. (eds.) Living Machines 2015. LNCS, vol. 9222, pp. 391–402. Springer, Heidelberg (2015)

Cameron, D., Collins, E.C., Chua, A., Fernando, S., McAree, O., Martinez-Hernandez, U., Aitken, J.M., Boorman, L., Law, J.: Help! i can't reach the buttons: facilitating helping behaviors towards robots. In: Wilson, S.P., Verschure, P.F.M.J., Mura, A., Prescott, T.J. (eds.) Living Machines 2015. LNCS, vol. 9222, pp. 354–358. Springer, Heidelberg (2015a)

Cameron, D., Fernando, S., Collins, E., Millings, A., Moore, R., Sharkey, A., Evers, V., Prescott, T.: Presence of life-like robot expressions influences childrens enjoyment of human-robot interactions in the field. In: Salem, M., Weiss, A., Baxter, P., Dautenhahn, K. (eds.) 4th International Symposium on New Frontiers in Human-Robot Interaction, Canterbury, UK (2015b)

Cameron, D., Fernando, S., Millings, A., Moore, R., Sharkey, A., Prescott, T.: Children's age influences their perceptions of a humanoid robot as being like a person or machine. In: Wilson, S.P., Verschure, P.F.M.J., Mura, A., Prescott, T.J. (eds.) Living Machines 2015. LNCS, vol. 9222, pp. 348–353. Springer, Heidelberg (2015c)

Cameron, D., Fernando, S., Collins, E., Millings, A., Moore, R., Sharkey, A., Prescott, T.: Impact of robot responsiveness and adult involvement on childrens social behaviours in human-robot interaction. In: Salem, M., Weiss, A., Baxter, P., Dautenhahn, K. (eds.) 5th International Symposium on New Frontiers in Human-Robot Interaction, Sheffield, UK (2016a)

Cameron, D., Loh, E., Chua, A., Collins, E., Aitken, J., Law, J.: Robot-stated limitations but not intentions promote user assistance. In: Salem, M., Weiss, A., Baxter, P., Dautenhahn, K. (eds.) 5th International Symposium on New Frontiers in Human-Robot Interaction, Sheffield, UK (2016b)

Cameron, D., Millings, A., Davison, D., Reidsma, D., Fernando, S., Prescott, T.: A framework for affect-led human-robot symbiotic interaction (submitted)

Charisi, V., Davison, D., Wijnen, F., Van Der Meij, J., Reidsma, D., Prescott, T., Van Joolingen, W., Evers, V.: Towards a child-robot symbiotic co-development: a theoretical approach. In: New Frontiers in Human-Robot Interaction, AISB (2015)

Charisi, V., Davison, D., Reidsma, D., Evers, V.: Evaluation methods for user-centered child-robot interaction. In: Proceedings of the 25th IEEE International Symposium on Robot and Human Interactive Communication, RO-MAN 2016, New York City, United States, August 26-31 (to appear, 2016)

Coleman, E.B., Brown, A.L., Rivkin, I.D.: The effect of instructional explanations on learning from scientific texts. J. Learn. Sci. **6**(4), 347–365 (1997)

Davison, D., Schindler, L., Reidsma, D.: Physical extracurricular activities in educational child-robot interaction. In: Salem, M., Weiss, A., Baxter, P., Dautenhahn, K. (eds.) 5th International Symposium on New Frontiers in Human-Robot Interaction, Sheffield, UK (2016)

Dillenbourg, P.: What do you mean by collaborative learning? Collaborative-Learning: Cognitive and Computational Approaches, Chap. 1, pp. 1–19 (1999)

Dweck, C.: Mindset: how you can fulfill your potential. Constable and Robinson (2012). ISBN: 9781780333939

Holmes, J.: Designing agents to support learning by explaining. Comput. Educ. **48**, 523–547 (2007)

Lazzeri, N., Mazzei, D., Zaraki, A., De Rossi, D.: Towards a believable social robot. In: Lepora, N.F., Mura, A., Krapp, H.G., Verschure, P.F.M.J., Prescott, T.J. (eds.) Living Machines 2013. LNCS(LNAI), vol. 8064, pp. 393–395. Springer, Heidelberg (2013). doi:10.1007/978-3-642-39802-5-45

Mazzei, D., Zaraki, A., Lazzeri, N., De Rossi, D.: Recognition and expression of emotions by a symbiotic android head. In: 2014 14th IEEE-RAS International Conference on Humanoid Robots (Humanoids), pp. 134–139. IEEE (2014)

Povey, D., Ghoshal, A., Boulianne, G., Burget, L., Glembek, O., Goeln, N., Hannemann, M., Motlicek, P., Qian, Y., Schwarz, P., Silovsky, J., Stemmer, G., Vesely, K.: The Kaldi speech recognition toolkit. In: IEEE Workshop on Automatic Speech Recognition and Understanding (2011). ISBN 978-1-4673-0366-8

ter Maat, M., Heylen, D.: Flipper: An Information State Component for Spoken Dialogue Systems. In: Vilhjálmsson, H.H., Kopp, S., Marsella, S., Thórisson, K.R. (eds.) IVA 2011. LNCS, vol. 6895, pp. 470–472. Springer, Heidelberg (2011). doi:10.1007/978-3-642-23974-8_67

van Welbergen, H., Reidsma, D., Kopp, S.: An incremental multimodal realizer for behavior co-articulation and coordination. In: Nakano, Y., Neff, M., Paiva, A., Walker, M. (eds.) IVA 2012. LNCS, vol. 7502, pp. 175–188. Springer, Heidelberg (2012). ISBN: 978-3-642-33196-1, ISSN: 0302-9743

Verschure, P.F.: Distributed adaptive control: a theory of the mind, brain, body nexus. Biol. Inspired Cogn. Architectures **1**, 55–72 (2012). doi:http://dx.doi.org/10.1016/j.bica.2012.04.005. ISSN: 2212-683X

Vouloutsi, V., Lallée, S., Verschure, P.F.M.J.: Modulating behaviors using allostatic control. In: Lepora, N.F., Mura, A., Krapp, H.G., Verschure, P.F.M.J., Prescott, T.J. (eds.) Living Machines 2013. LNCS, vol. 8064, pp. 287–298. Springer, Heidelberg (2013)

Vouloutsi, V., Munoz, M., Grechuta, K., Lallee, S., Duff, A., Puigbo, J.Y.L., Verschure, P.F.M.J.: A new biomimetic approach towards educational robotics: the distributed adaptive control of a synthetic tutor assistant. In: Salem, M., Weiss, A., Baxter, P., Dautenhahn, K. (eds.) 4th International Symposium on New Frontiers in Human-Robot Interaction, Canterbury, UK (2015)

Vouloutsi, V., Blancas, M., Zucca, R., Omedas, P., Reidsma, D., Davison, D., Charisi, V., Wijnen, F., van der Meij, J., Evers, V., Cameron, D., Fernando, S., Moore, R., Prescott, T., Mazzei, D., Pieroni, M., Cominelli, L., Garofalo, R., Rossi, D.D., Verschure, P.F.: Towards a synthetic tutor assistant: the EASEL project and its architecture. In: Proceedings of the 5th International Conference on Biomimetic and Biohybrid Systems, Living Machines 2016, Edinburgh, Scotland, July 2016 (2016)

Vygotsky, L.: Mind in Society: The Development of Higher Psychological Processes. Harvard University Press, Cambridge (1978). ISBN: 0674076680

Zaraki, A., Mazzei, D., Giuliani, M., De Rossi, D.: Designing and evaluating a social gaze-control system for a humanoid robot. IEEE Trans. Hum. Mach. Syst. **44**(2), 157–168 (2014)

Wasp-Inspired Needle Insertion with Low Net Push Force

Tim Sprang, Paul Breedveld, and Dimitra Dodou[✉]

Department of BioMechanical Engineering, Faculty of Mechanical,
Maritime and Materials Engineering, Delft University of Technology,
Delft, the Netherlands
d.dodou@tudelft.nl

Abstract. This paper outlines the development of a four-part needle prototype inspired by the ovipositor of parasitic wasps. In the wasp ovipositor, three longitudinal segments called valves move reciprocally to gain depth in the substrate. It has been suggested that serrations located along the wasp ovipositor induce a friction difference between moving and anchoring valves that is needed for this reciprocal motion. Such an anchoring mechanism may not be desired in a medical setting, as serrations can induce tissue damage. Our aim was to investigate whether a multipart needle can penetrate tissue phantom material with near-zero net push force while using needle parts devoid of surface gripping textures or serrations. Accordingly, a four-part needle prototype was developed and tested in gelatine substrates. The performance of the prototype was assessed in terms of the degree of slipping of the needle with respect to the gelatine, with less slip implying better performance. Slip decreased with decreasing gelatine concentration and increasing offset between the needle parts. Motion through gelatine was achieved with a maximum push force of 0.035 N. This study indicates the possibility of needle propagation into a substrate with low net push force and without the need of serrations on the needle surface.

Keywords: Percutaneous interventions · Wasp ovipositor · Biomimetics

1 Introduction

1.1 Percutaneous Needles: A Brief State-of-the-art

Blood sampling, biopsies, regional anaesthesia, neurosurgery, and brachytherapy, all rely on percutaneous needles [1]. Most needles used in these procedures are rigid and follow straight trajectories. Accessibility of targets located deep inside the body is inhibited, and deviations from the desired path due to organ deformation are common [2]. These functional limitations hamper medical treatment when designated areas cannot be reached [2] and compromise safety when undesired areas are penetrated [3].

Several flexible steerable needles have been developed in an effort to overcome the path planning limitations encountered with rigid straight needles (e.g., [2, 4–9]). Some of these newly developed needles rely on concentric axial insertion of multiple pre-bent needle parts, whereas others rely on reaction forces from the tissue to control the steering curvature (for reviews see [10, 11]). One drawback of flexible needles is that the axial

© Springer International Publishing Switzerland 2016
N.F. Lepora et al. (Eds.): Living Machines 2016, LNAI 9793, pp. 307–318, 2016.
DOI: 10.1007/978-3-319-42417-0_28

load applied at the back of a needle increases with resistive forces on the needle tip and shaft when penetrating deeper into the tissue. The tissue can only support this axial load up to a certain penetration depth; further needle advancement can lead to buckling of the needle and/or lateral slicing and damage of the tissue.

1.2 The Wasp Ovipositor and Biologically Inspired Steerable Needles

The wasp ovipositor is a needle-like structure, with no musculature. The ovipositor extends from the last abdominal segments of the wasp and serves the main purpose of egg deposition. The ovipositor consists of three longitudinal segments called valves: two ventral valves and one dorsal valve. The dorsal valve is connected to the two ventral valves along its length by means of a tongue-and-groove mechanism that allows for relative movement of the valves in the direction of the ovipositor while preventing their separation. Abdominal musculature actuates the valves independently from each other [12, 13]. The wasp penetrates the substrate by antagonistically moving the ventral valves while the dorsal valve acts as sliding support. Serrations at the ovipositor tip are likely to allow the valves anchor against the substrate (Fig. 1). It has been hypothesized that the wasp applies a pull force on the anchored valve, allowing to load the penetrating valve with a push force higher than its critical buckling load [14, 15].

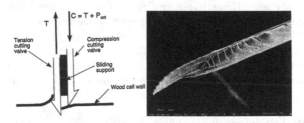

Fig. 1. Schematic representation of the hypothesized working principle of penetration of a wasp ovipositor (left). The SEM image (right) shows the tip of the ovipositor of *Megarhyssa nortoni nortoni*, an Ichneumonoid parasitoid wasp (figures from [14]).

The wasp ovipositor has been used as an inspiration source for designing steerable percutaneous needles [3, 15–19]. For example, Oldfield et al. [18] described a 6-mm diameter four-part needle prototype with reciprocally moving segments. Tissue transversal with no net axial force by means of microstructures on the needle surface has been reported in [6, 16, 19]. Similarly, inspired by insertion methods observed in mosquito proboscis, Aoyagi et al. [20] indicated that substrate penetration with a microneedle comprising serrated stylets is possible without applying a net push force (see also [21] for a description of buckling prevention in mosquitos and wasps).

1.3 Goal of this Study

It has been suggested that the serrations located on the wasp ovipositor induce a friction difference between moving and anchoring valves [19]. Serrations may not be desired in medical settings, as they can induce tissue damage. Our aim was to investigate whether a multipart needle can penetrate tissue phantom material with low net push force while being devoid of any surface gripping textures or serrations.

2 Design of a Multipart Needle Prototype

A multipart needle prototype was developed, comprising a low-friction cart that suspended a four-part needle assembly, four linear actuators, and driving electronics to move a needle assembly back and forth (Fig. 2). The needle assembly consisted of four stainless steel square needle parts and a stainless steel guide tube. The needle assembly had a total diameter of 2 mm (cf. diameter range of common hypodermic needles: 0.2– 4.3 mm [22]) and a length of 200 mm (in line with the penetration depths in [6]). The needle tip was conical (angle about 20°) and sharp. The needle parts had a smooth surface finish ($R_a = 0.2$); here we differentiate from [19] where saw-tooth microstructures were used to induce friction.

Fig. 2. Image of the prototype, showing the needle (1), needle assembly (2), linear actuator assembly (3), and the low friction cart with driving electronics (4).

Each needle part ended in a fin mounted to a leadscrew, each leadscrew being translated back and forth by a linear stepper motor (resolution: 1.8°, corresponding to 0.0061 mm/step of leadscrew translation). The motors were mounted in an aluminium bracket. Two alignment pins, extending from the bottom of the bracket, aligned the needle and actuator assembly with the low-friction cart. The cart was supported by two axes, each containing two ball bearings with metal shields (624-ZZ, 4 × 13 × 5 mm).

Four stepper drivers were mounted behind the linear actuator module. The drivers were controlled with a microcontroller. A two-row LCD display showed a timer and cycle counter during actuation, and settings when standby. The settings were controlled by two push buttons and a rotary encoder. The system was battery powered.

The embedded code was written in C++ including Arduino libraries. The software consisted of a user interface part containing a rotational encoder with a push button, a green button, a red button, and the display driver, and an embedded part containing the register, stepper sequence controller, stepper controller, and stepper driver.

Needle penetration was realized in cycles, each of which consisted of a protrusion and a retraction phase. In the protrusion phase the needle parts were protruded one-by-one (or two-by-two, depending on the sequence setting) with respect to the cart over a distance equal to a predefined offset. In the retraction phase the needle parts were retracted altogether with respect to the cart, over a distance equal to the offset.

3   Hypotheses

In a four-part needle, zero net push force penetration relying on a friction difference due to a difference between the number of stationary and protruding needle parts would require the advancement of one needle part at a time. Previous measurements in porcine liver showed that friction force on a needle with a diameter of 1.27 mm increased linearly with penetration depth, whereas cutting forces remained about constant along the penetration depth [23, 24]. It can be therefore expected that, when protruding one needle part at a time, an equilibrium between the friction force on the stationary needle parts and the resistive force (i.e., sum of cutting force F_c and friction force F_f) on the protruding needle part is reached at a certain penetration depth:

$$\left(\frac{1}{4}F_c + \frac{1}{4}F_f\right)_{protruding} = \left(\frac{3}{4}F_f\right)_{stationary} \tag{1}$$

From this depth onwards, penetration with zero net push force is theoretically possible, since the friction force on the stationary needle parts is higher than the resistive force on the protruding needle part.

The performance of the prototype was assessed in terms of the degree of slipping of the needle with respect to the gelatine, with less slip implying better performance. The following hypotheses were investigated:

- *Protrusion sequence:* When moving one needle part at the time, performance is not expected to be affected by the protrusion sequence, as the surface area difference between moving and stationary needle parts is independent from the protrusion sequence. A two-by-two actuation is expected to result in no penetration, as in that case the friction force on the stationary needle parts is smaller than the resistive forces on the protruding needle parts.
- *Needle-part velocity:* Friction force increases with insertion velocity more steeply than the corresponding cutting force [25]. Following Eq. 1, it is thus expected that the needle performs better at higher velocities.
- *Gelatine concentration:* In very low gelatine concentrations no penetration is expected, as there is no sufficient needle-substrate friction to overcome the bearing friction and inertia of the car. Resistive forces on the protruding needle parts are likely to also be dominant in high concentrations, leading to poor performance. An optimum in gelatine concentration is thus expected for best performance.

- *Offset:* The friction force on the protruding needle part is expected to be higher for a larger offset, due to the larger surface area compared to the area for lower offsets.

4 Methods

Two experiments were conducted. In the first experiment (called henceforth prototype experiment), the effect of gelatine concentration, needle-part offset, needle-part velocity, and needle-part sequence on the performance of the prototype was measured. In the second experiment (force experiment), force measurements were conducted during the insertion of the needle into gelatine to estimate the relation between cutting and friction force as a function of needle velocity and gelatine concentration.

4.1 Experimental Setup

Prototype Experiment. A platform consisting of an aluminium frame and a $500 \times 500 \times 8$ mm acrylic plate formed a level surface for the low-friction cart to travel back and forth between a proximity sensor and a gelatine sample. Adjustment screws ran through river nuts on the four corners of the frame for levelling. Gelatine substrates were produced in $434 \times 344 \times 107$-mm trays. One panel of each tray had a 20×2 grid of 15-mm diameter holes, each 40 mm apart. The tray was set to height with spacers, and oriented so that the panel with holes was positioned against the platform. A laser proximity sensor (Micro-Epsilon optoNCDT1302-200; range: 200 mm; resolution: 0.1 mm) measured the travelled distance of the prototype. The sensor was mounted on a bracket and placed against the platform to ensure a constant distance between the sensor and gelatine tray. Sensor data were sampled at 50 Hz using LabVIEW 2010 and a National Instruments NI USB-6210 16-bit data acquisition system.

Force Experiment. The four needle parts and guide tube were clamped to a force sensor assembly with the tip directed downwards. The needle was translated by means of a linear stage. The axial force was measured during the insertion and retraction movement of the needle in a gelatine container positioned underneath the needle. The force sensor assembly consisted of a mechanical decoupler holding a force transducer. The sampling rate was set at 1 kHz. The decoupler assembly was statically calibrated with balance weights ranging between 0.25 and 1.0 kg in steps of 0.25 kg.

4.2 Variables

Independent Variables. The following variables were manipulated:

- *Needle-part offset* [mm]: The length of the protruding part of the moving needle parts relative to the stationary needle parts varied between 3, 10, and 20 mm.
- *Needle-part velocity* [mm/s]: The protrusion and retraction velocity of the needle parts with respect to the cart, varied between 4, 8, and 13.5 mm/s, where 13.5 mm/s is the maximum velocity that the linear actuators were able to generate.

- *Gelatine concentration* [wt%]: The % weight of gelatine powder in water, varied between 3, 8, and 13 wt%, to simulate a variety of soft tissues, from brain and fat to breast and liver tissue [26, 27].
- *Needle-part sequence* [no units]: The sequence of protruding needle parts, varied between circular, diagonal, and a two-by-two manner of actuation.

Dependent Variables (Measured). The following variables were measured:

- S_a [mm]: The distance travelled by the cart (prototype experiment).
- *Number of cycles* (N_c) [no units]: The number of cycles performed by the cart to travel a given distance (prototype experiment).
- F_i [N]: The axial force acting on the needle during insertion, measured with the force transducer. F_i represents the sum of the needle cutting force and the needle-substrate friction force (force experiment).
- F_f [N]: The axial force acting on the needle during retraction. F_f represents the needle-substrate friction force (force experiment).

Dependent Variables (Derived). The following variables were calculated:

- S_t [mm]: The number of cycles multiplied by the offset representing the theoretical distance that the cart would travel if there were zero slip (prototype experiment).
- $Slip_{pro}$ [mm]: The slip during protrusion, equalling the backwards travelled distance of the cart during the protrusion of the needle in one cycle (prototype experiment).
- $Slip_{ret}$ [mm]: The slip during retraction, equalling the difference between the offset and the actual travelled distance of the cart during retraction (prototype experiment).
- SR_{tot} [no units]: The slip ratio over an entire measurement, defined as $SR_{tot} = 1 - S_a/S_t$. Less slip means less dependence of the needle penetration on the net axial push force on the needle (prototype experiment).
- SR_{pro} [no units]: The slip ratio in the protrusion phase, defined as: $SR_{pro} = Slip_{pro}/Offset$ (prototype experiment).
- SR_{ret} [no units]: The slip ratio in the retraction phase, defined as: $SR_{ret} = Slip_{ret}/Offset$ (prototype experiment).
- F_c [N]: The cutting force defined as the axial force on the needle as a result of the subtraction of the retraction force from the insertion force (force experiment).
- *Depth of equilibrium* [mm]: The depth at which F_c and F_f are in equilibrium according to Eq. 1 (force experiment).

4.3 Experimental Procedure

Prototype Experiment. Gelatine was prepared in trays one day in advance of the experiment. The needle was positioned in front of the allocated hole of the tray and a machinist square was used to align the cart perpendicular to the platform. Before the start of each measurement, the needle was manually pushed 35 mm into the gelatine (this initial depth was defined in pilot measurements as the minimum distance for which needle advancement was possible; for smaller distances, the contact area of the needle with the gel was not sufficient and the needle slipped instead of advancing). Next, the

Table 1. Parameter variation in prototype experiment.

Sequence [no unit]	Offset [mm]	Gelatine concentration [%wt]	Velocity [mm/s]
Circular	3	3	**4**
Diagonal	**10**	**8**	8
Two-by-two	20	13	13.5

Note. The baseline condition is annotated in bold.

proximity sensor was placed against the base of the platform, aligned with the cart, and switched on. About 1 s later, the actuation of the needle parts started and continued until the needle had travelled about 120 mm inside the gelatine.

All measurements were performed over a time span of one day. The four independent variables were varied from a common baseline, see Table 1. Measurements were repeated five times for each combination of parameters. Measurement allocation to the tray holes was done quasi-randomly, so that each set of five measurements in 8 wt% gelatine was distributed over the upper and lower rows of three trays. All measurements in 3 wt% and 13 wt% gelatine were performed in a single tray and distributed randomly between the upper and lower rows of the tray.

Force Experiment. Gelatine samples were prepared one day in advance of the experiment. A drilling spot along the edge of the gelatine container at least 30 mm apart from other drillings and from the edge of the container was chosen and the needle was positioned 1 mm above the gelatine. The waiting time between needle insertion and retraction was set to 3 s, to allow the gelatine around the needle to settle.

All force measurements were performed over a time span of one day. Needle velocity (4, 8, and 13.5 mm/s) and gelatine concentration (3, 8, and 13 % wt) were varied in a fully crossed manner. Each combination of parameters was repeated five times. Temperature of the gelatine was kept between 4 and 8 C° in both experiments, as the insertion force of a needle in gelatine decreases for temperatures above 8°C [28].

4.4   Data Processing

Prototype Experiment. The raw signal of the cart position was filtered using a moving average filter over five samples. S_a was calculated by subtracting the mean value of the last 10 data samples from the first 10 data samples. N_c was determined by the number of local maxima in the signal. $Slip_{pro}$ was determined by subtracting the cart position after the protrusion phase of a given cycle from the cart position before the protrusion phase of the cycle. $Slip_{ret}$ was determined by subtracting the difference in cart position before and after the retraction phase from the offset.

Force Experiment. The raw force signal was filtered using a moving average filter over 100 samples and shifted so that the first 1,000 data points (i.e., 1 s) averaged at zero. The retraction phase was defined as the part of the curve from maximum retraction to substrate exit. The protrusion phase was defined as the part of the curve until maximum protrusion, with the same length as the retraction part. The protrusion and retraction force profiles were aligned at their maximum force value. The resultant of the retraction and protrusion force is the cutting force. The depth at which cutting and friction forces were in equilibrium (Eq. 1) was determined by the intersection between a linearly fitted retraction force and the mean cutting force.

4.5 Statistical Analysis

For the prototype experiment, a one-way ANOVA was performed to compare the means of SR_{pro}, SR_{ret}, and SR_{tot} as a function of the parameters in Table 1. For the force experiment, a one-way ANOVA was performed to compare the means of the maximum friction and average cutting force as a function of needle velocity and gelatine concentration. Data analysis was performed in MATLAB R2013a (The MathWorks, Inc., Natick, Massachusetts).

5 Results

5.1 Prototype Experiment

SR_{tot} was not significantly different between circular and diagonal actuation in 8 wt% gelatine at 4 mm/s ($F(1,8) = 0.56$, $p = .476$; Fig. 3). Two-by-two actuation of needle parts resulted in no penetration at 4 mm/s and in penetration with a SR_{tot} higher than 0.9 at 13.5 mm/s. A 4-mm offset was associated with a higher SR_{tot} as compared to SR_{tot} for 10-mm and 20-mm offsets ($F(2,12) = 160.54$, $p < .001$; the latter two also being significantly different from each other; post-hoc paired t-test: $p = .009$). SR_{tot} increased with gelatine concentration ($F(2,12) = 154.71$, $p = 2.70.10^{-9}$). SR_{tot} did not differ between the three tested velocities ($F(2,12) = 1.43$, $p = .278$). SR_{pro} decreased with offset and increased with gelatine concentration, in line with the effects observed for SR_{tot}, whereas SR_{ret} decreased with offset and gelatine concentration.

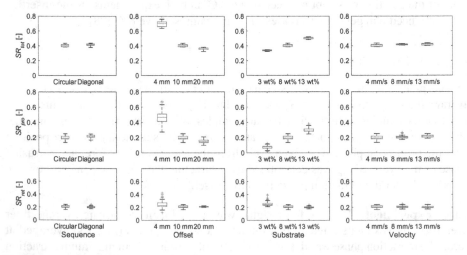

Fig. 3. Prototype experiment. Slip ratio as a function of each of the independent variables (from left to right: sequence, offset, gelatine concentration, velocity) (top row) per cycle (SR_{tot}); (middle row) in the protrusion phase (SR_{pro}); and (bottom row) in the retraction phase (SR_{ret}).

5.2   Force Experiment

The maximum friction force increased with needle velocity for all three gelatine concentrations (3 wt%: $F_{(2,12)} = 17.6$, $p < .001$; 8 wt%: $F_{(2,12)} = 10.33$, $p = .003$; 13 wt%: $F_{(2,12)} = 47.83$, $p = < .001$). Also the average cutting force increased with needle velocity (3 wt%: $F_{(2,12)} = 40.62$, $p < .001$; 8 wt%: $F_{(2,12)} = 131.3$, $p < .001$; 13 wt%: $F_{(2,12)} = 50.91$, $p < .001$). The equilibrium depth increased with velocity and decreased with gelatine concentration.

6   Discussion

We tested the principle of needle advancement through tissue phantom material with low net push force while using a four-part needle devoid of surface gripping textures. We found that slip ratio decreased with decreasing gelatine concentration and increasing offset. Friction force increased with velocity, gelatine concentration, and penetration depth, whereas cutting force increased with velocity and gelatine concentration, and remained constant with penetration depth.

6.1   Main Findings

Gelatine Concentration. We found a proportional relation between SR_{tot} and gelatine concentration, which is not in line with our hypothesis and Eq. 1 but is consistent with Parittotokkaporn et al. (2009) who reported substrate traversal with slip in 6 wt% gelatine and no substrate traversal at all in 8 wt% gelatine. A parameter possibly contributing to the observed proportional relation is the elastic deformation of the gelatine: when all four needle parts are protruded, the cart moves slightly backwards due to the gelatine springing back. It is possible that this effect increases with gelatine concentration, as friction increases with gelatine concentration more than elasticity does [29–31]. Cart inertia and bearing friction possibly also contribute to the observed proportional relation between SR_{tot} and gelatine concentration. Specifically, the resistance against movement of the cart may act as a pushing force on the protruding needle part, preventing the cart from moving backwards. Cutting force increases with gelatine concentration (in line with [25, 28, 30]), leaving us to believe that the effect of cart inertia and bearing friction on $Slip_{pro}$ decreases with gelatine concentration.

Offset. Whereas the absolute amount of slip *per cycle* was proportional to the offset in agreement with our hypothesis, the total amount of slip decreased with offset, meaning that the best overall performance was achieved with the largest offset. The elastic deformation described in the previous paragraph could contribute to the relatively high SR_{tot} for small offsets. As the absolute effect of elasticity should be equal for the three offsets, the relative contribution to the SR_{pro} is higher for small offsets. This is supported by an exploratory test with an offset of 1 mm, which resulted in no penetration at all, possibly because in that case only elastic deformation was achieved.

Sequence. No performance difference between one-by-one circular and diagonal actuation was found, which is expected, as the surface area subjected to resistive forces is independent from the actuation order of the needle parts. Two-by-two actuation yielded no penetration at 4 mm/s, suggesting that the difference in friction between protruding and stationary needle parts is the dominant factor for needle penetration. Two-by-two actuation at 13.5 mm/s resulted in penetration with a slip ratio over 0.9, probably due to cart inertia: due to the relatively large mass of the cart, the protruding needle part accelerated forwards faster than the cart was able to accelerate backwards.

Velocity. No slip reduction was observed for increasing needle-part velocity, contradicting our hypothesis. Force measurements indicated that the cutting force increased more steeply with needle velocity than the friction force did, opposite to [25]. The difference in maximum acceleration for different velocities in the prototype experiment makes the validation of the performance results difficult. Future experiments with varying velocity require a trade-off between maximum acceleration and acceleration time to reach a certain needle-part velocity, as a reduction of acceleration results in a longer acceleration time, whereas for high acceleration differences between velocity settings, inertia confounds the effect of velocity on performance.

6.2 Limitations

Cart inertia and bearing friction possibly prevented slip in the protrusion phase in which the resultant force on the cart was directed out of the gelatine. The cart weighed 0.73 kg and accelerated with a maximum of 27 mm/s^2 in the 4-mm/s velocity setting, therefore requiring a force of about 0.02 N to overcome its inertia. To overcome bearing friction, a force of about 0.015 N would be needed (assuming a friction coefficient of 0.002 for lubricated ball bearings; [32]). Thus, in the protrusion phase, where the resultant force on the needle is likely to be lower than 0.035 N, an effect of inertia and bearing friction cannot be ruled out.

To validate the method used to define friction and cutting force, supplementary measurements were conducted, in which friction force was measured when the entire needle was inserted 5 times in the same hole in 3, 8, and 13 wt% gelatine. Results showed that in all three concentrations the retraction force declined after the first insertion by about 10 %. These results indicate that friction force both in our force experiments and in the literature is generally underestimated and therefore the cutting force is overestimated, due to the fact that the retraction (thereby friction) force is measured *after* the hole is already made during the insertion movement of the needle.

In this work, we only focused on straight trajectories. Frasson et al. [7] showed that steering with a multipart needle can be achieved by controlling the relative offset between needle parts. Asymmetric forces acting from the substrate on the (bevelled) tip cause the needle to cut a curved trajectory. Wasps are also capable of bending their ovipositors in several ways (e.g., [33]). Further research is required to find the most suitable way of steering a multipart needle.

Acknowledgements. This research is supported by the Dutch Technology Foundation STW, which is part of the Netherlands Organization for Scientific Research (NWO), and which is partly funded by the Ministry of Economic Affairs (project 12712, STW Perspectief Program iMIT-Instruments for Minimally Invasive Techniques).

References

1. Abolhassani, N., Patel, R., Moallem, M.: Needle insertion into soft tissue: a survey. Med. Eng. Phys. **29**, 413–431 (2007)
2. Misra, S., Reed, K.B., Schafer, B.W., Ramesh, K.T., Okamura, A.M.: Observations and models for needle-tissue interactions. In: IEEE International Conference on Robotics and Automation, pp. 2687–2692 (2009)
3. Frasson, L., Parittotokkaporn, T., Davies, B.L., Rodriguez y Baena, F.: Early developments of a novel smart actuator inspired by nature. In: 15th International Conference on Mechatronics and Machine Vision in Practice, pp. 163–168 (2008)
4. Ebrahimi, R., Okazawa, S., Rohling, R., Salcudean, S.E.: Hand-held steerable needle device. In: Ellis, R.E., Peters, T.M. (eds.) MICCAI 2003. LNCS, vol. 2879, pp. 223–230. Springer, Heidelberg (2003)
5. Engh, J.A., Podnar, G., Khoo, S.Y., Riviere, C.N.: Flexible needle steering system for percutaneous access to deep zones of the brain. In: IEEE 32nd Annual Northeast Conference on Bioengineering, pp. 103–104 (2006)
6. Frasson, L., Ko, S.Y., Turner, A., Parittotokkaporn, T., Vincent, J.F., Rodriguez y Baena, F.: STING: a soft-tissue intervention and neurosurgical guide to access deep brain lesions through curved trajectories. J. Eng. Med. **224**, 775–788 (2010)
7. Frasson, L., Ferroni, F., Ko, S.Y., Dogangil, G., Rodriguez y Baena, F.: Experimental evaluation of a novel steerable probe with a programmable bevel tip inspired by nature. J. Robot. Surg. **6**, 189–197 (2012)
8. Walsh, C.J., Franklin, J.C., Slocum, A., Gupta, R.: Material selection and force requirements for the use of pre curved needles in distal tip manipulation mechanisms. In: Design of Medical Devices Conference, 3913 (2010).
9. Webster, R.J., Okamura, A.M., Cowan, N.J.: Toward active cannulas: miniature snake-like surgical robots. In: IEEE/RSJ International Conference on Intelligent Robots and Systems, pp. 2857–2863 (2006)
10. Scali, M., Pusch, T.P., Breedveld, P., Dodou, D.: Needle-Like Instruments for Steering Through Solid Organs: a Review of the Scientific and Patent Literature (submitted)
11. Van de Berg, N.J., Van Gerwen, D.J., Dankelman, J., Van den Dobbelsteen, J.J.: Design choices in needle steering—a review. IEEE-ASME T. Mech. **20**, 2172–2183 (2015)
12. Rahman, M.H., Fitton, M.G., Quicke, D.L.: Ovipositor internal microsculpture in the braconidae (insecta, hymenoptera). Zool. Scr. **27**, 319–332 (1998)
13. Scudder, G.G.E.: Comparative morphology of insect genitalia. Annu. Rev. Entomol. **16**, 379–406 (1971)
14. Vincent, J.F.V., King, M.J.: The mechanism of drilling by wood wasp ovipositors. Biomimetics **3**, 187–201 (1995)
15. Burrows, C., Secoli, R., Rodriguez y Baena, F.: Experimental characterisation of a biologically inspired 3D steering needle. In: 13th IEEE International Conference on Control, Automation and Systems, pp. 1252–1257 (2013)
16. Frasson, L., Parittotokkaporn, T., Schneider, A., Davies, B.L., Vincent, J.F.V., Huq, S.E., Degenaar, P., Rodriguez y Baena, F.R.: Biologically inspired microtexturing: investigation into the surface topography of next-generation neurosurgical probes. In: 30th IEEE Annual International Conference of the Engineering in Medicine and Biology Society, pp. 5611–5614 (2008)
17. Leibinger, A., Oldfield, M., Rodriguez y Baena, F.R.: Multi-objective design optimization for a steerable needle for soft tissue surgery. In: 15th International Conference on Biomedical Engineering, pp. 420–423 (2014)

18. Oldfield, M.J., Burrows, C., Kerl, J., Frasson, L., Parittotokkaporn, T., Beyrau, F., Rodriguez y Baena, F.: Highly resolved strain imaging during needle insertion: results with a novel biologically inspired device. J. Mech. Beh. Biomed. Mat. **30**, 50–60 (2014)

19. Parittotokkaporn, T., Frasson, L., Schneider, A., Huq, E., Davies, B.L., Degenaar, P., Biesenack, J., Rodriguez y Baena, F.R.: Soft tissue traversal with zero net push force: feasibility study of a biologically inspired design based on reciprocal motion. In: IEEE International Conference on Robotics and Biomimetics, pp. 80–85 (2009)

20. Aoyagi, S., Takaoki, Y., Takayanagi, H., Huang, C.-h., Tanaka, T., Suzuki, M., Takahashi, T., Kanzaki, T., Matsumoto, T.: Equivalent negative stiffness mechanism using three bundled needles inspired by mosquito for achieving easy insertion. In: 2012 IEEE/RSJ International Conference on Intelligent Robots and Systems, pp. 2295–2300 (2012)

21. Sakes, A., Dodou, D., Breedveld, P.: Buckling prevention strategies in nature as inspiration for improving percutaneous instruments: a review. Bioinsp. Biomim. **11**, 021001-1–021001-26 (2016)

22. Syringe Needle Gauge Chart (2014). http://www.sigmaaldrich.com/chemistry/stockroom-reagents/learning-center/technical-library/needle-gauge-chart.html

23. Hing, J.T., Brooks, A.D., Desai, J.P.: Reality-based needle insertion simulation for haptic feedback in prostate brachytherapy. IEEE International Conference on Robotics and Automation, pp. 619–624 (2006)

24. Simone, C., Okamura, A.M.: Modeling of needle insertion forces for robot-assisted percutaneous therapy. In: IEEE International Conference on Robotics and Automation, vol. 2, pp. 2085–2091 (2002)

25. Van Gerwen, D.J., Dankelman, J., Van den Dobbelsteen, J.J.: Needle-tissue interaction forces-a survey of experimental data. Med. Eng. Phys. **34**, 665–680 (2012)

26. Nguyen, M.M., Zhou, S., Robert, J.L., Shamdasani, V., Xie, H.: Development of oil-in-gelatin phantoms for viscoelasticity measurement in ultrasound shear wave elastography. Ultrasound Med. Biol. **40**, 168–176 (2014)

27. Moreira, P., Peterlik, I., Herink, M., Duriez, C., Cotin, S., Misra, S.: Modelling prostate deformation: SOFA versus experiments. Mech. Eng. Res. **3**, 64–72 (2013)

28. Ng, K.W., Goh, J.Q., Foo, S.L., Ting, P.H., Lee, T.K.: Needle insertion forces studies for optimal surgical modeling. Int. J. Bioscience Biochem. Bioinformatics **3**, 187–191 (2013)

29. Zhang, X., Qiang, B., Greenleaf, J.: Comparison of the surface wave method and the indentation method for measuring the elasticity of gelatine phantoms of different concentrations. Ultrasonics **51**, 157–164 (2011)

30. De Lorenzo, D.: Force Sensing and Display in Robotic Driven Needles for Minimally Invasive Surgery. Doctoral dissertation, Politecnico di Milano, Milano (2012)

31. Crouch, J.R., Schneider, C.M., Wainer, J., Okamura, A.M.: A velocity-dependent model for needle insertion in soft tissue. In: Duncan, J.S., Gerig, G. (eds.) MICCAI 2005. LNCS, vol. 3750, pp. 624–632. Springer, Heidelberg (2005)

32. Van Beek, A.: Advanced Engineering Design: Lifetime Performance and Reliability. Delft University of Technology, Delft (2006)

33. Quicke, D.L.: Ovipositor mechanics of the braconine wasp genus zaglyptogastra and the ichneumonid genus pristomerus. J. Nat. Hist. **25**, 971–977 (1991)

Use of Bifocal Objective Lens and Scanning Motion in Robotic Imaging Systems for Simultaneous Peripheral and High Resolution Observation of Objects

Gašper Škulj[✉] and Drago Bračun

Faculty of Mechanical Engineering, University of Ljubljana,
1000 Aškerčeva 6, Ljubljana, Slovenia
{gasper.skulj,drago.bracun}@fs.uni-lj.si
http://www.fs.uni-lj.si/

Abstract. Imaging systems are widely used in robotic systems to detect features of their surroundings and their position in order to guide the robot's motion. In the paper, a variant of an artificial imaging system based on the unique bifocal eye of sunburst diving beetle (Thermonectus marmoratus) larvae is proposed. The biologically inspired imaging system of a single sensor and a coaxial lens form a superposition of two focused narrow and wide view angle images. The output image contains a high resolution area of interest and its periphery. The scanning motion of the bifocal imaging system is also imitated and provides positional relations between objects. Acquired images are used for distance assessment. The intended use of the proposed imaging system is in a guidance system of an autonomously moving robot with biologically inspired locomotion.

Keywords: Biomimetics · Imaging system · Bifocal lens · Scanning motion · Image disparity

1 Introduction

Imaging systems are useful for autonomous robots in order for them to observe their surroundings and navigate around obstacles. An advantage in comparison to other sensors is that imaging systems can work on considerable distance and produce an output with plenty of information. Robotic imaging systems need to be able to observe region of interest with high resolution and wide surrounding area with lower resolution. This is usually realised with complex combination of several imaging systems with different view angles. To reduce the complexity of autonomous robots new design of an imaging system is needed that provides the desired capabilities without the use for numerous cameras or special costly optical components.

Nature provides us with many examples of different eye designs. The mammalian eye for example has an adaptable deformable lens that focuses the image

© Springer International Publishing Switzerland 2016
N.F. Lepora et al. (Eds.): Living Machines 2016, LNAI 9793, pp. 319–328, 2016.
DOI: 10.1007/978-3-319-42417-0_29

on the retina on the curved surface on the back of the eye. Mimicking the mammalian eye design with standard optical components is problematic because artificial lenses are rigid and image sensors are flat with uniform pixel density [1,2]. Insects, on the other hand, usually have a rigid compound eye design that is also studied and imitated, but again has to be created with non-standard optical parts [3,4].

A possible inspiration that could solve the need for simultaneous high resolution observation of an area of interest and a peripheral area is the recently discovered bifocal eye design of sunburst diving beetle larvae. The larvae frontal eyes have a fully featured bifocal function. The lenses are rigid and because of asymmetry focus the two images onto distinct distal and proximal retinas. The larvae scan their surroundings with these principal eyes by oscillating their heads dorso-ventrally as they approach potential prey [5,6]. Based on the sunburst diving beetle larvae eye and the idea of two images in one eye a robotic imaging system design of standard optical components is proposed.

2   Imaging System Design

The proposed imaging system (Fig. 1) that mimics the larvae bifocal eye functions is composed of a Galilean telescope that is mounted in front of a machine vision camera lens. The Galilean telescope is assembled from a convex lens 1 and concave lens 2. The telescope diameter is smaller than the camera lens diameter. This way the light can pass through the telescope and further on through the camera lens as it is marked with A light rays in Fig. 1. However, the light can also pass besides the telescope directly to the camera lens, as it is marked with B light rays. A combination of A and B light paths forms a bifocal optical system, that creates two images with a different magnification on the same image detector. The light path A has a greater magnification because of the Galilean telescope.

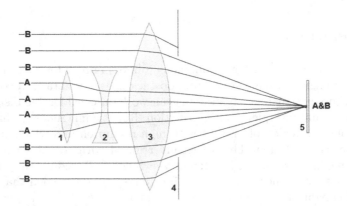

Fig. 1. Bifocal optical system configuration with a fully opened iris. The Galilean telescope (1 and 2), camera lens (3) with iris (4), image detector (5).

Let us denote the light path A as the central vision and light path B as the peripheral vision. Switching between these two systems is carried out by changing the iris opening in the camera lens. If the iris is fully opened (Fig. 1), the light passes through both optical systems to the image detector and two differently magnified images overlap. Consequently the complete imaging system has two different view angles. If the iris is fully closed as is demonstrated in Fig. 2, the peripheral vision is switched off.

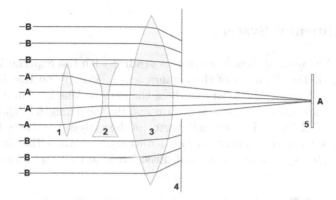

Fig. 2. Bifocal optical system configuration with fully closed iris. The Galilean telescope (1 and 2), camera lens (3) with iris (4), image detector (5).

In order to avoid overlapping problem at the fully opened iris, and to achieve a smooth transition between only the central or only the peripheral or both images, a suitable ratio of telescope and camera lens diameters should be selected. The camera lens should have a much greater light power capacity then the telescope. This way, when the iris is fully opened, the peripheral image is much brighter than the central image. Consequently the image detector acquires mostly peripheral image. The central image is in this case barely noticeable. If the iris is precisely tuned, both equally bright images can be acquired. However, if the iris is fully closed, the peripheral image vanishes and only the central narrow angle image remains on the image detector. The telescope magnification defines the magnification of the central image.

As a possibility, a similar imaging system can also be achieved if the telescope is reversed. In this case the light first encounters the concave lens 2 and then the convex lens 1 of the telescope. The complete magnification of the light path A is therefore smaller than the one of the light path B. The light path A thus becomes the wide angle peripheral vision and the light path B the narrow angle central vision. The iris effect is consequently also turned around. When the iris is fully closed only the peripheral wide angle view remains visible.

The difference between the proposed imaging system and the larvae bifocal eye in terms of the visual function are image A & B alignment and the number of image sensors. The larvae eye has two distinct retinas that correspond to two image sensors where images are focused. The proposed imaging sensor uses only a single image sensor. The larvae eye also produces images A and B that have an offset equal to the distance between the two retinas. Images A and B produced in the proposed system are centrally aligned and completely overlap. The offset of the two images could be achieved by offsetting the telescope axis to the camera imaging system axis. The proposed system employs the function of the iris that is not present in the larvae eye, to compensate for these differences by possible selection between the central or the peripheral images.

3   Experimental System

The proposed imaging system concept was studied with the experimental setup shown in Fig. 3. The diameter of the camera lens is 45 mm and the diameter of the telescope is 15 mm. The focal length of the camera lens is 50 mm, the focal length of the telescope convex lens is 25 mm and the focal length of the telescope concave lens is -12.7 mm. The image detector of the system has 744 × 480 pixels. The imaging system is mounted on two rotational actuators that are used to point the imaging system in the desired direction with the resolution of 0.01°.

Fig. 3. The experimental setup.

The scene consists of two bolts of different sizes (Fig. 4). Bolts are used because of their simple but at the same time recognisable shape. And also when the bolt is observed in high enough resolution its thread is visible. A bolt as an observed object therefore offerers a lot of distinct features that can be used in image analysis. The bolts are positioned 120 cm from the imaging system and spaced apart approximately 5 cm. The smaller bolt is the main observed object in the experiment. The bigger bolt serves as a reference point in the scene and is positioned further away form the imaging system and to the right of the smaller bolt.

First of all, the function of the iris is demonstrated. Figure 5 (a) shows a scene with a fully opened iris. The image where the light passes besides the telescope

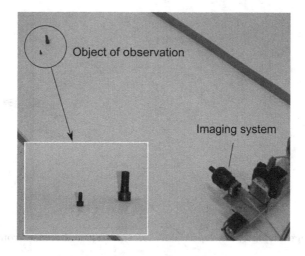

Fig. 4. The experimental scene.

directly to the camera lens prevails. This is the wide angle view image with lower magnification. The image formed with the light path A is barely visible. In Fig. 5 (b) the iris is partially closed and both images are sufficiently represented. In the last case shown by the Fig. 5 (c), the iris is fully closed and only the narrow angle view image with higher magnification is visible. The image formed with the light path B is not visible. It was also found that the telescope works as a central iris for a light path B when the camera lens iris is fully opened. The image formed by a light path B is consequently high-pass filtered. This enhances small details, which can be an advantage in disparity searching algorithms.

Fig. 5. Effect of the iris settings. Fully open iris (a); partially closed iris (b); fully closed iris (c).

Secondly, the use of image disparity to estimate distance to the object is verified. Image disparity of close objects is larger than the disparity of far objects. Figure 6 shows the difference in disparity between two distant points L1 and

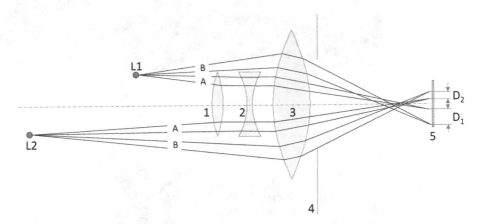

Fig. 6. The image disparity in dependence of the object distance from the imaging system.

L2. The point L1 closer to the imaging system has a greater disparity D1 than the more distant point L2 with disparity D2. In general, disparity exponentially decreases with larger distances. If the observed point has no disparity, it is positioned directly in the imaging system optical axis or it is very far away.

Image disparity in the proposed bifocal imaging system results in two overlapping images with different magnifications. The magnification for the image A is denoted as M_A and for the image B as M_B. Ratio between M_A and M_B is denoted as Z. With the use of equations for thin lens optics (1, 2 and 3) the relation between Z and the observed object's distance L from the image detector is derived. A condition for the given equations is that the image of the observed object must be focused.

$$L = S_1 + S_2 \tag{1}$$

$$\frac{1}{S_1} + \frac{1}{S_2} = \frac{1}{F} \tag{2}$$

$$M = -\frac{I}{O} = -\frac{S_2}{S_1} \tag{3}$$

Distance L is the sum of distances S_1 and S_2. S_1 is the distance between the object and the lens. S_2 is the distance between the lens and the image detector. Focal distance F for the optical system A is denoted as F_A and focal distance for the optical system B as F_B. Magnification is the ratio between the image size I and the object size O. The derived theoretical relation Z(L) is shown in (4).

$$Z(L) = \frac{I_A(L)}{I_B(L)} = \frac{M_A(L)}{M_B(L)} = \frac{\left(L - \sqrt{L(L - 4F_A)}\right)\left(L + \sqrt{L(L - 4F_B)}\right)}{\left(L + \sqrt{L(L - 4F_A)}\right)\left(L - \sqrt{L(L - 4F_B)}\right)} \tag{4}$$

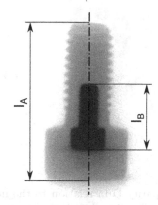

Fig. 7. The image used to measure ratio Z at position 1200 mm.

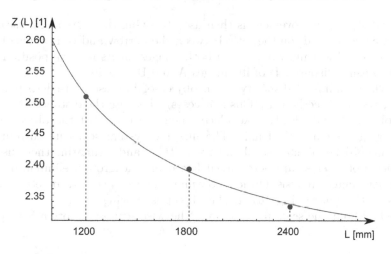

Fig. 8. Comparison of measured values (red dots) with theoretical model for Z(L). (Color figure online)

During the experiment ratio Z was determined on three locations 1200 mm, 1800 mm and 2600 mm. The measurements (Fig. 7) are compared to the theoretical model Z(L) in Fig. 8. The measurements sufficiently correspond to the model to confirm that the proposed imaging system can be used for distance assessment of an object. However, for a precise distance measurement the experimental system is not yet sufficient. The images taken are not sharp enough to precisely determine the edges of the object. This can be improved with better image focusing and with an image detector of higher resolution.

Lastly, the use of the scanning motion of the imaging system is explained. The scanning motion is achieved with a rotational actuator that locally swings the whole imaging system predominantly around one axis. Alternatively the scanning motion can also be achieved by rigidly mounting the imaging system

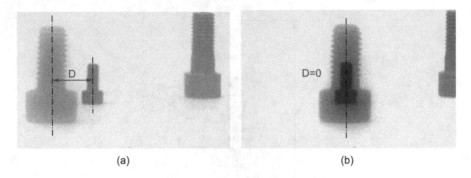

(a)   (b)

Fig. 9. Example of image disparity D in relation to the imaging system rotation (a) left of the object and (b) centrally aligned with the object.

on the robot, whose movement is then used to swing the system. The scanning rotation angle depends on the ratio between the narrow and wide angle of view. The observed object must appear on both images for its relative position to be estimated using disparity D of its images A and B (Fig. 9).

Search for minimal disparity of an object of interest is used to point the imaging system directly at it. This is necessary because the distance assessment of an object should ideally take place when the centre of the object lies on the imaging system optical axis. The imaging system scans its surroundings identifying objects of interest, aligning with them and estimating their distance.

Objects of interest are determined based on background elimination using optical flow image analysis. Optical flow image analysis is used to detect moving objects in the image (Fig. 10). If the object is moving on its own, it can be detected without the scanning motion. Otherwise scanning motion is employed

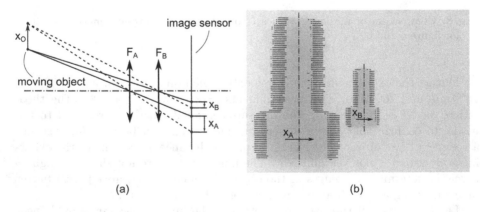

(a)   (b)

Fig. 10. (a) Projection of the same point of the scanned object at different locations on to the sensor of the bifocal imaging system. (b) Optical flow image analysis of the scanned object. (Color figure online)

to move the image of the stationary object across the out-of-focus background. An example of optical flow algorithm use is shown in Fig. 10 (b). The result of the optical flow algorithm are movement vectors shown as red lines that depict the direction and the amplitude of the detected movement. It can be seen that the movement of the object xo is detected on both parts of the image (for A and B light path). The resulting vectors for the object's image A and B are marked as xA and xB, respectively. The amplitude ratio between the movement vectors corresponds to ratio Z(L). The information provided by the optical flow analysis can therefore be used to estimate the distance of the observed object.

4 Discussion

The proposed imaging system design successfully mimics the main functionality of the bifocal eye of sunburst diving beetle (Thermonectus marmoratus) larvae. The imaging system was realized as a working experimental prototype. The bifocal imaging system produces two images on a single imaging sensor that are used to determine the object's distance from the system. The scanning motion of the system is used to locate the object of interest and distinguish them from the background. With the bifocal imaging system it is possible to simultaneously observe the object in relation to its surroundings and in high resolution.

Future work will continue the development based upon the bifocal eye inspiration with the goal to improve image quality and increase the system's distance measurement resolution. To enable an efficient use of the system, decision algorithms based on image processing will be developed. It is also desirable to reduce the size of the imaging system, because of the intention to integrate the imaging system on a robot with biologically inspired locomotion. The integration will provide an opportunity for co-evolution of individual robot's subsystems and further development of the artificial bifocal imaging system. For a biologically inspired robotic system to be truly successful, all of its parts have to be in synergy and build upon their strengths.

An example that can provide an insight into future development of the proposed system is a snake like robot. An advantage of a snake like robot is its large length-to-width ratio. A small width and flexible body enable it to slither through tight spaces and corners. The improved version of the proposed imaging system would be placed centrally on the robot's head in order to preserve its dimensions and at the same time provide depth perception. Mimicry of a snake's motion would be used in synergy with the imaging system to passively scan the surroundings and detect objects of interest. The distance assessment function of the imaging system would provide the robot with enough data to construct a map of its local environment and enable it to autonomously navigate towards the desired location. A potential use for a snake-like robot would be autonomous surveillance on rough terrain or underwater. A specially interesting case would be a search for survivors in demolished buildings after natural disasters.

References

1. Lee, L.P., Szema, R.: Inspirations from biological optics for advanced photonic systems. Science **310**, 1148–1150 (2005)
2. Rim, S.B., Catrysse, P.B., Dinyari, R., Huang, K., Peumans, P.: The optical advantages of curved focal plane arrays. Opt. Express **16**, 4965–4971 (2008)
3. Jeong, K.H., Kim, J., Lee, L.P.: Biologically inspired artificial compound eyes. Science **312**, 557–561 (2006)
4. Song, Y.M., Xie, Y., Malyarchuk, V., Xiao, J., Jung, I., Choi, K.J., Li, R.: Digital cameras with designs inspired by the arthropod eye. Nature **497**, 95–99 (2013)
5. Stowasser, A., Rapaport, A., Layne, J.E., Morgan, R.C., Buschbeck, E.K.: Biological bifocal lenses with image separation. Curr. Biol. **20**, 1482–1486 (2010)
6. Bland, K., Revetta, N.P., Stowasser, A., Buschbeck, E.K.: Unilateral range finding in diving beetle larvae. J. Exp. Biol. **217**, 327–330 (2014)

MantisBot Uses Minimal Descending Commands to Pursue Prey as Observed in *Tenodera Sinensis*

Nicholas S. Szczecinski[✉], Andrew P. Getsy, Jacob W. Bosse,
Joshua P. Martin, Roy E. Ritzmann, and Roger D. Quinn

Case Western Reserve University, Cleveland, OH 44106, USA
nss36@case.edu
http://biorobots.case.edu/

Abstract. Praying mantises are excellent models for studying directed motion. They may track prey with rapid saccades of the head, prothorax, and legs, or actively pursue prey, using visual input to modulate their walking patterns. Here we present a conductance-based neural controller for MantisBot, a 28 degree-of-freedom robot, which enables it to use faux-visual information from a head sensor to either track or pursue prey with its prothorax and appropriate movements of one of its legs. The controller can switch between saccades and smooth tracking, as seen in pursuit, modulating only two neurons in its model thoracic ganglia via descending commands. Similarly, the neural leg controller redirects the direction of locomotion, and automatically produce reflex reversals seen in other insects when they change direction, via two simple descending commands.

Keywords: Real-time neural control · Praying mantis · CPG · Descending commands

1 Introduction

Praying mantises are ambush hunters who primarily use ballistic, visually-guided saccades of the head and body to track prey while standing still [7,9,12]. Starving them, however, will cause them to actively pursue prey, smoothly adjusting their gaze with their body joints as well as changes in their walking patterns. This behavior is of particular interest to roboticists, to understand how visual information affects the rhythms and reflexes that produce walking.

Walking is made of two phases, stance and swing (See [2] for a thorough review of how animals coordinate their walking). In stance phase, muscles must activate to support the body and move it in the intended direction. Depending on the linear and angular velocity of the animal in the ground plane, reflexes

---

N.S. Szczecinski—This work was supported by a NASA Office of the Chief Technologists Space Technology Research Fellowship (Grant Number NNX12AN24H).

N.F. Lepora et al. (Eds.): Living Machines 2016, LNAI 9793, pp. 329–340, 2016.
DOI: 10.1007/978-3-319-42417-0_30

are modulated to excite the appropriate muscles [1,8]. The networks that control locomotion are distributed throughout the nervous system, so presumably the brain does not directly modulate these reflexes. Populations in the central complex (CX) code for intended direction [6], and stimulating these same neurons will modulate inter-joint reflexes [8]. But it is also known that descending neurons in the ventral nerve cord carry information about visual targets, presumably to the thoracic networks [16]. This suggests that the intended direction of motion is communicated as descending commands to the thoracic ganglia, which use this information to modulate reflexes locally.

In this paper we present a conductance-based neural controller for MantisBot, shown in Fig. 1 and described in [14], that can reproduce visually-guided saccades like those seen in the animal, actively "pursue" prey with appropriate movements of one leg, and adjust the direction and load-specificity of walking based on simple descending commands. The robot autonomously switches between these modes depending on the distance and location of the prey in its visual field. The walking controller requires only minimal parameter tuning based on the kinematics of the leg. The tuning is completely automated and takes less than five minutes to design the entire robot's controller. The resulting controller is a hypothesis of how low-level locomotion control networks may be structured to readily implement the reflex reversals observed in animals via only minimal descending commands.

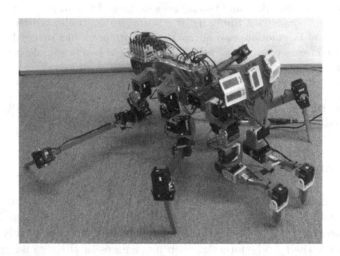

Fig. 1. MantisBot supporting its own weight on four legs, shown with its head sensor.

2   Methods

2.1   Animal Experimental Procedures

Adult female praying mantises, *Tenodera sinensis*, were starved for 3 to 5 days prior to the experiment. Individual animals were placed in a clear acrylic arena, which was then placed on top of a computer monitor. High speed video (120 fps) was taken as the mantis was presented with an artificial stimulus, a 1×2 cm black oval on a white background, which moved at 4 cm/s through a 60° arc in the center of the praying mantis's visual field. The animal was presented with this stimulus two times each at 3, 6, 9, 12, 15, and 18 cm away from the head.

2.2   Animal Data Analysis

Using the software WINanalyze (Mikromak, Berlin, Germany), the angle of the artificial stimulus and the praying mantis's head were tracked. For the artificial stimulus this angle was the angle between a line going from the center of the arc to the origin (line A), and another line from the origin to the stimulus's position. For the praying mantis's head, this angle was between line A and a line projecting perpendicularly out of the center of the animal's, representing the direction that the animal was focusing in any given frame of video. If the animal did not move significantly from its starting position, then these angles were used verbatim, but if the animal moved closer to the stimulus, they were adjusted according to a factor that varied with the distance that the praying mantis moved from its starting position to account for the fact that the visual angle to the stimulus will change if the origin moves. The animal's behaviors were also quantified by observing the video and tracking the frames in which the behavior occurred.

2.3   Robot Experimental Procedures

MantisBot's thorax was fixed in all six dimensions by bolting it to a rigid arm suspended from a wheeled frame. When the trial began, all leg joints were moved to their zero positions, and the strain gage in the leg was tared. Once the leg moved from its initial posture, an 11.4 cm block topped with a sheet of Teflon^{TM} was placed under the foot. This low-friction surface enabled the leg to register leg strain when in stance phase, but allowed the foot to slip laterally. A 1600 lm, 23 W compact fluorescent light bulb was then presented in front of the robot as a faux-visual prey-like signal. An array of solar cells is used to detect the angular position and distance of the light source [5].

3   Animal Behavior

Mantises track prey with saccadic motions of the head, prothorax and thorax [7,9,12]. When starved, they may actively pursue prey, walking toward the target. Figure 2 shows the orientation of the head when presented with a prey-like stimulus. When the stimulus starts far from the mantis, it approaches, smoothly adjusting its heading over time. Once the mantis comes within reach of the prey, it strikes, and later tracks with saccades. Starting the stimulus nearer to the mantis, as in the second trial, evokes typical saccades.

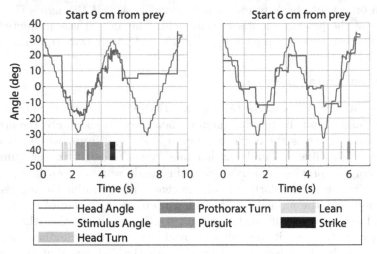

Fig. 2. Traces of mantis gaze compared to an artificial stimulus. The mantis's state is encoded by the bars below the traces. (Color figure online)

4   Network Structure

All neurons in this controller are modeled as nonspiking leaky integrators. Most neurons have no additional ion channels, but those in the CPGs possess a persistent sodium channel to provide additional nonlinear dynamics [4]. A neuron's membrane voltage V fluctuates according to

$$C_{mem}\frac{dV}{dt} = G_{mem}\cdot(E_{rest}-V)+\sum_{i=1}^{n} g_{syn}\cdot(E_{syn}-V)+G_{Na}m_{\infty}h\cdot(E_{Na}-V) \quad (1)$$

in which C is a constant capacitance, G is a constant conductance, g is a variable conductance, E is a constant voltage, and the subscript mem stands for membrane, syn stands for synaptic, and Na stands for persistent sodium. m and h are dynamical variables that describe the sodium channel activation and deactivation, respectively. m is fast, and thus is simulated as its steady state value m_{∞}. h changes according to

$$\dot{h} = (h_{\infty} - h)/\tau_h(V). \quad (2)$$

Both m_{∞} and h_{∞} are sigmoids, and have the same form,

$$z_{\infty}(V) = \frac{1}{1 + A_z \cdot \exp(S_z \cdot (V - E_z))}, \quad (3)$$

the primary difference being that $S_m > 0$ and $S_h < 0$. The full simulation, including all parameter values, is available for download at http://biorobots. case.edu/projects/legconnect/mantisbot/.

4.1   Head Networks

The robot uses the position of the light to decide whether to remain stationary and track or actively pursue it. It then uses this information to generate motion that moves the prey toward the center of its field of view. The hierarchical network in Fig. 3 enacts these behaviors.

The top section in Fig. 3 models the function of networks in the lobula and other visual processing centers, identifying the distance, angular position, and velocity of visual targets. The *Left* neuron receives a current proportional to the head sensor's left/right comparison. The *Right* neuron rests at a voltage that signifies centered prey. The comparison between the *Right* and *Left* neurons activates the *Error Right* and *Error Left* neurons. When in the *Track* state, the *Down Bias* neuron is tonically active, inhibiting the *Error Right* and *Error Left* neurons. This ensures that no error is detected as the target moves within 10° of center, mimicking the "dead zone" observed in mantises as they saccade toward prey [12]. If either *Error* neuron becomes active, it will excite both the *Error Large* and *Slow* neurons. These neurons are identical, except that the *Slow* neuron has a larger capacitance, and thus larger time constant. The *Trigger* neuron then computes the difference between these, forming a Reichardt detector [11]. Thus *Trigger* detects when the prey is leaving the "dead zone," requiring corrective action.

The visual center also computes the distance to the target by mapping the head's center panel reading to a $1/r^2$ relationship, which controls *Dist.*'s activity. When the distance to the target is small enough, *Too Close* is disinhibited, halting pursuit. The visual center also computes the velocity of the visual input via another Reichardt detector [11].

Saccades are by their nature brief bursts of motion. Therefore the *Sacc.* neuron is configured to produce brief, stereotypical pulses when *Trigger* becomes active. *Sacc.* and *No Sacc.* include persistent sodium channels, which are frequently used to model CPGs in locomotion models [4]. By biasing *No Sacc.* upward with a 1 nA tonic current, the equilibrium state of the *Sacc./No Sacc.* system is for *No Sacc.* to be tonically active, and *Sacc.* to be inactive. When *Trigger* inhibits the interneuron, it disinhibits *Sacc.*, causing it to temporarily inhibit *No Sacc.*, producing a temporary pulse which disinhibits the *Sacc. Right* and *Sacc. Left* neurons. These pulses can be observed in Fig. 6.

Flashing targets at mantises reduces the latency of subsequent saccades in the same direction [7]. This suggests that there is some memory of the previous saccade direction, which primes the system to move in that direction again. Thus the *Left Mem.* and *Right Mem.* neurons mutually inhibit one another such that exactly one is always tonically active while the other is inactive. The *Error Right* and *Error Left* neurons can change the state of this memory, which then gate the *Sacc. Right* and *Sacc. Left* neurons. These, in turn, gate the *Right Vel.* and *Left Vel.* neurons, which ultimately stimulate the motor neurons of the prothorax and leg joints.

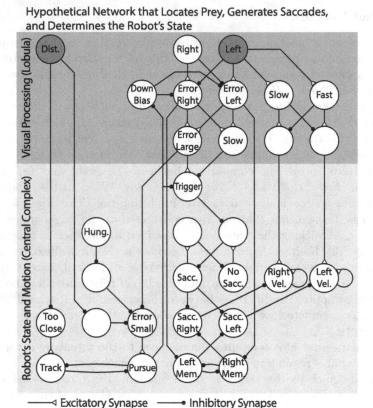

Fig. 3. Network that processes visual information for MantisBot and communicates descending commands to the thoracic ganglia model, partially shown in Fig. 4.

4.2   Thoracic Networks

Insects' nervous systems are highly distributed. How, then, are individual joints controlled to direct posture or locomotion based on descending input? Signals from load sensors signal the onset of stance phase, causing supportive and propulsive muscles to contract. Which muscles are activated depends on the direction of travel [1]. Signals from movement sensors signal the end of stance phase, causing the leg to lift from the ground and return to its position prior to stance [3]. Figure 4 shows a control network for one joint based on these observations, which underpins the leg controller in this work. Every joint's controller is identical, except for the weight of the four synapses drawn in bold, and the resting potential of the *Des. PEP* and *Des. AEP* neurons. The network's function directly informs the tuning of these values, as shown below.

Our joint network receives descending commands about the position and velocity of the prey signal, as well as the state of the robot, whether statically tracking or actively pursuing prey. The labels PEP (posterior extreme position)

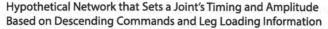

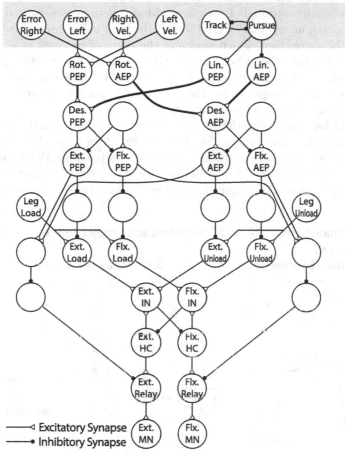

Fig. 4. Network that determines the timing and amplitude of the motion of each joint. With only minimal tuning and limited descending commands, the network produces walking motion and reflex reversals to change the travel.

and AEP (anterior extreme position) refer only to the intended position of the joint at the end of stance and swing, respectively. When *Track* is active, the transition to swing is inhibited, and the PEP is simply the desired position of that joint. If *Pursue* is active, then the AEP and PEP are different, establishing the beginning of stance and swing, respectively. This is manifested by the excitation of *Lin. PEP* and the inhibition of *Lin. AEP*. Visual information affects the rotation of the PEP (*Rot. PEP*) and AEP (*Rot. AEP*). Even though this paper only presents results from one leg of MantisBot, our design method is general, and is intended to control all the legs simultaneously. Here we describe how to tune the network in Fig. 4 for a single joint. If we were to translate and rotate some point on the robot distance p and angle θ, as shown in Fig. 5, then a particular foot would have to move distance d at angle ϕ from straight ahead to stay in the original position. For each ϕ, the Moore-Penrose generalized inverse of the leg's Jacobian matrix can be used to find the joint velocities required to move the foot in the $\hat{d}$ direction, which minimize the norm of the joint velocity vector [10]. The ratio between joint velocities in the leg specify the relative proportion of motor neuron activation required to move the foot in a straight line in the direction specified. If negative position feedback is present at each joint, then scaling this proportion will specify an equilibrium position.

Each joint has a map of its deflection from rest as a function of the body's rotation and translation (for example, Fig. 5C). We can draw a line showing the body motions that would require no change in the joint's position. Points above that line represent body motions that would require the joint to flex, and those below correspond to motions that require extension. Because this map is specifying the

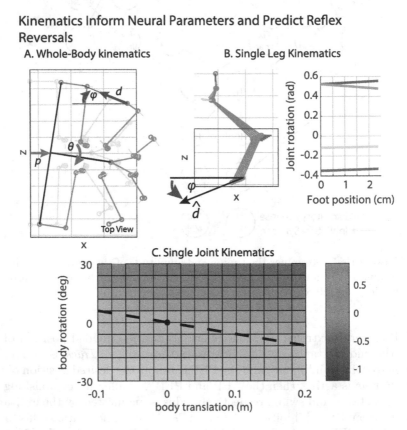

Fig. 5. Inverse kinematics used to design the joint network. To move the robot's body in a particular direction (A), each leg must move in a different direction (B). The inverse kinematics reveal at which body rotations and translations the joint must change its direction of travel (C). (Color figure online)

PEP of the leg, the line predicts at what locomotion speeds and turning radii a reflex reversal should take place, a consideration of prior studies [8,13].

Using this method, the four synapses in bold in Fig. 4 are designed such that the descending commands, which encode intended *body* translation and rotation, can be used to command the *joint's* rotation. This network is duplicated, once to command the PEP and once to command the AEP. If the desired PEP, encoded by *Des. PEP* is greater than the resting posture, then the joint needs to extend in stance phase, and the *Ext. PEP* neuron will be active. When active, it disinhibits the pathway from the leg's strain gage, exciting *Ext. IN* and inhibiting *Flx. HC* during stance, causing the joint to extend. Changing the descending commands may instead cause *Flx. PEP* to be active, which will route load information to the opposite half-center of the CPG. The AEP network works the same way, but with a signal that the leg is unloaded. In this way the kinematics of the robot are used to not only cause desired foot trajectories, but also enforce known reflex reversals with minimal descending commands.

5   Results

Figure 6 shows that the network in Fig. 3 can accurately orient the robot toward prey using the same strategy as the animal. The target is tracked smoothly when in active pursuit, and with saccades when tracking. Locomotion also halts when this change takes place, and produces saccadic motions afterward. When the target strays from the center of view, the *Error Left* or *Error Right* neuron's

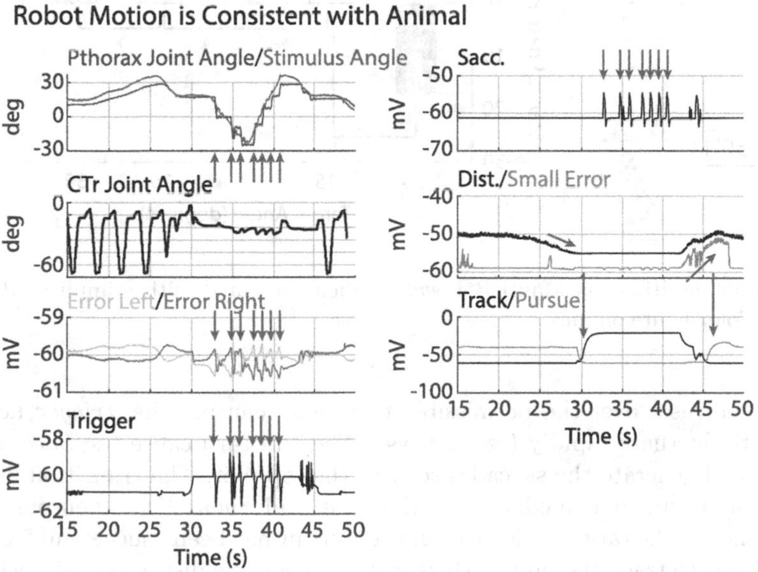

Fig. 6. Traces from MantisBot tracking a stimulus and autonomously changing between tracking and pursuit. (Color figure online)

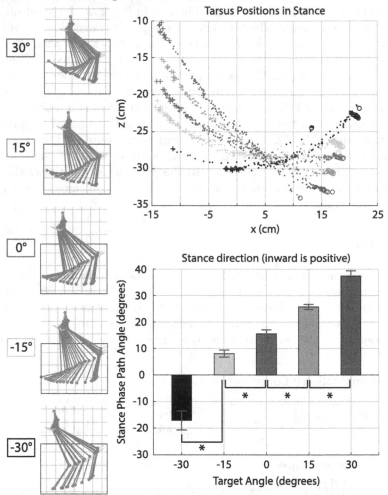

Fig. 7. Foot positions as MantisBot walks when presented with stimuli at different angles. (Color figure online)

activity increases over the permissible threshold, causing the *Trigger* neuron's activity to fluctuate rapidly (red arrows). These in turn cause the *Sacc.* neuron to pulse and generate the saccades seen in the top plot. The transition between tracking and pursuit is mediated by the *Dist.* and *Small Error* neurons. When the distance to the target is decreased, the pursuit has been "successful," and the robot begins to track the prey with saccades. When the distance again decreases and the prey is located within the field of view, the robot again pursues.

Reflex Reversals Naturally Occur from the Hypothetical Leg Controller

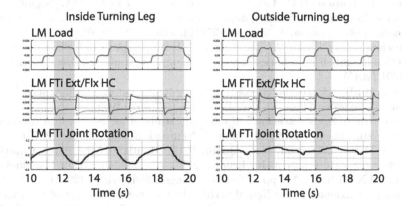

Fig. 8. The network in Fig. 4 automatically produces the reflex reversals observed in other insects.

Figure 7 shows that the network in Fig. 4 can control the direction of the foot in stance. Leg kinematics were recorded as the robot took 12 steps toward the light in five different positions. The joint angles read from the servomotors were fed through the kinematic model to create 3D plots of the leg motion over time, as well as the foot placement on the ground. Calculating the angle of the line that passes through the AEP and PEP with respect to straight ahead for each step shows that the stance direction is statistically significantly different for each target presented ($p < 0.001$, 1 sided ANOVA). This includes both inward and outward stance directions, suggesting that reflex reversals must take place to change the stepping so drastically. Figure 8 shows that this is the case; the sign of the load information fed to the femur-tibia joint is reversed by the network in Fig. 4. Rather than guessing when this transition should occur, our network automatically changes the sign of this reflex with minimal descending commands.

6 Conclusion

The data shows that this joint controller is capable of producing basic stance and swing motions, and modifying them with minimal descending commands. Insects use a host of reflexes to counteract external forces and transition from stance to swing, and vice versa (for a review, see [2]). Reflexes in response to decreasing or excessive load, as well as searching for missing footholds, will be added in the future [15]. In the future, this controller will be expanded to use all of the legs of MantisBot to orient towards and pursue faux-visual targets.

References

1. Akay, T., Ludwar, B.C., Göritz, M.L., Schmitz, J., Büschges, A.: Segment specificity of load signal processing depends on walking direction in the stick insect leg muscle control system. J. Neurosci. **27**(12), 3285–3294 (2007)
2. Buschmann, T., Ewald, A., Twickel, A.V., Büschges, A.: Controlling legs for locomotion insights from robotics and neurobiology. Bioinspiration Biomimetics **10**(4), 41001 (2015)
3. Cruse, H.: Which parameters control the leg movement of a walking insect? II: the start of the swing phase. J. Exp. Biol. **116**, 357–362 (1985)
4. Daun-Gruhn, S.: A mathematical modeling study of inter-segmental coordination during stick insect walking. J. Comput. Neurosci. **30**(2), 255–278 (2010)
5. Getsy, A.P., Szczecinski, N.S., Quinn, R.D.: MantisBot: the implementation of a photonic vision system. In: Lepora, N.F., et al. (eds.) Living Machines 2016. LNCS(LNAI), vol. 9793, pp. 429–435. Springer, Switzerland (2016)
6. Guo, P., Ritzmann, R.E.: Neural activity in the central complex of the cockroach brain is linked to turning behaviors. J. Exp. Biol. **216**(Pt 6), 992–1002 (2013)
7. Lea, J.Y., Mueller, C.G.: Saccadic head movements in mantids. J. Comp. Physiol. A **114**(1), 115–128 (1977)
8. Martin, J.P., Guo, P., Mu, L., Harley, C.M., Ritzmann, R.E.: Central-complex control of movement in the freely walking cockroach. Curr. Biol. **25**(21), 2795–2803 (2015)
9. Mittelstaedt, H.: Prey capture in mantids. In: Scheer, B.T. (ed.) Recent Advances in Invertebrate Physiology, pp. 51–72. University of Oregon, Eugene (1957)
10. Murray, R.M., Li, Z., Sastry, S.S.: A Mathematical Introduction to Robotic Manipulation. CRC Press, Boca Raton (1994)
11. Reichardt, W.: Autocorrelation, a principle for the evaluation of sensory information by the central nervous system. In: Rosenblith, W.A. (ed.) Sensory Communication, pp. 303–317. MIT Press, Cambridge, MA (1961)
12. Rossel, S.: Foveal fixation and tracking in the praying mantis. J. Comp. Physiol. A Neuroethology Sens. Neural Behav. Physiol. **139**, 307–331 (1980)
13. Szczecinski, N.S., Brown, A.E., Bender, J.A., Quinn, R.D., Ritzmann, R.E.: A neuromechanical simulation of insect walking and transition to turning of the cockroach blaberus discoidalis. Biol. Cybern. **108**(1), 1–21 (2013)
14. Szczecinski, N.S., Chrzanowski, D.M., Cofer, D.W., Moore, D.R., Terrasi, A.S., Martin, J.P., Ritzmann, R.E., Quinn, R.D.: MantisBot: a platform for investigating mantis behavior via real-time neural control. In: Wilson, S.P., Verschure, P.F.M.J., Mura, A., Prescott, T.J. (eds.) Living Machines 2015. LNCS, vol. 9222, pp. 175–186. Springer, Heidelberg (2015)
15. Szczecinski, N.S., Chrzanowski, D.M., Cofer, D.W., Terrasi, A.S., Moore, D.R., Martin, J.P., Ritzmann, R.E., Quinn, R.D.: Introducing MantisBot: hexapod robot controlled by a high- fidelity, real-time neural simulation. In: IEEE International Conference on Intelligent Robots and Systems. pp. 3875–3881. Hamburg, DE (2015)
16. Yamawaki, Y., Toh, Y.: A descending contralateral directionally selective movement detector in the praying mantis tenodera aridifolia. J. Comp. Physiol. A **195**(12), 1131–1139 (2009)

Eye-Head Stabilization Mechanism for a Humanoid Robot Tested on Human Inertial Data

Lorenzo Vannucci[1]([✉]), Egidio Falotico[1], Silvia Tolu[2], Paolo Dario[1], Henrik Hautop Lund[2], and Cecilia Laschi[1]

[1] The BioRobotics Institute, Scuola Superiore Sant'Anna,
Viale Rinaldo Piaggio 34, 56025 Pontedera, Pisa, Italy
{l.vannucci,e.falotico,p.dario,c.laschi}@sssup.it
[2] Department of Electrical Engineering, The Center for Playware,
Technical University of Denmark, Elektrovej Building 326,
2800 Kongens Lyngby, Copenhagen, Denmark
{stolu,hhl}@elektro.dtu.dk

Abstract. Two main classes of reflexes relying on the vestibular system are involved in the stabilization of the human gaze: the vestibulocollic reflex (VCR), which stabilizes the head in space and the vestibulo-ocular reflex (VOR), which stabilizes the visual axis to minimize retinal image motion. Together they keep the image stationary on the retina.

In this work we present the first complete model of eye-head stabilization based on the coordination of VCR and VOR. The model is provided with learning and adaptation capabilities based on internal models. Tests on a simulated humanoid platform replicating torso disturbance acquired on human subject performing various locomotion tasks confirm the effectiveness of our approach.

Keywords: Head stabilization · VOR · VCR · Eye-head coordination · Humanoid robotics

1 Introduction

Several neuroscientific studies focus on eye-head behaviour during locomotion. Results about head and eyes during walking mostly come from two-dimensional studies on linear overground, turning, treadmill locomotion, running and walking on compliant surface [1–6]. These studies have shown that the body, head, and eyes rotate in response to the up-down and side-to-side motion to maintain stable head pointing and gaze in space. This is achieved through the joint effect of two main classes of reflexes, which rely on the output of the inertial system: 1. the vestibulo-ocular reflex (VOR), which stabilizes the visual axis to minimize retinal image motion; 2. the vestibulocollic reflex (VCR), which stabilizes the head in space through the activation of the neck musculature in response to vestibular inputs. The VOR compensates for head movements that would perturb vision

© Springer International Publishing Switzerland 2016
N.F. Lepora et al. (Eds.): Living Machines 2016, LNAI 9793, pp. 341–352, 2016.
DOI: 10.1007/978-3-319-42417-0_31

by turning the eye in the orbit in the opposite direction of the head movements [7]. Several approaches have been used to model the VOR depending on the aim of the study. In robotics literature we found some controllers inspired by the VOR [8–11]. The VCR stabilizes the head based on the inertial input space by generating a command that moves the head in the opposite direction to that of the current head in space displacement. When the head is rotated in the plane of a semicircular canal, the canal is stimulated and the muscles are activated. This stimulation produces a compensatory rotation of the head in the same plane. If more than one canal is activated, then an appropriate reflex response is produced. Unlike the VOR, the VCR controls a complex musculature. The VOR involves six extraocular muscles, each pair acts around a single rotation axis. On the other hand, the neck has more than 30 muscles controlling pitch, roll and yaw rotations.

In robotics, some head stabilization models already exist implemented on humanoid robots. Gay et al. [12] proposed a head stabilization system for a bipedal robot during locomotion controlled by the optical flow information. It is based on Adaptive Frequency Oscillators to learn the frequency and phase shift of the optical flow. Although the system can successfully stabilize the head of the robot during its locomotion, it does not take in consideration the vestibular inputs. The most close to the neuroscientific findings of the VCR are the works proposed by Kryczka et al. [13–15]. They proposed an inverse jacobian controller [13,14] based on neuroscientific results [16] and an adaptive model based on a feedback error learning (FEL) [15] able to compensate the disturbance represented by the trunk rotations. All the presented models try to reproduce specific aspects of the gaze stabilization behaviour, but none of them can provide a comprehensive model of gaze stabilization, integrating eye stabilization (VOR) together with head stabilization (VCR).

By considering the analysis of neuroscience findings, we can conclude that in order to replicate eye-head stabilization behaviours found in humans it is necessary to be able to replicate the joint effect of VCR for the head and VOR for the eye. This work goes in this direction by presenting a model that replicates the coordination of VCR and VOR and is suitable for the implementation on a robotic platform. We used, as a disturbance motion, inertial data acquired on a human subject performing various locomotion tasks (straight walking, running, walking a curved path on normal and soft ground) and replicated by the torso of a humanoid robot. The purpose of these tests is to assess the effectiveness of the stabilization capabilities of the proposed model rejecting the torso disturbance measured in real walking tasks through the joint stabilizing effect of head and eye of the simulated iCub robot.

2   Eye-Head Stabilization Model

In order to implement the VOR-VCR system, a bio-inspired feed-forward control architecture was used. The model uses classic feedforward controllers that generate motor commands purely based on the current error. Each controller is coupled with a learning network that generates predictions based on internal models that are used to fine tune motor commands. An overview of the model can be seen in Fig. 1.

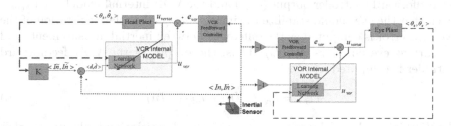

Fig. 1. The proposed model of eye-head stabilization. Dashed lines represent encoder readings, dotted lines show inertial values.

2.1 Head Stabilization System

Inside the head stabilization system, the output of the VCR internal model (u_{vcr}) is added to the output of the feedforward controller (e_{vcr}) in order to generate motor commands that stabilized the head against the disturbance originating from the torso movements. The VCR Feedforward Controller is implemented as a PD controller, and its output is computed as a function of the inertial readings ($In, \dot{In}$):

$$e_{vcr} = k_p \cdot In + k_d \cdot \dot{In}. \tag{1}$$

The inputs to the learning network are the current and desired position and velocity of the robotic head, and the network is trained with newly generated motor commands. In order to provide a proper reference to the VCR internal model, the current value of the external disturbance must be estimated. Using the readings coming from the inertial measurement unit and the encoder values, the disturbance vector (d) can be estimated using only direct kinematics functions, by computing $d = In - \tilde{In}$, i.e. by subtracting the expected angular rotations given by the encoder values ($\tilde{In}$) from the inertial readings (In). $\tilde{In} = [\varphi, \vartheta, \psi]$ are the Euler angles for the rigid roto-translation matrix $K(\theta_h)$ from the root reference frame to the inertial frame, computed as:

$$\varphi = atan2(-K(\theta_h)_{2,1}, K(\theta_h)_{2,2}), \tag{2}$$
$$\vartheta = asin(K(\theta_h)_{2,0}), \tag{3}$$
$$\psi = atan2(-K(\theta_h)_{1,0}, K(\theta_h)_{0,0}). \tag{4}$$

Likewise, the same procedure can be followed in order to estimate the velocity of the disturbance:

$$\dot{d} = \dot{In} - \dot{\tilde{In}} = \dot{In} - J(\theta_h) \cdot \dot{\theta}_h, \tag{5}$$

where J is the geometric Jacobian from the root reference frame to the inertial frame.

2.2 Eye Stabilization System

The eye stabilization system implements the VOR and, similarly to the head stabilization system, produces a motor command for the eyes that is the sum of the feedforward controller output (e_{vor}) and the VOR internal model one (u_{vor}). Given that the eye should stabilize the image against the relative rotation of the head, the error is computed as the difference between inertial measurements and the current eye encoders (θ_e, $\dot{\theta}_e$). Thus, the output of the VOR feedforward controller is computed as

$$e_{vor} = k_p \cdot (-In) + k_d \cdot (-\dot{In}). \tag{6}$$

The VOR internal model receives in input the head position and velocity signal as references, acquired through the vestibular system, along with the proprioceptive feedback, and uses the generated motor command as a training signal.

2.3 Learning Network

Prediction of the internal model is provided by a learning network that is implemented with a machine learning approach, Locally Weighted Projection Regression (LWPR) [17]. This algorithm has been proved to provide a representation of cerebellar layers that in humans are responsible for the generation of predictive motor signals that produce more accurate movements [18,19]. The LWPR spatially exploits localized linear models at a low computational cost through an online incremental learning. Therefore, the prediction process is quite fast, allowing real-time learning. LWPR incrementally divides the input space into a set of receptive fields defined by a centre c_k and a Gaussian area characterized by a positive definite distance matrix D_k. The activation on each receptive field k in response to an input x is expressed by

$$p_k(x) = exp\left(-\frac{1}{2}(x - c_k)^T D_k (x - c_k)\right), \tag{7}$$

while the output is $y_k(x) = w_k \cdot x + \epsilon_k$, where w_k and ϵ_k are the weight vector and bias associated with the k-th linear model. For each iteration, the new input, x, is assigned to the closest RF based on its weight activation, and consequently, the centre, the weights and the kernel width are updated proportionally to a training signal. Moreover, the number of local models increases with the complexity of the input space.

The global output of the LWPR is given by the weighted mean of all the predictions y_k of the linear local models created:

$$u(x) = \frac{\sum_{k=1}^{N} p_k(x) y_k(x)}{\sum_{k=1}^{N} p_k(x)}. \tag{8}$$

3   Experimental Procedure

In order to collect human inertial data relative to locomotion tasks, experiments were conducted on a human subject with no visual and vestibular impairments. An inertial measurement unit (IMU) was placed on the back of the subject, near T10, the tenth vertebra of the thoracic spine, as depicted in Fig. 2.

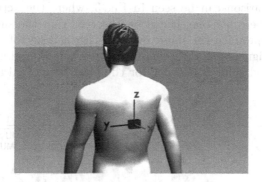

Fig. 2. Placement of the inertial measurement unit on the subject.

The IMU used was an Xsens MTi orientation sensor[1], that incorporates an on-board sensor fusion algorithm and Kalman filtering. The inertial unit is able to produce the current orientation of the torso at a frequency of 100 Hz.

Three different tasks were performed by the subject: straight walking (25 m), circular walking and straight running (25 m). The circular walking was carried out by asking the subject to walk with a circular pattern, without any indication of the pattern on the ground. Such task was executed both on normal and soft ground, provided by placing a foam rubber sheet on the ground. The foam had a density of $40 \, kg/m^3$ and the sheet measured $103 \times 160 \times 15$ cm. All tasks were performed with bare feet.

Due to the fact that the inertial readings relative to the yaw rotational axis (rotation around z) can often be inaccurate because of drifting, we decided not to use such readings. Moreover, in order to prevent drifts of the sensor measurements on the other two rotational axis (pitch and roll, rotations around y and x respectively), each trial lasted less than one minute with a reset of the rotational angle at the beginning of the trial [20].

4   Robotic Platform

The proposed model was implemented for the iCub robot simulator [21], a software included with the iCub libraries. The iCub head contains a total of 6 degrees of freedom: 3 for the neck (pan, tilt and swing) and 3 for the eyes (a common

---

[1] https://www.xsens.com/products/mti/.

tilt, version and vergence), while the torso has 3 degrees of freedom (pan, tilt and swing). The visual stereo system consists of 2 cameras with a resolution of 320×240 pixels.

In order to assess the repeatability of the experiments on the iCub simulator, first test were conducted to evaluate whether the measurements of the simulated robot inertial rotations were compatible with the collected data. Thus, the collected torso rotations were given as motor commands to the robot torso. A graphical comparison can be seen in Fig. 3, where the actual IMU data is shown alongside the robot one. It can be observed that the simulation is accurate enough to reproduce the data, even if with a delay of 50 ms. The error between the two signals was then computed after a temporal alignment and its Root Mean Squared value is 0.21 deg for the pitch rotational axis and 0.12 deg for the roll.

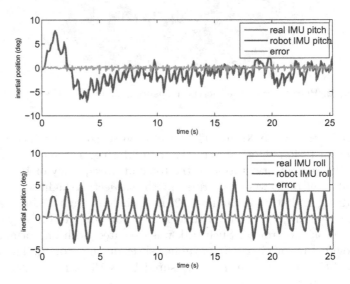

Fig. 3. Comparison between collected IMU readings and simulated ones, on pitch rotational axis (top) and roll (bottom). (Color figure online)

5   Results

The stabilization model was tested on the data coming from the three different locomotion tasks (straight walking, circular walking and straight running). Due to the fact that the collected inertial data related to the yaw rotational axis was not considered, the eye-head stabilization model has been simplified, so that no stabilization on the yaw axis was performed. Moreover, given that the robot eyes cannot influence stabilization on the roll rotational axis, due to the fact that only tilt and pan motors are present, only disturbance on the pitch axis was compensated by the VOR model.

The main measure of error during a stabilization task is the movement of the camera image. In particular, human vision is considered stable if the retinal slip (the speed of the image on the retina) is under 4 deg/s [22]. In order to compute the error from the camera image, a target was placed in front of the simulated robot and its position was tracked from the camera images via a colour filtering algorithm during the execution of the task. Another measure of performance considered is the inertial orientation and speed of the head. As already stated before, no movement on the yaw rotational axis was considered, thus only the camera error on the vertical axis is relevant for the evaluation.

For each task, a comparison between the same task performed with and without the stabilization model will be presented. The values of the gains of the PD controllers were set to $k_p = 5.0, k_d = 0.1$ for the VCR model and to $k_p = 1.0, k_d = 0.1$ for the VOR model, for all trials.

5.1   Straight Walking

Results for the compensation of the disturbance of straight walking inertial data can be found in Table 1, where the Root Mean Square (RMS) values for inertial readings and target position and speed are presented. In this and subsequent tables, $In_p, \dot{In}_p$ are the inertial readings for rotation (deg) and rotation speed (deg/s) on the pitch axis, $In_r, \dot{In}_r$ are the inertial readings for rotation (deg) and rotation speed (deg/s) on the roll axis, $v, \dot{v}$ are the position of the target on the camera image (deg) and its speed (retinal slip, deg/s).

Table 1. Results for straight walking data.

Trial	In_p	$\dot{In}_p$	In_r	$\dot{In}_r$	v	$\dot{v}$
No stabilization	2.06	11.36	2.05	11.56	3.66	8.88
Stabilization	0.84	14.71	0.22	3.45	2.71	3.30

Figures 4, 5 and 6 show the behaviour of the task, showing the target position and retinal slip, inertial data for the pitch rotational axis and inertial data for the roll axis, respectively. From these results it can be noticed that while the roll disturbance is almost completely compensated by the VCR model, the magnitude of the rotational velocity on the pitch axis is too high to be fully compensated by the said model, that only provides an improvement in the position space. Nevertheless, the VOR subsystem is still able to maintain the camera image stable, with a mean vertical retinal slip lower than 4 deg/s. Moreover, Fig. 4 also shows a comparison between the full stabilization model and a simplified model with only the PD controllers. While the PD only implementation is able to reduce the error on the camera, it is outperformed by the complete model, thus proving the effectiveness of the latter.

5.2 Circular Walking

Two sets of data were collected for circular walking tasks: one for normal ground and one for soft ground. Results for both cases are presented in Table 2, where it can be observed that walking on soft ground produces a greater disturbance, especially in the velocity space. Despite the higher disturbance the model is still able to stabilize the head and the camera image, achieving stable vision in both cases. As in the straight walking case, the disturbance on the pitch axis cannot be

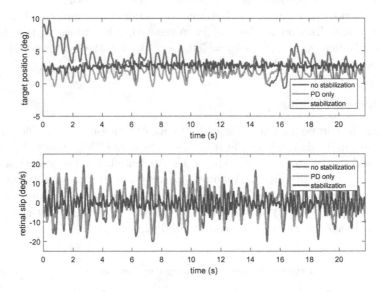

Fig. 4. Stabilization task with data from a straight walking task, target position (top) and retinal slip (bottom). (Color figure online)

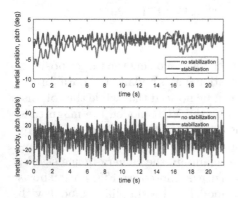

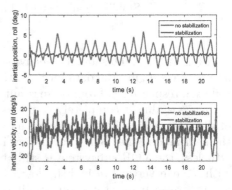

Fig. 5. Stabilization task with data from a straight walking task, inertial position data (top) and inertial velocity (bottom), pitch axis. (Color figure online)

Fig. 6. Stabilization task with data from a straight walking task, inertial position data (top) and inertial velocity (bottom), roll axis. (Color figure online)

fully compensated by the VCR alone, but thanks to the VOR module, the vision remains stable. The behaviour on the soft ground task can be observed in Fig. 7.

Table 2. Results for circular walking data on normal and soft ground.

Ground	Trial	In_p	$I\dot{n}_p$	In_r	$I\dot{n}_r$	v	$\dot{v}$
Normal	no stabilization	3.35	14.03	2.11	6.36	3.61	12.86
Normal	stabilization	0.49	6.34	0.17	2.84	0.88	1.85
Soft	no stabilization	3.09	16.00	3.89	10.95	4.23	15.15
Soft	stabilization	0.46	6.35	0.24	3.07	2.74	3.56

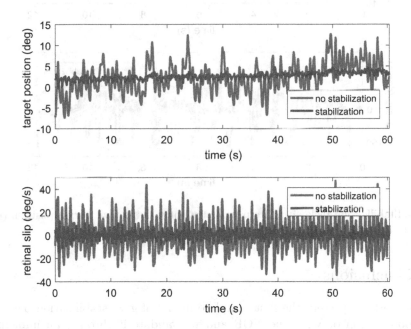

Fig. 7. Stabilization task with data from a circular walking task on soft ground, target position (top) and retinal slip (bottom). (Color figure online)

5.3   Straight Running

During the last experiment, data from the straight running was used to move the robot torso. From Table 3 it can be observed that the model is not able to achieve a complete compensation of the disturbance, due to the high rotational velocities on the two axes. Nevertheless, the mean retinal slip is reduced to a quarter of the one of the trial with no stabilization. Thus, the model provides a viable solution even for disturbances of this magnitude, as it is also shown in Fig. 8.

Table 3. Results for straight running data.

Trial	In_p	$I\dot{n}_p$	In_r	$I\dot{n}_r$	v	$\dot{v}$
No stabilization	3.14	45.36	1.77	16.28	3.49	25.96
Stabilization	1.06	43.03	0.53	10.78	1.61	6.36

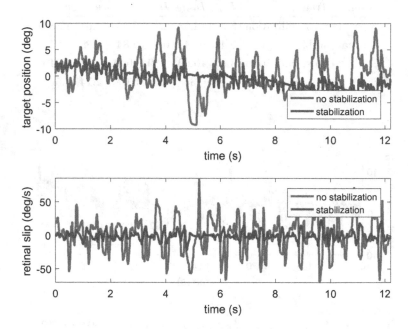

Fig. 8. Stabilization task with data from a straight running task, target position (top) and retinal slip (bottom). (Color figure online)

6   Conclusions

In this work we present the first complete model of gaze stabilization based on the coordination of VCR and VOR and we validate it through an implementation on a simulated humanoid robotic platform. We tested the model using, as a disturbance motion, inertial data acquired on a human subject performing various locomotion tasks that we replicated with the torso of the simulated iCub robot. Results show that the model is able to perform well in almost all trials, with the exception of the straight running task, by reducing the retinal slip below 4 deg/s, thus achieving stable vision. In the running task, the model was still able to improve the stabilization by reducing the retinal slip to a quarter of the one from the task were no stabilization was present. As such, this model has proven suitable to be used on humanoid robotic platforms, where it could help during visually guided locomotion tasks by stabilizing the camera view against the disturbance produced by walking.

Acknowledgment. The research leading to these results has received funding from the European Union Seventh Framework Programme (FP7/2007-2013) under grant agreement no. 604102 (Human Brain Project). The authors would like to thank the Italian Ministry of Foreign Affairs, General Directorate for the Promotion of the "Country System", Bilateral and Multilateral Scientific and Technological Cooperation Unit, for the support through the Joint Laboratory on Biorobotics Engineering project.

References

1. Imai, T., Moore, S.T., Raphan, T., Cohen, B.: Interaction of the body, head, and eyes during walking and turning. Exp. Brain Res. **136**(1), 1–18 (2001)
2. Hirasaki, E., Moore, S.T., Raphan, T., Cohen, B.: Effects of walking velocity on vertical head and body movements during locomotion. Exp. Brain Res. **127**(2), 117–130 (1999)
3. Pozzo, T., Berthoz, A., Lefort, L., Vitte, E.: Head stabilization during various locomotor tasks in humans. Exp. Brain Res. **85**(1), 208–217 (1991)
4. Nadeau, S., Amblard, B., Mesure, S., Bourbonnais, D.: Head and trunk stabilization strategies during forward and backward walking in healthy adults. Gait Posture **18**(3), 134–142 (2003)
5. Hashimoto, K., Kang, H.J., Nakamura, M., Falotico, E., Lim, H.O., Takanishi, A., Laschi, C., Dario, P., Berthoz, A.: Realization of biped walking on soft ground with stabilization control based on gait analysis. In: IEEE International Conference on Intelligent Robots and Systems, pp. 2064–2069 (2012)
6. Kang, H.J., Hashimoto, K., Nishikawa, K., Falotico, E., Lim, H.O., Takanishi, A., Laschi, C., Dario, P., Berthoz, A.: Biped walking stabilization on soft ground based on gait analysis. In: Proceedings of the IEEE RAS and EMBS International Conference on Biomedical Robotics and Biomechatronics, pp. 669–674 (2012)
7. Barnes, G.: Visual-vestibular interaction in the control of head and eye movement: the role of visual feedback and predictive mechanisms. Prog. Neurobiol. **41**(4), 435–472 (1993)
8. Shibata, T., Schaal, S.: Biomimetic gaze stabilization based on feedback-error-learning with nonparametric regression networks. Neural Netw. **14**(2), 201–216 (2001)
9. Viollet, S., Franceschini, N.: A high speed gaze control system based on the vestibulo-ocular reflex. Robot. Auton. Syst. **50**(4), 147–161 (2005)
10. Porrill, J., Dean, P., Stone, J.V.: Recurrent cerebellar architecture solves the motor-error problem. Proc. Roy. Soc. Lond. B **271**(1541), 789–796 (2004)
11. Franchi, E., Falotico, E., Zambrano, D., Muscolo, G., Marazzato, L., Dario, P., Laschi, C.: A comparison between two bio-inspired adaptive models of vestibulo-ocular reflex (VOR) implemented on the iCub robot. In: 2010 10th IEEE-RAS International Conference on Humanoid Robots, Humanoids 2010, pp. 251–256 (2010)
12. Gay, S., Santos-Victor, J., Ijspeert, A.: Learning robot gait stability using neural networks as sensory feedback function for central pattern generators. In: 2013 IEEE/RSJ International Conference on Intelligent Robots and Systems (IROS), pp. 194–201, November 2013
13. Kryczka, P., Falotico, E., Hashimoto, K., Lim, H., Takanishi, A., Laschi, C., Dario, P., Berthoz, A.: Implementation of a human model for head stabilization on a humanoid platform. In: Proceedings of the IEEE RAS and EMBS International Conference on Biomedical Robotics and Biomechatronics, pp. 675–680 (2012)

14. Kryczka, P., Falotico, E., Hashimoto, K., Lim, H.O., Takanishi, A., Laschi, C., Dario, P., Berthoz, A.: A robotic implementation of a bio-inspired head motion stabilization model on a humanoid platform. In: IEEE International Conference on Intelligent Robots and Systems, pp. 2076–2081 (2012)

15. Falotico, E., Cauli, N., Hashimoto, K., Kryczka, P., Takanishi, A., Dario, P., Berthoz, A., Laschi, C.: Head stabilization based on a feedback error learning in a humanoid robot. In: Proceedings - IEEE International Workshop on Robot and Human Interactive Communication, pp. 449–454 (2012)

16. Falotico, E., Laschi, C., Dario, P., Bernardin, D., Berthoz, A.: Using trunk compensation to model head stabilization during locomotion. In: IEEE-RAS International Conference on Humanoid Robots, pp. 440–445 (2011)

17. Vijayakumar, S., Schaal, S.: Locally weighted projection regression: incremental real time learning in high dimensional space. In: ICML 2000: Proceedings of the Seventeenth International Conference on Machine Learning. Morgan Kaufmann Publishers Inc., San Francisco, pp. 1079–1086 (2000)

18. Tolu, S., Vanegas, M., Luque, N.R., Garrido, J.A., Ros, E.: Bio-inspired adaptive feedback error learning architecture for motor control. Biol. Cybern. **106**(8–9), 507–522 (2012)

19. Tolu, S., Vanegas, M., Garrido, J.A., Luque, N.R., Ros, E.: Adaptive and predictive control of a simulated robot arm. Int. J. Neural Syst. **23**(3), 1350010 (2013)

20. Bergamini, E., Ligorio, G., Summa, A., Vannozzi, G., Cappozzo, A., Sabatini, A.: Estimating orientation using magnetic and inertial sensors and different sensor fusion approaches: accuracy assessment in manual and locomotion tasks. Sens. (Switz.) **14**(10), 18625–18649 (2014)

21. Tikhanoff, V., Cangelosi, A., Fitzpatrick, P., Metta, G., Natale, L., Nori, F.: An open-source simulator for cognitive robotics research: the prototype of the icub humanoid robot simulator. In: Proceedings of the 8th Workshop on Performance Metrics for Intelligent Systems, PerMIS 2008, pp. 57–61. ACM, New York (2008)

22. Collewijn, H., Martins, A., Steinman, R.: Natural retinal image motion: origin and change. Ann. N. Y. Acad. Sci. **374**(1), 312–329 (1981)

Towards a Synthetic Tutor Assistant: The EASEL Project and its Architecture

Vasiliki Vouloutsi[1]([✉]), Maria Blancas[1], Riccardo Zucca[1], Pedro Omedas[1], Dennis Reidsma[3], Daniel Davison[3], Vicky Charisi[3], Frances Wijnen[3], Jan van der Meij[3], Vanessa Evers[3], David Cameron[4], Samuel Fernando[4], Roger Moore[4], Tony Prescott[4], Daniele Mazzei[5], Michael Pieroni[5], Lorenzo Cominelli[5], Roberto Garofalo[5], Danilo De Rossi[5], and Paul F.M.J. Verschure[1,2]

[1] SPECS, N-RAS, DTIC, Universitat Pompeu Fabra (UPF), Barcelona, Spain
{vicky.vouloutsi,maria.blancas,riccardo.zucca,pedro.omedas,
paul.verschure}@upf.edu
[2] Catalan Institute of Advanced Studies (ICREA), Barcelona, Spain
[3] Human Media Interaction/ELAN, University of Twente, Enschede, Netherlands
{d.reidsma,d.p.davison,v.charisi,f.m.wijnen,j.vandermeij,
v.evers}@utwente.nl
[4] Sheffield Robotics, University of Sheffield, Sheffield, UK
{d.s.cameron,s.fernando,r.k.moore,t.j.prescott}@sheffield.ac.uk
[5] University of Pisa, Pisa, Italy
{daniele.mazzei,michael.pieroni,roberto.garofalo,
danilo.derossi}@centropiaggio.unipi.it, lorenzo.cominelli@for.unipi.it

Abstract. Robots are gradually but steadily being introduced in our daily lives. A paramount application is that of education, where robots can assume the role of a tutor, a peer or simply a tool to help learners in a specific knowledge domain. Such endeavor posits specific challenges: affective social behavior, proper modelling of the learner's progress, discrimination of the learner's utterances, expressions and mental states, which, in turn, require an integrated architecture combining perception, cognition and action. In this paper we present an attempt to improve the current state of robots in the educational domain by introducing the EASEL EU project. Specifically, we introduce the EASEL's unified robot architecture, an innovative Synthetic Tutor Assistant (STA) whose goal is to interactively guide learners in a science-based learning paradigm, allowing us to achieve such rich multimodal interactions.

Keywords: Education · Robotic tutor assistant · Pedagogical models · Cognitive architecture · Distributed Adaptive Control

1 Introduction

Robot technology has become so advanced that automated systems are gradually taking on a greater role in our society. Robots are starting to make their

N.F. Lepora et al. (Eds.): Living Machines 2016, LNAI 9793, pp. 353–364, 2016.
DOI: 10.1007/978-3-319-42417-0_32

mark in many domains ranging from health-care and medicine, to automotive technologies, to service robots and robot companions.

Here our interest is in educational robots and, specifically, how automated technologies can assist learners with their intellectual growth in the classroom.

Despite being still disputed, the use of robots in education has been shown to positively affect students' concentration, learning interest and academic achievements [1]. The introduction of assisting robots as part of a course has been proved effective for integration, real-world issues, interdisciplinary work as well as critical thinking [2]. Moreover, educational robots can be flexible, as they can assume the role of mere tools [3], peers [4,5] or even tutors [6]. Although the preferred characterization is not yet conclusive [7], when used as a tutor or a peer, thus implying a continuous robot–learner interaction, the robot's design and behavior become a crucial aspect [6,8,9]. Typically, robots in education are employed in scenarios ranging from technical education [10] (usually related to robotics or technology), to science [11] and learning of a foreign language [5], just to cite a few – see also [12] for a recent review and discussion about the use of robots in education.

Although a comprehensive examination of pedagogical theories in educational robotics is still lacking, two main approaches have been mostly influential. On one side, Piaget's theory of *constructivism*, which defines knowledge as the process in which an individual creates a meaning out of his own experiences [13]. On the other side, Papert's theory of *constructionism*, stating that learning is the result of building knowledge structures through progressive internalization of actions and conscious engagement through making [14]. More recently, the work of the Russian psychologist Lev Vygotsky gained more and more attention, as it introduced the principle of *scaffolding* and the one of *Zone of Proximal Development (ZPD)* [15]. The former refers to the usage of tools or strategies providing help, whereas the latter corresponds to the distance between what a learner can do by himself and what he may do under the guidance of an effective mediator. All these pedagogical approaches are highly relevant to robotic applications in education, and a review of related studies can be found in [16].

The work we present here constitutes an effort to move one step forward in the domain of robots in education by introducing the EASEL EU–project. EASEL is the acronym of *Expressive Agents for Symbiotic Education and Learning* and it is a collaborative project aimed to explore and develop a theoretical understanding of the Human-Robot Symbiotic Interaction (HRSI) realized in the domain of tutoring. The final outcome of EASEL is the delivery of an innovative Synthetic Tutor Assistant (STA), whose goal is to interactively guide learners (e.g., children in the age 8–11) using a science-based learning paradigm. The main focus of this paper is in describing the underlying STA's architecture that will allow the agent to interact with a human user in an educational task, whereas the theoretical approach and expected impact can be found in an accompanying paper (Reidsma et al. [in preparation]).

The paper's structure is as follows: in Sect. 2 we present the Distributed Adaptive Control (DAC) model of human and animal learning that serves as both a pedagogical model and the control architecture of the STA. In Sect. 3 we present the developed educational scenarios that serve as both the test case of the architecture as well as an evaluation method. Finally, in Sect. 4 we introduce the overall architecture and present its individual components.

2   The DAC Architecture and Pedagogical Model

The STA's main goal is to guide the learner through the science–based educational scenario in order to maximize learning. Based on the perception of the social, communicative and educational context, the robot will respond accordingly within the educational scenario and the specific learning task. The STA's reasoning and memory components need to continuously extract relevant knowledge from sequences of behavioural interactions over prolonged periods to learn a model of the user and adapt its actions to the partner. To successfully interact with the user, the STA thus requires an integrated architecture combining perception, cognition, and action. Here, we adopt the Distributed Adaptive Control (DAC) cognitive architecture (Fig. 1) as the basis to control the STA's behavior.

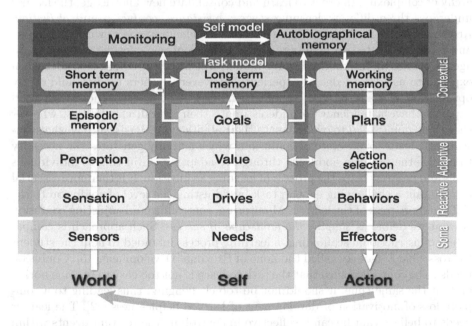

Fig. 1. Schematic illustration of the Distributed Adaptive Control architecture (DAC) and its four layers: somatic, reactive, adaptive and contextual. Across the layers we can distinguish three main functional columns of organization: world related (exosensing, red), self related (endosensing, blue) and the interface to the world through action (green). The arrows indicate the flow of information. Adapted from [17]. (Color figure online)

DAC [17–19] is a robot-based neuronal model of perception, cognition and behavior that is a standard in the domains of new artificial intelligence and behavior-based robotics. It is biologically constrained and fully grounded since it autonomously generates representations of its primary sensory input [18,20]. The DAC architecture is organized around four tightly coupled layers: Soma, Reactive, Adaptive and Contextual. Across these layers, three functional columns of organization can be distinguished: exosensing, defined as the sensation and perception of the world, endosensing which detects and signals the states of the self, and the interface to the world through action.

The Soma represents the body itself and the information acquired from sensation, needs and actuation. In the Reactive Layer, predefined sensorimotor reflexes are triggered by low complexity signals and are coupled to specific affective states of the agent. The Adaptive Layer extends these sensorimotor loops with acquired sensor and action states, allowing the agent to escape the predefined reflexes through learning. Finally, the Contextual Layer develops the state–space acquired by the Adaptive Layer to generate behavioral plans and policies that can be expressed through actions.

Furthermore, DAC postulates that learning is organized along a similar hierarchy of complexity. In order to learn and consolidate new knowledge, the learner undergoes three different learning stages: *resistance*, *confusion* and *abduction*. Resistance is the mechanism that results from defending one's own world-model and is highly related to maintaining the balance of one's feeling of agency. All agents possess an internalized world-model, which they need to reconsider when exposed to new knowledge or experience. However, learners tend to hold overly optimistic and confused views about their level of knowledge: those with good expertise have a tendency to underestimate their overall capabilities, whereas those who don't, tend to overestimate their abilities [21]. Resistance is what consequently leads to the state of confusion, which actually generates the necessity to resolve the problem and learn through re-adapting. Adjusting to individual's skills and progress helps the process of learning acquisition: it is therefore essential to maintain a challenging enough task by adjusting the level of confusion to the skills and progress of the learner. Monitoring, controlling and adjusting confusion is what we define as *shaping the landscape of success*. Such approach is comparable to instructional scaffolding, a learning process intended to help the student to cross what Vygotsky called the Zone of Proximal Development [15]. Confusion needs to be controlled such that the task to learn is not too easy to become boring and, at the same time, it should not be too challenging, thus leading to a complete loss of motivation or development of learned helplessness [22]. The learner needs to believe that he can be effective in controlling the relevant events within the learning process [23]. Confusion is needed in order to discover and generate theories to asses them later, that is, to be able to perform abduction, which is the very process of acquiring the new knowledge.

In the EASEL project, the role of DAC is thus twofold: on the one hand, it acts as a control system to define and drive the STA's actions; on the other hand, it serves as the pedagogical model through which learning can be achieved.

3   Science–Based Educational Scenarios

For the EASEL project we designed two interaction scenarios in real–life settings based on inquiry–based learning tasks. Typically, inquiry–based learning tasks involve active exploration of the world, asking questions, making discoveries and testing hypotheses. The first scenario aims at teaching children about physics concepts based on the Piagetian balance-beam experiments. The second scenario is meant to help children learn about healthy living and physical exercise. These two situations allow to exploit two different perspectives: a formal teaching scenario compared to a more "voluntary free exploration", and a language oriented versus more bodily involvement.

3.1   The Balance Beam

The *balance beam problem* was first described by Inhelder and Piaget to characterize and explain children's stages of cognitive development [24]. Following the Piagetian work, Siegler developed a methodology which allowed him to classify children's cognitive developmental stages on the base of four rules of increasing complexity that children of different ages would apply while solving the balance beam task (Fig. 2) [25,26].

Briefly, in the balance beam scenario different numbers of weights are placed at varying distances from the fulcrum on the equally spaced pegs positioned on both arms of the scale. Children explore the physics of the balance problem using tangible materials and guided by an artificial agent (e.g., a robot or a virtual avatar) that serves as the bodily manifestation of the STA. Children are then asked to predict the behavior of the beam given the configuration provided: if it will stay in equilibrium, tip to the left or tip to the right. To succeed in this task children have to identify the relevant physical concepts (i.e., weight and distance) and understand the underlying multiplicative relation between the two variables (i.e., the "torque rule"). The goal of the interaction is that the child learns about balance and momentum by going through a series of puzzle tasks with the balance beam. The artificial agent is there to encourage the students, to help them get through the different tasks and to provide feedback; thus, learning improves by constantly monitoring the learner's progresses.

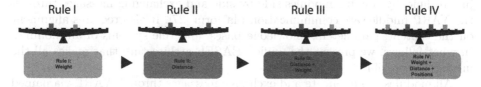

Rule I   Rule II   Rule III   Rule IV

Fig. 2. Schematic illustration of the four rules assessed by Siegler [25]. At each developmental stage one or both dimensions (i.e., weight and distance) are considered. For instance, Rule I exclusively considers the weight, whereas Rule IV considers both weights and distance from the fulcrum.

3.2 Healthy-Living

The second interaction scenario designed is that of healthy-living. This scenario involves a synthetic agent assisting learners in an inquiry-based learning task about the benefits of physical exercise. Specifically, the children will investigate the effects of different types of exercise on a number of physiological variables (e.g., heart rate, blood pressure, respiration rate, etc.). Two types of sessions are considered: in the first one, the artificial agent will encourage the children to perform exercises at varying speeds. It will then provide information about the outcome of the exercise in a friendly, accessible way, either via voice or through a video display allowing the children to have immediate feedback about their own state. In the second type of session, the child sits down with a prepared worksheet about healthy-living, which is used to prompt a spoken dialogue with the artificial agent. The child reads through the text on the worksheet, which comprises instructions and questions to the artificial agent about the previous interaction. For instance, it explains how the physiological values are related to other forms of energy, such as the calories contained in different foods, and how much energy is burnt during an exercise. The worksheet prompts the child to start off the interaction providing cues about the kind of things that can be said to the artificial agent.

4 Architecture Overview

The proposed interaction scenarios consist of a humanoid robot, a handheld device (e.g., tablet) called the EASELscope and in the case of the balance beam task, the Smart Balance Beam (SBB), a motorized scale equipped with sensors to detect the object's weight and position. Both scenarios imply a detailed interaction between the learner and the robot which create a series of challenges such as effective social behaviour to get a good social relation with the child, proper modelling of the students' learning progress, various kinds of perception of the childs utterances, expressions, bodily and mental states, as well as bodily and facial expressions from the robot.

Rich multimodal interactions require an integrated architecture which combines perception, reasoning, and action (Sect. 2). Such integration is more than just technically running modules side-by-side and exchanging messages through the YARP middleware communication platform [27]. It also requires alignment on the content, parameters, knowledge bases, and rule systems of all modules. In what follows, we present the unified EASEL architecture that makes all the interaction possible.

All modules communicate and exchange messages through YARP via named communication channels. This middleware platform allows us to not only distribute the system over different machines, but also permits abstraction from different Operating Systems.

4.1   Modules and Mapping to the DAC Architecture

The integrated EASEL architecture is a practical incarnation of the conceptual architecture of the Distributed Adaptive Control (DAC) theory of the design principles underlying perception, cognition and action. Each implemented module in the EASEL architecture can be mapped to one or more of the core components of the DAC model that they embody and the specific modules are schematically illustrated in Fig. 3.

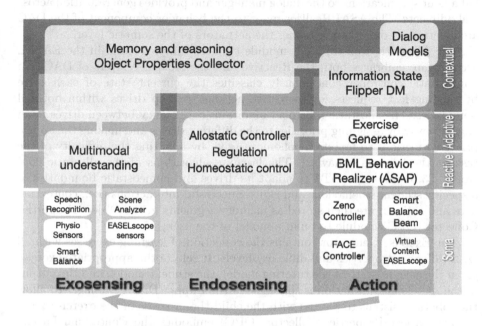

Fig. 3. Overview of the EASEL architecture, where each implemented module is mapped to the core components of the DAC architecture.

The Speech recognition (ASR module), the SceneAnalyzer, the PhysioReader and the EASELScope sensors embed the exosensing component of the Soma layer through which the states of the world are acquired and the internal drives are established. More precisely, the ASR module is based on the open-source Kaldi speech recognition toolkit [28] with an EASEL specific vocabulary, language model and recognition grammar.

The SceneAnalyzer builds upon several other libraries to deliver integrated recognition of multimodal features of the users and their behaviour [29,30]. The physiological signal acquisition module uses non-obtrusive and robust methods for obtaining information about the users physiological state: by integrating sensors in the robot or in the EASELscope tablet, information can be unobtrusively obtained without sensors worn or strapped to the body of user. The EASELScope sensors allow detection of the current state of the balance beam allowing the EASEL system to respond to the actions of the user with respect to the learning materials.

In the reactive layer, ASAPRealizer module [31,32] is responsible for the choreography of the behavior (verbal and non-verbal) of the STA using the generic robot-independent Behavior Markup Language (BML). We also use an easily configurable XML binding between the BML and the motion primitives of each robotic platform (that serves as the physical instantiation of the STA). Such approach abstracts away from specific motor control by exposing more general behaviour specifications to the dialog manager and provides generalization across embodiments. The ASAPRealizer maps to the behavior component of the DAC architecture by directly controlling the actuators of the somatic layer.

The Allostatic Control (AC) module currently implemented in the EASEL architecture embraces both the Reactive and the Adaptive layers of DAC. An homeostatic controller continuously classifies the current state of each drive by sending fast requests for corrective actions to keep drives within optimal boundaries. The allostatic controller maintains consistency between drives in an adaptive way by assigning priorities to the different drives and making the appropriate corrections to maintain coherence (e.g., by adapting the difficulty of the task to the learners behavior). The learning algorithms of the allostatic controller [33,34] allow the STA to adapt its drives and homeostatic boundaries to a specific student's behaviour and skills. Successful interactions (i.e., contextual cues and actions) are then stored as memory segments in the Object Properties Collector (OPC) module to build a model of the user.

The Exercise Generator contains the collection of learning exercises with all their different properties and difficulty levels. It selects the appropriate exercise given the current state of the tutoring model, the student model, and the output of the AC. This information is shared with the Flipper Dialog Manager to allow the robotic assistant to discuss with the child the progress of the exercise.

The Object Properties Collector (OPC) embodies the Contextual Layer's memory components of DAC. At this stage of development, the OPC implements the memory for events that can be stored and distributed to the other STA's modules as well as its short-term memory component. At each instant of the interaction, the OPC can temporarily retain ongoing perceptions, actions and values (i.e., outcomes of the current interaction) as segments of memory (relations). These relations allow the definition of rules for specific interactions that can be further stored as long-term memories if a high-level goal is successfully achieved. Between the OPC and the sensors lies the multimodal understanding module, a light and simple interpreter of speech, emotions and speaker's probabilities, which simplifies the requirements for the Flipper Dialog Manager's scripts. Flipper offers flexible dialog specification via information state and rule-based templates to trigger information state changes as well as behaviour requests [35].

The robots (Zeno (Hanson Robotics, Hong Kong) and FACE [36]), the virtual robot avatars and the EASEL-scope hidden state visualizer correspond to the DAC effectors and represent the main interface of the STA with the world

(Fig. 4). The EASELscope offers an augmented reality (AR) interface that allows the learner to interact with the task materials. It can be used to present extra information to the child about the learning content. This allows the system to vary between different ways of scaffolding the learning of the user. For instance, using the EASELscope we can present "hidden information" about the balance beam, such as the weights of the pots, or the forces acting on the arms of the beam; both are types of scaffolds in learning that would not be possible without the scope (Fig. 4).

Fig. 4. The Zeno robot, the EASELscope and the Smart Balance Beam (SBB). Zeno acts as the embodiment of the STA and continuously interacts with the learner. Information about the task is displayed as augmented reality on top of the physical SBB.

5   Discussion and Conclusions

In this paper we presented the DAC based EASEL architecture, designed to guide learners through learning in two science-based educational scenarios. Each module within the framework has been integrated in a cohesive setup and the configuration options, models, behaviour repertoires and dialog scripts will allow us to validate the EASEL system through specific experiments with child-robot interaction in the proposed learning scenarios.

The way the architecture is organized gives us three key advantages: scalability, configurability and abstraction. This allows us to easily add sensory components with negligible changes to the main core of the system: it is sufficient to add the input to the multimodal understanding module that, in turn, will store the new information with an appropriate format in the OPC module. Furthermore, all modules are fully configurable: for instance, we can add new behaviors (ASAPRealizer), drives (Allostatic Control) as well as dialogues (Flipper) in an easy way with the usage of configuration files such as XML scripts. Thus, any additional implementation for the needs of the EASEL architecture (in terms of scenarios or sensory inputs) can be done in a flexible way. Finally, the proposed architecture permits abstraction from the physical manifestation of the STA, in a way that using the same scenario, we can choose the robotic platform (or even avatar) with small changes in the main core of the system.

At this stage, we are now ready to start validating our educational architecture and focus on concrete long-term studies on human-robot symbiotic interactions in learning tasks. For instance, by looking at the types of hypotheses that the child frames and the outcome in solving exercises, the STA can continuously build and refine a model of the student's understanding of the task. By varying the exercises, the scaffolding provided by the robot or the hidden-states information provided through the EASELScope, the STA can explore the most effective learning strategies for a specific task. By modifying the social and conversational strategies of the robot, the STA can extract the best "personality" and style of the robot leading to a good relationship between the child and the robot itself and their impact on the overall learning process.

Acknowledgments. This work is supported by grants from the European Research Council under the European Union's 7th Framework Programme FP7/2007-2013/ERC grant agreement n. 611971 (EASEL) and n. 341196 (CDAC) to Paul F. M. J. Verschure.

References

1. Han, J.-H., Jo, M.-H., Jones, V., Jo, J.-H.: Comparative study on the educational use of home robots for children. J. Inf. Process. Syst. **4**(4), 159–168 (2008)
2. Beer, R.D., Chiel, H.J., Drushel, R.F.: Using autonomous robotics to teach science and engineering. Commun. ACM **42**(6), 85–92 (1999)
3. Mondada, F., Bonani, M., Raemy, X., Pugh, J., Cianci, C., Klaptocz, A., Magnenat, S., Zufferey, J.-C., Floreano, D., Martinoli, A.: The e-puck, a robot designed for education in engineering. In: Proceedings of the 9th Conference on Autonomous Robot Systems, Competitions, vol. 1, pp. 59–65. Instituto Politécnico de Castelo Branco (2009)
4. Wijnen, F., Charisi, V., Davison, D., van der Meij, J., Reidsma, D., Evers, V.: Inquiry learning with a social robot: can you explain that to me? In: Heerink, M., de Jong, M. (eds.) Proceedings of New Friends 2015: The 1st international conference on social robotics in therapy and education, pp. 24–25, Windesheim Flevoland, Almere (2015)
5. Kanda, T., Hirano, T., Eaton, D., Ishiguro, H.: Interactive robots as social partners and peer tutors for children: a field trial. Hum.-Comput. Interact. **19**(1), 61–84 (2004)
6. Saerbeck, M., Schut, T., Bartneck, C., Janse, M.D.: Expressive robots in education: varying the degree of social supportive behavior of a robotic tutor. In: Proceedings of the SIGCHI Conference on Human Factors in Computing Systems, pp. 1613–1622. ACM (2010)
7. Shin, N., Kim, S.: Learning about, from, with robots: students' perspectives. In: The 16th IEEE International Symposium on Robot and Human Interactive Communication, RO-MAN 2007, pp. 1040–1045. IEEE (2007)

8. Vouloutsi, V., Munoz, M.B., Grechuta, K., Lallee, S., Duff, A., Llobet, J.-Y.P., Verschure, P.F.M.J.: A new biomimetic approach towards educational robotics: the distributed adaptive control of a synthetic tutor assistant. New Frontiers in Human-Robot Interaction, p. 22 (2015)

9. Blancas, M., Vouloutsi, V., Grechuta, K., Verschure, P.F.M.J.: Effects of the robot's role on human-robot interaction in an educational scenario. In: Wilson, S.P., Verschure, P.F.M.J., Mura, A., Prescott, T.J. (eds.) Living Machines 2015. LNCS, vol. 9222, pp. 391–402. Springer, Heidelberg (2015)

10. Balch, T., et al.: Designing personal robots for education: hardware, software and curriculum. IEEE Pervasive Comput. **7**(2), 5–9 (2008)

11. Highfield, K., Mulligan, J., Hedberg, J.: Early mathematics learning through exploration with programmable toys. In: Proceedings of the Joint Meeting of PME 32 and PME-NA, pp. 169–176. Citeseer (2008)

12. Mubin, O., Stevens, C.J., Shahid, S., Al Mahmud, A., Dong, J.-J.: A review of the applicability of robots in education. J. Technol. Educ. Learn. **1**, 209–215 (2013)

13. Piaget, J., Inhelder, B.: The Psychology of the Child. Basic Books, New York (1972)

14. Papert, S., Harel, I.: Situating constructionism. Constructionism **36**, 1–11 (1991)

15. Vygotsky, L.S.: Mind in Society: The Development of Higher Psychological Processes. Harvard University Press, Cambridge (1980)

16. Charisi, V., Davison, D., Wijnen, F., Van Der Meij, J., Reidsma, D., Prescott, T., Van Joolingen, W., Evers, V.: Towards a child-robot symbiotic co-development: a theoretical approach. In: AISB Convention 2015, The Society for the Study of Artificial Intelligence and the Simulation of Behaviour (AISB) (2015)

17. Verschure, P.F.M.J.: Distributed adaptive control: a theory of the mind, brain, body nexus. Biologically Inspired Cogn. Architectures **1**, 55–72 (2012)

18. Verschure, P.F.M.J., Voegtlin, T., Douglas, R.J.: Environmentally mediated synergy between perception and behaviour in mobile robots. Nature **425**, 620–624 (2003)

19. Verschure, P.F.M.J., Pennartz, C.M., Pezzulo, G.: The why, what, where, when and how of goal-directed choice: neuronal and computational principles. Phil. Trans. R. Soc. B **369**(1655), 20130483 (2014)

20. Maffei, G., Santos-Pata, D., Marcos, E., Sánchez-Fibla, M., Verschure, P.F.M.J.: An embodied biologically constrained model of foraging: from classical and operant conditioning to adaptive real-world behavior in DAC-X. Neural Netw. **72**, 88–108 (2015)

21. Kruger, J., Dunning, D.: Unskilled and unaware of it: how difficulties in recognizing one's own incompetence lead to inflated self-assessments. J. Pers. Soc. Psychol. **77**(6), 1121 (1999)

22. Abramson, L.Y., Seligman, M.E., Teasdale, J.D.: Learned helplessness in humans: critique and reformulation. J. Abnorm. Psychol. **87**(1), 49 (1978)

23. Seligman, M.E.: Learned helplessness. Annu. Rev. Med. **23**(1), 407–412 (1972)

24. Inhelder, B., Piaget, J.: The Growth of Logical Thinking from Childhood to Adolescence: An Essay on the Construction of Formal Operational Structures. Basic Books, New York (1958)

25. Siegler, R.S.: Three aspects of cognitive development. Cogn. Psychol. **8**(4), 481–520 (1976)

26. Siegler, R.S., Strauss, S., Levin, I.: Developmental sequences within and between concepts. Monogr. Soc. Res. Child Dev. **46**, 631–683 (1981)

27. Metta, G., Fitzpatrick, P., Natale, L.: Yarp: yet another robot platform. Int. J. Adv. Rob. Syst. **3**(1), 43–48 (2006)

28. Povey, D., Ghoshal, A., Boulianne, G., Burget, L., Glembek, O., Goel, N., Hannemann, M., Motlicek, P., Qian, Y., Schwarz, P., et al.: The kaldi speech recognition toolkit. In: IEEE 2011 workshop on automatic speech recognition and understanding, no. EPFL-CONF-192584. IEEE Signal Processing Society (2011)
29. Zaraki, A., Mazzei, D., Giuliani, M., De Rossi, D.: Designing and evaluating a social gaze-control system for a humanoid robot. IEEE Trans. Hum.-Mach. Syst. **44**(2), 157–168 (2014)
30. Zaraki, A., Mazzei, D., Lazzeri, N., Pieroni, M., De Rossi, D.: Preliminary implementation of context-aware attention system for humanoid robots. In: Lepora, N.F., Mura, A., Krapp, H.G., Verschure, P.F.M.J., Prescott, T.J. (eds.) Living Machines 2013. LNCS, vol. 8064, pp. 457–459. Springer, Heidelberg (2013)
31. Reidsma, D., van Welbergen, H.: AsapRealizer in practice - a modular and extensible architecture for a bml realizer. Entertainment Comput. **4**(3), 157–169 (2013)
32. van Welbergen, H., Yaghoubzadeh, R., Kopp, S.: AsapRealizer 2.0: the next steps in fluent behavior realization for ECAs. In: Bickmore, T., Marsella, S., Sidner, C. (eds.) IVA 2014. LNCS, vol. 8637, pp. 449–462. Springer, Heidelberg (2014)
33. Vouloutsi, V., Lallée, S., Verschure, P.F.M.J.: Modulating behaviors using allostatic control. In: Lepora, N.F., Mura, A., Krapp, H.G., Verschure, P.F.M.J., Prescott, T.J. (eds.) Living Machines 2013. LNCS, vol. 8064, pp. 287–298. Springer, Heidelberg (2013)
34. Lallée, S., Vouloutsi, V., Wierenga, S., Pattacini, U., Verschure, P.F.M.J: Efaa: a companion emerges from integrating a layered cognitive architecture. In: Proceedings of the 2014 ACM/IEEE International Conference on Human-Robot Interaction, pp. 105–105. ACM (2014)
35. ter Maat, M., Heylen, D.: Flipper: an information state component for spoken dialogue systems. In: Vilhjálmsson, H.H., Kopp, S., Marsella, S., Thórisson, K.R. (eds.) IVA 2011. LNCS, vol. 6895, pp. 470–472. Springer, Heidelberg (2011)
36. Lazzeri, N., Mazzei, D., Zaraki, A., De Rossi, D.: Towards a believable social robot. In: Lepora, N.F., Mura, A., Krapp, H.G., Verschure, P.F.M.J., Prescott, T.J. (eds.) Living Machines 2013. LNCS, vol. 8064, pp. 393–395. Springer, Heidelberg (2013)

Aplysia Californica as a Novel Source of Material for Biohybrid Robots and Organic Machines

Victoria A. Webster[✉], Katherine J. Chapin, Emma L. Hawley, Jill M. Patel, Ozan Akkus, Hillel J. Chiel, and Roger D. Quinn

Case Western Reserve University, 10900 Euclid Ave, Cleveland, OH 44106, USA
vaw4@case.edu
http://www.case.edu

Abstract. *Aplysia californica* is presented as a novel source of actuator and scaffold material for biohybrid robots. Collagen isolated from the *Aplysia* skin has been fabricated into gels and electrocompacted scaffolds. Additionally, the I2 muscle from the *Aplysia* buccal mass had been isolated for use as an organic actuator. This muscle has been characterized and the maximum force was found to be 58.5 mN with a maximum muscle strain of $12 \pm 3\%$. Finally, a flexible 3D printed biohybrid robot has been fabricated which is powered by the I2 muscle and is capable of locomotion at 0.43 cm/min under field stimulation.

Keywords: Biohybrid devices · Organic actuators · Biorobotics · *Aplysia californica*

1 Introduction

The field of biohybrid devices involves the combination of synthetic substrates with living tissue in order to produce a functioning device. A common application of such devices is in the area of biohybrid robotics. By seeding contractile cells on synthetic or organic polymers, researchers are able to produce devices using thin films for lab-on-chip style testing [1–4], or even devices capable of swimming [5,6], and crawling [7–11]. Recent efforts in the area of biohybrid robotics and the development of organic actuators have focused on mammalian [5–10] or avian [11] contractile cells to serve as actuators. However, both mammalian and avian cells require precise pH, temperature, osmotic balances, and CO_2 levels. Such requirements make these devices difficult to fabricate and maintain, and limit their applicability beyond the lab.

A few studies have investigated the possibility of using frog [12] or insect [13,14] muscle to power biohybrid devices. Instead, we propose *Aplysia californica* as a potential source for contractile tissue using the I2 muscle and scaffold material, using isolated collagen-like material from the skin. *Aplysia californica* is a species of sea slug native to the southern coast of California whose range extends into Mexico.

© Springer International Publishing Switzerland 2016
N.F. Lepora et al. (Eds.): Living Machines 2016, LNAI 9793, pp. 365–374, 2016.
DOI: 10.1007/978-3-319-42417-0_33

These animals live in the intertidal and sub-littoral zones of the photic zone and as a consequence can be subjected to large temperature changes as well as changes in the surrounding salt water environment [15]. As a result, *Aplysia californica* are very robust, making them an ideal candidate for use in biohybrid robotics.

In order to build biohybrid devices, suitable scaffolds are needed. Such scaffolds should be strong enough to allow handling and assembly, while being deformable enough to allow the muscle to deflect the substrate. In this study, we present a basic characterization of the I2 muscle as well as a protocol for isolating collagen from the skin. Additionally, biobots powered by the I2 muscle have been 3D printed using flexible polymeric resin.

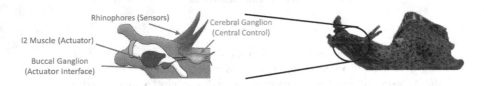

Fig. 1. Right: *Aplysia californica*. Left: a diagram of the relevant anatomy of the *Aplysia*. The I2 muscle (red) has been highlighted on the buccal mass (white). (Color figure online)

2   Methods

2.1   Materials

Aplysia californica were obtained from Marinus Scientific and South Coast Biomarine and maintained in artificial seawater (Instant Ocean, Aquarium Systems) in aerated aquariums at a temperature of 16°C. Animals weighing 200–450 g were anesthetized by injecting isotonic $MgCl_2$ (333 mM) at a volume in milliliters equal to 50 % of the animal's body weight in grams. For isolation of collagen like materials, the skin was harvested and stored at -20^0C until processed. A diagram of relevant sea slug anatomy is presented in Fig. 1.

2.2   I2 Muscle Isolation

Using tweezers and surgical scissors, a horizontal cut was made behind the rinophores, followed by a perpendicular cut towards the jaws of the slug. These cuts are only as deep as the skin. Any connective tissue limiting access to the buccal mass of the slug was cut away from the skin. The esophagus was identified, exposed and cut above the cerebral ganglia. Nerves or connective tissue connected to the buccal mass were removed and the scissors were used to cut the buccal mass at the jaws of the slug. The buccal mass was then placed in a small dish in order to hold it stationary for the removal of the I2 muscle. The buccal ganglia was removed to reveal the entirety of the I2 muscle which was then removed by carefully cutting it away from the tissue at both the origin and insertion lines (Fig. 2). The I2 muscle was stored in *Aplysia* saline (0.46 M NaCl, 0.01 M KCl, 0.01 M MOPS, 0.01 M $C_6H_{12}O_6$, 0.02 M $MgCl_2 \cdot 6H_2O$, 0.033 M $MgSO_4 \cdot 7H_2O$, 0.01 M $CaCl_2 \cdot 2H_2O$, pH 7.5) until used.

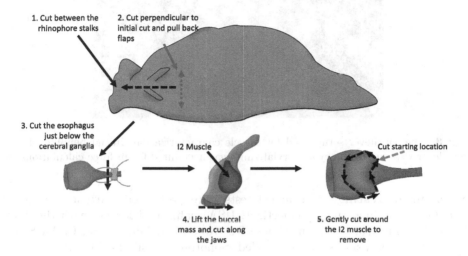

Fig. 2. A diagram of the animal dissection and I2 isolation process.

2.3   I2 Muscle Characterization

Cantilever-Based Actuator Tension Measurement: In order to characterize the muscle force and strain capabilities, flexible cantilevers were printed using a Formlabs Form 1+ stereolithographic (SLA) printer with photocurable resin. The I2 muscles were attached to the cantilever and base using gel based cyanoacrylate (Fig. 3). In order to produce contractions, the muscle was stimulated using two platinum electrodes spaced 3 cm apart. Stimulation patterns were based on the stimulation activity seen in *Aplysia* during biting behavior. The muscle was stimulated at 20 Hz for a duration of 3 s followed by a 3 s rest [16]. In order to investigate the effect of stimulation current on muscle force, the muscle was stimulated at 0.25 V/mm, 0.3 V/mm, and 0.42 V/mm. At each voltage, the muscle was stimulated 10 times and then allowed to rest for 5 min. The muscle force was then determined based on the deflection of the cantilever.

Muscle Dimensions: The dimensions of the Y-shaped I2 muscle (Fig. 4) were measured first across the two arms to measure the width of the Y, and second, along the base of the Y starting from the cleft between the two arms to the connective tissue at the center of the edge which connected to the I1/I3 complex (Fig. 4).

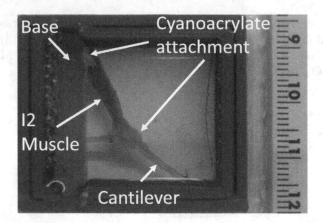

Fig. 3. The cantilever setup used for muscle characterization. Initial bending of the cantilever was due to the cyanoacrylate and was accounted for in force calculations.

Mechanical Testing: Mechanical tests were performed using a 5 lb. load cell (Omega Engineering, Connecticut, USA) calibrated for use with the Test Resources universal testing machine (Test Resources, Minne sota, USA). Samples were tested using strain controlled monotonic tensile loading at a rate of 1 mm/min. Failure load and displacement were recorded by Test Resources, and were used with width analyzed in ImageJ and thickness measurements taken using a FemtoTools MicroRobotics Testing and Handling Module with a FT-S1000 Microforce sensing probe to calculate modulus.

2.4 *Aplysia* Collagen Isolation and Scaffold Fabrication

The *Aplysia* collagen-like material isolation procedure was developed based on a previously reported procedure for isolating collagen from the oyster *Crassostrea gigas* [17]. *Aplysia* skin was thawed at room temperature and minced with scissors before being suspended in 0.1 M NaOH (10:1 volume:weight) for 24 h at 4°C. After 24 h, the mixture was centrifuged at 10000 x g for 20 min. All centrifuging was performed at 4°C. The supernatant was discarded and the remaining residue was washed by gently stirring in distilled water for 20 min. The washed solution was then centrifuged again and the supernatant discarded. The residue was then suspended in disaggregating solution composed of 0.1 M Tris-HCL (pH 8.0) containing 0.5M NaCl, 0.05 M EDTA, and 0.2 M 2-mercaptoethanol [18]. This extraction occurred over 7 days with gentle stirring at 4°C, after which the solution was centrifuged and the supernatant discarded. The residue was washed as before and resuspended in 4 M guanidine hydrochloride in 0.05 M Tris-HCL (pH 7.5) at a volume to weight ratio of 1/2 the original tissue weight. After 24 h at 4°C, the solution was again centrifuged, the supernatant was discarded and the residue was washed. The residue was further processed in 0.5 M acetic acid containing porcine pepsin (enzyme to substrate ratio of 1:20) for 48 h at 4°C. After 48 h, the solution was fractionated by salt precipitation by adding NaCl to a final concentration of 2 M.

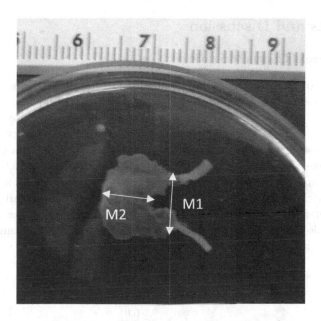

Fig. 4. A diagram of I2 muscle length measurements. The first measurement (M1) is made between the arms of the Y while the second measurement (M2) measures from the base of the Y to the cleft between the arms.

The solution was again centrifuged and the supernatant discarded. The collected collagen was dialyzed for 24 h against 0.02 M Na_2HPO_4.

Aplysia collagen was gelled by incubation at 25°C for 3 h. Additionally, collagen was dialyzed against distilled water for 18 h with water changes every 2 h for the first 6 h in order to produce stock for electrocompaction. This stock was electrocompacted into aligned threads using a dual wire setup [19] and into compacted, unaligned sheets using plate electrodes [11]. Compaction was achieved by application of 15 V for 30 s.

2.5   Flexible 3D-printed Devices

Flexible 3D-printed devices have been fabricated using a Formlabs Form 1+ SLA printer with photocurable resin. The resulting structures had a modulus of 2.29 - 2.76 MPa (as reported on the Formlabs datasheet), and are strong enough to be manipulated but flexible enough that the I2 muscle was able to deflect thin members. The I2 muscle was attached to each arm half way along the length and to the rear of the body using cyanoacrylate (Fig. 8 A and B). In order to provide surface traction, devices were tested in a dish with a sandpaper base.

3   Results and Discussion

3.1   I2 Muscle Characterization

The Effect of Animal Size on Muscle Dimensions: Muscle measurements were made and compared to the mass of the animal for animals weighing 280–420 g. Neither the muscle width nor the base length correlated strongly to the animal mass (Fig. 5) for the range tested. This presents one possible drawback to the use of the I2 muscle as an actuator because it is not possible to predict the size of the muscle with a high degree of certainty prior to dissection. However, all animals tested were mature adult animals and only a narrow size range was checked. If the size range was widened to include small juveniles as well as larger adults a correlation may emerge. However, for the purposes of designing a biohybrid device, the observed variability can be accommodated in the design of the structure.

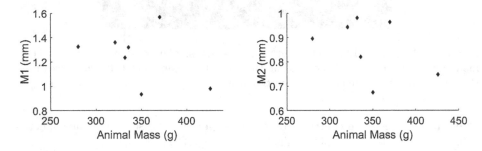

Fig. 5. Measurements for M1 (left) and M2 (right). Neither measurement corresponded strongly to the animal mass.

Mechanical Testing: Tensile testing of 4 I2 muscle samples in the muscle fiber direction resulted in a Young's modulus of 0.49 ± 0.99 MPa.

The Effect of Stimulation Voltage on Muscle Tension: During this study the I2 muscle had a maximum actuating force of 58.5 mN which occurred when the muscle was stimulated at 0.3 V/mm. However, previous literature indicates that the muscle may be capable of an active tension of up to 120 mN [20]. The lower force output may be due to the use of field stimulation, rather than suction electrodes as were used by Yu et al. [20]. However, suction electrodes are stiff and may interfere with muscle motion as well as biohybrid device performance, making field stimulation preferable.

Noticeable deflection did not occur at or below 0.2 V/mm (data not shown). Noticeable deflection began at 0.25 V/mm and increased with increasing voltage. Additionally, no noticeable difference in active tension was recorded between 0.3

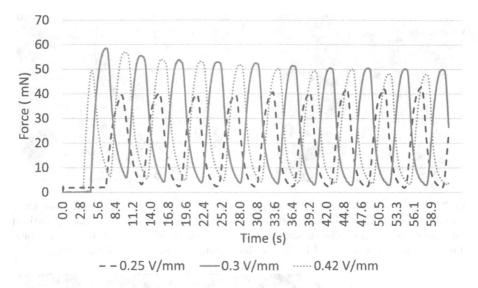

Fig. 6. The effect of stimulation voltage on muscle force. 0.25 V/mm resulted in the lowest muscle force while there was no appreciable difference between 0.3 and 0.42 V/mm.

and 0.42 V/mm (Fig. 6). Therefore a voltage of 0.3 V/mm is recommended for stimulation in order to maximize actuator force while minimizing fatigue and power requirements.

Muscle Strain: The I2 muscle was able to reach considerable strains when contracting with an average maximum strain of $12 \pm 3\%$.

3.2 *Aplysia* Collagen

Isolated collagen was capable of forming robust gels which retained their molded shape (Fig. 7.A). Additionally, the compacted threads showed strong alignment along the thread axis (Fig. 7.B). These structures could be suitable for use as scaffolds for cell based organic machines [11].

3.3 Flexible 3D-printed Devices

Flexible 3D-printed devices were capable of locomotion under the actuating force of the I2 muscle. Devices designed with frictionally anisotropic legs (Fig. 8.A) were capable of locomotion at speeds of approximately 0.5 cm/min. A characteristic example of locomotion is presented in Fig. 8 B and C. All stimulation trials began with 30–35 s without stimulation in order to verify that the device was not moving due to residual fluid flow in the surrounding medium. During each step the device experienced some degree of slip, indicating that ground contact and frictional anisotropy could be improved (Fig. 8.D). Over the course

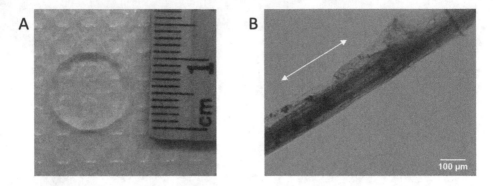

Fig. 7. A: *Aplysia* collagen gel. Gels were robust to manipulation and held their molded shape. B: Electrocompacted *Aplysia* collagen thread imaged using polarized light microscopy. The collagen shows strong alignment along the thread axis (blue coloration, direction indicated by the double headed arrow) (Color figure online)

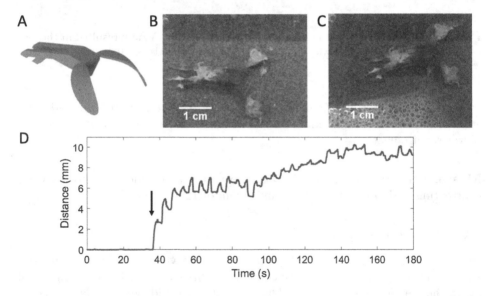

Fig. 8. A: A rendering of the 3D printed biohybrid robot. The muscle was attached to the two arms and between the legs. Angled legs served to provide frictional anisotropy to allow forward motion. B: The starting position of the biohybrid device. C. The position of the biohybrid device after 2 min of stimulation. The device moved approximately 1 cm. D: A plot of the devices motion with respect to time. Electrical stimulation began as indicated by the black arrow.

of each trial, fatigue is observed in the device, with initial steps producing more forward motion than later steps. In order to improve the device lifetime, alternative stimulation techniques, such as including the buccal ganglia, which contains the motor neurons, and using chemical stimulation, should be investigated.

Additionally, the device was observed to become caught on larger features of the rough dish, which stalled motion until later steps freed the device. In order to reduce the occurrence of such events, the geometry of the underside of the device can be improved.

4   Conclusion

In this study we have demonstrated several ways in which *Aplysia californica* can be used as a novel source of materials for biohybrid robots. Additionally, we have provided one application for the use of the I2 muscle to powered 3D printed biohybrid robots. The collagen-like material and muscle from the *Aplysia* offer a robust alternative to mammalian or avian sources. In the future, ganglia or individual neurons may be isolated from the animal to provide an organic control system. Such devices could have applications in environmental sensing and aquatic surveillance.

Acknowledgments. This material is based upon work supported by the National Science Foundation Graduate Research Fellowship under Grant No. DGE-0951783 and a GAANN Fellowship (Grant No. P200A150316). This study was also funded in part by grants from the National Science Foundation (Grant No. DMR-1306665), and the National Institute of Health (Grant No. R01 AR063701). Any opinion, findings, and conclusions or recommendations expressed in this material are those of the authors and do not necessarily reflect the views of the National Science Foundation.

References

1. Feinberg, A.W., Feigel, A., Shevkoplyas, S.S., Sheehy, S., Whitesides, G.M., Parker, K.K.: Muscular thin films for building actuators and powering devices. Science **80**(317), 1366–1370 (2007)
2. Grosberg, A., Alford, P.W., McCain, M.L., Parker, K.K.: Ensembles of engineered cardiac tissues for physiological and pharmacological study: Heart on a chip. Lab Chip **11**(24), 4165 (2011)
3. Vannozzi, L., Ricotti, L., Cianchetti, M., Bearzi, C., Gargioli, C., Rizzi, R., Dario, P., Menciassi, A.: Self-assembly of polydimethylsiloxane structures from 2D to 3D forbio-hybrid actuation. Bioinspir. Biomim. **10**(5), 056001 (2015)
4. Nagamine, K., Kawashima, T., Sekine, S., Ido, Y., Kanzaki, M., Nishizawa, M.: Spatiotemporally controlled contraction of micropatterned skeletal muscle cells on a hydrogel sheet. Lab Chip **11**(3), 513–7 (2011)
5. Nawroth, J.C., Lee, H., Feinberg, A.W., Ripplinger, C.M., McCain, M.L., Grosberg, A., Dabiri, J.O., Parker, K.K.: A tissue-engineered jellyfish with biomimetic propulsion. Nat. Biotechnol. **30**(8), 792–797 (2012)
6. Williams, B.J., Anand, S.V., Rajagopalan, J., Saif, M.T.A.: A self-propelled biohybrid swimmer at low Reynolds number. Nat. Commun. **5**, 3081 (2014)
7. Xi, J., Schmidt, J.J., Montemagno, C.D.: Self-assembled microdevices driven by muscle. Nat. Mater. **4**, 180–184 (2005)
8. Chan, V., Park, K., Collens, M.B., Kong, H., Saif, T.A., Bashir, R.: Development of miniaturized walking biological machines. Sci. Rep. **2**, 857 (2012)

9. Cvetkovic, C., Raman, R., Chan, V., Williams, B.J., Tolish, M., Bajaj, P., Sakar, M.S., Asada, H.H., Taher, A., Taher A Saif, M., Bashir, R.: Three-dimensionally printed biological machines powered by skeletalmuscle. PNAS **111**(28), 10125–10130 (2014)

10. Kim, J., Park, J., Yang, S., Baek, J., Kim, B., Lee, S.H., Yoon, E.-S., Chun, K., Park, S.: Establishment of a fabrication method for a long-term actuated hybrid cell robot. Lab Chip **7**, 1504–1508 (2007)

11. Webster, V.A., Hawley, E.L., Akkus, O., Chiel, H.J., Quinn, R.D.: Fabrication of electrocompacted aligned collagen morphs for cardiomyocyte powered living machines. In: Wilson, S.P., Verschure, P.F.M.J., Mura, A., Prescott, T.J. (eds.) Living Machines 2015. LNCS, vol. 9222, pp. 429–440. Springer, Heidelberg (2015)

12. Herr, H., Dennis, R.G.: A swimming robot actuated by living muscle tissue. J. Neuroeng. Rehabil. **1**, 6 (2004)

13. Baryshyan, A.L., Woods, W., Trimmer, B.A., Kaplan, D.L.: Isolation and maintenance-free culture of contractile myotubes from Manduca sexta embryos. PLoS One **7**(2), e31598 (2012)

14. Uesugi, K., Shimizu, K., Akiyama, Y., Hoshino, T., Iwabuchi, K., Morishima, K.: Contractile performance and controllability of insect muscle-powered bioactuator with different stimulation strategies for soft robotics. Soft Robot. **3**(1), 13–22 (2016)

15. Kandel, E.: Behavioral Biology of Aplysia: Contribution to the Comparative Study of Opistobranch Molluscs. W.H. Freeman and Company, San Francisco (1979)

16. Hurwitz, I., Neustadter, D., Morton, D.W., Chiel, H.J., Susswein, A.J.: Activity patterns of the B31/B32 pattern initiators innervating the I2 muscle of the buccal mass during normal feeding movements in Aplysia californica. J. Neurophysiol. **75**(4), 1309–26 (1996)

17. Mizuta, S., Miyagi, T., Yoshinaka, R.: Characterization of the quantitatively major collagen in the mantle of oyster Crassostrea gigas. Fish. Sci. **71**(1), 229–235 (2005)

18. Matsumura, T.: Shape, size and amino acid composition of collagen fibril of the starfish asterias amurensis

19. Cheng, X.: Umut a Gurkan, Christopher J Dehen, Michael P Tate, Hugh W Hillhouse, Garth J Simpson, and Ozan Akkus.: An electrochemical fabrication process for the assembly of anisotropically oriented collagen bundles. Biomaterials **29**(22), 3278–3288 (2008)

20. Yu, S.N., Crago, P.E., Chiel, H.J.: Biomechanical properties and a kinetic simulation model of the smooth muscle I2 in the buccal mass of Aplysia. Biol. Cybern. **81**, 505–513 (1999)

A Soft Pneumatic Maggot Robot

Tianqi Wei[1]([✉]), Adam Stokes[2], and Barbara Webb[1]

[1] School of Informatics, Institute of Perception, Action and Behaviour,
University of Edinburgh, Edinburgh EH8 9AB, UK
chitianqilin@163.com, B.Webb@ed.ac.uk
[2] School of Engineering, Scottish Microelectronics Centre, University of Edinburgh,
Edinburgh EH9 3FF, UK
A.a.stokes@ed.ac.uk

Abstract. *Drosophila melanogaster* has been studied to gain insight into relationships between neural circuits and learning behaviour. To test models of their neural circuits, a robot that mimics *D. melanogaster* larvae has been designed. The robot is made from silicone by casting in 3D printed moulds with a pattern simplified from the larval muscle system. The pattern forms air chambers that function as pneumatic muscles to actuate the robot. A pneumatic control system has been designed to enable control of the multiple degrees of freedom. With the flexible body and multiple degrees of freedom, the robot has the potential to resemble motions of *D. melanogaster* larvae, although it remains difficult to obtain accurate control of deformation.

1 Introduction

We have designed a robot to mimic *Drosophila melanogaster* larvae (maggots), as a platform to test and verify their learning and chemotaxis models. *Drosophila* as a model system has a useful balance between relatively small number of neurons yet interestingly complex behaviours [10]. Many genetic techniques, such as GAL4/UAS systems developed by Brand and Perrimon [2], facilitate research on the connectivity and dynamics of the circuits. As a result, a number of necessary components of neural circuits for sensorimotor control and learning are being found and modelled. Currently, the models are tested by comparing between wildtype and genetic mutation lines, or using simulations of neural circuits and comparing output with biological experimental recordings. To test models in a wider environment, more similar to a larva, a physical agent that copies properties of the larval body is important.

Larvae have high degrees of freedom (DOFs) and flexible bodies. As a result, they are able to do delicate and spatially continuous motion. Simplified in mechanics, a larval body consists of body wall attached to the muscles and body fluids inside the body wall. The 2 parts works together as a hydrostatic skeleton [5]. The skin has regular repeating symmetrical folds, which are essential for its deformation and friction, forming its segments. The muscles of Drosophila larvae are in 3 orientations: dorso-ventral, anterioro-posterior and

© Springer International Publishing Switzerland 2016
N.F. Lepora et al. (Eds.): Living Machines 2016, LNAI 9793, pp. 375–386, 2016.
DOI: 10.1007/978-3-319-42417-0_34

oblique. Antero-posterior muscles are located nearer the interior than dorso-ventral muscles. The body wall muscles are segmentally repeated, and in each abdominal half segment there are approximately 30 of them ([1]) (Fig. 1).

Fig. 1. A Drosophila larva expressing mCherry (a type of photoactivatable fluorescent proteins [14]) in its muscles. From Balapagos (2012).

Based on the property of their bodies, Drosophila larvae are able to do several motions, such as peristaltic crawling, body bending and rolling. Forward peristaltic motion is best described. In the centre of the body, viscera suspended in hemolymph is essential for limiting body wall deformation and produces piston motion. During the 'piston phase' of peristalsis, muscles on the tail contract and push the viscera forward. The second 'wave phase' involves a wave of muscle contraction travelling through the bodywall segments from tail to head [4]. To mimic various and motions of a Drosophila larval, it is important to utilize this anatomical structure and avoid oversimplifying the high DOFs.

Some soft robots have been developed as bionic robots. The main materials are silicone, rubber, or other flexible and stretchable materials. They are usually actuated by Shape-Memory Alloy (SMA) or pneumatically, such as Biomimetic Miniature Robotic Crawler [7], GoQBot [6], Multigait soft robot [13], and a fluidic soft robot [11]. These robots only have several degrees-of-freedom (DOFs) and usually only have one type of motion, which is not sufficient to mimic larval motion. Although SMA is widely applied on soft robots, it has a significant shortcoming. As SMAs deform according to temperature, their response is limited by control of temperature. Because soft robots are usually not sufficient in heat dissipation, heat accumulates inside the robots, and response times of SMAs get too long so that continuous actuation is infeasible. The shortcoming does not exist on pneumatic actuation. Hence, pneumatic actuators are a feasible option as they have a faster response and longer effective working time. However, the main action most of soft pneumatic actuators is off-axis bending, and the axial elongation and contraction are only side effects. For examples: Micro Pneumatic Curling Actuator- Nematode Actuator [9], Pneu-net [13]), and Robot Air Muscles made from Oogoo [8]. As axial contraction is necessary for some motion (such as peristalsis), we designed a new type of pneumatic actuators.

2   Methods

The robot is made from soft silicone rubber, instead of rigid material, because: (1) motions of Drosophila larva are based on continous body deformation; (2)

soft materials have more similar properies to biological tissue than rigid material, such as nolinear elasticity and hysteresis, which are suitable to simulate dynamic characteristics of the muscle; and (3) defomation of Drosophila larval body wall is one method to control friction between body and contacted surface.

Figure 2 shows a sketch of a possible structure of a maggot robot. The robot has repeating modular body wall segments, with a water bag or air bag inside. Here we described the construction and control of 4 body segments. At present, the control system and pneumatic system are placed off board because of limited space and load.

Fig. 2. Sketch of the soft maggot robot. A central bag of fluid is surrounded by muscle segments.

2.1 Design of the Actuator and Body Wall of the Soft Robot

Pneu-nets (Fig. 3(a) and (b)) are usually made from 2 different soft materials: (1) flexible and stretchable material, such as Ecoflex, to form chambers to inflate and expand; (2) flexible but less or not stretchable material, such as Polydimethyl-siloxane (PDMS). Thus, when pneu-nets are inflated, the actuator bends to the side made from less stretchable material. Pneu-nets are not suitable for tubular body wall because the stretchable layer limits axial bending, hence we have modifeid the design to produce a new actuator type, which we called Extensible Pneu-nets.

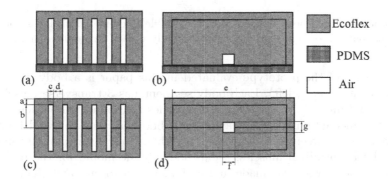

Fig. 3. Structure of Pneu-nets and Extensible Pneu-nets (a) and (b) are longitudinal and transverse sections of Pneu-nets, respectively; (c) and (d) are longitudinal and transverse sections of Extensible Pneu-nets, respectively.

Extensible Pneu-nets (Fig. 3(c) and (d)) use only 1 stretchable material. Small air chambers are connected by air tunnels to form a muscle. Different muscles are isolated. When an air chamber is inflated, it expands in all directions, and the direction with maximum expansion is the direction with the maximum cross sectional area. To limit deformation in the unwanted direction, thickness of the inner walls between chambers and thickness of the outer walls are carefully selected and tested. As stretchable material allows not only bending but also expansion along the surface, tubular body wall based on Extensible Pneu-nets are possible to axially bend.

To make the air chambers and tunnels inside, the actuator is divided into 2 layers which are cast separately. The moulds can be manufactured in conventional machining process or by 3D printing. Then the 2 layers are glued together with the same material. Finally, tubes for injecting pressed compressed air are inserted and glued. By including more air chambers and tunnels on a model, a body wall with multiple pneumatic actuators can be cast.

The first attempt at a muscle pattern was designed according to real muscle pattern on dissected and flattened body wall of Drosophila larva (Fig. 4). Dorsal oblique (DO) muscles, lateral transverse (LT) muscles, oblique lateral (LO) muscles, ventral longitudinal (VL) muscles and ventral acute (VA) muscles are simplified and mapped on the muscle pattern of the body wall. However, the adjacent muscles limited each others motions, especially when they have different orientations. The cause of limitation is that inner walls between air chambers limit transverse deformation, which is the direction that the adjacent muscles are designed to deform. Thus adjacent muscles should either be parallel, or should not be contiguous.

Fig. 4. Body wall of a body segment with Extensible Pneu-nets designed according to real muscle pattern on dissected and flattened body wall of Drosophila larva.

The design of the prototype evaluated in this paper is a body wall with 4 body segments (Fig. 5, left). Each body segment has 3 transverse muscles and 3 longitudinal muscles. These 2 types of muscles are connected perpendicularly and only connected on corners, leaving gaps between them to avoid limitation of deformation between each other (Fig. 5, right). Body segments are connected in series by longitudinal muscles. Figure 6 shows the mould for the body wall. After a flat body wall was made, it was folded end to end and clamped by 2 specially cut boards. Through the window of the board, the end was carefully aligned and glued together. By this process, the flat body wall is formed into a hollow cylinder shape (Fig. 7).

In this 4 body segment version, because the limited resolution of the 3D printer we use (Wanhao Duplicator with 0.4 mm nozzle) and resistance of air flow in tube, the dimensions of air chamber, as shown in Fig. 3(c) and (d), are: $a = 1.2\,\text{mm}$, $b = 3.0\,\text{mm}$, $c = 0.8\,\text{mm}$, $d = 1.2\,\text{mm}$, $e = 18\,\text{mm}$ (longitudinal muscles) or $28\,\text{mm}$ (transverse muscles), $f = 2.0\,\text{mm}$, $g = 3.0\,\text{mm}$. In a curved single body segment, the longitudinal length is $40\,\text{mm}$, the diameter is of the robot is about $50\,\text{mm}$. The total length of the 4 body segment body wall is about $175\,\text{mm}$.

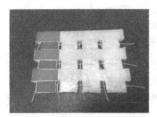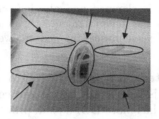

Fig. 5. (left) Prototype design of a body wall with perpendicular arrangement of muscles. Transverse muscles and longitudinal muscles of the first body segment are highlighted in red and green, respectively. (centre) A closer view of the flatten body wall shows gaps and spaces between the muscles to allow expansion. (right) The gaps and spaces when the body wall curved. (Color figure online)

Fig. 6. The 3D printed mould for body wall casting.

2.2 Pneumatic Actuation and Control System

The pneumatic actuation and control system controls the robot by controlling air pressures of air chambers. Air pressure sensors measures pressure in every muscle, pumps and valves control the air flow.

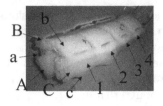

Fig. 7. (left) The body wall is clamped and glued. (centre) Formed into a hollow cylinder. (right) Names of muscles: body segments are numbered, longitudinal muscles named in capital letters, transverse muscles named in lower case letters.

Pneumatic Control System. A pneumatic control system has been designed for the robot. The system is located off board and connects to the robot with rubber tubes. As the robot has more DOFs than previous pneumatic soft robots mentioned above, the size of the pneumatic control system is designed to be compact.

The main component of the system is a valve island with 24 pairs of miniature 2 way solenoid valves (Fig. 8). The size of solenoid valves is 10 mm × 11 mm × 23 mm. Overall, the size of the valve island is 120 mm × 91 mm × 60 mm. The valves are installed the 3D main structure by interference fit. The main structure of the valve island consists of layers of 3D printed parts. The upper layer made form Acrylonitrile Butadiene Styrene (ABS), which offers Mechanical strength to fix valves, and lower layer made from Thermoplastic Elastomer (TPE), which has build in air channels with air-tightness. Every channel connects 4 ways, which are 2 valves, a pressure sensor, and an air chamber on the robot. The other 2 ways of each pair of valves are connected to compressed air and open to air, respectively. As the solenoid valves speed up to 100 Hz, the air flow can be finely controlled.

Fig. 8. (left) A valve and pump in the system. (centre) Structure of the 3D printed valve island. (right) Pneumatic valve island with 24 pairs of valves

Embedded Control System. An Embedded Control system has been designed for control and actuation. The control system is a hierarchical control system consisting of 1 main controller and 3 slave controllers. Their micro controllers are

STM32F411RE by STMicroelectronics. They are based on Cortex-M4 by ARM with digital signal processor (DSP) and floating-point unit (FPU). The main controller receives commands from a computer, and distributes them among the slave controllers by Universal Synchronous/Asynchronous Receiver/Transmitters (USART). On each of the slave controllers, 16 hardware Pulse-width modulation (PWM) channels and 8 Analog-to-digital converters (ADC) are configured to control 8 muscles. The PWMs control Darlington transistor arrays (ULx2003 by Texas Instruments). On each slave control board, 3 of them are adopted to drive valves. MPS20N0040D-D, which is an air pressure sensor to measure pressure in air chambers, is adopted to measure the pressures.

Algorithm. At present stage, the robot is controlled by feedforward preprogrammed motion. According to a approximate linearization between deformation and pressure at the initial state of equilibrium, the pressure is utilized as feedback of motion of muscle.

3   Experiments

The robot was tested for individual control of every muscle and coordination between them. Three motions are programmed and tested on the robot (A video of the experiments: https://youtu.be/aFE9dANHowk). The muscles are named based on their location. As show in (Fig. 7), body segments are numbered, longitudinal muscles named in capital letters, transverse muscles named in lower case letters.

Turn. In this motion, muscle a and B on every body segment was actuated at the same time, then pressure released. To minimize friction and show relevance between pressure and deformation, the robot is tested while floating on water. Figure 9 shows the motion of the robot. The pressure of the muscles is shown in Fig. 10.

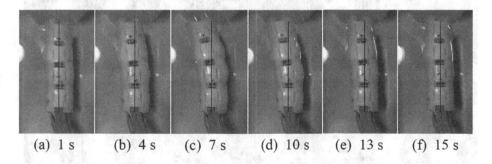

(a) 1 s   (b) 4 s   (c) 7 s   (d) 10 s   (e) 13 s   (f) 15 s

Fig. 9. Turning left. The black lines show the initial central axis.

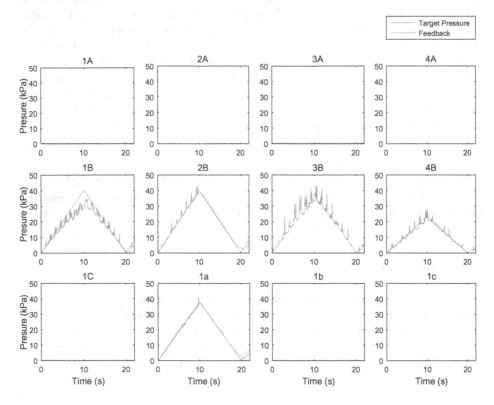

Fig. 10. Pressure of muscles in the first body segment, muscles A and muscles B during turning. (Color figure online)

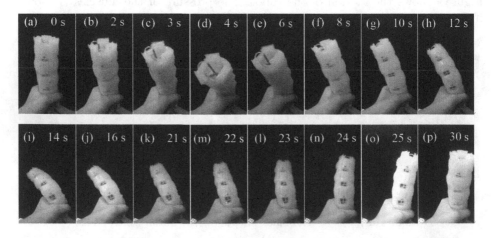

Fig. 11. Roll.

Roll. In this motion, muscle a and B, b and C, c and A, are inflated alternately. As the bundle of the tubes flowing the robot impact the rolling on a surface in water, the robot is hold on its tail vertically during test. Figure 11 shows the motion of the robot. The pressure of muscles show in Fig. 12.

Peristalsis. In this simplified peristalsis, all the muscles on a body segment inflate at same time and muscles of different segments inflate alternately.

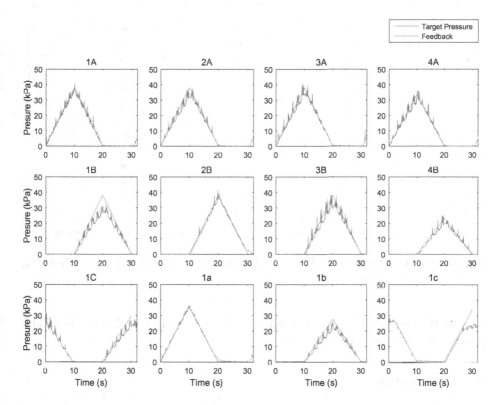

Fig. 12. Pressure of muscles in the first body segment, muscles A and muscles B during rolling. (Color figure online)

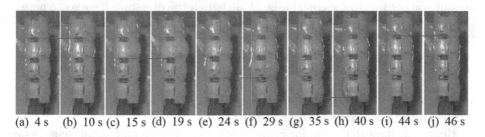

(a) 4 s   (b) 10 s (c) 15 s (d) 19 s (e) 24 s (f) 29 s (g) 35 s (h) 40 s (i) 44 s (j) 46 s

Fig. 13. Peristalsis. The parts on the blue lines were expanding. (Color figure online)

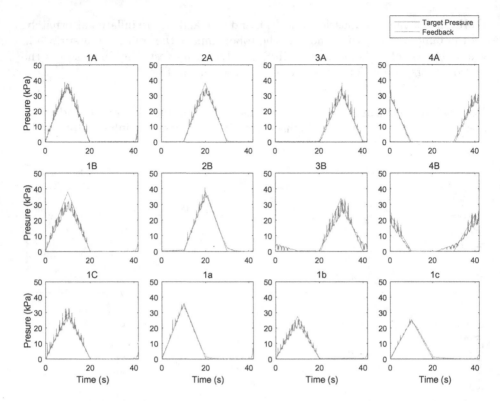

Fig. 14. Pressure of muscles in the first body segment, muscles A and muscles B during peristalsis. (Color figure online)

The motion is tested on water. Figure 13 shows the motion of the robot. The pressure of muscles show in Fig. 14.

4   Discussion

In the tests above, the robot produced three different motions from muscles actuated individually in different orders. The system was able to control the pressures according to the control signal, although the pressures have some noise. However, the three motions are not accurate. Deformation for the same pressure is different between the muscles. That is because muscles are slightly different and the relationship between deformation and pressure is not ideally linear. When an air chamber is inflated to a given range, the pressure does not change much even with obvious deformation. Hence, applying the same pressure to different muscles can result in different deformations. Thus, deformation sensors will be important to precise control of the robot.

Hence our immediate aim for future work is to develop and install sensors on the robot for deformation feedback. As the sampling density of the deformation sensors is limited, different deformations may map to the same output, hence a model or method to learn the relationship between sensor outputs and posture is necessary. We should then be able to explore more thoroughly the movement capabilities of the current design. Some additional redesign of the pneumatic muscle and body wall may be necessary, for example, surface processes to mimic denticles on Drosophila larval skin which generate asymmetric friction so that peristalsis produces forward locomotion [12].

5 Conclusion

Our longer term aim for this robot is to use it as a platform to test neural circuit models of *Drosophila* larvae. Initially this could focus on the motor circuits that generate and control peristalsis and bending. In particular these circuits could form the basis of an adaptive method for learning the control signals needed to adjust to the irregularities and non-linearities in the actuators and their interactions, in the same way that maggots are able to adapt to rapid change and growth in their body while maintaining efficient locomotion. Ultimately we would like to add sensors for environmental signals and investigate the sensorimotor control and associative learning involved in, e.g., odour search [3].

References

1. Bate, M.: The mesoderm and its derivatives. The Development of Drosophila Melanogaster, pp. 1013–1090. Cold Spring Harbor Laboratory Press, New York (1993)
2. Brand, A.H., Perrimon, N.: Targeted gene expression as a means of altering cell fates and generating dominant phenotypes. Development (Cambridge, England) **118**(2), 401–415 (1993)
3. Gomez-Marin, A., Louis, M.: Multilevel control of run orientation in Drosophila larval chemotaxis. Front. Behav. Neurosci. **8**, 38 (2014)
4. Heckscher, E.S., Lockery, S.R., Doe, C.Q.: Characterization of Drosophila larval crawling at the level of organism, segment, and somatic body wall musculature. J. Neurosci. **32**(36), 12460–12471 (2012)
5. Kohsaka, H., Okusawa, S., Itakura, Y., Fushiki, A., Nose, A.: Development of larval motor circuits in Drosophila. Dev. Growth Differ. **54**(3), 408–419 (2012)
6. Lin, H.T., Leisk, G.G., Trimmer, B.: GoQBot: a caterpillar-inspired soft-bodied rolling robot. Bioinspir. Biomim. **6**(2), 026007 (2011)
7. Menciassi, A., Accoto, D., Gorini, S., Dario, P.: Development of a biomimetic miniature robotic crawler. Auton. Robots **21**(2), 155–163 (2006)
8. mikey77: Soft robots: Making robot air muscles. Webpage (2012). http://www.instructables.com/id/Soft-Robots-Making-Robot-Air-Muscles/
9. Ogura, K., Wakimoto, S., Suzumori, K., Nishioka, Y.: Micro pneumatic curling actuator - Nematode actuator. In: 2008 IEEE International Conference on Robotics and Biomimetics, pp. 462–467 (2009)
10. Olsen, S.R., Wilson, R.I.: Cracking neural circuits in a tiny brain: new approaches for understanding the neural circuitry of Drosophila. Trends Neurosci. **31**(10), 512–520 (2008)

11. Onal, C.D., Rus, D.: Autonomous undulatory serpentine locomotion utilizing body dynamics of a fluidic soft robot. Bioinspir. Biomim. **8**(2), 026003 (2013). http://www.ncbi.nlm.nih.gov/pubmed/23524383
12. Ross, D., Lagogiannis, K., Webb, B.: A model of larval biomechanics reveals exploitable passive properties for efficient locomotion. In: Wilson, S.P., Verschure, P.F.M.J., Mura, A., Prescott, T.J. (eds.) Living Machines 2015. LNCS, vol. 9222, pp. 1–12. Springer, Heidelberg (2015)
13. Shepherd, R.F., Ilievski, F., Choi, W., Morin, S.A., Stokes, A.A., Mazzeo, A.D., Chen, X., Wang, M., Whitesides, G.M.: From the cover: multigait soft robot. Proc. Nat. Acad. Sci. **108**(51), 20400–20403 (2011)
14. Subach, F.V., Patterson, G.H., Manley, S., Gillette, J.M., Lippincott-Schwartz, J., Verkhusha, V.V.: Photoactivatable mCherry for high-resolution two-color fluorescence microscopy. Nat. Methods **6**(2), 153–159 (2009)

Short Papers

On Three Categories of Conscious Machines

Xerxes D. Arsiwalla[1]([✉]), Ivan Herreros[1], and Paul Verschure[1,2]

[1] Synthetic Perceptive Emotive and Cognitive Systems (SPECS) Lab,
Center of Autonomous Systems and Neurorobotics, Universitat Pompeu Fabra,
Barcelona, Spain
x.d.arsiwalla@gmail.com
[2] Institució Catalana de Recerca i Estudis Avançats (ICREA), Barcelona, Spain

Abstract. Reviewing recent closely related developments at the cross-roads of biomedical engineering, artificial intelligence and biomimetic technology, in this paper, we attempt to distinguish phenomenological consciousness into three categories based on embodiment: one that is embodied by biological agents, another by artificial agents and a third that results from collective phenomena in complex dynamical systems. Though this distinction by itself is not new, such a classification is useful for understanding differences in design principles and technology necessary to engineer conscious machines. It also allows one to zero-in on minimal features of phenomenological consciousness in one domain and map on to their counterparts in another. For instance, awareness and metabolic arousal are used as clinical measures to assess levels of consciousness in patients in coma or in a vegetative state. We discuss analogous abstractions of these measures relevant to artificial systems and their manifestations. This is particularly relevant in the light of recent developments in deep learning and artificial life.

Keywords: Conscious agents · Artificial intelligence · Complex systems · Bioethics

1 Introduction

The topic of this discussion addresses the question of what it takes to engineer conscious processes in an artificial machine and measures useful for describing such processes. By conscious processes we mean phenomenological aspects of consciousness. This concerns an epistemically objective description of a problem that may well have an ontologically subjective element. Drawing from what is known (howsoever little) about biological systems, we build a parallel argument for artificial and collective systems.

2 Biological Consciousness

In patients with disorders of consciousness ranging from coma, locked-in syndrome to those in vegetative states, levels of consciousness are assessed through

© Springer International Publishing Switzerland 2016
N.F. Lepora et al. (Eds.): Living Machines 2016, LNAI 9793, pp. 389–392, 2016.
DOI: 10.1007/978-3-319-42417-0_35

a battery of behavioral tests as well as physiological recordings. More generally, these measurements are grouped into those referring to awareness and those to arousal [5,6]. In fact, for clinical purposes, closely associated states of consciousness can be grouped into clusters on a plane with awareness and arousal representing orthogonal axes [5,6]. Scales of awareness target both higher- and lower-order cognitive functions enabling complex behavior (see [1,2] for a discussion on measures of consciousness). Arousal results from biochemical homeostatic mechanism regulating survival drives and is clinically measured in terms of glucose metabolism in the brain. Clinical consciousness is thus assessed as a bivariate measure on these two axes. In fact, in all known organic life forms, biochemical arousal is a necessary precursor supporting the hardware necessary for cognition. In turn, evolution has shaped cognition in such a way so as to support the organism's basic survival as well as higher-order drives associated to cooperation and competition in a multi-agent environment. Awareness and arousal thus form a closed-loop, resulting in phenomenological consciousness.

How can we then abstract and map arousal and awareness to non-biological system to set-up a comparative measure of consciousness? As noted above, arousal results from autonomous homeostatic mechanisms necessary for the self-preservation of an organism's germ line in a given environment. In other words, arousal results from self-sustaining life processes necessary for basic survival, whereas awareness can be functionally abstracted as general forms of intelligence, necessary for higher-order functions in a complex social world. If biological consciousness as we know it, is phenomenologically an amalgamation of life and intelligence, how can we use this argument to conceive a functional notion of consciousness in artificial systems?

3   Artificial Consciousness

The reason why the aforementioned question has now become very relevant is due to some remarkable recent advances in two seemingly unrelated fields: artificial intelligence and synthetic life. The most recent example of the former is Google DeepMind's AlphaGo system [8] and recent examples of the latter include synthesis of artificial DNA with six nucleotide bases (the naturally occurring adenine (A), cytosine (C), guanine (G) and thymine (T) plus a new synthetic pair), which was engineered into the cell of the bacterium Escherichia coli and successfully passed from one generation to the next [7]; synthetic proto-cells capable of replicating themselves [4]; and a fully functioning synthetic genome implanted into the cell of the bacterium Mycoplasma genitalium that converts it into a novel bacterium species [3].

Of course the abstract functional criterion we have defined above for consciousness is not even remotely satisfied by any of these systems as they either have some limited form of intelligence or life but not yet both. However, AlphaGo's feat in beating the top human Go champion was noteworthy for a couple of reasons. Unlike Chess, possible combinations in Go run into millions and when played using a timer any brute-force algorithm trying to scan the entire search space would simply run out of computational capacity or time. Hence, a pattern recognition algorithm mimicking intuition was crucial to the development of AlphaGo, where the machine learned complex strategies by playing itself over and over again while reinforcing successful sequence of plays through the weights of its deep neural networks [8]. Most remarkably, it played counterintuitive moves that shocked the best human players and the only game that Lee Sedol, the human champion won out of five, was possible after he himself adopted a brilliant counterintuitive strategy.

Hence asking whether a machine can think like a human is similar to asking whether a submarine can swim. It just does it differently. If the goal of a system is to learn and solve complex tasks close to human performance or better, current machines are already doing that in specific domains. However, these machines are still far from learning and solving problems in generic domains and in ways that would couple its problem solving abilities to its own survival drives. On the other side, neither have the above-mentioned synthetic cells yet been used to build complex structures with computing or cognitive capabilities. Nevertheless, this does suggest that the synthesis between artificial life forms and AI would satisfy the same criterion of functional consciousness as that used by known biological systems. This form of consciousness, if based on a life form with different survival drives/mechanisms and non-human forms of intelligence, would also likely lead to non-human behavioral outcomes.

4   Collective Consciousness

The notion of collective consciousness is not really a distinct type of consciousness by itself when one thinks of biological consciousness as a manifestation of this collective phenomenon where individual cells making up an organism are not considered conscious by themselves even though the organism as a whole is. As such collective intelligence is a phenomenon widely studied in systems ranging from ant colonies to social networks to the internet. But these are generally not regarded as conscious systems. As a whole they are not considered to be life forms with survival drives that compete or cooperate with other similar agents. But these considerations begin to get blurred at least during transient epochs when collective survival comes under threat. For example, when a bee colony comes under attack by hornets, it demonstrates a prototypical survival drive, similar to lower-order organisms and which is transiently implemented alongside its existing social intelligence. This does suggest a third category of conscious machines, which possess a distributed architecture along with collective survival drives and learning abilities, needed for competition or cooperation in a multi-agent environment.

5   Societal Impact, Ethics and Conclusions

Both, the societal impact and ethics of any form of intelligent machine, especially conscious machines, constitutes a very serious issue. For example, the impact of medical nanobots for removing tumors, attacking viruses or non-surgical organ reconstruction has the potential to change medicine forever. Or intelligent systems to clear pollutants from the atmosphere or the rivers are absolutely essential for some of the biggest problems that humanity faces. However, as discussed above, purely increasing the performance of a machine along the intelligence axis will not constitute consciousness as along as this intelligence is not accessible by the system to autonomously regulate or enhance its survival drives. On the other hand, whenever the latter is indeed made possible, issues of societal interactions of machines with humans and the ecosystem, becomes an imminent ethical responsibility. It becomes important to understand the kind of cooperation-competition dynamics that a futuristic human society will face. The early stages of designing these machines are probably the best times to regulate future behavioral outcomes of their interaction dynamics. This analogy might not be surprising to any parent that has a child. Hence, a serious effort into understanding the evolution of complex social traits is crucial alongside the engineering that goes into building such systems.

Acknowledgments. This work has been supported by the European Research Council's CDAC project: "The Role of Consciousness in Adaptive Behavior: A Combined Empirical, Computational and Robot based Approach" (ERC-2013-ADG 341196).

References

1. Arsiwalla, X.D., Verschure, P.: Computing information integration in brain networks. In: Wierzbicki, A., Brandes, U., Schweitzer, F., Pedreschi, D., et al. (eds.) NetSci-X 2016. LNCS, vol. 9564, pp. 136–146. Springer, Heidelberg (2016). doi:10. 1007/978-3-319-28361-6_11
2. Arsiwalla, X.D., Verschure, P.F.: Integrated information for large complex networks. In: The 2013 International Joint Conference on Neural Networks (IJCNN), pp. 1–7. IEEE (2013)
3. Hutchison, C.A., Chuang, R.Y., Noskov, V.N., Assad-Garcia, N., Deerinck, T.J., Ellisman, M.H., Gill, J., Kannan, K., Karas, B.J., Ma, L., et al.: Design and synthesis of a minimal bacterial genome. Science 351(6280), aad6253 (2016)
4. Kurihara, K., Okura, Y., Matsuo, M., Toyota, T., Suzuki, K., Sugawara, T.: A recursive vesicle-based model protocell with a primitive model cell cycle. Nat. Commun. 6, 8352 (2015)
5. Laureys, S.: The neural correlate of (un) awareness: lessons from the vegetative state. Trends Cogn. Sci. 9(12), 556–559 (2005)
6. Laureys, S., Owen, A.M., Schiff, N.D.: Brain function in coma, vegetative state, and related disorders. Lancet Neurol. 3(9), 537–546 (2004)
7. Malyshev, D.A., Dhami, K., Lavergne, T., Chen, T., Dai, N., Foster, J.M., Corrêa, I.R., Romesberg, F.E.: A semi-synthetic organism with an expanded genetic alphabet. Nature 509(7500), 385–388 (2014)
8. Silver, D., Huang, A., Maddison, C.J., Guez, A., Sifre, L., van den Driessche, G., Schrittwieser, J., Antonoglou, I., Panneershelvam, V., Lanctot, M., et al.: Mastering the game of go with deep neural networks and tree search. Nature 529(7587), 484–489 (2016)

Gaussian Process Regression for a Biomimetic Tactile Sensor

Kirsty Aquilina[1,2(✉)], David A.W. Barton[1,2], and Nathan F. Lepora[1,2]

[1] Department of Engineering Mathematics, University of Bristol, Bristol, UK
{ka14187,david.barton,n.lepora}@bristol.ac.uk
[2] Bristol Robotics Laboratory, University of Bristol, Bristol, UK

Abstract. The aim of this paper is to investigate a new approach to decode sensor information into spatial information. The tactile fingertip (TacTip) considered in this work is inspired from the operation of dermal papillae in the human fingertip. We propose an approach for interpreting tactile data consisting of a preprocessing dimensionality reduction step using principal component analysis and subsequently a regression model using a Gaussian process. Our results are compared with a classification method based on a biomimetic approach for Bayesian perception. The proposed method obtains comparable performance with the classification method whilst providing a framework that facilitates integration with control strategies, for example to perform controlled manipulation.

Keywords: Tactile sensors · Gaussian process regression · Bayesian perception

1 Introduction

Touch is essential in everyday life as it provides the required information to properly perform manipulation tasks [1]. This kind of sensory input should thereby also be provided to robots in order to make them capable of performing complex manipulation tasks. In consequence, it is important to have robust tactile sensors and algorithms that are able to decode all the important sensor information.

The aim of this paper is to investigate a new approach to decode tactile sensor information into spatial information. The standard approach is to use a classification framework in which percepts are discrete classes and then the best class is chosen, for example via Bayesian inference as in the biomimetic active perception method of refs [2, 3]. In the present paper, another Bayesian approach is considered where the input signal is preprocessed using Principal Component Analysis (PCA) and then given as input to a Gaussian Process (GP) [4] that performs regression.

The sensor used here is a biologically-inspired tactile fingertip, the TacTip [5]: a dome-shaped sensor with internal artificial papillae inspired by dermal papillae, i.e. protrusions on the inside of the sensor skin. These artificial papillae will be here referred to as taxels (the sensing elements) or pins. The motion of each

© Springer International Publishing Switzerland 2016
N.F. Lepora et al. (Eds.): Living Machines 2016, LNAI 9793, pp. 393–399, 2016.
DOI: 10.1007/978-3-319-42417-0_36

individual pin is tracked using an internally-encased camera and image processing techniques. The algorithm proposed in this work also has a biological inspiration. The dimensionality reduction preprocessing step is analogous to human vision, where retinal cone photoreceptors reduce the dimensionality of the incoming light to three dimensions depending on the wavelength range [6]. Additionally, second-order statistics such as PCA have been shown to have similarities with receptive fields in the primary visual cortex [7]. The idea of using a regression model is also rooted in human behaviour where it has been proposed that people learn continuous-valued functions by a regression framework [8]. Furthermore, Griffiths et al. [9] show that GPs provide a possible explanation to how humans learn continuous functions.

This paper is organised as follows: the Methods will provide a description of the decoding of the sensor data, then the Results will compare this decoding with that of the Bayesian perception method from [2,3] that has been previously used for interpreting data from the TacTip sensor.

2   Methods

The tactile data analysed in this paper was collected from the TacTip mounted on a 6-DOF robotic arm (ABB IRB120) that performed vertical taps on an edge stimulus as shown in Fig. 1. The sensor is rotated over this edge, between taps, by 1° collecting data ranging from 0–359°. The sensor readings for 0° and 180° are opposites of each other, as in each case a different subset of the pins is in contact with the object, thereby permitting to distinguish between all the angles over the considered range. The pin arrangement of the TacTip can be seen in Fig. 2a. This version of the TacTip [10] has 127 pins leading to a total of 254 dimensions, as each pin (taxel) gives measurements in the x and y direction. Furthermore, each taxel produces a time series of pixel deflections in the x direction and in the y direction during each tap.

Two datasets were collected, one being used as a training set and the other one used as a test set. Each set contained a time series of taxel positions for each dimension for each tap, having in total 360 taps covering the whole range of interest. In order to obtain one value for each of these 254 dimensions for every tap, the first principal component of each time series is computed using PCA. This was done by collecting all the data over all 360 taps for each taxel dimension, then finding the most dominant eigenvector and projecting the data on it. The result of doing this dimensionality reduction can be seen in Fig. 2b which is a scaled down version (0.4 times) of the principal components superimposed on the position of each taxel. This means that the time series provided by each tap is mapped to one single value per taxel dimension computed using the whole time series. This transformed the training dataset into a matrix of 360 × 254 which was then further preprocessed by obtaining the principal components that explain the motion of the taxels when different edge orientations were sensed. The first two principal components were considered for the rest of the analysis which can be seen in Fig. 2c. The image shows that considering these two dimensions is a valid approach as this representation encodes each orientation differently but similar orientations are related to each other. This PCA was performed so as to reduce the computational expense incurred in this approach by optimising 4 hyperparameters instead of 256 hyperparameters.

The training set was composed of 2 dimensions and 360 observations. This data was then used to optimise the GP hyperparameters built using a squared exponential (SE) covariance function. Two GPs were used, as angles which are close to 0° or 360° are actually the same angle. Therefore in order not to cause interference between these two extremes a GP was used to learn the mapping for angles between 0° and 179° and another for the rest of the range to 359°. The use of multiple local GPs was inspired from the approach presented in [11]. However, in the present work the training points for each model are chosen before the model learning and the Euclidean distance is used as a measure to choose which model is to be associated with a new test point. The predictions are made using only one model and are not a merged version of both models.

Fig. 1. The setup used for data gathering.

The hyperparameters were optimised using the MATLAB optimisation toolbox by minimizing the negative log marginal likelihood of the GP [4] thereby obtaining values for the signal noise standard deviation σ_n, the signal standard deviation σ_f, and the length-scale l (one for each dimension) required by the SE covariance. These are the respective values for model 1 (0°–179°) 0.26, 66.23, 121.57, 144.5 and model 2 (180°–359°) 0.21, 59.56, 151.78, 95.98. The length-scale can be seen to be related to the circle depicted in Fig. 2c which has a radius of approximately 106 in the principal components dimension.

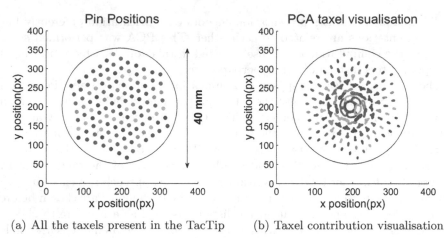

(a) All the taxels present in the TacTip   (b) Taxel contribution visualisation

(c) Overall Principal Components visualisation

Fig. 2. The pins or taxels of the sensor are seen in (a). These are then displaced over the range of interest (b) showing the first principal component in the x and y direction scaled by 0.4. Finally the first and second principal components over all taxels are shown in (c) where 'start' denotes 0° and 'end' denotes 359°.

3   Results

The framework described in the Methods section was tested using the test dataset. The regression results can be seen in Fig. 3. The mean predicted values had an average orientation error of 0.33° as shown in Fig. 4a. It is interesting to note that the point that had the largest error also showed a larger variance implying that the model is less sure about the predicted value.

The passive Bayesian perception approach explained in [2] having 1° per class and using one tap to make a decision was then used to analyse the same two datasets. When using these parameters the passive Bayesian perception is

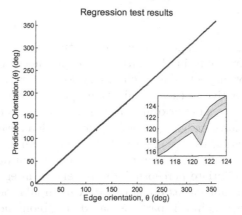

Fig. 3. Regression results for the two joined GP models, along with a zoomed in plot section. The blue line depicts the computed mean and the grey bands show ± 3 standard deviations. (Color figure online)

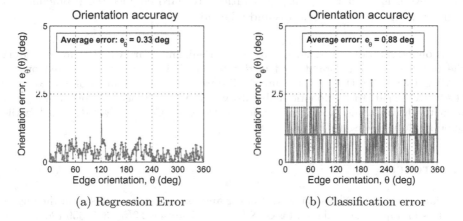

Fig. 4. Comparing the error obtained using the proposed regression method and the currently used classification method.

equivalent to using maximum likelihood which resulted in an average error of $0.88°$ as shown in Fig. 4b. Therefore the proposed method managed to obtain results comparable to the classification method and achieved an improvement in accuracy due to the averaging implicit in a GP. Additionally, the proposed method has the advantage of providing a measure of variance on the prediction, which might be useful for a decision making process. Another advantage is that a regression approach requires fewer training points while still covering the whole range. For instance using 92 training points over the whole range which include $0°$, $179°$, $180°$ and $359°$ lead to the same average error of $0.33°$. Most importantly, having a regression method to infer sensor readings is important in order to facilitate the integration of control strategies in tactile robots.

4   Conclusions

The results show that the proposed decoding method using PCA and a GP can learn the required mapping in the particular experiment considered here (an edge over a range of angles). This preliminary work showed that this method can provide an alternative to the classification approach with discrete percepts. Additionally, it provides more flexibility as continuous predictions are computed due to the interpolation capabilities of a GP, along with a measure of uncertainty via the computed variances. We expect that fewer training points will also be required. Obtaining a good regression model to infer sensor data is important as this can give an appropriate perceptual component to integrate within control loops, which is important for attaining complex manipulation behaviour.

The work investigated in this paper is based on a biomimetic sensor inspired from the human tactile sensory system and the considered inference approach is biologically inspired. Natural organisms such as humans are a great source of inspiration for artificial systems as they have been perfected over many millennia of evolution, which can inspire solutions to hard engineering problems in a manner with superior robustness and adaptability.

Acknowledgements. I would like to thank Benjamin Ward-Cherrier for providing the training and test dataset used in this paper. This work was supported by the EPSRC Centre for Doctoral Training in Future Autonomous and Robotic Systems (FARSCOPE) at Bristol Robotics Laboratory. This work was also supported by a grant from the Engineering and Physical Sciences Research Council (EPSRC) on 'Tactile Superresolution Sensing' (EP/M02993X/1).

References

1. Johansson, R.S., Flanagan, J.R.: Coding and use of tactile signals from the fingertips in object manipulation tasks. Nat. Rev. Neurosci. **10**(5), 345–359 (2009)
2. Lepora, N.F., Ward-Cherrier, B.: Superresolution with an optical tactile sensor. In: 2015 IEEE/RSJ International Conference on Intelligent Robots and Systems (IROS), pp. 2686–2691, September 2015
3. Lepora, N.F.: Biomimetic active touch with fingertips and whiskers. IEEE Trans. Haptics **9**(2), 170–183 (2016)
4. Rasmussen, C., Williams, C.: Gaussian Processes for Machine Learning. Adaptative Computation and Machine Learning Series. University Press Group Limited, New Era Estate (2006)
5. Chorley, C., Melhuish, C., Pipe, T., Rossiter, J.: Development of a tactile sensor based on biologically inspired edge encoding. In: International Conference on Advanced Robotics, ICAR 2009, pp. 1–6, June 2009
6. Gegenfurtner, K.R., Kiper, D.C.: Color vision. Annu. Rev. Neurosci. **26**(1), 181–206 (2003)
7. Baddeley, R.J., Hancock, P.J.B.: A statistical analysis of natural images matches psychophysically derived orientation tuning curves. Proc. R. Soc. Lond. B: Biol. Sci. **246**(1317), 219–223 (1991)

8. Koh, K., Meyer, D.E.: Function learning: induction of continuous stimulus-response relations. J. Exp. Psychol. Learn. Mem. Cogn. **17**(5), 811 (1991)
9. Griffiths, T.L., Lucas, C., Williams, J., Kalish, M.L.: Modeling human function learning with gaussian processes. In: Koller, D., Schuurmans, D., Bengio, Y., Bottou, L. (eds.) Advances in Neural Information Processing Systems 21, pp. 553–560. Curran Associates Inc. (2009)
10. Ward-Cherrier, B., Cramphorn, L., Lepora, N.F.: Exploiting sensor symmetry to generalize biomimetic touch. In: Ohwada, H., Yoshida, K. (eds.) Living Machines 2016. LNCS, vol. 9793, pp. 540–544. Springer, Switzerland (2016)
11. Nguyen-Tuong, D., Peters, J.: Local gaussian process regression for real-time model-based robot control. In: IEEE/RSJ International Conference on Intelligent Robots and Systems, IROS 2008, pp. 380–385, September 2008

Modulating Learning Through Expectation in a Simulated Robotic Setup

Maria Blancas[1]([envelope]), Riccardo Zucca[1],
Vasiliki Vouloutsi[1], and Paul F.M.J. Verschure[1,2]

[1] Laboratory of Synthetic Perceptive Emotive Cognitive Systems (SPECS), DTIC,
N-RAS, Universitat Pompeu Fabra, Barcelona, Spain
{maria.blancas,riccardo.zucca,vicky.vouloutsi,paul.verschure}@upf.edu
[2] Catalan Institute of Advanced Studies (ICREA), Barcelona, Spain

Abstract. In order to survive in an unpredictable and changing environment, an agent has to continuously make sense and adapt to the incoming sensory information and extract those that are behaviorally relevant. At the same time, it has to be able to learn to associate specific actions to these different percepts through reinforcement. Using the biologically grounded Distributed Adaptive Control (DAC) robot-based neuronal model, we have previously shown how these two learning mechanisms (perceptual and behavioral) should not be considered separately but are tightly coupled and interact synergistically via the environment. Through the use of a simulated setup and the unified framework of the DAC architecture, which offers a pedagogical model of the phases that form a learning process, we aim to analyze this perceptual-behavioral binomial and its effects on learning.

Keywords: Distributed Adaptive Control · Autonomous synthetic agents · Rule learning

1 Introduction

You wake up at night and you want to reach a small (turned on) lamp at the other side of the room. As you see the light, reaching it will not be a problem. But, what would happen if you were in complete darkness and wanted to turn on that light? You could first try a random path and probably crash with the objects in the room. You would try to find another path, and then another one, and so on, until you finally find the lamp. This strategy, although probably leading you to a correct solution at the end, would not be really optimal. An optimal strategy would be to use the objects you run into as cues and use the distance between them to learn the whole sequence that allows you to arrive from your bed to the lamp. In order to correctly reach your final goal, you have to make accurate predictions of the position of the objects in your path.

This example illustrates the critical interplay occurring between perceptual and behavioral learning. If you learned the room only through perception, you

© Springer International Publishing Switzerland 2016
N.F. Lepora et al. (Eds.): Living Machines 2016, LNAI 9793, pp. 400–408, 2016.
DOI: 10.1007/978-3-319-42417-0_37

would have to perceive as many objects in the room as possible, to form a general idea of the room. That would provide you with a large amount of information of the surrounding environment, but the most of it would not be needed for your goal. If you learnt to reach the lamp only through behavior, you would have to try different paths until finding the right one, but it would have been easier if you had some references of the environment. Thus, perceptual learning gives you cues of the environment, the sensory events; and behavioral learning allows you to build associations between perceptions and actions, thus reshaping your internal model of the world and improve your accuracy. In that sense, when you have enough information of the environment that allows you to form quasi-correct expectations of the cues you are going to receive, you can start to learn from context the most optimal solution that leads you to reach your goal.

The positive effect of the interplay between perceptual and behavioral learning on accuracy and performance has been previously investigated in a robot foraging task in the context of the Distributed Adaptive Control (DAC) theory of mind, brain and body nexus [1–3]. Similarly to the example described previously, in the foraging task, the agent first explores the world to learn its regularities using the simple reflexes it is equipped with; then, through adaptation, those behaviors can be reshaped and internal representations can be formed. When the difference between the stimulus expected (prediction) and the stimulus received (perception) is low enough, that is, when the outcome of expectations is close to reality, the agent begins to learn from context by learning the sequences of actions that lead to a goal. This form of learning does not focus only on the appropriate actions but also on the sequences that lead to a goal state, shaping the world model of the agent and leading to a stabilization of the agent's behavior and internal representations.

The use of the DAC cognitive architecture offers an already defined framework to monitor and control this behavior. Furthermore, DAC predicts that learning is organized along a hierarchy of complexity and in order to learn and consolidate new material, the learner undergoes a sequence of learning phases: resistance to new information when that threatens an agent's inner model of the world, confusion that creates the necessity to resolve the problem and learn through re-adapting and abduction which is the process of acquiring the new knowledge. Given the interplay between perceptual and behavioral learning and the three phases of learning according to DAC, we propose a first approach to map these two ideas. That is: how, on the one hand, the accuracy of one's expectations allows for a more optimal learning (perceptual) while contextual learning stabilizes behavior and internal representations and on the other hand, how learning occurs based on the phases suggested by DAC. Here, we focus on introducing the idea of stage transitions, and how the different layers of the DAC architecture contribute to the stabilization of behavior. This paper will later serve as the basis for our main goal: map these stage transitions to the phases of learning and understand how they occur.

In the following sections, we first present the DAC architecture (see Sect. 1.1) as a model of cognitive development. In Sect. 2 we describe the experimental setup (simulated environment) used to test the proposed stage transition (Sect. 3). Finally, in the last section we discuss how we plan to proceed in order to map the current stage transition to the phases of learning.

1.1   DAC as a Model of Cognitive Development

The Distributed Adaptive Control (DAC) architecture views the learner as a complex individual whose behavior derives from the interplay of sensation, reaction, adaptation and contextualization. This biologically grounded cognitive architecture is defined in the context of classical and operant conditioning and represents the association between mind, brain and behavior [4]. DAC is organized across four different layers: Soma, Reactive, Adaptive and Contextual. The Soma Layer represents the body itself and the information acquired from sensation, needs and actuation. The Reactive Layer enables the interaction of the system with its environment through a set of pre-wired behaviors in response to specific stimuli. The Adaptive Layer uses the cues provided by the Reactive Layer to form sensory representations and their associated behaviors. These internal representations of the world, or *prototypes*, are later used by the Contextual Layer, a behavioral learning system based on short and long memory systems, to construct goal-oriented behavioral plans based on mechanisms of operant conditioning.

The Reactive and Adaptive layers of the DAC architecture follow the paradigm of classical conditioning, where a stimulus that was initially neutral becomes conditioned (CS) by relating it to an intrinsically motivational or unconditioned stimulus (US), allowing the CS to trigger actions, or conditioned responses (CR). In this sense, the acquired sensor and motor states are associated through the valence states triggered by the Reactive Layer. The US is defined through internal states and the CS, through sensory modalities.

The Contextual Layer is a behavioral learning system that uses the sensorimotor representations acquired by the Adaptive Layer as inputs to construct goal-oriented behavioral plans. It comprises systems for memorizing and recalling behavioral sequences, consisting of two structures: short-term memory (STM) and long-term memory (LTM), which allow for the formation (conditional on the goal achievement of the agent as signaled by the Reactive and the Adaptive Layers) of sequential representations of states of the actions generated by the agent and of the environment. What triggers the activation of the Contextual Layer is the Discrepancy (D) measure. When this measure, calculated as the discrepancy between expected and actual CS events, reaches a value below a pre-specified Discrepancy threshold (θ_D), the Contextual Layer is enabled.

Learning is defined as a process where the learner moves through a hierarchically organized progression of stages to acquire new knowledge. Therefore, an agent's actions result from merging the perceived events coming from a constantly changing environment and the agent's own internal needs. DAC predicts that learning is organized along a hierarchy of complexity. The Reactive Layer allows the organism to explore the world and gain experiences based on which the Adaptive Layer would learn the states of the world and their associations.

Only after the states are well consolidated the Contextual Layer can extract consistent rules and regularities.

DAC predicts that in order to learn and consolidate new material the learner undergoes a sequence of learning phases: *resistance, confusion,* and *abduction* that a learner has to go through in order to acquire and consolidate new knowledge. *Resistance* occurs when the learner overestimates his knowledge and skills by defending his own world's model when facing a new condition [5]. This resistance to accept new knowledge drives the learner to *confusion.* When feeling confused, the learner has the need to readjust his internal model to cope with the new one. Nevertheless, this confusion has to stay in a controllable range so that he does not feel frustrated or not challenged enough. Lastly, the development of new theories to overcome *confusion* and their posterior assessment is what is called *abduction.* This is what allows the learner to acquire new knowledge. The agent begins forming his inner world model in the Reactive Layer, through reflexes, but it is through adaptation and experience that this world model is shaped, to reach its maximum accuracy in the Contextual Layer.

This process is not completely parallel to the layers of the DAC architecture, but results from their combination.

2   Materials and Methods

2.1   Experimental Setup and Procedure

We tested the DAC architecture using a simulated robot solving a foraging task in a restricted arena as in [6]. The arena contains a light source used as a target (unconditioned stimulus, US) and six patches of different color on the floor serving as contextual cues (conditioned stimuli, CSs) (Fig. 1). Colored patches on the floor are distinguished by their hue value. On each trial, the robot is randomly placed in one of the three starting positions (grey patches on the bottom of the arena) and the goal of the robot is to maximize the number of times it can correctly find the light. To learn the task, the robot has to form adequate representations of the environmental cues and stable associations of sensory events and actions (CS-UR associations).

We considered two experimental conditions in which the DAC contextual layer was either "Diasbled" or "Enabled". For each condition the experimental session is divided in two main cycles, each of them beginning with a stimulation period (2000 time steps), where the US signal from the target is activated, and followed by a recall period (5000 time steps), where the US signal from the target is deactivated.

The foraging task is implemented in the Gazebo robot simulator (v4.0) [7] whereas the DAC neuronal model is implemented with the *iqr* neuronal network simulator (v4.2) [8] running on a Linux machine.

2.2   Predictive Correlative Subspace Learning Rule

In DAC the adaptive layer forms internal representations of the environment and learns sensory-motor contingencies based on the classical conditioning paradigm

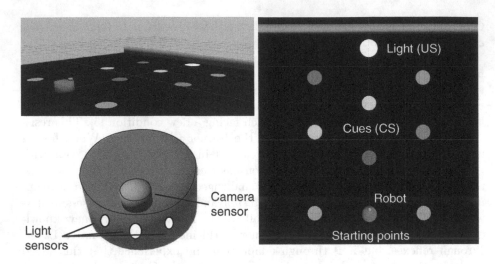

Fig. 1. The experimental setup. The restricted arena foraging task used for the experiments. The unconditioned stimulus target (US, white circle) corresponds to the light source, colored patches on the floor represent the contextual cues (conditional stimuli, CSs) whereas grey patches correspond to the starting positions. The wheeled simulated robot is equipped with a set of light sensors to detect the source of light (US sensors) and a camera to detect the patches on the floor (CS sensor).

[1,3,6]. The slow dynamics describing the change of synaptic weights (W) follow the predictive correlative subspace learning rule (PCSL) [2].

$$\Delta W = \eta((x - W_y)((1 - \zeta)y + (1 + \zeta)r)^T + \rho e(r - y)^T)$$

The parameter η represents the learning rate, the ρ parameter varies the influence of the behavioral error term on learning. The ζ parameter regulates the balance between perceptual ($\zeta \sim -1$) and behavioral learning ($\zeta \sim 1$), that is, how much the system relies on perception or behavior to define the prototypes. An extended study on the individual effects of these components can be found in [2].

The contextual layer relies on the internal representations formed by the adaptive layer and is activated only when the CS prototypes approximate the CS. This transition is controlled by a discrepancy measure D, which, whenever falls below 0.2, enables the contextual layer. D is a running average of the difference between the perceived and the expected CS (or prototype) defined as:

$$d(x, e) = \frac{1}{N} \sum_{j=1}^{N} \left| \frac{x_j}{max(x)} - \frac{e_j}{max(e)} \right|$$

3   Results

In this study, we tested the differences in the behavior of a simulated robot during a foraging task. In this task, the robot had to navigate in an environment and learn to associate sensory events (CS) to actions (UR), based on classical conditioning. Two conditions (with 10 repetitions each) were tested: the "Disabled" condition (where the only DAC layers that are enabled are the Reactive and the Adaptive), and the "Enabled" condition (where, as soon as D falls below 0.2, the Contextual layer is enabled). We analysed the temporal evolution of the discrepancy measure across simulations for the two conditions. In the Enabled condition, the activation of the Contextual Layer makes the Discrepancy fall below the value observed for the Disabled condition (Fig. 2 left). This is caused by the reduction of the discrepancy between the predicted and actual CS events and the stabilization of the synaptic weights inside the Adaptive Layer, which leads to a transition to the Contextual Layer. The graph at the right represents the distribution of the changes in discrepancy, where we can see that the discrepancy of the "Enabled" condition is lower than the one of the "Disabled" condition.

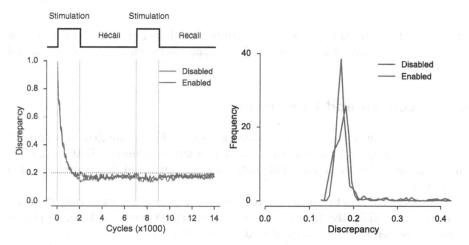

Fig. 2. (Left) Evolution of the discrepancy measure (D). The dashed vertical lines indicate the beginning of the periods inside each cycle. The dashed horizontal line indicates the transition value of D. (Right) Distribution of discrepancy changes for the two experimental conditions (enabled and disabled contextual layer) (Color figure online)

Figure 3 represents the evolution of the change in the weights. During the second stimulation and recall periods, there is a significantly lower value of the average absolute change in synaptic weights: enabled $|\Delta W| = 1.6x10^{-3}$, disabled $|\Delta W| = 1.8x10^{-3}$ (Kolmogorov-Smirnov $p \ll 0.001$).

Verschure et al. [3] showed that enabling the Contextual Layer results in a less variable behavior, what, in turn, lowers the variability of the perceived sensory inputs that the system has to classify. This leads to a more reduced set of input states to adjust the structures for perceptual learning; a change that, at the same time, can be seen in the reduction and stabilization of the Discrepancy

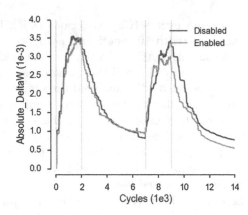

Fig. 3. Evolution of the change in synaptic weights during the experimental session). (Color figure online)

measure. Thus, the transition to the Contextual Layer may be facilitated by behavioral feedback, as it lowers the discrepancy between expected and actual CS events.

4   Discussion and Conclusion

In this study, we implemented a learning task with a simulated autonomous robotic system controlled by the Distributed Adaptive Control (DAC) architecture, a neuronal based model that includes perceptual and behavioral learning mechanisms.

Our aim was to define a first approach that would allow us to operationally identify and explain the transition between the three stages of learning postulated by the pedagogical model of DAC.

The first time the robot faces the arena, it does not have any knowledge of it either. It only has the schemes coming from its reflexes, provided from the Reactive Layer to interact with the environment. These schemes are similar to the neonatal schemes postulated by Piaget which come from the cognitive structures underlying innate reflexes, even when they have not had much experience about the world [9].

This system performs through a balanced combination between perceptual and behavioral learning. On the one hand, perceptual learning permits to acquire more knowledge about the environment, but can lead to the acquisition of erroneous behaviors. On the other hand, behavioral learning allows for a better performance but drives the system to an acquisition of a smaller state space. Only when the confidence between perceived and expected stimuli is high (i.e., the discrepancy variable D is lower than a specified threshold), the Contextual Layer is enabled and begins to store the correct sequences in long-term memory. Enabling the Contextual Layer, then, ensures the generation of actions most supported by the knowledge stored in the LTM.

In this setup, the balance measure, responsible for the interplay between perceptual and behavioral learning, was set to achieve a balanced combination of the two kinds of learning. Nevertheless, if this variable was set to enable only one of the two kinds of learning, the behavior of the robot would be affected. Duff et al. [6] shown how, on the one side, enabling only perceptual learning causes the robot to associate arbitrary actions to the patches, what complicates its possibilities to reach the target. On the other side, when there is only behavioral learning, the robot learns to associate patches and correct actions that lead it to the light. When there is an interplay between perceptual and behavioral learning, the robot not only associates some of the patches to the correct actions that lead it to the light, but also learns to associate some others to the actions that do not lead to it. Moreover, the results of Verschure et al. [3] confirmed that behavioral feedback directly enhances performance.

Considering Peirce theory [10], every agent has his own idea of the world (defined as *ego*), in contrast with the real way the world is (defined as *non-ego*). When the agent is faced with a situation that does not fit his inner world model (*ego*), he would be surprised. DAC proposes an equivalent paradigm, where this difference between the inner world of the learner derived from his previous knowledge (similar to Peirce's *ego*) and the world itself, which presents him new knowledge (Peirce's *non-ego*) causes confusion, what the learner has to overcome in order to abduct and acquire knowledge. A similar difference can be found in the Discrepancy measure, that is, the difference between the expected CS and the actual one. Whenever this measure has reached a pre-defined value, the confusion generated by the difference between expectation and reality disappears.

Moreover, following DAC model, it has been previously demonstrated that is not the specific use of perceptual or behavioral learning what leads to the best results, but the interplay among them [3]. In the same way, we can extract from Peirce that the best results of this difference between *ego* and *non-ego* would be coming from interplay between Perception and Imagination.

4.1   Future Work

Further steps will be focused on continuing exploring the similarities between the phases of learning based on DAC's pedagogical model and the dichotomy of perceptual and behavioral learning. One possibility would be the study of the *resistance* phase as being reflected on the system when it is more dependent of perceptual learning and less prone to adapt to changes into the environment.

Moreover, another plausible experiment would be testing the behavior of the simulated robot when the distribution of the cue patches is changed after the Contextual Layer has been enabled and how it affects to its behavior and learning, with the aim of studying the modulation of the transition from *confusion* phase to the *abduction* one.

Acknowledgments. This work is supported by the EU FP7 project WYSIWYD (FP7-ICT-612139) and EASEL (FP7-ICT- 611971).

References

1. Marcos, E., Sánchez-Fibla, M., Verschure, P.F.M.J.: The complementary roles of allostatic and contextual control systems in foraging tasks. In: Doncieux, S., Girard, B., Guillot, A., Hallam, J., Meyer, J.-A., Mouret, J.-B. (eds.) SAB 2010. LNCS(LNAI), vol. 6226, pp. 370–379. Springer, Heidelberg (2010)
2. Duff, A.: The dynamics of perceptual and behavioral learning in executive control. Ph.D. thesis, Diss., Eidgenössische Technische Hochschule ETH Zürich, Nr. 18766, 2009 (2009)
3. Verschure, P.F.M.J., Voegtlin, T., Douglas, R.J.: Environmentally mediated synergy between perception and behaviour in mobile robots. Nature **425**(6958), 620–624 (2003)
4. Verschure, P.F.M.J.: Distributed adaptive control: a theory of the mind, brain, body nexus. Biologically Inspired Cogn. Architectures **1**, 55–72 (2012)
5. Kruger, J., Dunning, D.: Unskilled and unaware of it: how difficulties in recognizing one's own incompetence lead to inflated self-assessments. J. Pers. Soc. Psychol. **77**(6), 1121 (1999)
6. Duff, A., Verschure, P.F.M.J.: Unifying perceptual and behavioral learning with a correlative subspace learning rule. Neurocomputing **73**(10–12), 1818–1830 (2010)
7. Koenig, N., Howard, A.: Design and use paradigms for gazebo, an open-source multi-robot simulator. In: Proceedings. 2004 IEEE/RSJ International Conference on Intelligent Robots and Systems, (IROS 2004). vol. 3, pp. 2149–2154. IEEE (2004)
8. Bernardet, U., Verschure, P.F.M.J.: iqr: A tool for the construction of multi-level simulations of brain and behaviour. Neuroinformatics **8**(2), 113–134 (2010)
9. Piaget, J.: The Origins of Intelligence in Children, vol. 8(5), pp. 18–1952. International Universities Press, New York (1952)
10. Peirce, C.S., Turrisi, P.A.: Pragmatism as a Principle and Method of Right Thinking: The 1903 Harvard Lectures on Pragmatism. SUNY Press, New York (1997)

Don't Worry, We'll Get There: Developing Robot Personalities to Maintain User Interaction After Robot Error

David Cameron[✉], Emily Collins, Hugo Cheung, Adriel Chua,
Jonathan M. Aitken, and James Law

Sheffield Robotics, University of Sheffield, Sheffield, UK
{d.s.cameron,e.c.collins,mcheung3,
dxachua1,jonathan.aitken,j.law}@sheffield.ac.uk
http://www.sheffieldrobotics.ac.uk/

Abstract. Human robot interaction (HRI) often considers the human impact of a robot serving to assist a human in achieving their goal or a shared task. There are many circumstances though during HRI in which a robot may make errors that are inconvenient or even detrimental to human partners. Using the ROBOtic GUidance and Interaction DEvelopment (ROBO-GUIDE) model on the Pioneer LX platform as a case study, and insights from social psychology, we examine key factors for a robot that has made such a mistake, ensuring preservation of individuals' perceived competence of the robot, and individuals' trust towards the robot. We outline an experimental approach to test these proposals.

Keywords: Human-robot interaction · Design · Guidance · Psychology

1 Background

Human-robot interaction (HRI) research typically explores interactions in which the robot plays a supportive or collaborative role for the human user [4]. However, there are circumstances in which robots may fail to meet these requirements, either through errors in processing the interaction scenario, or failure to adapt to changing HRI scenario circumstances. Furthermore, reliability and error rates of robots have both been identified as important factors in user trust towards robots [4]. Recent work has explored the social impact of a robot's fault or error, in terms of user cooperation [9], and whether an apology from a robot can mitigate the negative impact of the mistake [10]. However, there still remains much to be explored: first in terms of the negative impact even simple robot mistakes can have on user trust and their willingness to engage in HRI; and second, if the methods by which a robot acknowledges a mistake, and then potentially corrects for it, have differential impacts on HRI.

Recent work identifies that the means by which a socially adaptive robot asks for help can impact on: users' attitudes towards the robot, clarity in the support the robot needs, and people's willingness to use the robot in collaborative

© Springer International Publishing Switzerland 2016
N.F. Lepora et al. (Eds.): Living Machines 2016, LNAI 9793, pp. 409–412, 2016.
DOI: 10.1007/978-3-319-42417-0_38

tasks [2]. The context for that interaction does not concern robot error but rather robots requiring user intervention (completing a task outside of the robot's capability) to progress towards a goal. Nevertheless, the principles on which that interaction is based can be drawn upon to identify means by which robots can use particular social interactions to recover, in part, from mistakes.

The synthetic personality a robot exhibits can have substantial impact on the user's experience of, and their engagement with, HRI [3]. The relatively new social domain of HRI is unfamiliar to many, but principles of social psychology have been applied to social HRI scenarios with promising results. These include: accurate recognition of synthetic personality types, even in non-humanoid robots [6]; development of classification of social 'rules' for robots to adhere to [3]; and participant response towards synthetic personalities corresponding with theoretical models of interpersonal cooperation [2]. This paper draws on social psychological theories to develop a model of social factors for robots that support recovery from robot error and maintain user interaction.

2   ROBO-GUIDE Interactive Scenario

To explore the social factors that support a robot in recovery from error, and maintain user interaction, it is useful to consider an interactive scenario in which these circumstances might arise. The ROBOtic GUidance and Interaction DEvelopment (ROBO-GUIDE) project [1,8] is an ideal scenario to consider the impact of such social factors as it requires humans to place trust in a robot, even in the instance where an error might occur.

ROBO-GUIDE is embodied on the Pioneer LX mobile platform. The platform is capable of autonomously navigating a multistory building and leading users from their arrival point to their desired destination. Our focus here is a critical point in building navigation: floor determination whilst using the elevator to navigate between floors. Each floor in the building is similar and ROBO-GUIDE relies on subtle cues to differentiate between them; in noisy environments errors can occur [8]. For example, during busy periods the corridor structural features or floor-indication signs may be obscured, noise may mask lift announcements, or the ROBO-GUIDE might be misinformed or misled by a member of the public.

We identify the disembarkation of the elevator on the wrong floor as a simple and, critically, natural type of error for users to encounter (for the purposes of the experiment, errors would be staged). This tour guide scenario, in which individuals use a robot to navigate an unfamiliar building, means any error would mildly inconvenience the user. Moreover, it gives opportunity for the robot to recover from the error and allows testing of different means for the robot to socially communicate error and attempts to correct it.

3   Social-Recovery and Experimental Proposal

Previous HRI work has identified a social psychological model of cooperation [7] as a useful framework for exploring the impact different personalities can have on user willingness to engage with robots [1]. This model, when applied in an HRI context, contrasts the impact of friendly-oriented statements to build user-liking, and goal-oriented statements to suggest the robot's task competency. Findings indicate that individuals are more willing to use a robot they like than a robot suggesting task competency; they also regard the interaction with a friendly robot to be less ambiguous [2]. Individuals further report trust towards the robot across the dimensions offered in the model: affective (from personable interactions) and cognitive (from evidence of competency). While results do not show substantive differences in trust between conditions, this may be due to a ceiling effect because there was no challenge made to the competency of the robot.

The error scenario (Sect. 2) provides such a challenge to the robot's apparent competency; it further provides opportunity for the robot to attempt to socially recover from the error and work to restore use perceptions of competency. Again, using the framework developed in [1], friendly-oriented and goal-oriented synthetic personalities, as means to socially communicate error, may result in differences in individuals' views towards the robot.

The framework may be further enhanced by considering social psychological understanding on the impact of acknowledging and apologising for error. Apologies for errors in competency are observed to raise an individual's trustworthiness but not their apparent competency [5]. Simple apologies may therefore support a user's affective trust towards a robot but not their cognitive trust. In contrast, identification of the error and communication of means to resolve it, to maintain progress towards a goal, may restore users cognitive trust.

We outline a brief experimental proposal to test the impact of statements promoting affective and cognitive trust for the user in the error scenario (Sect. 2). Acknowledgments of the error by the robot are planned to be manipulated in a 2×2 experimental between-subjects design: inclusion or absence of the competency-oriented apology-oriented statements following error. These will be communicated using the on-board speech synthesizer[1]. We anticipate that these statements will impact on participant willingness to use the robot through the key channels of affective and cognitive trust.

Participants will experience one of the four conditions: (1) the control condition comprising simple instructions for the user to follow after making an error, (e.g., 'Follow me back to the lift'); (2) inclusion of competency-oriented statements that emphasise the robot's ability to recognise the environment, identifying the cues used to orient, and reaffirming the goal (e.g., 'That sign said we are on C floor and we need to go to B floor. Follow me back to the lift'); (3) inclusion of apology-oriented statements that emphasise attempts to relate to users but *do not* indicate competency (e.g., 'Sorry about the error; we can all make mistakes sometimes. Follow me back to the lift'); (4) inclusion of both the competency- and apology-oriented statements.

Outcomes are measured using the prior measures of affective and cognitive trust, clarity of interaction and willingess to use the robot [2]. We anticipate that

---

[1] The viable alternatives of pre-recorded spoken phrases or an on-screen display are acknowledged.

the apology and competency statements will support affective and cognitive trust respectively, although affective trust better predict willingness to use the robot in future. Findings from this work will support the development of a socially adaptive robots for HRI and further reveal the social models users draw upon when interacting with socially engaging robots.

Acknowledgments. This work was supported by European Union Seventh Framework Programme (FP7-ICT-2013-10) under grant agreement no. 611971.

References

1. Cameron, D., et al.: Help! I cant reach the buttons: facilitating helping behaviors towards robots. In: Wilson, S.P., Verschure, P.F.M.J., Mura, A., Prescott, T.J. (eds.) Living Machines 2015. LNCS(LNAI), vol. 9222, pp. 354–358. Springer, Heidelberg (2015)
2. Cameron, D., Loh, E.J., Chua, A., Collins, E.C., Aitken, J.M., Law, J.: Robot-stated limitations but not intentions promote user assistance. In: Salem, M., Weiss, A., Baxter, P., Dautenhahn, K. (eds.) Proceedings of the 5th International Symposium on New Frontiers in Human-Robot Interaction. AISB (In Press)
3. Dautenhahn, K.: Socially intelligent robots: dimensions of human-robot interaction. Philos. Trans. R. Soc. Lond. B Biol. Sci. **362**, 679–704 (2007)
4. Hancock, P.A., Billings, D.R., Schaefer, K.E., Chen, J., De Visser, E., Parasuraman, R.: A meta-analysis of factors affecting trust in human-robot interaction. J. Hum. Factors Ergon. Soc. **53**, 517–527 (2011)
5. Kim, P., Ferrin, D., Cooper, C., Dirks, K.: Removing the shadow of suspicion: the effects of apology versus denial for repairing competence- versus integrity-based trust violations. J. Appl. Psychol. **89**, 104–118 (2004)
6. Lee, K.M., Peng, W., Jin, S.A., Yan, C.: Can robots manifest personality?: An empirical test of personality recognition, social responses, and social presence in human-robot interaction. J. Communication **56**, 754–772 (2006)
7. McAllister, D.J.: Affect-and cognition-based trust as foundations for interpersonal cooperation in organizations. Acad. Manage. J. **38**, 24–59 (1995)
8. McAree, O., Aitken, J.M., Boorman, L., Cameron, D., Chua, A., Collins, E.C., Fernando, S., Law, J., Martinez-Hernandez, U.: Floor Determination in the operation of a lift by a mobile guide robot. In: Proceedings of the European Conference on Mobile Robots (2015)
9. Salem, M., Lakatos, G., Amirabdollahian, F., Dautenhahn, K.: Would you trust a (faulty) robot? Effects of error, task type and personality on human-robot cooperation and trust. In: Proceedings of the 10th ACM/IEEE International Conference on Human-Robot Interaction (HRI 2015), Portland (2015)
10. Snijders, D.: Robots recovery from invading personal space. In: 23rd Twente Student Conference on IT, Enschede, The Netherlands (2015)

Designing Robot Personalities for Human-Robot Symbiotic Interaction in an Educational Context

David Cameron[✉], Samuel Fernando, Abigail Millings, Michael Szollosy,
Emily Collins, Roger Moore, Amanda Sharkey, and Tony Prescott

Sheffield Robotics, University of Sheffield, Sheffield, UK
{d.s.cameron,s.fernando,a.millings,m.szollosy,
e.c.collins,r.k.moore,a.sharkey,t.j.prescott}@sheffield.ac.uk
http://easel.upf.edu/project

Abstract. The Expressive Agents for Symbiotic Education and Learning project explores human-robot symbiotic interaction with the aim to understand the development of symbiosis over long-term tutoring interactions. The final EASEL system will be built upon the neurobiologically grounded architecture - Distributed Adaptive Control. In this paper, we present the design of an interaction scenario to support development of the DAC, in the context of a synthetic tutoring assistant. Our humanoid robot, capable of life-like simulated facial expressions, will interact with children in a public setting to teach them about exercise and energy. We discuss the range of measurements used to explore children's responses during, and experiences of, interaction with a social, expressive robot.

Keywords: Human-robot interaction · Humanoid · Psychology · Symbiosis · Facial expression

1 Background

A key challenge in Human-Robot symbiotic interaction (HRSI) is to develop robots that can adapt to users during interactions. The Expressive Agents for Symbiotic Education and Learning (EASEL) project seeks to develop a biologically-grounded robot [9] that is responsive to users and capable of adaptation to user needs within, and across, interactions. One particular challenge for long-term or repeat human-robot interactions is maintaining successful social engagement with the user. A means to address this issue is ensuring the synthetic personality for a social robot is engaging and responsive to users [1]. A cornerstone of the EASEL project is the development of a synthetic personality for a social robot, which promotes *sustained* user engagement.

Robots that can effectively achieve the process of sustained social engagement and further act to positively shape user behavior (such as communicating to inform further user interactions) could be considered to behave symbiotically. HRSI extends beyond standard human-robot interaction by considering the: '[D]ynamic process of working towards a common goal by responding and

© Springer International Publishing Switzerland 2016
N.F. Lepora et al. (Eds.): Living Machines 2016, LNAI 9793, pp. 413–417, 2016.
DOI: 10.1007/978-3-319-42417-0_39

adapting to a partners actions, while affording a partner to do the same. This term suggests a mutually beneficial relationship between various parties' [5]. The EASEL project explores HRSI in the context of one-to-one tutoring interactions between a socially-adaptive humanoid robot and children.

A tutoring scenario (described in Sect. 2) gives context for the robot to adapt in *response* to a child's behaviour, updating its communication based on the developing interaction. The scenario also serves to direct children's interactions with the robot and gives goals for children to work in concert with the Synthetic Tutoring Assistant (STA). Effectiveness of the STA as an engaging personality, educational device, and model for an HRSI theoretical framework can be assessed through development of field interaction scenarios.

2   Field Interaction Scenario

2.1   The STA

The STA is to be developed upon the neurobiologically grounded architecture *Distributed Adaptive Control* (DAC) [9]. The DAC is a tiered, self-regulation system, structured to manage low- and high-level behaviours for synthetic agents, comprising: allostatic control, adaptation to sensory information, and the acquisition and expression of contextual plans. Behavioural outputs of the DAC are guided through synthetic motivation and emotion states. Behavioural interaction is scheduled using the AsapRealiser (e.g., [8]) interaction manager, which determines processing of sensory inputs to the robot and how content and feedback are delivered to the user.

This complete system is embodied in the Hanson Robokind Zeno R25 [7] (Fig. 1). A distinctive feature of the Zeno model robot is the platform's realistic face, capable of displaying a range of life-like simulated facial expressions. This model enables the conveying of synthetic emotions from the DAC in immediately recognizable ways that are minimally obtrusive to ongoing interactions [6].

Fig. 1. The Robokind Zeno R25 platform (humanoid figure approximately 60 cm tall)

2.2 Tutoring

HRSI between the user and STA will be explored in the context of a healthy-living tutoring scenario. Children will be invited to interact with the STA at a two-week special exhibit on robotics at a natural history museum.

The interaction comprises of four key stages. First, the robot briefly introduces itself and the activities for the interaction; in this period, the robot also calibrates its automatic speech recognition (ASR) to brief prompted responses from the child. Second, children engage in robot-led physical activities of different intensities and duration. After each activity, the robot provides responsive feedback about the energy used by the child to perform each action. Third, children complete a quiz based on their activity (i.e., which exercise used the most/least energy) and a similar quiz based on exercise and energy in general (i.e., identifying low- and high-energy activities). Finally, children can give Zeno voice commands to perform facial expressions or other physical actions. The interaction lasts approximately ten minutes in total.

The scenario will be delivered autonomously by the STA[1]. Development of the robot's personality is drawn from prior HRSI research to best ensure user engagement throughout the interaction. This includes regular confirmation from the robot that it is responsive to the *individual user* [4] and showing context-appropriate simulated facial expressions when delivering feedback [2]. Within the context of a tutoring scenario, supportive interactions, such as these, are anticipated to not just encourage interaction with the robot but also promote user engagement with the learning activity.

3 Exploring User Engagement, Learning, and HRSI

User Engagement with the scenario and the robot will be measured across multiple means, including: video recording, motion tracking, self-report and parental-report of child. User engagement will be examined using a between-subjects design. The standard scenario (Sect. 2.2) will be contrasted against an 'enhanced' scenario, which includes a brief rapport-building introduction by the robot (experimental condition) or researcher (active-control condition). We anticipate that the rapport-building introduction by the robot will promote user engagement, particularly at the start of the interaction.

The interaction scenario will be filmed throughout and children's facial expressions coded, as done in a prior study [2]. Video data will be further used to determine children's gaze throughout interaction; our prior work indicates that children turn towards a socially-responsive robot, in anticipation of its speaking, more often than towards a less-responsive robot [4]. Children turning their attention towards parents or the researcher will be coded as measures of comfort seeking and clarity seeking respectively. Children looking towards the STA,

---

[1] Wizard of Oz style control is available should the ASR fail to adapt to an individual child's voice.

particularly with positive facial expressions, rather than towards other figures, will be regarded as an indication of engagement with the STA.

Physical activity (in terms of joules) can be calculated automatically by the STA. This provides an additional measure of engagement; children more engaged in a prior scenario performed more physical activity at the robot's direction [4]. Interpersonal distance between the user and the robot during the physical activity will further be used as an index of social engagement with the robot: a smaller distance between user and robot indicating greater comfort in the interaction.

After the interaction, children will complete a brief survey on their affect felt when working with Zeno and their enjoyment of the scenario. Parents will also complete a brief survey on their views of their child's engagement and enjoyment of the scenario.

User learning will be examined using between-subjects design experiment. The standard scenario schedules general questions on the relationship between energy usage and both exercise intensity and duration *after* children have completed (1) physical activity of varying intensities and duration and (2) questions on their activity consolidating this relationship. Baseline measures of children's understanding of this relationship can be determined by scheduling the critical questions before the physical activity and consolidation questions. The number of questions answered correctly by children in each condition will be compared as a measure of learning.

The learning activity is more appropriate for seven-year-old children because it addresses topics covered in the UK National Curriculum for Lower Key Stage Two (ages seven to nine). We anticipate that the supportive social interaction and learning activity provided will provide scaffold for children to understand the relationship between exercise and energy. Given that the interaction is part of a public exhibit and open to all ages, a wide age range is anticipated in our data collection. Children's age will be co-varied in analysis to make account for children's prior education on the subject of exercise and energy.

HRSI development will be informed by the above factors. It will further be explored both in terms of children's perceptions of the robot and the individual differences in personality that may facilitate interaction. After interaction, children will complete a brief survey of their perceptions of Zeno's 'status' as being like a machine or person [3] and familiarity (i.e., like a stranger, acquaintance, friend, best friend). They further rate Zeno's 'knowledge' about exercise and skill at teaching. The above measures are anticipated to correlate with user engagement and user learning and further inform our understand of the *social context* that children use to understand HRSI (e.g., working with a teacher versus co-operation with a fellow friendly learner).

This study offers a potentially large-scale sample so that we may examine individual differences in personality and the impact this can have on children's perceptions of, and interaction with, a socially-responsive humanoid robot. Parents will complete a brief five-factor personality questionnaire about their child. We anticipate that openness and agreeableness will positively correlate with user engagement with the robot, conscientiousness with scenario engagement and learning, and extroversion with positive expression towards the robot.

This interaction scenario enables the further development of an engaging STA and extension of HRSI research, so that robot personalities can better adapt to user requirements throughout interaction and be tailored to meet individual needs.

Acknowledgments. This work was supported by European Union Seventh Framework Programme (FP7-ICT-2013-10) under grant agreement no. 611971. We thank Theo Botsford and Jenny Harding for their assistance in scenario development.

References

1. Breazeal, C., Scassellati, B.: How to build robots that make friends and influence people. In: IEEE/RSJ International Conference on Intelligent Robots and Systems, IROS Proceedings, vol. 2, pp. 858–863 (1999)
2. Cameron, D., Fernando, S., Collins, E.C., Millings, A., Moore, R.K., Sharkey, A., Evers, V., Prescott, T.: Presence of life-like robot expressions influences children's enjoyment of human-robot interactions in the field. In: Salem, M., Weiss, A., Baxter, P., Dautenhahn, K. (eds.) 4th International Symposium on New Frontiers in Human-Robot Interaction, pp. 36–41 (2015)
3. Cameron, D., Fernando, S., Millings, A., Moore, R.K., Sharkey, A., Prescott, T.: Children's age influences their perceptions of a humanoid robot as being like a person or machine. In: Wilson, S.P., Verschure, P.F.M.J., Mura, A., Prescott, T.J. (eds.) Living Machines 2015. LNCS, vol. 9222, pp. 348–353. Springer, Heidelberg (2015)
4. Cameron, D., Fernando, S., Collins, E.C., Millings, A., Moore, R.K., Sharkey, A., Prescott, T.: Impact of robot responsiveness and adult involvement on children's social behaviours in human-robot interaction. In: Salem, M., Weiss, A., Baxter, P., Dautenhahn, K. (eds.) Proceedings of the 5th International Symposium on New Frontiers in Human-Robot Interaction. AISB (in press)
5. Charisi, V., Davison, D., Wijnen, F., van der Meij, J., Reidsma, D., Prescott, T., van Joolingen, W., Evers, V.:Towards a child-robot symbiotic co-development: a theoretical approach. In: Salem, M., Weiss, A., Baxter, P., Dautenhahn, K. (eds.) in 4th International Symposium on New Frontiers in Human-Robot Interaction, pp. 30-35 (2015)
6. Costa, S., Soares, F., Santos, C.: Facial expressions and gestures to convey emotions with a humanoid robot. In: Herrmann, G., Pearson, M.J., Lenz, A., Bremner, P., Spiers, A., Leonards, U. (eds.) ICSR 2013. LNCS, vol. 8239, pp. 542–551. Springer, Heidelberg (2013)
7. Hanson, D., Baurmann, S., Riccio, T., Margolin, R., Dockins, T., Tavares, M., Carpenter, K.: Zeno: A cognitive character. In: AI Magazine, and Special Proceedings of AAAI National Conference, Chicago (2009)
8. Reidsma, D., van Welbergen, H.: AsapRealizer in practice - a modular and extensible architecture for a BML Realizer. Entertainment Comput. 4, 157–169 (2013)
9. Verschure, P.F.: Distributed adaptive control: a theory of the mind, brain, body nexus. Biologically Inspired Cogn. Architectures 1, 55–72 (2012)

A Biomimetic Fingerprint Improves Spatial Tactile Perception

Luke Cramphorn[1,2(✉)], Benjamin Ward-Cherrier[1,2], and Nathan F. Lepora[1,2]

[1] Department of Engineering Mathematics, University of Bristol, Bristol, UK
{1114468,bw14452,n.lepora}@bristol.ac.uk
[2] Bristol Robotics Laboratory, University of Bristol, Bristol, UK

Abstract. The function of the human fingertip has been often debated. There have been studies focused on how the fingerprint affects the perception of high temporal frequencies, such as for improved texture perception. In contrast, here we focus on the effects of the fingerprint on the spatial aspects of tactile perception. We compare two fingertips, one with a biomimetic fingerprint and the other having a smooth surface. Tactile data was collected on a sharp edged stimulus over a range of locations and orientations, and also over a smooth (cylindrical) object. The perceptual capabilities of both (fingerprint and smooth) sensor types were compared with a probabilistic classification method. We find that the fingerprint increases the perceptual acuity of high spatial frequency features (edges) by 30–40% whilst not influencing the tactile acuity of low spatial frequency features (cylinder). Therefore the biomimetic fingerprint acts as an amplifier of high spatial frequencies, and provides us with evidence to suggest that the perception of high spatial frequencies is also one of the functions of the human fingertip.

Keywords: Tactile sensors · Biomimetic

1 Introduction

The human sense of touch defines our ability as a species to manipulate the world in unparalleled ways. There are many morphological aspects that aid in the performance of this sense. In particular, the fingerprint subtly enhances the temporal frequency of stimuli, providing us with the ability to detect and identify fine textures [1,2].

Here we look at the effects of the fingerprint for spatial perception with a biomimetic fingertip. Our inspiration for the design comes from the mechanical coupling between the surface of the skin and the touch receptor cells, created by the dermal ridges that make up the human fingerprint (Fig. 1a). This coupling occurs through the dermal papillae, protrusions of the dermis into the epidermis in which the touch receptor cells are found [3]. These papillae create a non-planar layer in the dermis that transfers the kinetic energy from the surface to the sensory cell. The exact function of the fingerprint is uncertain in the biological community. Hypotheses include that the fingerprint exists to assist grasping by acting as a high friction

© Springer International Publishing Switzerland 2016
N.F. Lepora et al. (Eds.): Living Machines 2016, LNAI 9793, pp. 418–423, 2016.
DOI: 10.1007/978-3-319-42417-0_40

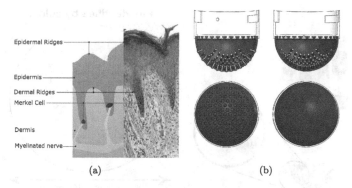

(a)   (b)

Fig. 1. (a) The epidermal ridges overlay protrusion in the dermis (inner layer of skin) called dermal ridges creating a mechanical coupling between the surface and the touch receptor cells (Merkel cells) that sit within the dermal ridges. (b) Cross sectional comparison of the two TacTip tips compared in this paper. The fingertip enhanced TacTip (left) has solid plastic cores that run from the inside of the artificial dermal ridges (nodules) to the top of the pins. The original structure (right) has a featureless skin and no plastic cores.

surface [4], and that the fingerprint aids tactile perception by acting as a small mechanical lever which magnifies the response to tactile contact [5]. It was also shown later that a lensing effect amplifies the signal/noise ratio due to the presence of a fingerprint [6]. Due to previous biomimetic studies, there is evidence that supports the idea that the fingerprint aids in perception of textures (and other high temporal frequency profiles) [1,2]. Finite element modelling of the effect of dermal ridges shows that there is an increase in stress contrast between adjacent measurement points, and thus better accuracy and consistency in edge encoding [7]. Here we present the evidence for the fingerprint also aiding in the perception of high spatial frequency (measured from the components of a spatial Fourier transform) such as edges. To demonstrate this effect of the fingerprint, a biomimetic optical tactile fingertip is used with two different tips, one with the fingerprint and one without. By comparing the accuracy of these tips on an edge (high spatial frequency) via localisation and orientation and over a cylinder (low spatial frequency) we show that the sensitivity and accuracy of the sensor is improved by the inclusion of the fingerprint over the large spatial frequency with no change to the accuracy or sensitivity on the low spatial frequencies. Thus we demonstrate that the fingerprint helps with the perception of high spatial features as well as the previously demonstrated high temporal features.

2   Methods

The sensor used is the TacTip optical tactile sensor developed at Bristol Robotics Laboratory [8–11]. The TacTip sensor is relativity cheap and easy to modify as it is 3D printed. The input of the sensor uses an off-the-shelf CCD (LifeCam Cinema HD, Microsoft) where an array of pins on the internal side of a 3D

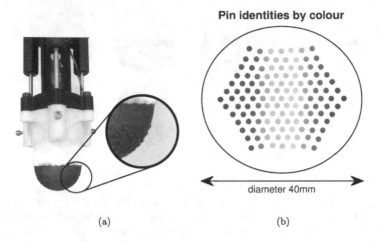

Fig. 2. (a) Image of the TacTip sensor used for the experiment, highlighting the nodules, that simulate the fingerprint, on the surface of the sensor. (b) Graphical representation of the pin layout on the interior of the sensor. Each pin is coloured and can be matched to deflections on data sets in Fig. 3.

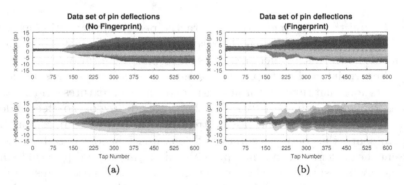

Fig. 3. The displayed data shows the time-series recordings over 600 taps spanning a 20 mm location range across the edge. Pin displacements are shown in units of pixels, where each colour represents the displacement of a corresponding pin (shown in Fig. 2b). A sharpening of the features (highlighted as an increase in rate of pin deflection) due to the fingerprint (b) in comparison to the set without (a) reduces ambiguity and thus improves classification.

printed rubber-like skin are tracked from frame to frame. The pins are inspired by the dermal papillae that are linked to tactile sensing in the human finger. As the sensor uses a plug and play web cam, the sensor is easy to install. The skin is filled with an optically clear gel, which gives the sensor compliance as well as structure. The sensor has interchangeable tips which can be produced at a relatively low cost and keep any contact far from any delicate electronics. This means the sensor is robust and any tip can be replaced with ease if damaged. The platform for the system was a six degree-of-freedom robot arm (IRB 120, ABB

Robotics), which precisely positions the TacTip with an absolute repeatability of 0.01 mm.

To produce a tip with the biomimetic fingerprint, domes are included on the exterior of the skin and mirror the interior locations of the pins. This creates mechanics akin to dermal ridges. Each dome has a plastic core that starts in the interior of the dome and connects to the pin tip providing direct coupling between the stimuli and the pin tips (Fig. 1b).

The validation of the two tips considered here was performed by analysing the accuracy in location and orientation of various stimuli. The task was performed collecting tactile data during taps on the surface of each stimulus. Each tap records a time series over the press (approximately 2 s), with ~50 recorded frames per series. The data used for the results in this paper was collected in two distinct sets, ensuring that the training and test sets are different, and that the validation of the results is based on sampling from an independent data set to that used to train the classifier.

3 Results

Varying sensor location (y) and orientation (θ) over the stimulating features changes the extent to which the pins displace, both in magnitude and direction, this creates unique time-series for each contact feature of the object. This uniqueness is what permits classification of the feature. Sets with lower ambiguity therefore have greater classification accuracy. The data series (Fig. 3) shows the s_x and s_y displacements of the recordings, where the key information from the contact is the region of increasing displacements in the central region.

The data over the edge feature shows a low rate of change in pin deflection for the smooth tip (Fig. 3a) contrasted with a high rate of change for the fingerprint tip (Fig. 3b). This contrast creates a less ambiguous representation of the feature in the direction of travel (y) for the tip with the fingerprint.

The validation and comparison of the effectiveness of the two different tips is done by using probabilistic methods for localization that have been described extensively in refs [9,12]. The collected training data sets are used such that each of the move increments are treated as distinct location classes. A likelihood model is constructed using the tactile data at each location using a histogram method. Given a sample of test data with an unknown location, this model can be used to determine the likelihood of which location class it originates from.

The two training sets taken from each of the stimuli are used with the Monte Carlo procedure where randomly sampled data from one of the sets are used to classify locations in the other (10000 iterations). Each increment of test data uses a maximum likelihood approach to classify the position classes. The maximum likelihood method means that the lower the ambiguity at each point the better the accuracy of the classification.

The errors in location acuity determined from validation show improvements in accuracy for the tip with the fingerprint: the correct class is classified more frequently. The average error with a fingerprint is $\bar{e}_{loc}(y) = 0.067$ mm and without is $\bar{e}_{loc}(y) = 0.091\,\text{mm}$ which represents a 32 % improvement in location acuity. The validation for orientation of the edge is classified in the same way. Once again the number of iterations classifying correctly is higher in the fingerprint version. The average error of the fingerprint version was $\bar{e}_{loc}(\theta) = 0.412$ and for the smooth version was $\bar{e}_{loc}(\theta) = 0.693$ which represents an improvement of 41 % in orientation accuracy. In the third experiment, the error in location perception is examined for a smooth cylinder, similar to the experiment on location acuity previously performed in [9]. The results show that the fingerprint neither added nor subtracted in any substantial accuracy with errors of $\bar{e}_{loc}(x) = 0.081$ for the fingerprint and $\bar{e}_{loc}(x) = 0.076$ without the fingerprint.

4   Conclusions

This study has demonstrated that the inclusion of a biomimetic fingerprint improves tactile acuity of a biomimetic tactile fingertip to features with a high spatial frequency: the accuracy of the location perception of an edge increased by 30–40 % with no notable decrease or increase in the accuracy of the location perception on the low spatial frequency features of a cylinder. This provides us with evidence to suggest that the perception of high spatial frequencies is also one of the functions of the human fingerprint.

Acknowledgment. We thank Sam Coupland, Gareth Griffiths and Samuel Forbes for their assistance with 3d-printing. LC was supported by the EPSRC Centre for Doctoral Training in Future Autonomous and Robotic Systems (FARSCOPE). BWC was supported by an EPSRC DTP studentship and NL was supported in part by an EPSRC grant on 'Tactile Superresolution Sensing' (EP/M02993X/1).

References

1. Scheibert, J., Leurent, S., Prevost, A., Debrégeas, G.: The role of fingerprints in the coding of tactile information probed with a biomimetic sensor. Science **323**(5920), 1503–1506 (2009)
2. Winstone, B., Griffiths, G., Pipe, T., Melhuish, C., Rossiter, J.: TACTIP - tactile fingertip device, texture analysis through optical tracking of skin features. In: Lepora, N.F., Mura, A., Krapp, H.G., Verschure, P.F.M.J., Prescott, T.J. (eds.) Living Machines 2013. LNCS, vol. 8064, pp. 323–334. Springer, Heidelberg (2013)
3. Dahiya, R., Metta, G., Valle, M., Sandini, G.: Tactile sensing from humans to humanoids. IEEE Trans. Rob. **26**(1), 1–20 (2010)
4. Jones, L., Lederman, S.: Human Hand Function. Oxford University Press, New York (2006)
5. Cauna, N.: Nature and functions of the papillary ridges of the digital skin. Anat. Rec. **119**(4), 449–468 (1954)
6. Fearing, R., Hollerbach, J.: Basic solid mechanics for tactile sensing. Int. J. Robot. Res. **4**(3), 40–54 (1985)

7. Gerling, G., Thomas, G.: The effect of fingertip microstructures on tactile edge perception. In: Eurohaptics Conference, 2005 and Symposium on Haptic Interfaces for Virtual Environment and Teleoperator Systems, World Haptics 2005, First Joint, pp. 63–72. IEEE (2005)
8. Chorley, C., Melhuish, C., Pipe, T., Rossiter, J.: Development of a tactile sensor based on biologically inspired edge encoding. In: ICAR International Conference Advanced Robotics, pp. 1–6. IEEE (2009)
9. Lepora, N., Ward-Cherrier, B.: Superresolution with an optical tactile sensor. In: IEEE/RSJ International Conference on Intelligent Robots and Systems (IROS), pp. 2686–2691 (2015)
10. Ward-Cherrier, B., Cramphorn, L., Lepora, N.: Tactile manipulation with a Tac-thumb integrated on the open-hand M2 gripper. Rob. Autom. Lett. $1(1)$, 169–175 (2016)
11. Cramphorn, L., Ward-Cherrier, B., Lepora, N.: Tactile manipulation with biomimetic active touch. In: IEEE International Conference on Robotics and Automation (ICRA), pp. 123–129
12. Lepora, N.: Biomimetic active touch with fingertips and whiskers. IEEE Trans. Haptics $9(2)$, 170–183 (2016)

Anticipating Synchronisation for Robot Control

Henry Eberle[✉], Slawomir Nasuto, and Yoshikatsu Hayashi

Brain Embodiment Lab, Reading University, Berkshire RG6 6UR, UK
nm010080@reading.ac.uk

Abstract. Anticipating synchronisation (AS) is an interesting phenomenon whereby a dynamical system can trace the future evolution of an analogous 'master' system using delayed feedback. Although some have theorised that AS (or a similar phenomenon) has a role in human motor control, research has primarily focused on demonstrating it in novel systems rather than its practical applications. In this paper we explore one such application: by coupling the dynamics of the joints in a simulated robot arm, we seek to demonstrate that AS can have a functional role in motor control by reducing instability during a reaching task.

Keywords: Anticipating synchronisation · Motor control · Joint synergy · Robotics

1 Introduction

Because sensing and actuation are not instantaneous (and cannot be), some degree of anticipation is necessary for an organism to execute any form of movement accurately. Typically this is considered the domain of computational forward models that accept the parameters of an action and calculate its result. An alternative formulation is that of strong anticipation where the prediction arises from the organism's continuous coupling with the environment rather than an internal model [1].

Anticipation Synchronisation (AS), the phenomenon where a 'slave' dynamical system can predict the behaviour of the qualitatively similar master driving it [2], is the closest realisation of this concept that yet exists. However, research has focused primarily on its applications in communication rather than motor control or cognition.

This paper offers a 1st step demonstrating that an AS coupling can play a stabilising role in motor control without the need for explicit kinematic models, using the example of a two-joint planar robot arm.

Closed-loop control of a robot arm generally requires the use of a kinematic model in order to transform target coordinates in the external work space into appropriate joint angles. However, the arm can also be treated as a dynamical system (and is, when analysing the stability of control laws). Kinematics is thus irrelevant so long as the robot's state space trajectory converges on the correct target.

© Springer International Publishing Switzerland 2016
N.F. Lepora et al. (Eds.): Living Machines 2016, LNAI 9793, pp. 424–428, 2016.
DOI: 10.1007/978-3-319-42417-0_41

Dispensing with kinematics means that the future course of the arm's motion cannot be computationally modelled. Since a deviation from a desired path must thus be sensed before it can be corrected, any delay in the joints' reactions will create delay in the control loop that can cause instability. Such a delay is inevitable due to the fact that the joints must move links with mass, and thus inertia, which prevents the reaction from being instantaneous.

However, because of the arm's control law being treated as a dynamical process, there is scope to utilise the phenomenon of anticipating synchronisation (AS) to counteract this instability.

2   The Problem of Delay

In order to demonstrate the tangible benefits of an AS coupling in motor control, we use visual control based on a linear transformation we term the joint relationship matrix (JRM) [3] (In preparation):

$$\boldsymbol{\tau} = \boldsymbol{JRM}(\boldsymbol{K_p}(\boldsymbol{x_d} - \boldsymbol{x}) - \boldsymbol{K_v}\dot{\boldsymbol{x}}) \tag{1}$$

$\boldsymbol{\tau}$ is the vector of joint torques, $\boldsymbol{x}$ is the work space position of the end effector. $\boldsymbol{x_d}$ is the target in the work space. $\boldsymbol{K_p}$ is a proportional gain matrix for the work space error, while $\boldsymbol{K_v}$ is a velocity gain matrix. $\boldsymbol{JRM}$ is a static transformation applied to the work space feedback, defining each joint's reaction each joint's response to the feedback terms.

The JRM is used to associate each joint in the simulated robot arm with a direction vector in the work space. Although this is not an accurate kinematic transformation, the resulting control law can correct any deviations from a straight motion as they occur. Since a deviation must be detected before it can be corrected, the correction always lags behind the error, causing unstable 'bumps' in the end effector's trajectory.

For the purposes of this paper $\boldsymbol{JRM}$ is fixed at the value $\left(\begin{smallmatrix} 0 & 1 \\ -1 & 0 \end{smallmatrix}\right)$, which relates the torque at the first and second joints to the vertical and horizontal work space error, respectively.

3   Anticipating Synchronisation

Instead of using explicit forward modelling, which would greatly increase the complexity of the control, we seek to introduce an element of anticipating synchronisation into Eq. 1.

The AS paradigm we base this paper on is that of anticipation via delayed self-feedback as described by Voss [2]: a 'slave' system anticipates the evolution of a 'master' through the addition of the coupling shown in Eq. 3.

$$\dot{x}(t) = f(x(t)) \tag{2}$$

$$\dot{y}(t) = f(y(t)) + K[x(t) - y(t - t_{delay})] \tag{3}$$

So long as $y(t)$ lags $x(t + t_{delay})$ the coupling drives the slave to evolve at a faster rate until it is in advance of the master. When $y(t)$ is equal to $x(t + t_{delay})$, the coupling disappears and the two systems become independent.

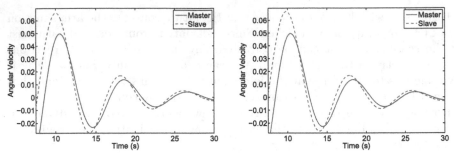

(a) Anticipation period = 0.425s, t_{delay} = 0.1s (b) Anticipation period = 0.455s, t_{delay} = 0.2s

Fig. 1. Comparison of anticipation periods where t_{delay} is varied and $k = 150$. $\boldsymbol{K_p} = \left(\begin{smallmatrix} 100 & 0 \\ 0 & 100 \end{smallmatrix}\right)$ and $\boldsymbol{K_v} = \left(\begin{smallmatrix} 75 & 0 \\ 0 & 75 \end{smallmatrix}\right)$. Master's angular velocity is multiplied by -1.

4 Applying Anticipating Synchronisation

In order to demonstrate the effect of adding an AS coupling to Eq. 1, the JRM is fixed at the value $\left(\begin{smallmatrix} 0 & 1 \\ -1 & 0 \end{smallmatrix}\right)$. This allows the arm's end effector perform a horizontal reaching movement towards a target in the work space, but does not produce straight trajectories as the two joints' reactions to each others' movement is delayed.

To reduce this effect, we designated the second joint the 'slave' and added an additional delay coupling term to Eq. 1. By causing the slave to anticipate the negative of the master's angular velocity, we aim to cause the end effector to

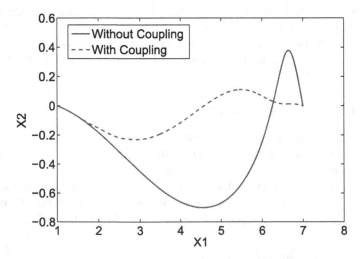

Fig. 2. Comparison of paths taken by the end effector taken between (1, 0) and (7, 0) in the non-coupled (controlled by Eq. 1) and coupled (controlled by Eq. 4 with $k = 250$ and $t_{delay} = 0.2$ s) cases. $\boldsymbol{K_p}$ and $\boldsymbol{K_v}$ are kept constant at $\left(\begin{smallmatrix} 100 & 0 \\ 0 & 100 \end{smallmatrix}\right)$ and $\left(\begin{smallmatrix} 150 & 0 \\ 0 & 150 \end{smallmatrix}\right)$, respectively.

trace a straight line. This results in a modified control law with the below form:

$$\tau = JRM(K_p(x_d - x) - K_v\dot{x}) - \binom{0}{k[\dot{x}_2(t) - \dot{x}_1(t - t_{delay})]} \tag{4}$$

where k is the coupling strength constant and t_{delay} is the coupling delay. This coupling creates a stable, anticipatory phase relationship between the angular velocities of the two joints, as measured by the time difference between the nearest peaks in the two joints' angular velocity time series (Fig. 1).

The addition of this coupling can also be used to significantly increase the straightness of the path taken by the end effector. As seen in Fig. 2, the master-slave coupling effectively counteracts the deviations caused by inertial delay, leading to a notably straighter path.

Similarly, the modified control law makes the arm much more robust to perturbation. As seen in Fig. 3, the deviation caused by a torque pulse applied to the master joint is quickly corrected, while the unmodified control law traces an unstable spiral when subject to the same disturbance.

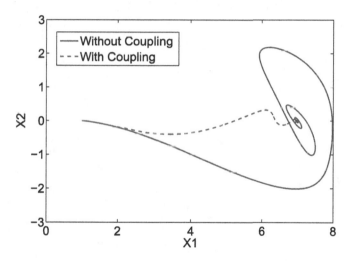

Fig. 3. Comparison of paths taken by the end effector taken between (1, 0) and (7, 0) in the non-coupled (controlled by Eq. 1) and coupled (controlled by Eq. 4 with $k = 250$ and $t_{delay} = 0.2$ s) cases. A torque pulse is applied to the master joint between 0.5 and 1 s. K_p and K_v are kept constant at $\begin{pmatrix} 100 & 0 \\ 0 & 100 \end{pmatrix}$ and $\begin{pmatrix} 150 & 0 \\ 0 & 150 \end{pmatrix}$, respectively.

5   Conclusions

Although delay would appear to be an unavoidable consequence of controlling a robotic arm without explicit kinematics, the phenomenon of AS allows us to run counter to this normal temporal relationship. This opens the possibility of effective motor control without the need for computational modelling. We believe this is a solid 1st step in showing that AS can perform an integral role in tasks such as reaching.

References

1. Stepp, N., Turvey, M.T.: On strong anticipation. Cogn. Syst. Res. **11**(2), 148–164 (2010)
2. Voss, H.U.: Anticipating chaotic synchronization. Phys. Rev. E **61**(5), 5115–5119 (2000)
3. Eberle, Y., Nasuto, S.J., Hayashi, Y.: Integration of visual and joint information to enable linear reaching motions (2016, forthcoming)

MantisBot: The Implementation of a Photonic Vision System

Andrew P. Getsy[✉], Nicholas S. Szczecinski, and Roger D. Quinn

Case Western Reserve University, Euclid Ave., 10900, Cleveland, OH 44106, USA
apg26@case.edu

Abstract. This paper presents the design of a vision system for MantisBot, a robot designed to model a male Tenodera sinensis. The vision system mounted on the head of the robot consists of five oriented voltaic panels that distinguish the location of the most dominant light source in reference to the head's location. The purpose of this vision system is to provide the neural control system of the robot with descending commands, allowing for the investigation of transitions between targeted motions. In order to mimic the vision system of the insect, the voltaic panels were positioned in a particular orientation that when combined with a designed electric circuit yields an output that can be used by the neural controller to control the robot's motion. This paper summarizes the design of the vision system and its outputs. It also presents calibration data revealing the system's ability to encode a light source's elevation and azimuth angle as well as its distance.

Keywords: Descending commands · Robot · Mantis

1 Introduction

Praying mantises are ambush predators that primarily rely on vision to orient their bodies toward prey before capture [3]. While large amounts of data have been collected regarding the kinematics of their motion prior to strike [2,6], the neural systems that control this motion are largely unknown. Our group has constructed hypothetical models of visually guided control of the legs [5], but simulations have the disadvantage of taking a long time to calculate, and are never a complete picture of the real world. Therefore, we constructed the robot MantisBot, which until now had no head or head sensors [4].

In this paper we present the design of a simple head sensor for MantisBot, which provides the descending commands needed to explore how these might modulate local control systems in the thoracic ganglia that orient the robot toward prey. The sensor is composed of five photovoltaic cells arranged in a convex pattern on a head made of balsa wood. The cell's output are measured, amplified, and compared to calculate the centroid and distance of a known light source in the robot's field of view. Here we describe the design and present calibration data.

© Springer International Publishing Switzerland 2016
N.F. Lepora et al. (Eds.): Living Machines 2016, LNAI 9793, pp. 429–435, 2016.
DOI: 10.1007/978-3-319-42417-0_42

2   Methods and Results

2.1   Mechanical Design

The design for the head, that holds the vision system, was constructed from 6.35 mm (0.25 in.) thick Balsa wood. Wood was chosen as the structural material due to its low density. The strength of this material is sufficient due to the lack of external and inertial loads present in the structure. The components of this structure were cut with an Epilog Legend 36EXT laser cutter, at the Larry Sears and Sally Zlotnick Sears think[box]. The general dimensions of the structure were chosen in order to fit on the MantisBot robot which was built with a 13.3:1 scale to the insect.

 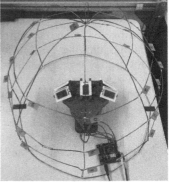

Fig. 1. Left: constructed head design with vision system positioned within the first testing fixture made from a cardboard box. Right: the second test fixture made of 1/16 in. steel rod.

Experiments were performed to calibrate the sensor. In all experiments, a 50 lumen light was shone onto the head from different known locations, and the locations were correlated with sensor readings. Two test fixtures were used for calibration. The first test fixture was a 45 cm by 45 cm by 30 cm cardboard box with twenty-five holes at measured locations, shown in Fig. 1. Testing was conducted by covering all holes except one and shining light, directed at the center panel, for a duration of five seconds through the only opening. The holes were utilized in a known order, and the location of each in all three dimensions was measured and recorded. Shadows and reflections within the box led to the development of a second test fixture, a cage constructed from 1.59 mm (1/16 in.) steel wire brazed together to form a 23 cm dome, shown in Fig. 1. This fixture allows for the vision system to be positioned at the center of the dome while, in a dark room, a light source is held at the measured locations. This test method eliminated the creation of unwanted reflections from the light source and external interferences.

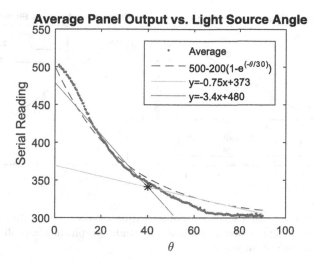

Fig. 2. Average solar panel readings with approximation function. Drop off angle (*) from linear fits of the segmented approximation function.

The collected readings were then analyzed in MATLAB (Mathworks, Natick, MA). Periods of zero volt readings from the center sensor were used to demarcate readings, and then the mean sensor values for each reading were recorded. Least squares curve fitting was conducted using a genetic algorithm and then a Newton optimizer. Initial test results showed that the head could only accurately and uniquely predict the position of the light source over a narrow range of angular positions due to the changing flux on each sensor (Fig. 3). Therefore, individual panels were tested by presenting a light to the panel at a variety of orientations

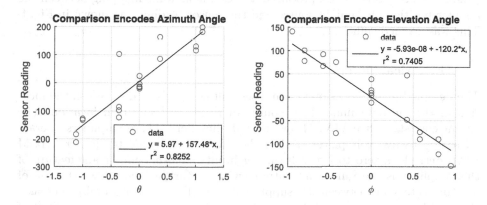

Fig. 3. Collected data using the cardboard box testing fixture. Left: graph showing collected reading versus light source azimuth angle. Right: graph showing collected reading versus light source elevation angle.

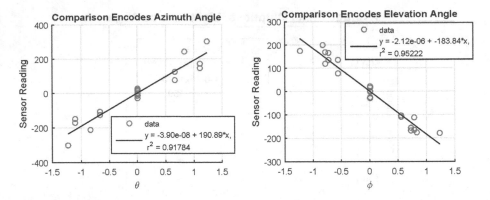

Fig. 4. Collected data using the steel dome testing fixture. Left: graph showing collected reading versus light source azimuth angle. Right: graph showing collected reading versus light source elevation angle.

ranging from orthogonal to parallel with respect to both the short and long axis of the panel while keeping the distance from the panel constant.

The collected data revealed that the solar panel reacted in a similar fashion with regard to the short and long axis. This allowed the average of the two trails to represent how the light source angle effects the solar panel's output. An approximation function was constructed to fit the average of the solar panel readings (Fig. 2). This function was then used to find the angle at which the panel's output began to drop off at a faster rate. Segmenting the function and performing linear fits on the two portions led to an intersection at an angle of forty degrees. This was the chosen angle to offset the panels in both the azimuth and elevation directions in order to achieve a field of view of one hundred eighty degrees (Fig. 5). Making this change produced a linear, monotonic mapping between the angular position of the light source and the sensor readings, as shown in Fig. 4. This was deemed sufficient.

2.2 Electrical Design

The designed photonic vision system uses voltaic panels to distinguish the location of the most dominant light source in reference to its own location. The circuit consists of five TLV274 operational amplifiers. These amplifiers operate on a zero to five volt power rail, so they were chosen due to the Arbotix-M Robocontroller's zero to five volt analog inputs. Another important feature of this amplifier is the rail-to-rail output swing feature, which allows the use of the complete span between the supply voltages [1]. This feature also increases the signal-to-noise ratio and dynamic range, which allows for a larger range of cleaner signals.

The circuit is connected to five Sundance Solar calibrated solar cells oriented with two on the left side of the head, two on the right side of the head, and one in the center, as seen in Fig. 5. These solar panels receive incoming light while

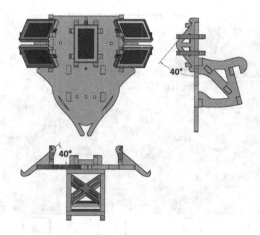

Fig. 5. 3D rendering of MantisBot's updated vision system with voltaic panel orientation angles.

outputting a maximum of 0.5 V at 200 mA based on the intensity of the light source. The objective of the circuit is to gather these signals and produce two outputs that reveal which of the panels are receiving the most light; the two up-most, down-most, left-most, or right-most panels, and one output revealing the general intensity of the light source. These outputs inform the control system when a change in direction needs to occur in order to keep the light source centered. The vision system also contains an isolated digital LED circuit (Fig. 6) which provides visual feedback to the operator showing which solar panels are receiving the most light.

The production of these visual signals was accomplished by the combination of three steps within the electronic circuit shown in Fig. 6. First, the signals of the four solar panels, not including the center panel, are each sent through a unity gain buffer to prevent the solar cells from loading the rest of the circuit (Fig. 6a). The signals are then added in four individual sums; the two left panels, two right panels, two top panels, and two bottom panels. These sums are performed by the operational amplifier summing circuits represented in Fig. 6b. These combine the voltages of the input signals (Fig. 6b Point I), which creates the average of the two signals, and then amplify that combined voltage with a gain of positive eight (Fig. 6b Point II). This is equivalent to the addition of the two signals with a gain of four. These four outputs are passed through individual unity gain buffers to ensure signal isolation (Fig. 6c).

The second step involves a unity gain summing amplifier which biases these signals upwards to ensure the output is between zero and five Volts. During this step a voltage divider is used to supply two and a half volts to the right and upward sum signals, shown in Fig. 6d Point III. This stage also introduces the center panel, whose signal is amplified by a factor of eleven (Fig. 6d Point IV). The increased gain for the center panel is necessary to distinguish light sources at farther distances once it was centered in the field of view. The gain also enhances

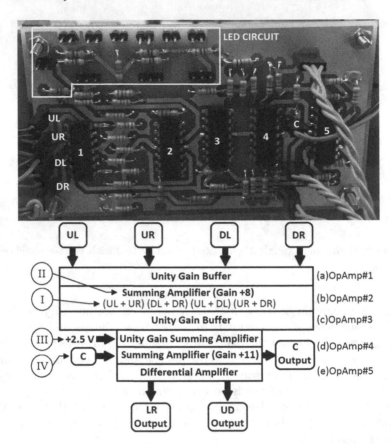

Fig. 6. Designed vision system circuit with labeled signal inputs UL (Up Left), UR (Up Right), DL (Down Left), DR (Down Right); LED circuit; and numbered TLV274 operational amplifiers.

the sensitivity of the signal making slight position changes of the light source more distinguished.

The final stage computes the centroid of the light source, and buffers the output. In this step the summed left signal is subtracted from the summed right signal, which has been offset by two and a half volts, and the summed down signal is subtracted from the offset summed upward signal (Fig. 6e). These signal manipulations result in two output analog signals ranging from 0.5 V to 4.5 V. Even though the TLV274 operational amplifier supports rail to rail output signals, a half volt safety region was incorporated for conservative operations.

These output signals are supplied to the Arbotix-M Robocontroller via three analog signal pins. The analog inputs are run through a ten-bit analog-to-digital converter, therefore the input signal voltage is divided into a range from zero to one thousand twenty three, which corresponds to the zero to five volt power supply. Since the supplied power is only positive voltage, a median voltage of zero to distinguish left from right and up from down would not be possible; it was for this reason that two and a half volts were added to the summed right and upward signals. The addition of this magnitude resulted in the median voltage being two and half volts, or five hundred twelve in the digital system. What this means is that when the centroid of the light source is left of the center of view, the first output signal will be some number larger than five hundred twelve, and when right the value will be below. When the light source is below the center of view, the output will be larger than five hundred twelve, and smaller when above the head. The center panel outputs a signal that will range from zero to one thousand twenty four. This signal decreases as the light source distance increases. The solar panel is also sensitive to the angle at which the light is held. The signal will be strongest when the light source is directly in front of the panel.

3   Conclusion

The vision system for MantisBot is capable of providing high level feedback to the robot's control system from an exterior light source. This feedback is in the form of serial outputs that reveal the location of a luminous target. This will allow the further investigation of how descending commands communicated to the local neural system can initiate reflexes that change the orientation of an animal or robot.

References

1. Family of 550-uA/Ch 3-MHz Rail-to-Rail Output Operational Amplifiers (Rev. D) (2004). http://www.ti.com/lit/ds/symlink/tlv274.pdf
2. Cleal, K.S., Prete, F.R.: The predatory strike of free ranging praying mantises, sphodromantis lineola (Burmeister). II: strikes in the horizontal plane. Brain Behav. Evol. **48**, 191–204 (1996)
3. Mittelstaedt, H.: Prey capture in mantids. In: Recent Advances in Invertebrate Physiology, pp. 51–72 (1957)
4. Szczecinski, N.S., Chrzanowski, D.M., Cofer, D.W., Terrasi, A.S., Moore, D.R., Martin, J.P., Ritzmann, R.E., Quinn, R.D.: Introducing MantisBot: hexapod robot controlled by a high-fidelity, real-time neural simulation. In: IEEE International Conference on Intelligent Robots and Systems, Hamburg, DE, pp. 3875–3881 (2015)
5. Szczecinski, N.S., Martin, J.P., Bertsch, D.J., Ritzmann, R.E., Quinn, R.D.: Neuromechanical model of praying mantis explores the role of descending commands in pre-strike pivots. Bioinspiration and Biomimetics **10**(6), 1–18 (2015). http://dx.doi.org/10.1088/1748-3190/10/6/065005
6. Yamawaki, Y., Uno, K., Ikeda, R., Toh, Y.: Coordinated movements of the head and body during orienting behaviour in the praying mantis Tenodera aridifolia. J. Insect Physiol. **57**(7), 1010–1016 (2011)

Force Sensing with a Biomimetic Fingertip

Maria Elena Giannaccini[1,2(✉)], Stuart Whyle[1,2], and Nathan F. Lepora[1,2]

[1] Department of Engineering Mathematics, University of Bristol, Bristol, UK
[2] Bristol Robotics Laboratory, University of Bristol, Bristol, UK
{maria.elena.giannaccini,n.lepora}@bristol.ac.uk

Abstract. Advanced tactile capabilities could help new generations of robots work co-operatively with people in a wider sphere than these devices have hitherto experienced. Robots could perform autonomous manipulation tasks and exploration of their environment. These applications require a thorough characterisation of the force measurement capabilities of tactile sensors. For this reason, this work focuses on the characterisation of the force envelope of the biomimetic, low-cost and robust TacTip sensor. Comparison with a traditional load cell shows that when identifying low forces and changes in position the TacTip proves significantly less noisy.

Keywords: Tactile sensing · Bioinspiration · Force sensing

1 Introduction

The sense of touch is one of the key human modalities. It is of vital importance in everyday tasks like object manipulation, grasp control and exploration. In order to develop an effective robotic sense of touch, inspiration is sought from the human tactile sensory system. This is chosen as a source of inspiration given its ability to process and transmit detailed information [1]. Here we consider the TacTip sensor, a biomimetic, dome-shaped device based on the human fingertip. Specifically, TacTip relies on the movement of internal pins inspired by human fingertip papillae: internal projections of the dermis that contain specialized nerve endings and are arranged in ridgelike lines.

Despite advancements in methods for tactile perception with the TacTip in localization [2–4], in shape perception [5,6] and texture discrimination [7], the characterisation of the force response of the sensor has not been tackled beyond an initial exploration [8]. In this work, we present a characterisation of the force sensing capability of the TacTip sensor, aimed at determining the force envelope the sensor can measure. This is a necessary step in order to identify the usability of TacTip in applications such as dexterous robot manipulation and high performance industrial quality control.

2 Method

Chorley et al. [8] were inspired to consider the behaviour of human glabrous skin. They built on previous research showing that the Meissner's Corpuscles

© Springer International Publishing Switzerland 2016
N.F. Lepora et al. (Eds.): Living Machines 2016, LNAI 9793, pp. 436–440, 2016.
DOI: 10.1007/978-3-319-42417-0_43

work in tandem with the dermal papillae, see Fig. 1A, to provide edge encoding of a touched surface. When the human finger makes contact with an object or surface, deformation results in the epidermal layers of the skin and the change is detected and reported by the mechanoreceptors. The human sensory response is one the TacTip, see Fig. 1B, device seeks to replicate. In the design of TacTip the epidermal layers are represented by internal papillae pins on the inside of its skin. These move when the device contacts an object and the resulting pin deformation is optically tracked by a camera.

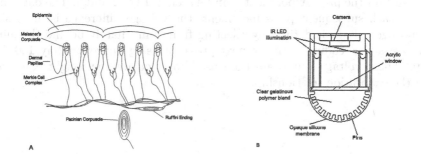

Fig. 1. A: Mechanoreceptors present on the top of the dermal papillae in the glabrous skin (cross section), B: The TacTip sensor with its pins, protrusions mechanically coupled with the outer skin surface, similarly to the dermal papillae.

For the force characterisation, a Tedea-Huntleigh, Model No. 1022 3 Kg beam load cell connected to a Novatech SY011 amplifier is utilised. In the experiment, the TacTip is pressed vertically against the load cell that outputs the magnitude of the pressing force. The calibration of the load cell is performed by placing weights on the load cell and then recording its voltage output.

In preparation for the experiments, the maximum pressing magnitude is identified to avoid damaging the sensor. It is found that beyond 7 mm the TacTip begins to deform irreversibly with its filling, a clear silicone elastomer gel, beginning to separate from the outer skin. This permanently degrades the TacTip performance. Hence, the maximum press magnitude is set, conservatively, at 5 mm from initial contact. At 5 mm the maximum output of the load cell is approximately 700 mV. A variety of weights are then used to give the range of outputs from 0 to 900 mV, which includes the 700 mV value we need.

After calibrating the beam load cell, the TacTip is fixed to the end-effector of an ABB IRB 120 robot arm and pressed onto the beam load cell for a range of 0.5–5 mm with 0.5 mm increments. In the interest of thoroughness it is important to clarify the effect of the load cell width on the calibration and experiment measurements. Due to its dome shape, the diameter of the TacTip increases gradually to its maximum (36 mm) while the width of the load cell is 25.4 mm. This implies that, due to the 25.4 mm limit, for final presses, the contact area magnitude is smaller than what it would be were the load cell as wide as the TacTip maximum diameter. Given that in the experiment the press displacement increments are kept constant (0.5 mm) and it takes less force to obtain the same penetration displacement given a smaller area of contact then the accuracy of the force measurements for the final presses could be affected. Nevertheless, it is important to keep in mind that given the conservative maximum press magnitude selected the effects of this phenomenon are very limited.

3   Results and Discussion

The results of the press experiments can be seen in Figs. 2 and 3. The data show that for each subsequent press increment, the voltage difference is increasing. The average voltage is derived by selecting from immediately behind the initial peak across the reading until the next press, utilising approximately 150 data points. These voltage values are then used to calculate the weight equivalent using the calibration relationship.

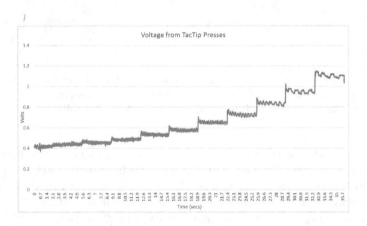

Fig. 2. The voltage output of the beam load cell generated from the TacTip pressing on the load beam, with 0.5 mm increments across a 0.5 to 5 mm range.

For the initial presses, Fig. 2, the voltage output from the load beam cell is similar to the noise level of the device. For the first press, the voltage change is barely discernible; whereas for the TacTip the step change at the initial presses is clearly evident and remains so throughout further presses, see Fig. 3. This shows that when identifying low forces and small changes in position the TacTip proves significantly less susceptible to noise than the load cell.

Inspection of the graph in Fig. 2 shows that, immediately after the press, the voltage peaks then subsides with a number of oscillations. The magnitude of the oscillation increases as the amount of presses increases. Silicone elastomers like the TacTip filling material do not have linear stress-strain behaviour. The initial peak in the load cell output is likely due to the viscoelastic response of the silicone elastomer. This would be dependent on the strain rate of the incremental end-effector loading.

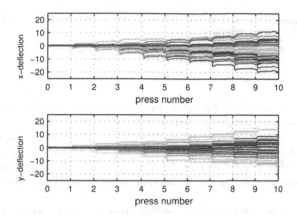

Fig. 3. Tactile data from TacTip, showing pin deflection against each press. Each press is a 0.5 mm increment across a range of 0.5–5 mm.

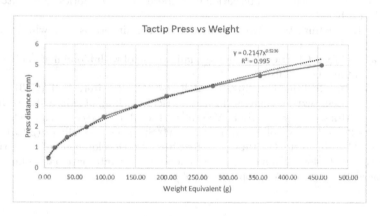

Fig. 4. Mapping the load beam outputs relative to the TacTip position. The dotted line shows the relationship to the function $y = 0.2147x^{0.5236}$. A common effect of viscoelasticity is strain stiffening at large stretches. Hence the fitted function may not be valid for larger deformations. At any rate, TacTip deformations are limited to avoid sensor damage.

Looking at Fig. 4 it can be seen that as the press magnitude increases, the relative voltage step and therefore the force, also increases. This covers a range of weights from 0 to 450 g which is equivalent to 0 to 4.413 N.

These results show that the TacTip is able to reliably detect changes in position of 0.5 mm. The experiments conducted yield a characterisation of the relationship between changes in position and force for the TacTip sensor along its full working range. In addition, the TacTip proves significantly less noisy than a load cell at identifying low forces and changes in position.

Acknowledgments. NL and MEG were supported by the EP/M02993X/1 EPSRC grant on Tactile Superresolution Sensing.

References

1. Dahiya, R.S., Valle, M.: Tactile sensing for robotic applications. In: Rocha, J.G., Lanceros-Mendez, J.G. (eds.) Sensors: Focus on Tactile Force and Stress Sensors, pp. 289–304 (2008). http://www.intechopen.com/books/sensors-focus-on-tactile-force-and-stress-sensors/tactile_sensing_for_robotic_applications
2. Lepora, N.F., Martinez-Hernandez, U., Evans, M., Natale, L., Metta, G., Prescott, T.J.: Tactile superresolution and biomimetic hyperacuity. Trans. Robot. **31**(3), 605–618 (2015)
3. Lepora, N.F., Ward-Cherrier, B.: Superresolution with an optical tactile sensor. In: Intelligent Robots and Systems (IROS), pp. 2686–2691, September 2015
4. Cramphorn, L., Ward-Cherrier, B., Lepora, N.: Tactile manipulation with biomimetic active touch. In: International Conference on Robotics and Automation (ICRA) (2016)
5. Lepora, L.: Biomimetic active touch with tactile fingertips and whiskers. IEEE Trans. Haptics (2016)
6. Lepora, N., Ward-Cherrier, B.: Tactile quality control with biomimetic active touch. Robot. Autom. Lett. (2016)
7. Winstone, B., Griffiths, G., Pipe, T., Melhuish, C., Rossiter, J.: TACTIP - tactile fingertip device, texture analysis through optical tracking of skin features. In: Lepora, N.F., Mura, A., Krapp, H.G., Verschure, P.F.M.J., Prescott, T.J. (eds.) Living Machines 2013. LNCS, vol. 8064, pp. 323–334. Springer, Heidelberg (2013)
8. Chorley, C., Melhuish, C., Pipe, T., Rossiter, J.: Development of a tactile sensor based on biologically inspired edge encoding. In: International Conference on Advanced Robotics (ICAR), pp. 1–6 (2009)

Understanding Interlimb Coordination Mechanism of Hexapod Locomotion via "TEGOTAE"-Based Control

Masashi Goda[1]([✉]), Sakiko Miyazawa[1], Susumu Itayama[1], Dai Owaki[1],
Takeshi Kano[1], and Akio Ishiguro[1,2]

[1] Research Institute of Electrical Communication, Tohoku University,
2-1-1 Katahira, Aoba-ku, Sendai 980-8577, Japan
{m.goda,saki,s.itayama,owaki,tkano,ishiguro}@riec.tohoku.ac.jp
[2] Japan Science and Technology Agency, CREST, 4-1-8 Honcho,
Kawaguchi, Saitama 332-0012, Japan
http://www.cmplx.riec.tohoku.ac.jp/

Abstract. Insects exhibit surprisingly adaptive and versatile locomotion despite their limited computational resources. Such locomotor patterns are generated via coordination between leg movements, *i.e.*, an interlimb coordination mechanism. The clarification of this mechanism will lead us to elucidate the fundamental control principle of animal locomotion as well as to realize truly adaptive legged robots that could not be developed solely by conventional control theory. In this study, we tried to model the interlimb coordination mechanism underlying hexapod locomotion on the basis of a concept called "TEGOTAE," a Japanese concept describing how well a perceived reaction matches an expectation. Preliminary experimental results show that our proposed TEGOTAE-based control scheme allows us to systematically design a decentralized interlimb coordination mechanism that can well-reproduce insects' gait patterns.

Keywords: Hexapod locomotion · Interlimb coordination · Local force feedback · Central pattern generator (CPG) · TEGOTAE

1 Introduction

Insects exhibit versatile gait patterns in response to their locomotion speed and environmental conditions [1,2]. These locomotor patterns are generated via the interlimb coordination mechanism. Although knowledge of these patterns at the behavioral and functional levels has been reported [3], the essential mechanism for interlimb coordination is still unknown. Thus, further clarification of this mechanism is required in order to design an adaptable and multifunctional legged robot as well as to understand the fundamental mechanism responsible for the remarkable abilities of legged animals.

© Springer International Publishing Switzerland 2016
N.F. Lepora et al. (Eds.): Living Machines 2016, LNAI 9793, pp. 441–448, 2016.
DOI: 10.1007/978-3-319-42417-0_44

To tackle this issue, we introduce a unique approach in this paper. We try to model the interlimb coordination mechanism underlying hexapod locomotion from the viewpoint of TEGOTAE, describing how well a perceived reaction matches an expectation. We introduce a TEGOTAE function, which is a function that quantitatively measures TEGOTAE, from which we can design a decentralized interlimb coordination mechanism in a systematic manner. Preliminary experimental results using an insect-like hexapod robot suggest that the designed interlimb coordination mechanism based on TEGOTAE well-reproduces the gait patterns of insects.

2 Model

Biological findings support that thoracic ganglia play a crucial role in the interlimb coordination in insect locomotion [4,5]. This strongly suggests that a decentralized control scheme, in which the coordination of simple individual components yields the nontrivial macroscopic behavior or functionalities, could be the key to understanding the interlimb coordination mechanism of insects. Thus far, various studies have been devoted to elucidating this mechanism [2,6] as well as to designing hexapod robots [7,8]. Here, we employ a "central pattern generator" (CPG)-based approach. For simplicity, we implement a phase oscillator into each leg as a decentralized controller.

The time evolution of the oscillator phase is described as follows:

$$\dot{\phi}_i = \omega + \frac{\partial T_i}{\partial \phi_i}, \tag{1}$$

where ω is the intrinsic angular velocity, ϕ_i is the phase of oscillator implemented into ith leg, and T_i is a TEGOTAE function. The ith leg is actively controlled according to ϕ_i such that the ith leg is in the swing phase when $0 \leq \phi_i < \pi$, i.e., $\sin\phi_i > 0$, and in the stance phase when $\pi \leq \phi_i < 2\pi$, i.e., $\sin\phi_i < 0$. T_i represents the key concept of our approach. T_i quantifies TEGOTAE, i.e., to what extent a perceived reaction matches an expectation, in real time. For convenience, we define T_i such that its value increases when good TEGOTAE is received at the ith leg. Thus, Eq. (1) means that ϕ_i is modified so that good TEGOTAE increases.

Now, the question is how to define T_i to well-reproduce the interlimb coordination observed in insect locomotion. In this study, we define T_i as follows:

$$T_i = \sigma_1 T_{i,1} + \sigma_2 T_{i,2}, \tag{2}$$

$$T_{i,1} = N_i(-\sin\phi_i), \tag{3}$$

$$T_{i,2} = \left(\frac{1}{n_L}\sum_{j \in L(i)}^{n_L} N_j\right)\sin\phi_i. \tag{4}$$

As Eq.(2) indicates, T_i consists of two TEGOTAE functions, $T_{i,1}$ and $T_{i,2}$, both of which are linearly coupled via the positive constants σ_1 and σ_2. N_i is the

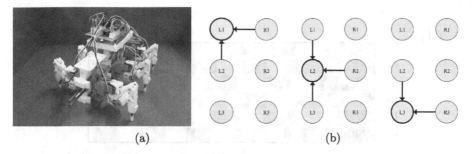

Fig. 1. (a) Developed hexapod robot. (b) The definition of $L(i)$, describing a set consisting of the legs neighboring the ith leg.

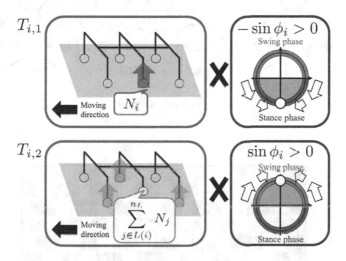

Fig. 2. Definition of the "TEGOTAE" functions.

ground reaction force acting on the ith leg. $L(i)$ denotes a set consisting of the legs neighboring the ith leg, and n_L is the number of elements of $L(i)$ (Fig. 1(b)). In what follows, we explain how we have designed these two TEGOTAE functions in detail.

$T_{i,1}$ quantifies TEGOTAE on the basis of the information only locally available at the corresponding leg; when the local controller intends to be in the stance leg $(-\sin\phi_i > 0)$, and results in receiving a ground reaction force $(N_i > 0)$ (Fig. 2 top), $T_{i,1}$ evaluates this situation as a good TEGOTAE, and returns a positive value. Note that the TEGOTAE function is described as the product of "what a local controller wants to do, $i.e.$, $-\sin\phi_i$," and "its resulting reaction, $i.e.$, N_i."

On the other hand, $T_{i,2}$ quantifies TEGOTAE on the basis of the relationship between the movements of the corresponding leg and its neighboring legs; when the local controller intends to be in the swing phase $(\sin\phi_i > 0)$ and its neighboring legs support the body well at that time $(\frac{1}{n_L}\sum_{j\in L(i)}^{n_L} N_j > 0)$ (Fig. 2

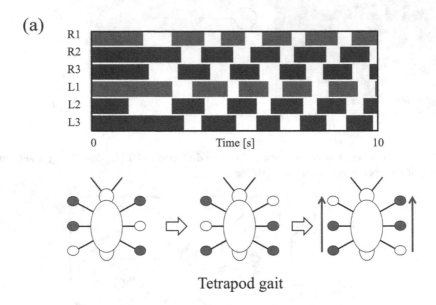

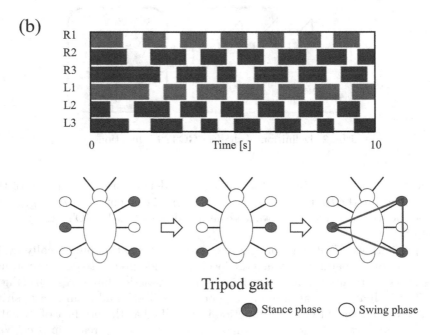

Fig. 3. Gait diagrams for (a) $\omega = 6.0$ and (b) $\omega = 7.5$. The colored regions in the upper graphs in (a) and (b) represent the stance phase, where $\sin \phi_i < 0$. The difference between these results and biological data [1] was the overlapping swing movements in the tetrapod gait. (Color figure online)

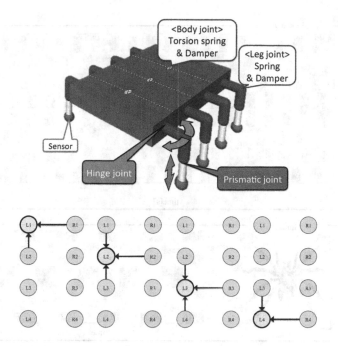

Fig. 4. Top: Octopod robot model. Bottom: The definition of $L(i)$, describing a set consisting of the legs neighboring the ith leg in the octopod model

bottom), $T_{i,2}$ evaluates that the corresponding leg well-establishes a relationship with its neighboring legs and returns a positive value.

By substituting Eqs. (2), (3) and (4) into Eq. (1), we obtain our CPG model as follows:

$$\dot{\phi}_i = \omega - \sigma_1 N_i \cos \phi_i + \sigma_2 \left(\frac{1}{n_L} \sum_{j \in L(i)}^{n_L} N_j \right) \cos \phi_i. \tag{5}$$

3   Preliminary Experimental Results

To verify the proposed control scheme in the real world, we developed a hexapod robot, as shown in Fig. 1(a). Each leg has two servo motors (Futaba, RS303MR) to generate the swing and stance phase motions according to the corresponding oscillator phase. We conducted experiments with the use of the parameter values $\omega = 6.0$ and 7.5. Here, the initial phases of all oscillators were $\phi_i = 3\pi/2$, and the parameters for the feedback gains were $\sigma_1 = 25$ and $\sigma_2 = 11$.

Figure 3 shows the gait diagrams for (a) $\omega = 6.0$ and (b) $\omega = 7.5$. In these graphs, the colored regions represent the stance phase, where $\sin \phi_i < 0$. The gait pattern observed in Fig. 3(a) corresponds to a tetrapod gait, where the feet touch the ground in the order of the hind, middle, and fore legs, whereas the gait

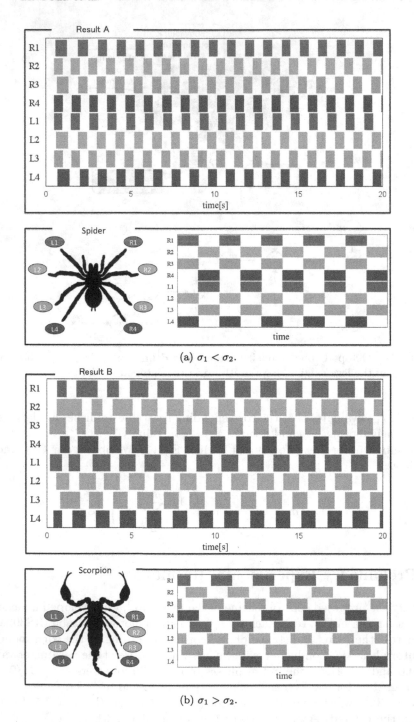

Fig. 5. Gait diagrams from the simulation (top) and in animal locomotion (bottom). The definitions of L1, L2, L3, L4, R1, R2, R3, and R4 are indicated at the bottom of each subfigure. The typical gait patterns for a (a) spider and (b) scorpion were drawn by authors after [9,10], respectively. (Color figure online)

pattern in Fig. 3(b) corresponds to a tripod gait, where the (L1, R2, L3) and (R1, L2, R3) feet alternately touch the ground in anti-phase. The results suggest that our model explains part of the observations of the gait patterns in insects. The difference between these results and biological data [1] is the overlapping swing movements in the tetrapod gait. Furthermore, we tested the effect of the variation in the initial oscillator phases on the gait patterns. As a result, we confirmed that more than 50 % of the initial conditions converged to the typical gait patterns in insects.

4   Applicability to Octopod Locomotion

Our long-term goal is to develop a systematic design scheme of interlimb coordination for multilegged robots by elucidating the interlimb coordination mechanism of various legged animals. No past studies tried to elucidate the mechanism underlying legged locomotion from a unified viewpoint: they focused on a specific locomotion generated via the particular number of legs. Toward the achievement of our goal, we investigate the applicability of the TEGOTAE-based control scheme to octopod locomotion, which is the locomotion observed in spiders or scorpions.

To this end, we modeled a simple octopod robot, as shown in Fig. 4. We conducted dynamical simulations using Open Dynamics Engine (ODE). To verify the applicability of the proposed control scheme, we implemented the decentralized controller designed in Sect. 2 in the octopod robot model. We compared the obtained gait patterns by changing the parameters of the feedback gains σ_1 and σ_2 in Eq. (5). Here, the initial phases of all oscillators were $\phi_i = 0$, and the parameter of the intrinsic angular velocity was $\omega = 2\pi$.

Figure 5 shows the simulation results for octopod locomotion. The upper graph in Fig. 5(a) shows the gait diagram in the case $\sigma_1 = 3 < \sigma_2 = 80$. The obtained gait pattern corresponds to a typical gait pattern in a spider [9], as shown in the lower graph in Fig. 5(a). On the other hand, the upper graph in Fig. 5(b) shows the gait diagram in the case $\sigma_1 = 217 > \sigma_2 = 70$. The gait pattern corresponds to a typical gait pattern in scorpions [10], as shown in the lower graph in Fig. 5(b). We found that the spider gait was observed when $T_{i,2}$ was dominant over $T_{i,1}$, whereas the scorpion gait was observed when $T_{i,1}$ was dominant over $T_{i,2}$. Surprisingly, our model also well-reproduced spider and scorpion locomotion with the same control scheme by only changing the magnitude of the local sensory feedback terms.

5   Conclusion and Future Work

In this study, we modeled the interlimb coordination mechanism underlying hexapod locomotion on the basis of the concept of TEGOTAE. Using a hexapod robot, we demonstrated that our proposed TEGOTAE-based control scheme well-reproduces insects' gait patterns. Furthermore, we confirmed that our model also well-reproduces spider and scorpion locomotion with the same control scheme. The applicability of the design scheme based on the TEGOTAE function for the locomotion of various animals seems to be of great interest and will also be studied in further investigations.

Acknowledgements. We acknowledge the support of a JSPS KAKENHI Grant-in-Aid for Scientific Research (B) (16H04381).

References

1. Graham, D.: A behavioural analysis of the temporal organisation of walking movements in the 1st instar and adult stick insect (Carausius morosus). J. Comp. Pysiol. **81**, 23–52 (1972)
2. Schilling, M., Hoinville, H., Schmitz, J., Cruse, H.: Walknet, a bio-inspired controller for hexapod walking. Biol. Cybern. **107**, 397–419 (2013)
3. Cruse, H.: What mechanisms coordinate leg movement in walking arthopods? Trends Neurosci. **13**, 15–21 (1990)
4. Jeffrey, D.: Leg coordination in the stick insect carausius morosus: effects of cutting thoracic connectives. J. Exp. Biol. **145**, 103–131 (1989)
5. Bässler, U., Büschges, A.: Pattern generation for stick insect walking movements-multisensory control of a locomotor program. Brain Res. Rev. **27**, 65–88 (1998)
6. Kimura, S., Yano, M., Shimizu, H.: A self-organizing model of walking patterns of insects. Biol. Cybern. **69**, 183–193 (1993)
7. Beer, R.D., Quinn, R., Chiel, H.: Biologically inspired approaches to robotics. Commun. ACM **40**, 30–38 (1997)
8. Ambe, Y., Nachstedt, T., Manoonpong, P., Worgotter, F., Aoi, S., Matsuno, F.: Stability analysis of a hexapod robot driven by distributed nonlinear oscillators with a phase modulation mechanism. In: IEEE/RSJ International Conference on Intelligent Robots and Systems (IROS 2013), pp. 5087–5092 (2013)
9. Biancardi, C.M., Fabrica, C.G., Polero, P., Loss, J.F., Minetti, A.E.: Biomechanics of octopedal locomotion: kinematic and kinetic analysis of the spider Grammostola mollicoma. J. Exp. Biol. **214**, 3433–3442 (2011)
10. Bowerman, R.F.: The control of walking in the Scorpion I. J. Comp. Physiol. A **100**, 183–196 (1975)

Decentralized Control Scheme for Myriapod Locomotion That Exploits Local Force Feedback

Takeshi Kano[1]([✉]), Kotaro Yasui[1], Dai Owaki[1], and Akio Ishiguro[1,2]

[1] Research Institute of Electrical Communication, Tohoku University,
2-1-1 Katahira, Aoba-ku, Sendai 980-8577, Japan
{tkano,k.yasui,owaki,ishiguro}@riec.tohoku.ac.jp
[2] Japan Science and Technology Agency, CREST, 4-1-8 Honcho,
Kawaguchi, Saitama 332-0012, Japan
http://www.cmplx.riec.tohoku.ac.jp/

Abstract. Legged animals exhibit adaptive and resilient locomotion through their inter-limb coordination. Our long-term goal of this study is to develop a systematic design scheme for legged robots by elucidating the inter-limb coordination mechanism of various legged animals from a unified viewpoint. As a preliminary step towards this, we here focus on millipedes. We performed behavioral experiments on a terrain with gap, and found that legs do not tend to move without the ground contact. Based on this qualitative finding, we proposed a decentralized control scheme using local force feedback.

Keywords: Millipede · Inter-limb coordination · Local force feedback

1 Introduction

In the animal kingdom, there are animals having various number of legs such as humans, quadrupeds, insects, and myriapods. What is interesting is that they can adapt to changes in the environment as well as changes in their own morphologies, *e.g.* leg amputation, in real time [1]. This adaptive and resilient locomotion is achieved by coordinating their limbs in a decentralized manner. Clarifying this inter-limb coordination mechanism will help develop legged robots that can move in harsh environments such as disaster areas.

Our challenge is to develop a systematic design scheme for legged robots by elucidating the inter-limb coordination mechanism of various legged animals from a unified viewpoint. Towards this goal, we started our research by focusing on individual animal species, particularly on animals with few number of legs, *i.e.*, two, four, and six legged animals [2–4]. However, animals having many legs have been less studied.

Accordingly, we here focus on millipedes. In fact, millipedes are suitable model because they have the largest number of legs on earth [5] and can move on unstructured terrain by propagating density waves of the leg tips from the tail to head appropriately [6]. In this study, we first performed behavioral experiments on a terrain with a gap. Then, based on the qualitative findings from

© Springer International Publishing Switzerland 2016
N.F. Lepora et al. (Eds.): Living Machines 2016, LNAI 9793, pp. 449–453, 2016.
DOI: 10.1007/978-3-319-42417-0_45

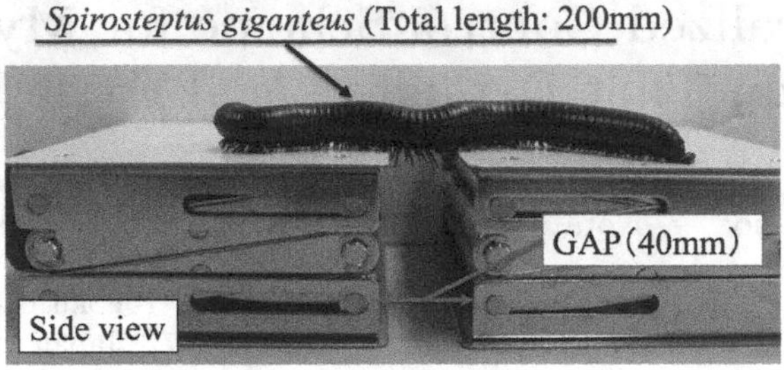

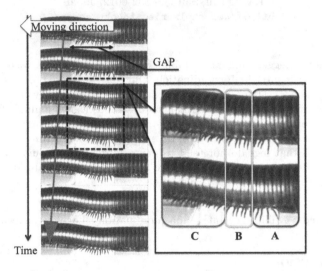

Fig. 1. Experiment setup.

Fig. 2. Millipede locomotion on a terrain with a gap.

the experiments, we proposed a simple decentralized control scheme using local force feedback.

2   Behavioral Experiments

We performed behavioral experiments using millipedes (*Spirosteptus giganteus*). To investigate the effect of the ground contact, we observed locomotion on a terrain with a gap (Fig. 1). The result is shown in Fig. 2. On the ground, the leg tips form density waves that propagate forward, as reported in the previous study [6]. However, this behavior changed on the gap: When a leg entered the gap, it exhibited searching movement [7] for a while (Fig. 2A), after which it stopped (Fig. 2B). It started to swing again when its several anterior legs contact the ground (Fig. 2C). Thus, each leg is likely driven by the ground contact of itself and its several anterior legs.

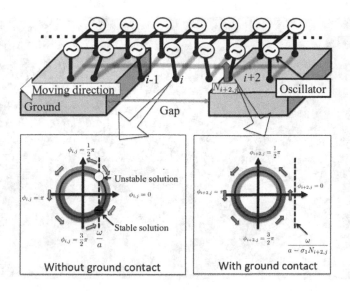

Fig. 3. Schematic for the proposed control scheme.

3   Model

We proposed a decentralized control scheme for millipede locomotion on the basis of the qualitative findings that the motion of each leg is likely triggered by the ground contact of itself and its anterior legs. The schematic is shown in Fig. 3. Legs are connected via the body trunk. Each leg can move both forward-backward and upward-downward. A phase oscillator is implemented in each leg. The motion of the ith leg on the right- and left-hand side is controlled according to the oscillator phase $\phi_{i,r}$ and $\phi_{i,l}$, respectively. Legs tend to be in the swing and stance phase when the oscillator phase is between 0 and π and between π and 2π, respectively.

The time evolution of the oscillator phase is described as follows:

$$\dot{\phi}_{ij} = \omega - (a - \sigma_1 N_{i,j}) \cos \phi_{i,j} + \sigma_2 N_{i-1,j} \cos \phi_{i,j}, \qquad (j = r, l) \qquad (1)$$

where ω is the intrinsic angular velocity, a, σ_1, σ_2 are positive constants, and $N_{i,j}$ is the ground reaction force acting on the leg.

The first and second terms on the right-hand side of Eq. (1) are based on the finding that the leg motion stops on the gap in the above behavioral experiment (Fig. 2B). When $\omega < a$ and the leg does not contact the ground ($N_{i,j} = 0$), $\omega - (a - \sigma_1 N_{i,j}) \cos \phi_{i,j} = 0$ has two solutions, where one is stable and the other is unstable (Fig. 3). Thus, the leg tends to stay at the position corresponding to the stable solution. When the leg contacts the ground and $N_{i,j}$ increases, the stable and unstable solutions vanish, and thus, the leg moves periodically.

The third term on the right-hand side of Eq. (1) is based on the finding that the leg begins to move when its anterior legs contact the ground (Fig. 2C).

Fig. 4. Myriapod-like robot we are now developing.

When the $i - 1$th leg contacts the ground, the ith oscillator phase converges to $\pi/2$, which makes the leg begin to move. Further, because the ith leg waits for its ground contact until the $i - 1$th leg lifts off the ground, it is expected that the density waves of the leg tips generate in a decentralized manner.

4   Conclusion and Future Work

We focused on millipedes and proposed a decentralized control scheme based on the results of the behavioral experiments. We are now developing a myriapod-like robot (Fig. 4) to investigate the validity of the proposed control scheme. The experimental results will be shown in the presentation.

Acknowledgments. This work was supported by Japan Science and Technology Agency, CREST. The authors would like to thank Kazuhiko Sakai of Research Institute of Electrical Communication of Tohoku University.

References

1. Schilling, M., Hoinville, T., Schmitz, J., Cruse, H.: Walknet, a bio-inspired controller for hexapod walking. Biol. Cybern. **107**, 397–419 (2013)
2. Owaki, D., Kano, T., Nagasawa, K., Tero, A., Ishiguro, A.: Simple robot suggests physical interlimb communication is essential for quadruped walking. J. Roy. Soc. Interface **10**, 20120669 (2013)
3. Owaki, D., Ishiguro, A.: CPG-based control of bipedal walking by exploiting plantar sensation. In: Proceedings of the CLAWAR 2014, pp. 335–342 (2014)

4. Ishiguro, A., Nakamura, K., Kano, T., Owaki, D.: Neural communication vs. physical communication between limbs: which is essential for hexapod walking? In: Dynamic Walking 2014, T 406 (2014)
5. Marek, P.E., Bond, J.E.: Rediscovery of the world's leggiest animal. Nature **441**, 707 (2006)
6. Kuroda, S., Tanaka, Y., Kunita, I., Ishiguro, A., Kobayashi, R., Nakagaki, T.: Common mechanics of mode switching in locomotion of limbless and legged animals. J. Roy. Soc. Interface **11**, 20140205 (2014)
7. Dürr, V.: Stereotypic leg searching-movements in the stick insect: kinematic analysis, behavioural context and simulation. J. Exp. Biol. **204**, 1589–1604 (2001)

TEGOTAE-Based Control Scheme
for Snake-Like Robots That Enables
Scaffold-Based Locomotion

Takeshi Kano[✉], Ryo Yoshizawa, and Akio Ishiguro

Research Institute of Electrical Communication, Tohoku University,
2-1-1 Katahira, Aoba-ku, Sendai 980-8577, Japan
{tkano,r-yoshi,ishiguro}@riec.tohoku.ac.jp
http://www.cmplx.riec.tohoku.ac.jp/

Abstract. Snakes exhibit "scaffold-based locomotion" wherein they actively utilize terrain irregularities and move effectively by pushing their body against the scaffolds that they encounter. Implementing the underlying mechanism in snake-like robots will enable them to work well in unstructured real-world environments. In this study, we proposed a decentralized control scheme for snake-like robots based on TEGOTAE, a Japanese concept describing how well a perceived reaction matches an expectation, to reproduce scaffold-based locomotion. A preliminary experimental result showed that reaction forces from environment are evaluated based on TEGOTAE in real time and those beneficial for propulsion of the body are selectively exploited.

Keywords: Snake-like robot · TEGOTAE-based control · Scaffold-based locomotion

1 Introduction

Snakes actively utilize terrain irregularities and move effectively by pushing their body against the scaffolds that they encounter [1,3]. This behavior, which we call "scaffold-based locomotion," is achieved by appropriately coordinating large number of bodily degrees of freedom in a decentralized manner. Clarifying the underlying mechanism will help develop robots that can work well in unstructured real-world environments. Unfortunately, however, previous models for the scaffold-based locomotion could not reproduce the innate behavior of snakes [2–4].

To address this issue, here we propose a decentralized control scheme for scaffold-based locomotion from the viewpoint of TEGOTAE, a Japanese concept describing how well a perceived reaction matches an expectation. In this control mechanism, reaction forces from environment are evaluated based on TEGOTAE in real time and those beneficial for propulsion of the body are selectively exploited. The validity of the proposed control scheme is investigated via a preliminary experiment using a snake-like robot developed (Fig. 2).

© Springer International Publishing Switzerland 2016
N.F. Lepora et al. (Eds.): Living Machines 2016, LNAI 9793, pp. 454–458, 2016.
DOI: 10.1007/978-3-319-42417-0_46

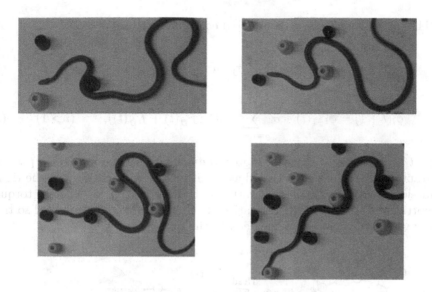

Fig. 1. Scaffold-based locomotion of real snake (*Elaphe climacophora*).

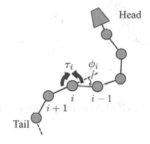

Fig. 2. Schematic of the proposed model.

2   Model

The schematic of the proposed model is shown in Fig. 1. Several segments are concatenated one-dimensionally via joints. Frictional anisotropy is implemented in each segment like real snakes, and the friction coefficient along the tangential direction is smaller than that along the normal direction. Each joint is controlled according to proportional control, and the target and real angle of the ith joint is denoted by $\bar{\phi}_i$ and ϕ_i, respectively. We assume that each segment can detect forces from the environment, and the contact forces detected at the ith segment from the right- and left-hand side are denoted by $f_{r,i}$ and $f_{l,i}$, respectively.

The target joint angle $\bar{\phi}_i$ is updated each time step as follows:

$$\bar{\phi}_1(t+1) = h_d(t) + \sigma \sum_{j=i-n_b}^{i+n_f} \tau_j(t)(f_{l,j}(t) + f_{r,j}(t)), \tag{1}$$

$$\bar{\phi}_i(t+1) = \phi_{i-1}(t) + \sigma \sum_{j=i-n_b}^{i+n_f} \tau_j(t)(f_{l,j}(t) + f_{r,j}(t)), \qquad (i > 1) \tag{2}$$

where t is a time step, $h_d(t)$ is the motor command from a operator, σ is a positive constant, $\tau_j(t)$ is a torque generated at the jth joint. The first term on the right-hand side of Eq. (2) denotes curvature derivative control [5] wherein torques proportional to the curvature derivative of the body curve are generated so that bodily waves propagate from the head to the tail.

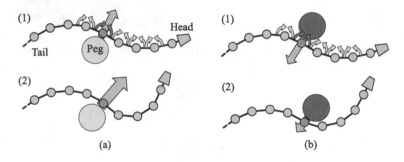

Fig. 3. Mechanism of TEGOTAE-based control. (Color figure online)

Fig. 4. Overview of the snake-like robot HAUBOT VI.

The second term on the right-hand side of Eqs. (1) and (2) denotes a sensory feedback based on TEGOTAE. Here, TEGOTAE is defined as a quantity that expresses "how well a perceived reaction matches an expectation" and described

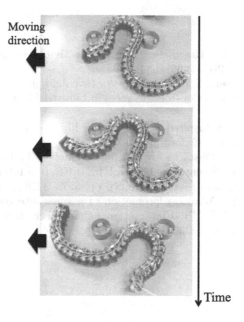

Fig. 5. Preliminary experiment.

as the product of the torque, *i.e.*, intended action, and the contact forces, *i.e.*, reaction. This feedback works as shown in Fig. 3. When a torque generated to bend the body leftward (red arrow in Fig. 3(a)(1)) resulted in receiving a contact force from the right (purple arrow in Fig. 3(a)(1)), the contact force assists propulsion, and the feedback works to generate further torques (yellow arrow in Fig. 3(a)(1)) so that the contact force increases (Fig. 3(a)(2)). On the other hand, when a torque generated to bend the body leftward (red arrow in Fig. 3(b)(1)) resulted in receiving a contact force from the left (purple arrow in Fig. 3(b)(1)), the contact force impedes propulsion, and the feedback works to generate further torques (yellow arrow in Fig. 3(b)(1)) so that the contact force decreases (Fig. 3(b)(2)). Thus, the TEGOTAE-based control scheme enables the robot to "selectively" exploit reaction forces from environment, unlike previous control schemes based on simple local reflexive mechanisms [2–4].

3   Preliminary Experiment

We developed a snake-like robot HAUBOT VI (Fig. 4) and performed a preliminary experiment on a terrain with two pegs as shown in Fig. 5. Then, the robot moved by selectively exploiting a peg that assists propulsion.

Acknowledgments. This work was supported by NEDO (Core Technology Development Program for Next Generation Robot). The authors would like to thank Daiki Nakashima of Research Institute of Electrical Communication of Tohoku University.

References

1. Moon, B.R., Gans, C.: Kinematics, muscular activity and propulsion in gopher snakes. J. Exp. Biol. **201**, 2669–2684 (1998)
2. Liljebäck, P., Pettersen, K.Y., Stavdahl, Ø., Gravdahl, J.T.: Snake Robots - Modelling, Mechatronics, and Control: Advances in Industrial Control. Springer, London (2012)
3. Hirose, S.: Biologically Inspired Robots (Snake-like Locomotor and Manipulator). Oxford University Press, Oxford (1993)
4. Kano, T., Ishiguro, A.: Obstacles are beneficial to me! Scaffold-based locomotion of a snake-like robot using decentralized control. In: IEEE/RSJ International Conference on Intelligent Robots and Systems (IROS), pp. 3273–3278 (2013)
5. Date, H., Takita, Y.: Adaptive locomotion of a snake like robot based on curvature derivatives. In: IEEE/RSJ International Conference on Intelligent Robots and Systems (IROS), pp. 3554–3559 (2007)

Modelling the Effect of Cognitive Load on Eye Saccades and Reportability: The Validation Gate

Sock C. Low[1(✉)], Joeri B.G. van Wijngaarden[1], and Paul F.M.J. Verschure[1,2]

[1] Synthetic Perceptive Emotive and Cognitive Systems Group (SPECS),
Universitat Pompeu Fabra, Barcelona, Spain
sockching.low@upf.edu
[2] Institució Catalana de Recerca i Estudis Avançats (ICREA), Barcelona, Spain

Abstract. Being able to selectively attend to stimuli is essential for any agent operating in noisy environments. Humans have developed, or are inherently equipped, with mechanisms to do so; these mechanisms are a rich source of debate. Here, we contribute to it by building a functional model to describe the findings of an existing psychophysiological experiment demonstrating how early and late saccades as well as "conscious report" are affected by varying levels of cognitive load. The model adheres to the established principles of neurophysiology. In a task focused on the monitoring of moving stimuli, where objects usually move predictably but randomly deviate from it every 0.2 s, change in cognitive load is reflected in the proportion of late saccades and behavioural reports in response to the task. It also provides evidence that the physiological structure of the brain is capable of implementing selective attention using a method other than the attentional spotlight.

Keywords: Attention · Model · Validation gate · Attentional mechanisms · Moving targets

1 Introduction

The human brain receives an overwhelming amount of sensory input and is unable to process all this information at a constant level of awareness. Instead, the brain chunks and selects specific parts of this sensory stream to attend to based on relevance and saliency. Over the years, many different theories for attention have been proposed, with one general consensus — that attention is not solely dependent on sensory input, commonly referred to as bottom-up attention (BU), but also relies heavily on contextual information (*e.g.*, memory or higher-level cognitive areas), known as top-down attention (TD). How these systems interact is, however, a matter of contention [1].

Currently, one of the most widely accepted views is that attention uses a so-called attentional spotlight: the saliency of objects from the bottom-up stream is modulated by top-down processes such that it boosts the saliency of goal- or task-relevant stimuli. This occurs either through an increase in their saliency, a

N.F. Lepora et al. (Eds.): Living Machines 2016, LNAI 9793, pp. 459–466, 2016.
DOI: 10.1007/978-3-319-42417-0_47

suppression of irrelevant stimuli, or a combination of both. This notion receives support on both a behavioural (*e.g.*, [2]) and neurophysiological (*e.g.*, [3]) level. However, this view is unable to comprehensively explain specific attentional shortcomings such as the attentional blink.

An alternative theory is that of the validation gate, one which originated from the field of aircraft position tracking. It proposes that rather than sampling the input for relevant stimuli and increasing their saliency respectively, higher-level cognitive regions use predictive coding in a feed-forward manner to predict upcoming states of objects. It then exerts an inhibitory influence on these predicted states such that only violations of the predictions become salient. With low cognitive load, there are sufficient cognitive resources to attend to a larger amount of objects, thus a person is able to observe small violations with high sensitivity. As cognitive load increases, and resources decline, this sensitivity to change decreases and the accuracy at which disruptions are observed diminish.

Mathews *et al.* designed a visual detection task to study this interaction between BU and TD influences [4]. During this task, ten objects (non-filled white circles) were presented on a screen, each moving at a constant speed in a randomly initialised direction and bouncing off the edges of the screen, giving them a predictable path. For each trial, the participant was allowed to freely look around the screen as a randomly chosen object is displaced from its linear trajectory, thus violating the next predicted state. Their task was to report the displacement as quickly as possible while simultaneously performing a secondary distractor task that modulates their cognitive load. Difficulty for this second task increased over three conditions: low (no second task), medium (reciting the alphabet normally) and high (reversed alphabet, skipping each second letter). The measures used were gaze movement, split into early (<250 ms) and late (250 ms–800 ms) saccades, and a button press to report a deviant translation.

The authors concluded that cognitive load did not affect the number of fast or slow saccades, but that the number of slow saccades followed by conscious detection did decrease as cognitive load increased. In this short paper, we developed a functional model of the visual circuit that can account for the validation gate and is consistent with existing literature. It aims to describe the findings of [4], specifically the interaction between cognitive load, early and late saccades, and behavioural reportability.

2   Method

2.1   Design

A series of frames were generated to simulate the visual input that the participants were exposed to. Assuming a frame rate of 50 Hz, 500 × 500 pixels frames were generated to last a duration of 20 s. For simplicity, the validation gate was taken to be a circular region around the stimuli rather than an oval. As this implies an equal probability of an object moving in any direction, instead of keeping a constant trajectory as in [4], objects were set to move randomly to one of eight points on a square four pixels in length that was centred on the circle (Fig. 1). In line with the heuristic of [5], eight filled circles with a radius of three pixels were used as stimuli. Jumps occurred every 10 frames (*i.e.* 0.2 s), and in order to generate them a random circle would use a square of random length between 20–100 pixels, while non-displaced objects continued with the original "rule" of four pixels. Simulations were performed over several runs, each with varying levels of cognitive load that ranged between 0–1 at intervals of 0.1; a value of 1 represented the maximal cognitive load, comparable to a person who is unable to pay attention to anything else other than the primary task.

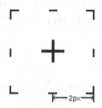

Fig. 1. Crosshair indicating centre of a circle, with black lines indicating the points where the circle may move to in the next time step.

2.2   Model

Of our five senses, the human visual system is arguably the most well-described. Influenced strongly by [6,7], we distil the circuitry into six main functional components (Fig. 2) — the retina, the thalamus, the visual cortices, the prefrontal cortex (PFC), the frontal eye fields (FEF) and the superior colliculus (SC).

Retina: provides input for the entire system and is thus represented by the images fed into the model. It has direct connections to the thalamus and the superior colliculus.

Thalamus: raw input from the retina invariably passes through the thalamus, specifically the lateral geniculate nucleus in the first stages of the cycle. This information is passed onwards to the primary visual cortex. The thalamus receives simultaneous inputs from the prefrontal cortex that are both excitatory (direct) and inhibitory (indirect). It is this indirect pathway that allows the PFC to selectively inhibit the relay of visual information to the primary visual cortex.

Visual cortices: uses recurrent loops from lower to higher visual cortices and the pulvinar to get increasingly abstract visual information from the input, and sends this to the prefrontal cortex to be integrated with contextual information.

Prefrontal cortex: generally associated with executive functions, decision-making and other high-level cognitive functions, with widespread connections from anterior to posterior regions of the cortex. Visual information that reaches the PFC has been processed and abstracted from the raw visual input in earlier stages of the circuit. The PFC's connections to the thalamus implies a structure suitable for retinotopic modulation in early stages of visual processing [8].

Frontal eye fields: an important component in the voluntary saccade, it has retinotopic structure. Although the frontal lobe is traditionally associated with higher-order cognitive functions, the FEF has been shown to play a part in low-level sensory processing as well [9].

Superior colliculus: as the gatekeeper for eye saccades, among other directing actions, it plays a key role in the visual circuit. Specifically, the SC is increasingly found to play an integrative role in mono- and multi-modal input for the control of attention and eye movements [10].

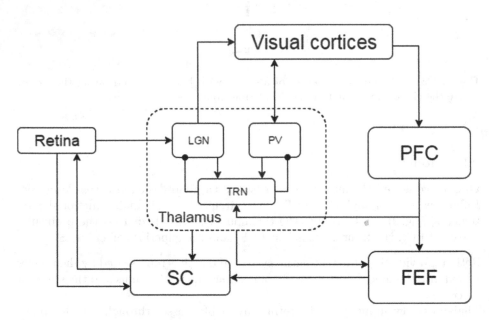

Fig. 2. Thalamocortical model showing the connections between the five major components of the visual circuit. Within the thalamus, the areas of interest are the thalamic reticular nucleus (TRN), the lateral geniculate nucleus (LGN) and the pulvinar nucleus (PV).

As the SC is the primary controller of eye movements, we consider its output to be the determinant of where a saccade is generated towards. Of the two types of saccades observed in humans, the instinctive, quick saccades are driven by BU saliency in the raw input signal from the retina. In the model, BU saliency is computed as the difference between two frames (Fig. 4) and can be said to take place in the SC; the greater the difference, the greater the bottom-up saliency. Normally, the object with the greatest saliency value is also the desired point towards which a saccade is directed based on the bottom-up circuit. However, if a generated number is less than a pre-set noise value, the jump has the same saliency as the other stimuli. In case of identical saliencies, a point is selected at random to saccade towards.

Subsequently, a late saccade in humans occurs due to the inherent latency of information passing through the thalamus and cortical regions before reaching the superior colliculus. This latency is simulated by having the top-down saliency of the circles calculated one frame after those used for the bottom-up saliency (*e.g.*, BU is calculated for frame 13 and 14 while TD is calculated for frame 12 and 13). The top-down system computes a circular validation gate (Fig. 3) around the circles of the earlier frame, the size (r_{VG}) of which is linearly related to the pre-determined cognitive load (CL) (Eq. 1). The next frame is then passed through this gate such that all stimuli within the gated regions are negated (*i.e.* ignored). Any remaining regions after this subtraction implies that one of the circles has moved outside of the predicted area, and is therefore important. In this case, the top-down system produces a saccade to be directed towards the location of this circle. When there is a conflict in positions for a saccade between BU and TD, TD always overrides BU.

Although the computations in the individual regions of the brain was not distinguished clearly in the model, it is expected that the calculation of the validation gate occurs in the PFC. Subsequently, the point to saccade towards from the TD system could come from either the FEF or the thalamus, albeit with differing latencies. This nuance is, however, outside the scope of this paper.

$$r_{VG} = 34 \times CL + 8 \tag{1}$$

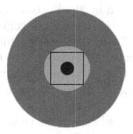

Fig. 3. The validation gate in the context of attending to translation of stimuli. Black circle: the individual stimulus' current position. Black square: possible locations that the stimulus will move to in the next frame. Blue and orange circles: with low and high cognitive loads respectively, region within which the stimulus is expected to be in the next frame. (Color figure online)

Finally, the model implies that a jump will only be reported when the PFC detects a change in the gated information. As cognitive load is introduced into the model mostly at the level of the PFC, this means the reporting of a jump is also affected by the cognitive load. Hence, a jump is detected only when there is a change outside of the validation gate *and* when it satisfies a probabilistic conditional that is dependent on the cognitive load value. The latter conditional has an additional constraint which prevents a very high cognitive load from completely blocking the reportability of the jumps. This is to provide some leeway which represents how humans are able to switch attention from one demanding task to another, even if they are unable to carry out both concurrently.

Fig. 4. Example of a difference map (inverted from the original black on white images) between two frames without a jump (LEFT) and with a jump (RIGHT) in the top-right corner.

3   Results and Discussion

Increasing levels of cognitive load influences specific components of the model's behaviour. Early saccades, driven by the bottom-up saliencies directly from the visual input, remain unaffected by any level of cognitive load (Fig. 5). Late saccades however do show a small negative relationship with cognitive load, where increasing levels of CL decrease the number late saccades elicited in the SC. Finally, the rate of reporting is most strongly coupled with cognitive load and ranges from near perfect response in the absence of a distractor to complete failure under maximal load.

These results are mostly consistent with those from [4], showing how it is possible for early saccades to be less effective at detecting changes depending on the noise present in the stimuli, and how reportability of the jumps is reduced as cognitive load increases. However, while [4] found the effect of cognitive load on late saccades to be insignificant, this was not the case in the model's data; for the model, the success of late saccades at detecting jumps is inversely proportional to cognitive load.

One possible explanation for this is that [4]'s task difficulty and cognitive loads had an interaction effect which masked the phenomenon. This could be due to the fact that ten stimuli were used in the experiment, instead of a lower number which would have made the task easier to complete and therefore highlighting the impact of the additional cognitive load. Also, the participants were allowed to freely move their gaze throughout the experiment. Humans have a

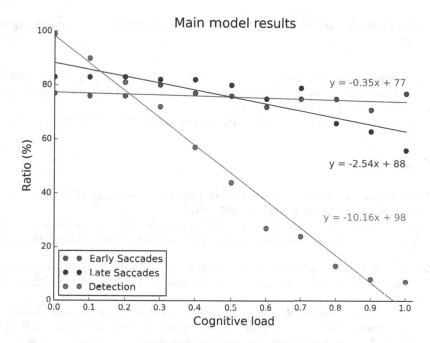

Fig. 5. Changes in the ratio of early (red) and late (green) saccades as well as detection resulting in a "behavioural response" (blue) due to increasing cognitive load. (Color figure online)

narrow region of acuity in the centre of their field of vision as that is the retinotopic location of their fovea. This would imply the participants would be less sensitive to changes outside of this region; when the field of acuity constantly changes, it is difficult to control for the distance between their visual point of focus and the stimulus that jumps at that same instant.

The current model was designed for functional replication of a physiological experiment, but does include all the main components reported in neuroanatomical literature. However, it remains difficult to draw generalised conclusions about the visual system and the neurological substrate of the validation gate from this alone. Further work on the model and psychophysiological experiments needs to be carried out to understand the phenomena, such as replacing functional components with more biophysically plausible spiking or mean-field models. It is also necessary to include other brain regions at some point, such as the posterior parietal cortex and basal ganglia which were neglected as they are mainly involved in voluntary saccades while our model does not distinguish voluntary and involuntary saccades. Nonetheless, this version already provides the main architecture to do so and is able to replicate and fit physiological data for a better understanding of the underlying processes. It is a plausible circuitry which forms the validation gate in humans, and demonstrates how the validation gate could be another form of attention selection mechanism in addition to the attentional spotlight.

Acknowledgements. The authors would like to thank all members of the SPECS group for their input during discussions, with a special mention for: Jordi-Ysard Puigbò, Riccardo Zucca and Clément Moulin-Frier.
This work is supported by the EU FP7 project WYSIWYD (FP7-ICT-612139).

References

1. Borji, A., Itti, L.: State-of-the-art in visual attention modeling. IEEE Trans. Pattern Anal. Mach. Intell. **35**(1), 185–207 (2013)
2. Kok, P., Rahnev, D., Jehee, J.F., Lau, H.C., de Lange, F.P.: Attention reverses the effect of prediction in silencing sensory signals. Cereb. Cortex **22**(9), 2197–2206 (2012)
3. Zikopoulos, B., Barbas, H.: Prefrontal projections to the thalamic reticular nucleus form a unique circuit for attentional mechanisms. J. Neurosci. **26**(28), 7348–7361 (2006)
4. Mathews, Z., Cetnarski, R., Verschure, P.F.: Visual anticipation biases conscious decision making but not bottom-up visual processing. Front. Psychol. **5** (2014)
5. Miller, G.A.: The magical number seven, plus or minus two: some limits on ourcapacity for processing information. Psychol. Rev. **63**(2), 81 (1956)
6. Urbanski, M., Coubard, O.A., Bourlon, C.: Visualizing the blind brain: brain imaging of visual field defects from early recovery to rehabilitation techniques. Neurovision: Neural bases of binocular vision and coordination and their implications in visual training programs (2014)
7. Büchel, C., Friston, K.: Modulation of connectivity in visual pathways by attention: cortical interactions evaluated with structural equation modelling and fMRI. Cereb. Cortex **7**(8), 768–778 (1997)
8. Dagnino, B., Gariel-Mathis, M.-A., Roelfsema, P.R.: Microstimulation of area V4 has little effect on spatial attention and on the perception of phosphenes evoked in area V1. J. Neurophysiol. **113**(3), 730–739 (2015)
9. Kirchner, H., Barbeau, E.J., Thorpe, S.J., Régis, J., Liégeois-Chauvel, C.: Ultrarapid sensory responses in the human frontal eye field region. J. Neurosci. **29**(23), 7599–7606 (2009)
10. Sparks, D.L.: Translation of sensory signals into commands for control of saccadic eye movements: role of primate superior colliculus. Physiol. Rev. **66**(1), 118–171 (1986)

Mutual Entrainment of Cardiac-Oscillators Through Mechanical Interaction

Koki Maekawa[1]([✉]), Naoki Inoue[1], Masahiro Shimizu[1], Yoshihiro Isobe[2],
Taro Saku[2], and Koh Hosoda[1]

[1] Graduate School of Engineering Science, Osaka University,
1-3 Machikaneyama-machi, Toyonaka, Osaka 560-8531, Japan
maekawa.koki@arl.sys.es.osaka-u.ac.jp
[2] Atree, Inc., 1-181 Sugi Machi, Yamatokoriyama City, Nara 639-1121, Japan

Abstract. This paper focus on bio-robots driven by cardiomyocyte-powered actuators. Towards biomaterial-based robotic actuators that exhibit co-operative motion, this study intends to deal with mutual entrainment between two cardiac-oscillators with mechanical interaction. The cardiac-oscillator is an oriented collagen sheet seeded cultured rat cardiomyocytes. Mechanically coupled cardiac-oscillator sysystem was developed consisting of two cardiac-oscillators connected mechanically via a PDMS film. Mechanically coupled cardiac-oscillator system and a single cardiac-oscillator was applied electric stimulation for investigating the function of mechanical structure. As a result of experiment, we observed mutual entrainment occurred in only mechanically coupled structure.

Keywords: Mutual entrainment · Cardiac-oscillator · Mechanical interaction

1   Introduction

Living things behave more adaptable on environmental changes than machines because living things are composed of biological devices that have significant abilities such as self-repair, self-assembly, and self-organization. Based on this circumstance, bio-robots consisting of biological materials are attracting a lot of attention [1].

Nawroth et al. [2] developed "medusoids" which imitates jellyfish and reproduces jellyfish propulsion. Akiyama et al. [3] developed iPAM, consisting of insect dorsal vessel tissue and a frame, which moved faster than other reported micro robots powered by mammalian cardiomyocytes. A great deal of effort had been made on developing bio-robot. What seems to be lacking, however, is driving method of bio-robot is limited to drive all actuators simultaneously. To implement co-operative motion such as driving actuators alternately, we have to consider a new locomotive method.

Then, we focused on mutual entrainment through mechanical interaction between cardiomyocytes. An important property of cardiomyocytes is cardiac

© Springer International Publishing Switzerland 2016
N.F. Lepora et al. (Eds.): Living Machines 2016, LNAI 9793, pp. 467–471, 2016.
DOI: 10.1007/978-3-319-42417-0_48

excitation-contraction coupling. Cardiac excitation-contraction coupling is the process from electrical excitation of the myocyte to heart [4]. The generation order is as follows: firstly, action potential occurs; secondly, Ca^{2+} flows; finally, cardiac contraction happen. Iribe et al. [5] confirmed that Ca^{2+} flows were prompted in cardiomyocytes by adopting mechanical stimulation. Therefore, there is a possibility of manipulating contraction rhythm of cardiomyocytes by constructing mechanical interaction.

This study describes analysis of mutual entrainment between two cardiomyocyte-powered actuators that have mechanical interaction. As the actuator, we fabricated a cardiac-oscillator that is an oriented collagen sheet contracted by cultured rat cardiomyocytes. We developed mechanically coupled cardiac-oscillator system consisting of two cardiac-oscillators connected via a PDMS film. Consequently, mutual entrainment was observed in mechanically coupled cardiac-oscillator system, whereas not observed in a single cardiac-oscillator.

2   Mechanically Coupled Cardiac-Oscillator System

As an actuator, a cardiac-oscillator was developed so that a collagen sheet contracted by cultured rat cardiomyocytes. The sheet was formed by aligning many collagen strings and dehydrating them [6]. As the characteristic of the sheet, it has orientation, therefore, a cardiac-oscillator always contract vertical to the oriented direction.

We developed mechanically coupled cardiac-oscillator system consisting of two cardiac-oscillators connected mechanically via PDMS film (see Fig. 1), the opposition was fixed on an acrylic mold. By continuing tissue culture, the cardiomyocytes on the collagen sheet were condensed and began contraction as a cardiac-oscillator. Generally, each cardiac-oscillator starts contracting in accordance with the endemic frequency of cardiomyocytes crowd. We expected, however, mutual entrainment occurs and co-operative contraction between two

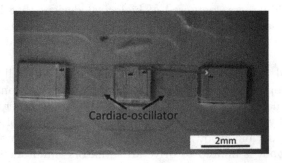

Fig. 1. Mechanically coupled cardiac-oscillator system consisting of acrylic resin, PDMS, and two cardiac-oscillators. We measure the displacement of red and blue lines. Then Contraction interval was calculated. (Color figure online)

cardiac-oscillators are voluntarily accrued because both cardiac-oscillators pull a PDMS film from both ends.

3   Verification of Mutual Entrainment

Electric stimulation was applied to mechanically coupled cardiac-oscillator system for investigating the effect of mutual entrainment. Same stimulation was applied to a single cardiac-oscillator for control experiment (see Fig. 2). The experimental results were expected that the mechanically coupled oscillator system is more stable against disturbance than a single cardiac-oscillator because mechanical coupling generates mutual entrainment and distinctive contraction rhythm, whereas a single cardiac-oscillator tends to be influenced by disturbance and changes the contraction rhythm owing to no mechanical coupling structure. Both mechanically coupled cardiac-oscillator system and the single cardiac-oscillator were paced at 0.75 Hz through an externally applied electric field of 5 V for 60 s.

The results are shown in the Figs. 3 and 4. The green line corresponds to electric stimulation's frequency in these figures. It seems that the single cardiac-oscillator contraction synchronized with the electric stimulation (see Fig. 3a), and after removing the stimulation, the single cardiac-oscillator restarted contraction with endemic frequency (see Fig. 3b). On the other hand, mechanically coupled cardiac-oscillator system was not attracted to electric stimulation (see Fig. 4a), and after removing the stimulation, the contraction behavior almost never changed (see Fig. 4b). From these results, we conformed that mechanically coupled cardiac-oscillator system had distinctive contraction rhythm. It is considered that the effect of mutual entrainment to contraction rhythm was stronger than electric stimulation, therefore, mechanically coupled cardiac-oscillator system was stable against disturbance.

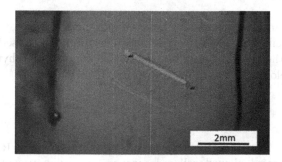

Fig. 2. Electric stimulation to a single cardiac-oscillator. The black line in picture is a silver wires for electric road.

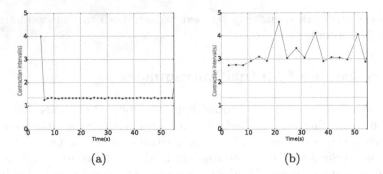

(a)   (b)

Fig. 3. The contraction interval of a single cardiac-oscillator, (a) applying electric stimulation and (b) after removing stimulation. (Color figure online)

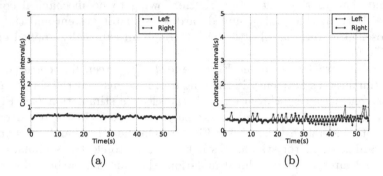

(a)   (b)

Fig. 4. The contraction interval of mechanically coupled cardiac-oscillator system, (a) Applying electric stimulation and (b) after removing stimulation. (Color figure online)

4   Conclusion

In this study, we made mechanically coupled cardiac-oscillator system consisting of two cardiac-oscillators and observed mutual entrainment caused by mechanical interaction. As the result, we found distinctive contraction rhythm. The next challenge is developing a bio-robot incorporated this system.

References

1. Webster, V.A., Hawley, E.L., Akkus, O., Chiel, H.J., Quinn, R.D.: Fabrication of electrocompacted aligned collagen morphs for cardiomyocyte powered living machines. In: Wilson, S.P., Verschure, P.F.M.J., Mura, A., Prescott, T.J. (eds.) Living Machines 2015. LNCS, vol. 9222, pp. 429–440. Springer, Heidelberg (2015)
2. Nawroth, J.C., Lee, H., Feinberg, A.W., Ripplinger, C.M., McCain, M.L., Grosberg, A., Dabiri, J.O., Perker, K.K.: A tissue-engineered jellyfish with bio-mimetic propulsion. Nat. Biotechnol. **30**, 792–797 (2012)

3. Akiyama, Y., Odaira, K., Sakiyama, K., Hoshino, T., Iwabuchi, K., Morishima, K.: Rapidly-moving insect muscle-powered microrobot and its chemical acceleration. Biomed. Microdevices **14**(6), 979–986 (2012)
4. Bers, D.M.: Cardiac excitation-contraction coupling. Nature **415**(6868), 198–205 (2002)
5. Iribe, G., Jin, H., Naruse, K.: Role of sarcolemmal BKCa channels in stretch-induced extrasystoles in isolated chick hearts. Circ. J. **75**(11), 2552–2558 (2011)
6. Isobe, Y., Kosaka, T., Kuwahara, G., Mikami, H., Saku, T., Kodama, S.: Oriented collagen scaffolds for tissue engineering. Materials **5**(3), 501–511 (2012)

"TEGOTAE"-Based Control of Bipedal Walking

Dai Owaki[1]([✉]), Shun-ya Horikiri[1], Jun Nishii[2], and Akio Ishiguro[1,3]

[1] Research Institute of Electrical Communication, Tohoku University,
2-1-1 Katahira, Aoba-ku, Sendai 980-8577, Japan
`owaki@riec.tohoku.ac.jp`
[2] Yamaguchi University, 1677-1 Yoshida, Yamaguchi 753-8512, Japan
[3] CREST, Japan Science and Technology Agency, 4-1-8 Honcho,
Kawaguchi 332-0012, Saitama, Japan
`http://www.cmplx.riec.tohoku.ac.jp`

Abstract. Despite the appealing concept of "central pattern generator" (CPG)-based control for bipedal walking, there is currently no systematic methodology for designing a CPG controller. To tackle this problem, we employ a unique approach: We attempt to design local controllers in the CPG model for bipedal walking based on the viewpoint of "TEGOTAE", which is a Japanese concept describing how well a perceived reaction matches an expectation. To this end, we introduce a TEGOTAE function that quantitatively measures TEGOTAE. Using this function, we can design decentralized controllers in a systematic manner. We designed a two-dimensional bipedal walking model using TEGOTAE functions and constructed simulations using the model to verify the validity of the proposed design scheme. We found that our model can stably walk on flat terrain.

Keywords: Bipedal walking · TEGOTAE · Plantar sensation · Central pattern generator (CPG)

1 Introduction

Humans and animals exhibit astoundingly adaptive and versatile locomotion given real-world constraints. To endow robots with similar capabilities, their bodies must have a significantly larger number of degrees of freedom than what they have at present. To successfully coordinate movement with many degrees of freedom in response to various circumstances, "autonomous decentralized control" plays a pivotal role and has therefore attracted considerable attention.

In fact, animals deftly coordinate the many degrees of freedom of their bodies using distributed neural networks called "central pattern generators" (CPGs), which are responsible for generating rhythmic movements, particularly locomotion [1,2]. Based on these biological findings, various studies have been conducted thus far to incorporate artificial CPGs into legged robots with the aim of generating highly adaptive locomotion [3–6]. However, there is currently no systematic methodology for designing a CPG controller; each individual CPG model has been designed on a completely ad hoc basis for a specific practical situation.

© Springer International Publishing Switzerland 2016
N.F. Lepora et al. (Eds.): Living Machines 2016, LNAI 9793, pp. 472–479, 2016.
DOI: 10.1007/978-3-319-42417-0_49

To tackle this issue, we herein introduce a unique approach: We attempt to design local controllers using the CPG model for bipedal walking based on the viewpoint of "TEGOTAE", which is a Japanese concept describing how well a perceived reaction, i.e., sensory information, matches an expectation, i.e., an intended motor command. To this end, we introduce the TEGOTAE function, which quantitatively measures TEGOTAE. This function can be described as the product of "what a local controller wants to do" and "its resulting reaction" on the basis of the concept of TEGOTAE. Thus, by only designing local sensory feedback such that each local controller increases "consistent" TEGOTAE in line with "expectation," or decreases "inconsistent" TEGOTAE otherwise, we can design decentralized controllers in a systematic manner. In this paper, we proposed a systematic design scheme of a decentralized CPG controller for bipedal walking based on TEGOTAE and verify the validity of the scheme through simulation.

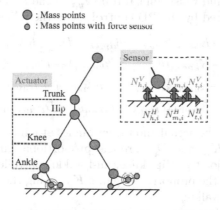

Fig. 1. Bipedal walking model.

2   Bipedal Walking Model

2.1   Musculoskeletal Structure

To validate the design scheme based on the TEGOTAE function, we conducted simulations using a two-dimensional bipedal walking model. Figure 1 shows the musculoskeletal structure of the bipedal walking model, whose movements are constrained in the sagittal plane for simplicity. We implemented seven actuators on the waist joint, two hip joints, two knee joints, and two ankle joints, at which each actuator generates torque based on proportional-derivative (PD) control, as explained in Sect. 2.2. Passive springs and dampers are implemented into the toe joints to passively generate an effective push-off force at the end of the stance phase. On the basis of findings in humans and animals, which have shown that cutaneous receptors in the foot play an essential role in the control of gait [7–9], we modeled plantar sensations that can detect vertical and horizontal ground reaction forces (GRFs) ($N_{x,i}^V$ and $N_{x,i}^H$) to the ground at heel ($x = h$), metatarsal ($x = m$), and toe ($x = t$) points on the feet. Here, the suffix i denotes each leg ($i = 0$: right and $i = 1$: left).

2.2 Systematic Design Scheme of a CPG Controller Based on the "TEGOTAE" Function

The proposed control system for bipedal walking consists of four components: (1) hip controllers, (2) knee controllers, (3) ankle controllers, and (4) a posture controller. The first three components coordinate the inter- and intra-limb movements via TEGOTAE functions for adaptive walking, and the forth component stabilizes the upper body using the waist actuator and vestibular sensor. Due to space limitations, the details of the posture control are not presented here.

Hip, knee, ankle, and waist joint torque $\tau_{y,i}$ in the ith leg (y indicates one of the joints) are generated by the PD control using the following equations:

$$\tau_{hip,i} = -K_{hip,i}(\theta_{hip,i} - \bar{\theta}_{hip,i}) - D_{hip,i}\dot{\theta}_{hip,i}, \tag{1}$$

$$\tau_{knee,i} = -K_{knee,i}(\theta_{knee,i} - \bar{\theta}_{knee,i}) - D_{knee,i}\dot{\theta}_{knee,i}, \tag{2}$$

$$\tau_{ankle,i} = -K_{ankle,i}(\theta_{ankle,i} - \bar{\theta}_{ankle,i}) - D_{ankle,i}\dot{\theta}_{ankle,i}, \tag{3}$$

$$\tau_{trunk} = -K_{trunk}(\theta_{trunk} - \bar{\theta}_{trunk}) - D_{trunk}\dot{\theta}_{trunk}, \tag{4}$$

where $\theta_{y,i}$ and $\bar{\theta}_{y,i}$ are the actual and target angles at joint y in the ith leg, respectively. Furthermore, $K_{y,i}$ and $D_{y,i}$ are the proportional and derivative gains of the PD controller at joint y. Hip, knee, and ankle controllers can modulate the target angles $\bar{\theta}_{y,i}$ and the proportional gains $K_{y,i}$ using the TEGOTAE function to generate adaptive walking.

The TEGOTAE function is a function formulated using the concept of TEGOTAE, which is described as the product of the (i) intended motor command of a controller $f(x)$, where x denotes a control variable, and (ii) resulting sensory information $g(S)$ obtained from sensor values S as follows:

$$T(x, S) = f(x)g(S), \tag{5}$$

where we design the TEGOTAE function such that it increases if we gain sensory information that is consistent with the intended motor command. Positive/negative values of the TEGOTAE function indicate consistency/inconsistency between the intended motor command and resulting sensory information.

Using TEGOTAE function $T(x, S)$, we can modulate the local control variable x as follows:

$$\dot{x} = h(x) + \frac{\partial T(x, S)}{\partial x}, \tag{6}$$

where the first term on the right denotes the intrinsic dynamics of the local controller, and the second term denotes the local sensory feedback for variable x

based on the TEGOTAE function. Using the sensory feedback described by the partial differential form of the TEGOTAE function, the controller modulates its control variable x such that it increases consistent TEGOTAE with expectation, or decreases inconsistent TEGOTAE otherwise. We now describe the local joint controllers at the hip, knee, and ankle joints designed using the TEGOTAE functions.

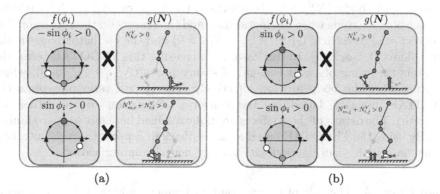

(a)   (b)

Fig. 2. Definition of the TEGOTAE function for the hip controller. (a) TEGOTAE function based on the corresponding leg's sensory information. (b) TEGOTAE function based on the other leg's sensory information.

Hip Control. The role of the hip joints in human waking is rhythmic motion generation for forward and backward leg swing [10]. To generate such rhythmic movements, we use phase oscillators to generate the target angle of the hip actuators (Eq. 1), which are described by the following equation:

$$\bar{\theta}_{hip,i} = -C_{1,hip} \cos \phi_i + C_{2,hip}, \tag{7}$$

where $C_{1,hip}$ and $C_{2,hip}$ [rad] denote the amplitude and offset angles of the hip target angle, respectively. According to the oscillator phases ϕ_i, legs are controlled to be in the swing phase for $0 \leq \phi_i < \pi$ and in the stance phase for $\pi \leq \phi_i < 2\pi$.

The dynamics of the phase oscillators with the local sensory feedback using the TEGOTAE function are as follows:

$$\dot{\phi}_i = \omega + \frac{\partial T_{hip,i}(\phi_i, \boldsymbol{N})}{\partial \phi_i}, \tag{8}$$

where ω [rad/s] denotes the intrinsic angular velocity of the oscillators. The TEGOTAE function for the hip control is defined as the following equation:

$$
\begin{aligned}
T_{hip,i}(\phi_i, \boldsymbol{N}) = \ & \sigma_{hip,1}\{N_{h,i}^V(-\sin\phi_i) + (N_{m,i}^V + N_{t,i}^V)(\sin\phi_i)\} \\
& + \sigma_{hip,2}\{N_{h,j}^V(\sin\phi_i) + (N_{m,j}^V + N_{t,j}^V)(-\sin\phi_i)\},
\end{aligned} \tag{9}
$$

where $\sigma_{hip,1}$ and $\sigma_{hip,2}$ [rad/Ns] denote the feedback gains. The suffix i and j denotes the corresponding and other legs, respectively. The first term on the right represents the TEGOTAE function based on the sensory information of the corresponding leg (Fig. 2(a)). The value of $N_{h,i}^V(-\sin\phi_i)$ becomes a positive value when the heel sensor on the corresponding leg detects a large vertical GRF ($N_{h,i}^V > 0$) and the oscillator phase is in the stance phase ($\pi < \phi_i < 2\pi$). By increasing this TEGOTAE term, the leg remains in the stance phase while supporting the body ($N_{h,i}^V > 0$). In contrast, the value of $(N_{m,i}^V + N_{t,i}^V)(\sin\phi_i)$ becomes positive value when the metatarsal and toe sensors on the corresponding leg detect a large vertical GRF ($N_{m,i}^V + N_{t,i}^V > 0$) and the oscillator phase is the swing phase ($0 < \phi_i < \pi$). In this case, by increasing this TEGOTAE term, the leg enters the swing phase at the end of stance phase ($N_{m,i}^V + N_{t,i}^V > 0$), which in turn pushes the body forward effectively. The second term represents the TEGOTAE function based on the sensory information of the other leg (Fig. 2(b)). The details of these effects are not explained here due to space limitation. By using the TEGOTAE-based local feedback in Eq. (8), the hip controllers enable "interlimb" coordination without any neural communication.

Knee Control. The role of a knee joint in human walking [10] is to support the body by increasing its stiffness in the stance phase and effective flexion by decreasing its stiffness in the swing phase. Thus, we define control variable χ_i, which denotes the control command that increases and decreases the stiffness of the knee joints. To implement such a stiffness control mechanism, we modify gain P in the knee controllers using χ_i as follows:

$$\tau_{knee,i} = -K_{knee,i}\theta_{knee,i} - D_{knee}\dot{\theta}_{knee,i}, \tag{10}$$

$$K_{knee,i} = \max[C_{1,knee}\tanh\chi_i, 0] + C_{2,knee}, \tag{11}$$

where $C_{1,knee}$ and $C_{2,knee}$ [Nm/rad] denote the variable range and offset value of gain P, respectively. In Eq. (10), the target angle $\bar{\theta}_{knee}$ of the knee controllers are set to 0 [rad], which indicates the state of the knee extension, allowing high/low stiffness to extend/flex the knee joints.

The dynamics of control variable χ_i with the local sensory feedback using the TEGOTAE function is as follows:

$$\dot{\chi}_i = -c_{knee,i}\chi_i + \frac{\partial T_{knee,i}(\chi_i, \boldsymbol{N})}{\partial \chi_i}, \tag{12}$$

where $c_{knee,i}$ denotes the time constant of the first order lag. The TEGOTAE function on the knee control is defined by the following equation:

$$T_{knee,i}(\chi_i, \boldsymbol{N}) = \sigma_{knee,1}N_i^V\chi_i + \sigma_{knee,2}N_j^V(-\chi_i), \tag{13}$$

where N_i^V and N_j^V [N] denote the sum of the vertical force sensor values on the heel, metatarsal, and toe, describing by, e.g., $N_i^V = N_{h,i}^V + N_{m,i}^V + N_{t,i}^V$, of the corresponding and other legs, respectively. Parameters $\sigma_{knee,1}$ and $\sigma_{knee,2}$ [1/N] denote the feedback gains. The first term on the right represents the TEGOTAE function based on the sensory information of the corresponding leg (Fig. 3(a) top). The value of $N_i^V \chi_i$ becomes a positive value when the foot sensors on the corresponding leg detect a large vertical GRF ($N_i^V > 0$) and the control command for the knee is increasing the stiffness ($\chi_i > 0$). Hence, by increasing this TEGOTAE term, the knee stiffness remains high while supporting the body ($N_i^V > 0$). The second term represents the TEGOTAE function based on the sensory information of the other leg (Fig. 3(a) bottom). Due to space limitation, the details of the effect is not explained here.

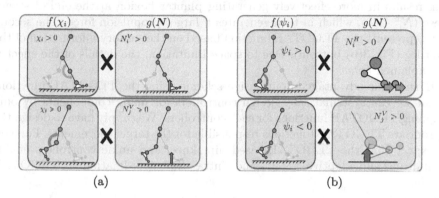

Fig. 3. Definition of the TEGOTAE function for the (a) knee and (b) ankle controller. Top: TEGOTAE function based on the corresponding leg's sensory information. Bottom: TEGOTAE function based on the other leg's sensory information.

Ankle Control. The role of an ankle joint in human walking [10] is to produce the push-off to generate the propulsion forces near the end of the stance phase and avoid colliding the foot with the ground during the swing phase. Thus, we define control variable ψ_i, which denotes the control command that increases or decreases the target angle of the ankle joints. We modify the target angle of the ankle controllers using ψ_i as follows:

$$\bar{\theta}_{ankle,i} = C_{1,ankle} \tanh \psi_i + C_{2,ankle}, \tag{14}$$

where $C_{1,ankle}$ and $C_{2,ankle}$ [rad] denote the variable range and offset value of the ankle target angle, respectively. The positive/negative value of ψ_i represents the plantar/dorsal flexion of an ankle joint.

The dynamics of control variable ψ_i with the local sensory feedback using the TEGOTAE function is as follows:

$$\dot{\psi}_i = -c_{ankle,i}\psi_i + \frac{\partial T_{ankle,i}(\psi_i, \boldsymbol{N})}{\partial \psi_i}, \tag{15}$$

where $c_{ankle,i}$ denotes the time constant of the first order lag. The TEGOTAE function on the ankle control is defined as follows:

$$T_{ankle,i}(\psi_i, \boldsymbol{N}) = \sigma_{ankle,1} N_i^H \psi_i + \sigma_{ankle,2} N_j^V (-\psi_i), \tag{16}$$

where N_i^H [N] denotes the sum of the horizontal force sensor values on the heel, metatarsal, and toe of the corresponding leg, described by $N_i^H = N_{h,i}^H + N_{m,i}^H + N_{t,i}^H$. In addition, N_j^V [N] denotes the sum of the vertical force sensor values of the other leg, described by $N_j^V = N_{h,j}^V + N_{m,j}^V + N_{t,j}^V$. Parameters $\sigma_{ankle,1}$ and $\sigma_{ankle,2}$ [1/N] denote the feedback gains. The first term on the right represents the TEGOTAE function based on the sensory information of the corresponding leg (Fig. 3(b) top). The value of $N_i^H \psi_i$ becomes a positive value when the foot sensors on the corresponding leg detects a large horizontal GRF ($N_i^H > 0$) and the command for the ankle is plantar flexion ($\psi_i > 0$). Increasing this TEGOTAE term results in more effectively generating plantar flexion at the end of stance phase ($N_i^H > 0$), which in turn generates a larger propulsion force. The second term represents the TEGOTAE function based on the sensory information of the other leg (Fig. 3(b) bottom). Due to space limitation, the details of the effect is not explained here.

In sum, the advantage of our design scheme using the TEGOTAE functions is that we can systematically design controllers for many components by only designing TEGOTAE functions for each controllers; We simply have to design the appropriate TEGOTAE functions responsible for the target movements. Further, we expect that the TEGOTAE-based hip, knee, and ankle controllers enable spontaneous and adaptive "inter"- and "intra"-limb coordination via TEGOTAE functions.

3   Simulation Result

Here, we describe the verification of our proposed design scheme using numerical simulation. Figure 4 shows a stick diagram plot (a) and time series data (b) (both hip angles, left knee and ankle angles, and stance phases of both legs) of steady walking motion. As shown in this figure, we achieved steady walking motion by designing each joint controller based on the TEGOTAE functions. Note that the time series data of the simulation were similar to human data of walking [10].

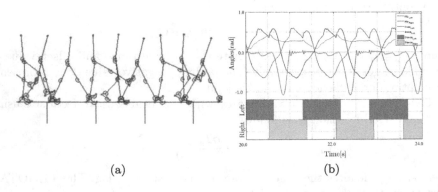

(a)   (b)

Fig. 4. Stick diagram (a) and time series data (b) of steady walking obtained over five periods. (Color figure online)

4 Conclusion and Future Work

The purpose of this study was to verify the validity of the proposed design scheme based on the TEGOTAE concept for bipedal walking. To this end, we constructed a bipedal walking model and applied our scheme to design joint controllers. We confirmed that the joint controllers designed using the TEGOTAE functions achieved stable bipedal walking on flat ground via spontaneous inter- and intra-limb coordination. The advantages of the proposed method over previous works [3,5] and the adaptability to environmental changes, which we did not verify, will be studied in future.

Acknowledgements. We acknowledge the support of a JSPS KAKENIII Grant-in-Aid for Young Scientists (A) (25709033) and Grant-in-Aid for Scientific Research on Innovative Areas "Understanding brain plasticity on body representations to promote their adaptive functions" (26120008).

References

1. Shik, M.L., Severin, F.V., Orlovskii, G.N.: Control of walking and running by means of electrical stimulation of the mesencephalon. Electroencephalogr. Clin. Neurophysiol. **26**, 549 (1969)
2. Grillner, S.: Neurobiological bases of rhythmic motor acts in vertebrates. Science **228**, 143–149 (1985)
3. Taga, G., Yamaguchi, Y., Shimizu, H.: Self-organized control of bipedal locomotion by neural oscillators. Biol. Cybern. **65**, 147–159 (1991)
4. Kimura, H., Akiyama, S., Sakurama, K.: Realization of dynamic walking and running of the quadruped using neural oscillator. Auton. Robots **7**, 247–258 (1999)
5. Aoi, S., Tsuchiya, K.: Locomotion control of a biped robot using nonlinear oscillators. Auton. Robots **19**, 219–232 (2005)
6. Righetti, L., Ijspeert, A.J.: Pattern generators with sensory feedback for the control of quadruped locomotion. In: Proceedings of ICRA 2008, pp. 819–824 (2008)
7. Duysens, J., Clarac, F., Cruse, H.: Load-regulating mechanisms in gait and posture: comparative aspects. Physiol. Rev. **80**, 83–133 (2000)
8. Dietz, V., Duysens, J.: Significance of load receptor input during locomotion: a review. Gait Posture **11**, 102–110 (2000)
9. Elis, E., Behrens, S., Mers, O., Thorwesten, L., Völker, K., Rosenbaum, D.: Reduced plantar sensation causes a cautious walking pattern. Gait Posture **20**, 54–60 (2004)
10. Perry, J., Burnfield, J.: Gait Analysis: Normal and Pathological Function, 2nd edn. Slack Inc., Thorofare (2010)

Tactile Vision – Merging of Senses

Nedyalka Panova[✉], Alexander C. Thompson[✉], Francisco Tenopala-Carmona[✉], and Ifor D.W. Samuel[✉]

SUPA, School of Physics and Astronomy, Organic Semiconductor Centre, University of St. Andrews, North Haugh, St. Andrews KY16 9SS, UK
{np44,at226,ftc2,idws}@st-andrews.ac.uk

Abstract. We developed a tactile vision sensor, by converting light intensity from narrowband reflection spectroscopy into different stimuli such as sound. This creates the perception of colours though alternative senses. The device extends work done to develop a wearable sensor of erythema and shows the outcome from the collaboration between scientists and artist. It discovers new applications for existing research combining technology and design.

1 Introduction

The absorption of light is an important process in nature and is often linked to transfer of energy. Sunlight is a thermal source and contains different wavelengths corresponding to different colours. Different materials can then selectively absorb specific wavelengths based on their molecular structure. In addition microstructure is sometimes used to give structural colours.

The carbon fixed by plants is used by mammals for nutrition and building body structures when on the other hand our vision relies on protein in the retina to enable us to interact with the environment. Additional scientific research on this topic results in advanced applications in medicine, sustainable energy, and sensing devices.

From an artistic perspective a combination of natural and man-made light absorbing materials and devices opens new territories for creative entanglement between art and science. The perception of the world around us and our interaction with it approaches limits of knowledge and representations. The new level of interaction that technology can offer for new senses and experience creates a dialogue about physiological aspect of what is to be a human.

For the 5th Conference on Living Machines we are presenting a specially engineered wearable device that is able to detect all visible colours. The development stage of the device is inspired by the same mechanism the human retina uses to detect colours and a device to measure skin redness (erythema) currently being investigated and studied in the Organic Semiconductor Centre, at the University of St. Andrews.

© Springer International Publishing Switzerland 2016
N.F. Lepora et al. (Eds.): Living Machines 2016, LNAI 9793, pp. 480–484, 2016.
DOI: 10.1007/978-3-319-42417-0_50

2   Erythema Measurement

Erythema, or skin redness, has a wide range of causes, such as UV irradiation, exposure to heat, intake of phototoxic drugs, allergens [1]. A current stream of research in the group is to develop a wearable monitor of erythema.

Existing devices to measure erythema are large benchtop devices, limiting the amount of temporal information that can be measured [2, 3]. However, a wearable device will allow continuous measurement and increase the temporal resolution of measurements. This improvement will provide a better understanding of erythema and the underlying causes of erythema from different triggers.

A range of techniques exist to measure erythema, including narrowband reflectance spectroscopy and analysis of RGB images [3, 4]. All rely on the difference in absorption between red and green light, as increased blood perfusion in tissue from erythema results in stronger absorption of green light (Fig. 1).

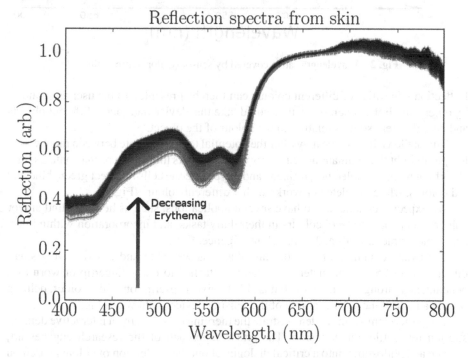

Fig. 1. Reflection spectra of skin with erythema, showing decreasing erythema (red to black). (Color figure online)

3   Tactile Vision

The ability of the erythema sensor to touch objects such as skin and discriminate between colours inspired the tactile vision device. This took the existing sensor and adapted it to measure reflection from nine different LEDs covering the visible range of light (Fig. 2).

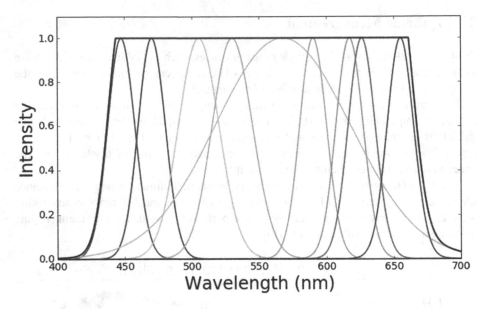

Fig. 2. Wavelength range covered by sensor (Color figure online)

Reflection intensities of different colours can then be presented to the user as sound or by other stimuli. For example, a user could take this device and touch different objects and hear different sounds relating to the colour of the object.

Our tactile vision device may offer the potential to discriminate between more wavelengths of light than humans are currently capable of. As the human retina contains light absorbing pigment molecules, retinene and 3 type of cone cells to detect green, blue and red colours, while our detector works with 9 different colours (Fig. 2).

We expect such a device to have several applications, such as helping blind and/or colourblind people to 'see' colours in their daily tasks, and incorporation within interactive electronic devices and artificial intelligence.

The main advantage of it is the fact that it is wearable and has relatively small dimensions of 25 mm diameter, so it can be attached to one's fingertip or worn as a pendant, and through a single touch it is able to give a description of the colour, helping people and machines. An example of the sensing device is shown in Fig. 3.

The poster/demo aims to demonstrate the merging of senses in an interactive demonstration for tactile colour vision. The artistic approach of the research engages art, science and philosophy into a critical dialogue about the perception of colours based on logic vs imagination. On one hand, the subjective vision of a gifted artist helps the public to see the world differently. On the other hand, if we strip the concept of colour from the human element and look at the primary colours as at primary numbers: the colours become a piece of data with no properties attached to them.

How could we describe colours to blind or colour blind people? Maybe they can learn the different smell of a green and yellow leaf, but can they smell reddish-green? They can learn different sounds of the ocean we associate with its colours, the calm blue, the windy grey and the crushing white. The difference in seeing and understanding

Fig. 3. Example of sensing device

colours is limited from our nature and nurture and as every concept is challenged by the experiment, the concept of colours is challenged by our sensory limitations.

"Science subjects the data of our experience to a form of analysis that we can never expect will be completed since there are no intrinsic limits to the process of observation" [5].

With the 'tactile vision' project we are looking at the perception of colours as at a closed system and the synthetic knowledge we achieve within is based not on uncertainty and appearance but on precise measurements inside the monitored area.

So we juxtapose the empirical measurements to the open end of colour variations associated with qualities such as cold, warm, dark and light and therefore, redesign the perception of the physical world to a digital one.

4   Conclusion

We are using the tactile vision workshop to present the biomimetic model of light absorbing mechanism studied within the Organic Semiconductor Centre (OSC) and especially for the development of a monitoring device that uses the light absorbing properties of blood to measure the redness of the patient's skin.

We see 5th Conference on Living Machines as a suitable platform to highlight the research of the OSC whose main focus is the use of organic (carbon based) semiconductors for solar cells, OLED displays, lasers and applications such as Li-Fi, sensing, medical applications (such as light-tissue interactions), optoelectronic design and engineering, computer simulations, and fluorescence imaging for Photodynamic Therapy (PDT) for skin lesions.

Collaboration between scientists and artists creates opportunities to further extend the possibilities and application of present research and to generate new tools for communicating science to larger audiences. The tactile vision device has been born out of discussion between an artist-in-residence and post-doctoral researchers, showing the benefits of merging science and art.

References

1. Wan, S., et al.: Quantitative evaluation of ultraviolet induced erythema. Photochem. Photobiol. **37**(6), 643–648 (1983)
2. Dolotov, L.E., et al.: Design and evaluation of a novel portable erythema-melanin-meter. Lasers Surg. Med. **34**, 127–135 (2004)
3. Diffey, B.L., Farr, P.M.: Quantitative aspects of ultraviolet erythema. Clin. Phys. Physiol. Meas. **12**(4), 311–325 (1991)
4. Stephen, S., et al.: Detection of skin erythema in darkly pigmented skin using multispectral images **22**(4), 172–179 (2009)
5. Merleau-Ponty, M.: The World of Perception, p. 44. Routledge, London (2004)

Tactile Exploration by Contour Following Using a Biomimetic Fingertip

Nicholas Pestell[1,2(✉)], Benjamin Ward-Cherrier[1,2], Luke Cramphorn[1,2], and Nathan F. Lepora[1,2]

[1] Department of Engineering Mathematics, University of Bristol, Bristol, UK
{np0877,bw14452,ll14468,n.lepora}@bristol.ac.uk
[2] Bristol Robotics Laboratory, University of Bristol, Bristol, UK

Abstract. Humans use a contour following exploratory procedure to estimate object shape by touch. Here we demonstrate autonomous robotic contour following with a biomimetic tactile fingertip, the Tac-Tip, using an active touch method previously developed for other types of touch sensors. We use Bayesian sequential analysis for perception and implement an active control strategy to follow an object contour. The technique is tested on a 110 mm diameter circle and yields results comparable with those previously achieved for other tactile sensors. We intend to extend the work onto a different robot platform with improved trajectory control to improve robustness, speed and match with human performance.

Keywords: Tactile sensors · Bayesian perception

1 Introduction

Touch is used extensively for various tasks including manipulation and exploration. Specifically, touch is implemented for characterising physical properties such as shape, texture, weight, hardness, temperature and size [1]. Lederman and Klatzky describe a set of distinct haptic exploratory procedures (EPs) which are employed separately depending on which physical property is being characterised. For example, lateral motion is associated with texture encoding, unsupported holding with encoding of weight, pressure with compliance, static contact with temperature, enclosure with volume and coarse shape and contour following with precise shape.

Shape encoding using robotic tactile perception for autonomous contour following has previously been achieved using the iCub fingertip [2]. Here we focus on achieving similar results using a novel biomimetic optical tactile sensor, the TacTip, developed at Bristol Robotics Laboratory [3,4]. The TacTip's novelty is informed by the structure of the human fingertip and the use of an optical system to process tactile information. We implement a perception-action cycle, where the TacTip perceives the edge or contour of an object and the subsequent action is a trajectory informed solely by the perception. The procedure is repeated so that the sensor traces a path around the object's edge.

© Springer International Publishing Switzerland 2016
N.F. Lepora et al. (Eds.): Living Machines 2016, LNAI 9793, pp. 485–489, 2016.
DOI: 10.1007/978-3-319-42417-0_51

2   Methods

2.1   Hardware

The TacTip is a 3D-printed, biomimetic optical tactile sensor with a soft deformable skin lined with a concentric array of pins on the inside layer. The pins are illuminated by a ring of LEDs and their positions are tracked by computer vision algorithms applied to the images from a CCD web-camera.

Novel features of the TacTip draw influence from the structure of the human fingertip. For example, dermal papillae are small pimple-like extensions of the dermis into the epidermis which form a strong interaction with tactile mechanoreceptors, enhancing receptor response to skin deformation. Papillae are analogues to the TacTip's pins which portray an amplified deformation response to skin-surface contact. Furthermore, the compliant gel separating the skin and the CCD camera has similar mechanical properties to the dermis and subcutaneous fat.

The TacTip is mounted as an end effector on a six degree-of-freedom robot arm (IRB 120, ABB Robotics). The arm can precisely and repeatedly position the sensor (absolute repeatability 0.01 mm) (Fig. 1).

Fig. 1. Hardware set up. The TacTip is mounted on an IRB 120 robot arm and makes successive taps on a test object.

2.2   Perception and Control

We implement a biomimetic approach to perception based on Bayesian sequential analysis for optimal decision making [5,6] which has been shown to achieve superresolved tactile sensing with the TacTip [7], where localization acuity was ∼ 0.1 mm as compared with a sensor resolution of 4 mm. Martinez-Hernandez *et al.*

successfully applies this Bayesian approach to the contour following problem with an iCub fingertip [2]. The iCub fingertip is a capacitive tactile sensor with 12 overlapping taxels (tactile elements) on a flexible printed circuit board. Here we adapt those techniques for use with the TacTip. Using Bayesian sequential analysis, the TacTip pin deflections are treated similarly to readings from the iCub taxels.

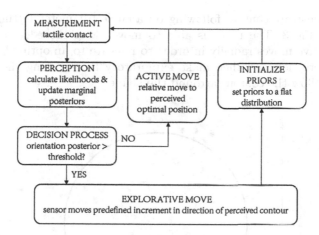

Fig. 2. Active contour following algorithm. Showing perception phase, decision process and active and exploratory moves.

Figure 2 depicts the contour following algorithm. Contact is made with an unknown object and likelihoods are estimated for each possible sensor orientation (angle) relative to the object edge. Bayes' rule is used, with an initially flat prior, after each successive contact to recursively update the posterior beliefs for each orientation hypotheses. The algorithm then evaluates a decision condition,

$$\text{any } P(c_n|z_{1:t}) > \theta_{dec} \tag{1}$$

where $P(c_n|z_{1:t})$ is the posterior for hypotheses c_n with evidence z accumulated after t taps and θ_{dec} is the decision threshold. If the output of the decision process is 'yes', the orientation is perceived as the class with the highest posterior. The robot then moves a predefined increment along a tangent to the perceived contour edge. If the output is no, the robot moves into an active process; it repositions the TacTip to the perceived most optimal position for orientation detection. The relative move is based on the marginal position posteriors. After relocation another tap is made, the posteriors are updated and the decision process (1) is re-implemented.

Martinez-Hernandez et al. [2] makes a distinction between active and passive perception for contour following. Active perception is when the sensor performs repositioning movements to more optimal locations for making perceptual decisions about its orientation. Alternatively, passive perception is when there is no repositioning and only the orientation is perceived. It was observed that active perception is far superior to the passive variant when perceiving edge orientation and is a key principle to achieve robust contour following [2]. Hence, in this work we only consider active contour following.

3   Results

Here we demonstrate contour following on a circle (diameter 110 mm). Results are shown in Fig. 3. The robot is able to move along the edge of the object by making active moves radially in order to relocate to an optimal position for orientation detection and then make exploratory moves along the tangent to successfully follow the contour arround the shape.

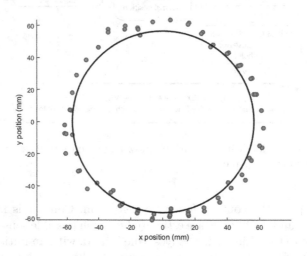

Fig. 3. Plot of sensor positions on one lap of 110 mm diameter circle. The actual shape edge is depicted as a solid black line, sensor positions at each tap are shown as smaller circles.

4   Conclusions and Future Work

Using the methods of Martinez-Hernandez et al. [2] we have demonstrated that the TacTip is an effective tool for identification of a previously unknown circular shape. The result achieved here suggests that the TacTip is comparable to the iCub fingertip for shape recognition, although an assessment on the same object shapes is needed for a more direct comparison.

It may be of benefit to introduce a short term memory factor into the control strategy which allows posteriors to influence priors between exploratory moves rather than initializing the priors to a flat distribution. This may improve both decision time and accuracy.

We intend to migrate the contour following tasks onto the UR5 robot arm. The UR5 is able to execute specified trajectories which, if informed by the object's edge shape, may improve speed and robustness of the tasks. In order to take advantage of this trajectory feature, the perception technique should be adapted so that posterior beliefs can be updated in a continuous fashion while the TacTip traces a smooth path around the object edge.

Acknowledgment. NP was supported by an EPSRC DTP studentship and NL was supported in part by an EPSRC grant on 'Tactile Superresolution Sensing' (EP/M02993X/1).

References

1. Ledermen, S., Klatzky, R.: Hand movements: a window into haptic object recognition. Cogn. Psychol. **19**, 342–368 (1987)
2. Martinez-Hernandez, U., Dodd, T., Natale, L., Metta, G., Prescott, T., Lepora, N.: Active contour following to explore object shape with robot touch. In: World Haptics Conference, pp. 341–346 (2013)
3. Chorley, C., Melhuish, C., Pipe, T., Rossiter, J.: Development of a tactile sensor based on biologically inspired edge encoding. In: International Conference on Advanced Robotics, pp. 1–6 (2009)
4. Assaf, T., Roke, C., Rossiter, J., Pipe, T., Melhuish, C.: Seeing by touch: evaluation of a soft biologically-inspired artificial fingertip in real-time active touch. Sensors **14**(2), 2561–2577 (2014)
5. Lepora, N.: Biomimetic active touch with tactile fingertips and whiskers. IEEE Trans. Haptics **9**(2), 170–183 (2016)
6. Lepora, N., Martinez-Hernandez, U., Evans, M., Natale, L., Metta, G., Prescott, T.: Tactile superresolution and biomimetic hyperacuity. IEEE Trans. Robot. **31**(3), 605–618 (2015)
7. Lepora, N., Ward-Cherrier, B.: Superresolution with an optical tactile sensor. In: IEEE/RSJ International Conference on Intelligent Robots and Systems (IROS), pp. 2686–2691 (2015)

Towards Self-controlled Robots Through Distributed Adaptive Control

Jordi-Ysard Puigbò[1](✉), Clément Moulin-Frier[1], and Paul F.M.J. Verschure[1,2]

[1] Laboratory of Synthetic, Perceptive, Emotive and Cognitive Science (SPECS),
DTIC, Universitat Pompeu Fabra (UPF), Barcelona, Spain
jordiysard.puigbo@upf.edu
[2] Catalan Research Institute and Advanced Studies (ICREA), Barcelona, Spain

Abstract. Robots, as well as machine learning algorithms, have proven to be, unlike human beings, very sensitive to errors and failure. Artificial intelligence and machine learning are nowadays the main source of algorithms that drive cognitive robotics research. The advances in the fields have been huge during the last year, beating expert-human performance in video games, an achievement that was unthinkable a few years ago. Still, performance has been assessed by external measures not necessarily fit to the problem to solve, what lead to shameful failure on some specific tasks. We propose that the way to achieve human-like *robustness* in performance is to consider the *self* of the agent as the real source of self-evaluated error. This offers a solution to acting when information or resources are scarce and learning speed is important. This paper details our extension of the cognitive architecture DAC to control embodied agents and robots, through self-generated signals, from needs, drives, self-generated value and goals.

1 Introduction

Since early history, human beings have dreamed of creating servants from inanimate mater. First examples can be found back to Jewish mythology with the concept of golem, a being magically animated from inanimate matter, like mud, or Greek mythology where the blacksmith god Hephaestus constructed several automatons to help on his workshop. Later on, technology allowed to construct automatons capable of simulating human behavior in an autonomous manner, but the level of autonomy was extremely limited to simple sets of actions, mechanically powered, and the capacity to adapt was very restricted, or absent. Robotics substituted automatons with expectations of greater autonomy and adaptability, together with big improvements on speed and precision. Still, the challenge remains on dealing with open, unconstrained and uncontrolled environments.

In parallel, also since the beginning of written history, evidence exist of the search for understanding the nature of human behavior, reasoning and other cognitive abilities. Back in ancient Greece, philosophers like Plato and Aristotle tried to understand which was the nature of human knowledge. The study of mind remained in the field of philosophy until the nineteenth century, with the

N.F. Lepora et al. (Eds.): Living Machines 2016, LNAI 9793, pp. 490–497, 2016.
DOI: 10.1007/978-3-319-42417-0_52

emergence of experimental psychology, until a few decades later, behaviorism dominated the field and practically denied the existence of mind. At that same time, the mid of 20th century, computer technology was enabling the study of Artificial Intelligence (AI), based on using mathematical reasoning to reproduce human cognitive abilities. Rapidly the field became dominated by the study of high level reasoning, applying to purely cognitive problems, detached from the physical world. It was then, around the 80's, that parallel advances on the fields of cognitive science, neuroscience and artificial intelligence, converged at highlighting the need to ground cognitive function on experience through embodied action. Theories of embodied cognition began to emerge at the hands of Rodney Brooks or Gerald Edelman.

Still, embodied cognition proposed the use of robots to understand cognitive function in a developmental perspective, but it was after two decades that the robotics community began to adopt developmental approaches to allow lifelong and open-ended learning of new skills and new knowledge on embodied agents. But still, the problem remains unsolved: embodiment seems a good paradigm, but embodied agents haven't outperformed non-embodied algorithms for the same tasks and still non-embodied artificial cognition have only outperformed human cognition in very constrained and abstract tasks, where average humans would fail, whereas can't compete with average human routine behavior.

We highlight the need of a central pillar of cognition that is usually ignored by computational approaches: the *self*. Whereas in general, sensory and motor modules are dissociated but communicating one with each other, we argue for the need of an active monitor or supervisor of the system that evaluates and selects the sensory and motor information that is used at every moment. Moreover, we ground this argument on the Distributed Adaptive Control cognitive architecture and theory of brain and body [11].

The following section will describe why subjectivity is an important feature of any cognitive system, tightly coupled to embodiment. Section 3 will introduce the theory of DAC and how it has been applied into our robotic cognitive architecture. Finally, Sect. 4 will detail the conclusions of previous discussion and propose future lines for improvement.

2   The Subjective Lens

All embodied agents are given sensors and actuators by default. This sensors and actuators suffer of physical constraints, what will define, in its essence, the robot capabilities. On top of this, in order to behave autonomously, an agent must have internal needs or drives that will define its general behavioral trends. We consider such needs as a characteristic of our robots and so, we can scale it to the next level and consider drives as the natural consequence of having a need, triggering a specific behavior due to a specific sensation. In this case, we must define sensations as predefined sensory processing of raw sensor data, and actions as predefined patterns of actuation. Later on, we will call this level of description *Reactive*.

We can imagine then, from an *Adaptive* level of description, perceptions as sensory information that is acquired through learning or adaptive processes; and behaviors as acquired and modulated action sequences. We argue that typical robotic approaches use the equivalent to reactive drives, without properly defined needs as the connection between adaptive processes, whereas the adaptive learning of drives and new sensory-motor associations *based on internal needs and motivations* is usually bypassed. We predict that this bypass is the main reason why state of the art robots are unable to robustly self-assess failure. Missing abstractions and experience based representations of an agent's own needs limits significantly the capacity to identify errors not foreseen by the designer.

Still, reinforcement learning has provided a framework capable of dealing with this problem. On the seminal work presented in [6], reinforcement learning was used as an adaptive process to learn sensory-motor associations to obtain the maximum possible score in a video-game. The drawback on their approach was twofold. First, a reactive layer of behavior was limited to random actions, whereas the appropriate reactive loops tend to speed up learning. Second, the internal needs of the agent were not carefully designed. The source of value in the algorithm was the scored obtained during game play and the mechanism to predict this reward. Choosing an inappropriate value measurement leads to undesired behavior and exploitation of the available rewards. The combination of both leaded to very low performances in games that had scarcity of reward and thus also required better exploration (reactive) mechanisms.

One of the arguments provided to justify the failures in some of the games, was the absence of a memory system. This brings us to a third level of description: *Contextual*. We can consider memory as a source of contextual information that is recalled under specific sensory, motor or internal conditions. In this sense, memory is the contextual version of sensory information, the motor equivalent could be understood as action plans and the subjective counterpart, the goals. We propose that, as resources become scarce and dangers become abundant, adaptive behavior is insufficient to ensure good performance. This can then be solved by the above mentioned contextual features: memory, planning and goals. Just one of them might be sufficient only in a limited set of tasks. In this sense, goals play again an important role in relating memories to form plans, allowing the acquisition of relevant memories and consequences from experience. As greater becomes the number of neurons in pallial regions of the brain, evolution has facilitated longer periods of guided experience after birth. We consider guided experience as the parental or social tutoring and the care off the offspring of an animal species. In these cases, guided experience reduces the ratio of danger and reward and facilitates learning through adaptive mechanisms. Nonetheless, contextual mechanisms serve the purpose of identifying, selecting, storing and recalling relevant information, where a goal oriented system would be in charge of regulating these processes.

3   From DAC to WR-DAC

3.1   Distributed Adaptive Control

We grounded this work on DAC theory of mind (Fig. 1), with the objective to construct a complete cognitive architecture that provides autonomy and cognitive abilities to nowadays robots. DAC is a theory of mind, brain and body nexus born in 1992, that has been gaining momentum in the past few years. DAC proposed the separation of cognitive processes in 2 dimensions. The first, separates cognitive processes in 3 columns: World, Self and Action.

World refers to all the sensory inputs that can be extracted from the world, from sensors themselves to stored memories.

Action is understood then as the column composed by all effecting mechanisms, from the simple actuators to the more abstract action plans.

Self is interpreted as all internal mechanisms, from needs, to goals through value assessment and attentional processes.

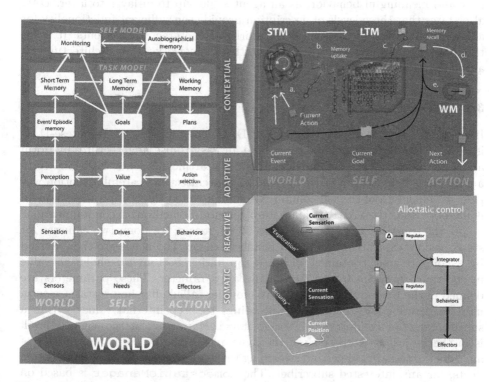

Fig. 1. Distributed Adaptive Control theoretical framework. On the left, the theoretical description of DAC is presented as described in the text [11]. On the right, the implementation of of the allostatic controller [10] (bottom) and the memory formation [4] (top). This figure was adapted from [11]

The second dimension describes cognitive elements in a layered, compositional structure.

Somatic: The somatic layer is formed by all sensors, actuators and internal needs given to the robot. This layer can be understood as the physical instance of the robot body.

Reactive: The reactive layer is then the first predefined loop for generating behavior that uses minimal sensory processing to trigger simple actions driven by the internal needs.

Adaptive: The adaptive layer provides the substrate for learning at the three columns. With this, an agent with an adaptive layer is able to change its behavioral loops according to environmental changes, acquiring new abstractions from sensory data and generating precise and adapted actions anticipating value functions.

Contextual: The contextual layer provides the substrate for memory, reasoning, planning and high level cognitive function.

DAC provides a framework that allows bottom-up generation of more complex and meaningful behavior, as an agent scales up from layer to layer. DAC also states that these levels of description provide the sufficient functionality for an agent to effectively survive in the world and interact with agents. It is for this reason that we chose to implement this framework as an architecture for autonomous robot self-control.

3.2   R-DAC

In contrast to other implementations of DAC along its history, Robotese-DAC is devised as the first attempt to provide an easy-to-use architecture, platform and task independent that provides robots or artificial agents with long term autonomous behavior. In order to build such architecture, careful attention was paid to the design of the self column. We implemented the first prototype on an iCub robot using the YARP middleware [5] for module facilitating communication.

At the somatic layer, the robot has 32 degrees of freedom, two cameras for stereo vision, capacitive skin and a Kinnect sensor. The needs are then defined by a comfort zone contained within a top and a bottom thresholds, a value in the range between 0 and 1 and a decay parameter. These parameters are defined in the `homeostasisManager` module, called after homeostasis, a term used to describe the tendency of life-beings to keep internal values within a specific range or comfort zone. `homeostasisManager`, at the somatic level, is in charge just to keep track of the dynamics of the needs, independently of any other external module, and streaming their information into communication ports, for any interested subscriber. The `homeostaticManager` is based on previous work in [10].

At the reactive layer, an `allostaticController` has been built that connects pre-processed inputs and pre-defined behaviors with the homeostatic manager. At this level, reactions are triggered automatically when the value of some homeostatic need goes over or below a specific threshold. We call this connection a *drive* drive to perform an action under specific sensory cues. Additionally, the allostatic controller provides additional functionality to define how sensory input will affect the dynamics of a specific need, which will still be computed in parallel in homesostaticManager.

As an adaptive approach, the allostaticController adds to the reactive function methods for self-regulation of multiple drives, in accordance with the name of the module. In this sense, active drives inhibit other drives from activating at the same time, while providing an interface for top-down regulation of drive dynamics. To become more adaptive, additional work is being done in the abstraction of new drives from old ones, what would lead to phenomenons like anticipation and modulation of the basic original ones.

Finally, the third loop is closed in the contextual layer with goal-oriented behavior. We propose with a first implementation to use goals as top-down modulators of the original drives. Top-down influence is achieved two-fold: from one side, a proper goal oriented module must be able to select what is the sensory

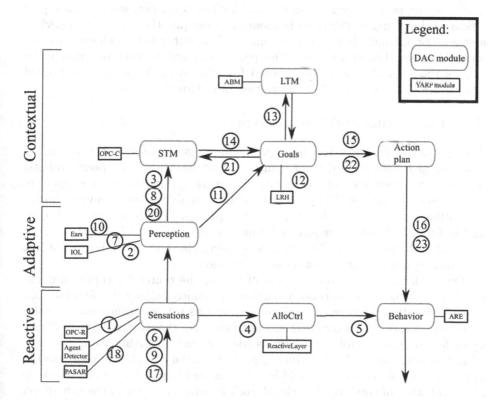

Fig. 2. Block description of WR-DAC. The DAC modules correspond to the boxes in the classical DAC diagram, with its columnar and layered distribution. These are usually wrappers for more complex sensory or motor modules or fully functional control modules dependent on inputs from sensations and behaviors.

information the agent must pay attention to; from the other side, it must be able to trigger behaviors and restrict other drives dynamics in order to gain priority over the other drives. The current implementation of the module includes just the last feature and already provides the necessary components to fulfill goals from past experiences stored in an AutoBiographical Memory (ABM) [7], while triggering actions from action plans generated by the system presented in [4], although the goal generation is still human driven.

In order to complete the architecture, the sensory and motor components must also be included. In this case, we created a wrapper that allows the inclusion of pre-processed sensations and pre-programmed behaviors as direct inputs and outputs to the `allostaticController`. The `sensatiuonsManager` is thought to contain modules that read information from sensory sources and publish both: contextual information for other behavioral modules, and driving information that directly modifies the drives dynamics. The `behaviorManager` receives triggering commands through a port, that can be sent through both `allostaticController` or `goalsManager` or any other module, while reading the relevant contextual information from the corresponding sensations port. The two wrappers are created on demand by the developer, whereas they provide a modular and configurable framework to use just the necessary behaviors and sensations from a large set, and connect them either to the self system or to any other general control module. This provides an architecture for autonomous control that can also easily be hijacked for Wizard of Oz or more controlled setups. See Fig. 2 for a detailed description of the architecture.

4   Conclusions and Future Work

A cognitive architecture has been proposed to generate robust autonomous behavior in robots and artificial agents. The architecture is a specific robotic oriented implementation of the DAC architecture, grounded on the sciences of biology and psychology. The proposed architecture differs from previous implementations by directly addressing the concept of self, self-generated behavior and self-interpretation of sensory information. It also differs from other cognitive architectures like SOAR or ACT-R by using also bottom-up instead of just top-down approaches to model cognitive function.

Our architecture proposes the use of self as the center for controlling and monitoring behavior, as opposed to engineering control algorithms for each conceivable situation. The current architecture exhibits robust, autonomous, continuous behavior, first demonstrated on a first version of the reactive layer in [3] and extended to pure control in [8]. Still, is the last that highlights the importance of adding adaptive mechanisms on triggering behaviors, as this, implemented in large scale set of behaviors, should lead to behavior scheduling in order to keep a general state of comfort. Additional work is being directed to the generation of new drives grounded on neuron dynamics on the limbic system, as well as anticipatory behaviors observed in conditioning experiments.

Special emphasis is being put in the contextual layer, that has the main framework ready, missing the mechanisms for the internal generation of goals, as now, can only be provided by a human being. The ability to use goals and idealized future states of the world has been shown to add robustness in presence of failure [9]. We are extending the architecture to have a working memory system that can be used it to compare desired and actual states of the world.

References

1. Anderson, J.R., Matessa, M., Lebiere, C.: ACT-R: a theory of higher level cognition and its relation to visual attention. Hum.-Comput. Interact. **12**(4), 439–462 (1997)
2. Laird, J.E., Newell, A., Rosenbloom, P.S.: Soar: an architecture for general intelligence. Artif. Intell. **33**(1), 1–64 (1987)
3. Lallee, S., et al.: Towards the synthetic self: making others perceive me as an other. Paladyn, J. Behav. Robot. **6**(1) (2015)
4. Marcos, E., Ringwald, M., Duff, Y., Sánchez-Fibla, M., Verschure, P.F.M.J.: The hierarchical accumulation of knowledge in the distributed adaptive control architecture. In: Baldassarre, G., Mirolli, M. (eds.) Computational and Robotic Models of the Hierarchical Organization of Behavior, pp. 213–234. Springer, Heidelberg (2013)
5. Metta, G., Fitzpatrick, P., Natale, L.: YARP: yet another robot platform. Int. J. Adv. Robot. Syst. **3**(1), 43–48 (2006)
6. Mnih, V., et al.: Human-level control through deep reinforcement learning. Nature **518**(7540), 529–533 (2015)
7. Petit, M., Fischer, T., Demiris, Y.: Lifelong augmentation of multi-modal streaming autobiographical memories. IEEE Trans. Cogn. Dev. Syst. **8**(3), 201–213 (2016)
8. Puigbò, J.-Y., Herreros, I., Moulin-Frier, C., Verschure, P.F.M.J.: Towards a two-phase model of sensor and motor learning. In: Wilson, S.P., Verschure, P.F.M.J., Mura, A., Prescott, T.J. (eds.) Living Machines 2015. LNCS, vol. 9222, pp. 453–460. Springer, Heidelberg (2015)
9. Puigbò, J.-Y., et al.: Using a cognitive architecture for general purpose service robot control. Connection Sci. **27**(2), 105–117 (2015)
10. Sanchez-Fibla, M., et al.: Allostatic control for robot behavior regulation: a comparative rodent-robot study. Adv. Complex Syst. **13**(03), 377–403 (2010)
11. Verschure, P.F., Pennartz, C.M., Pezzulo, G.: The why, what, where, when and how of goal-directed choice: neuronal and computational principles. Phil. Trans. R. Soc. B **369**(1655) (2014)

Discrimination-Based Perception for Robot Touch

Emma Roscow[1,4(✉)], Christopher Kent[2,4], Ute Leonards[2,4], and Nathan F. Lepora[3,4]

[1] School of Physiology, Pharmacology and Neuroscience, University of Bristol, Bristol, UK
emma.roscow@bristol.ac.uk
[2] School of Experimental Psychology, University of Bristol, Bristol, UK
[3] Department of Engineering Mathematics, University of Bristol, Bristol, UK
[4] Bristol Robotics Laboratory (BRL), University of Bristol, Bristol, UK

Abstract. Biomimetic tactile sensors often need a large amount of training to distinguish between a large number of different classes of stimuli. But when stimuli vary in one continuous property such as sharpness, it is possible to reduce training by using a discrimination approach rather than a classification approach. By presenting a biomimetic tactile sensing device, the TacTip, with a single exemplar of edge sharpness, the sensor was able to discriminate between unseen stimuli by comparing them to the trained exemplar. This technique reduces training time and may lead to more biologically relevant models of perceptual learning and discrimination.

Keywords: Robotics · Psychophysics · Tactile sensing · Biomimetics

1 Introduction

Almost all animals rely on tactile perception in order to successfully move around and interact with their environment: humans primarily use their hands, rats depend significantly on their whiskers, and cockroaches rely on their antennae. Many of these various tactile systems that exist in nature are now being used as inspiration in robot touch, with a range of tactile sensors available based on biologically-inspired principles [1].

But as demands are placed on these sensors to distinguish ever smaller differences in stimulus properties, they often require a very large, cumbersome amount of training to accurately discriminate between very similar classes. To make the training process more efficient, here we conjecture that when stimuli vary continuously over just one property (e.g. size or curvature or angle), the sensor can be trained on a single exemplar and subsequently a new previously unencountered stimulus can be compared with the exemplar. In tasks where there might be dozens of different stimuli, this new approach could reduce training time significantly. Moreover, this is likely to be a more biomimetic method of training, as humans can easily perceive object properties on continuous scales and make relative judgements about stimulus properties (e.g. bigger or smaller).

To test this conjecture we use the TacTip [2], a biologically-inspired optical tactile sensor designed to mimic responses to skin deformation in human fingertips, which has been shown to perform very well on a range of tactile perception and identification tasks [3–5]. Here we demonstrate that after training on just one exemplar of a stimulus

© Springer International Publishing Switzerland 2016
N.F. Lepora et al. (Eds.): Living Machines 2016, LNAI 9793, pp. 498–502, 2016.
DOI: 10.1007/978-3-319-42417-0_53

property (in this case, edge sharpness) the TacTip can generalise this property and distinguish between new, unseen stimuli that vary in sharpness with a level of sensitivity similar to human performance.

2   Methods

The TacTip is a biomimetic tactile sensor with a 3D-printed, hemispherical, compliant silicon outer membrane (the 'skin') with a diameter of 40 mm. 127 white-tipped pins ('papillae') cover the inside of the membrane in a hexagonal arrangement, such that when the TacTip comes into contact with an object the outer membrane is deformed and the pins are displaced (see Fig. 2). The membrane is fixed to a base with 6 LED lights which illuminate the white pins. A Microsoft Cinema HD webcam is mounted onto it which tracks the displacement of the pins, collecting tactile data in the form of images within the base (resolution 640 × 480 pixels, sampling rate 15 frames/sec), with computer vision techniques used to track the displacement of the pins [4].

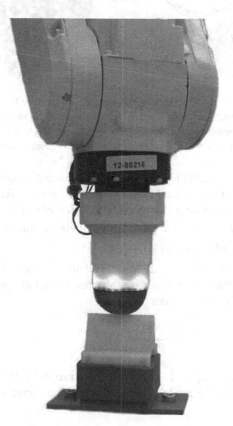

Fig. 1. The TacTip mounted on a robot arm and positioned above one of the stimuli. Data collection consisted of downward taps made onto the top of the stimuli. The apex angle of the stimuli varied from 35° to 85°.

We 3D-printed 11 triangular stimuli with apex angles ranging from 35° to 85° in 5° increments (see Fig. 1). The TacTip was mounted on a six-degree-of-freedom robot arm which allowed it to be brought into contact with each stimulus. Training consisted of a single tap moving 5 mm down onto the top edge of the sharpest stimulus (35°), and 5 mm back up. The displacement of each pin during the tap was measured in two dimensions to give 254 data dimensions. Log likelihoods were then obtained using a histogram method, whereby the sensor values for each data dimension were binned into 100 bins with equal intervals [6].

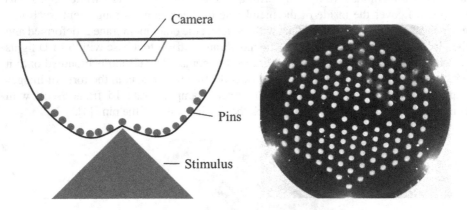

Fig. 2. An illustration of the displacement of the pins in the TacTip when in contact with a stimulus. Left: the flexible skin of the TacTip deforms around the stimulus, which displaces the pins. The camera is mounted directly above the pins to track this movement. Right: the displacement of the pins as captured by the camera, while in contact with a horizontal edge.

Testing consisted of trials in which the TacTip made a single tap sequentially onto a selected pair stimuli (any of the 11) and produced two log likelihoods, one for each stimulus: the probability of being identical to the exemplar (E) given the data from the first stimulus (S_1), $P(E|S_1)$, and the probability of being identical to the exemplar given the data from the second stimulus (S_2), $P(E|S_2)$.

The ratio of these two likelihoods was calculated such that a ratio less than 1 was taken to indicate that the first stimulus was sharper and a ratio greater than 1 indicated that the second stimulus was sharper (Eq. 1).

$$\frac{P(E|S_1)}{P(E|S_2)} \gtrless 1 \tag{1}$$

Every stimulus was compared against every other stimulus 10 times, giving 550 trials in total. After each trial, the model's judgement and the correctness of that judgement were recorded.

3 Results

On each trial, a pair of stimuli were tested and compared to the exemplar. A log likelihood ratio of their similarity to the exemplar was obtained for each pair, and whichever stimulus elicited the higher log likelihood was judged to be sharper. To assess the model's performance, we recorded the percentage of trials on which it gave the correct judgement when pairs of stimuli differed by different amounts, and fitted a logarithmic curve to the data, including the upper and lower 95 % confidence bounds (Fig. 3).

The just noticeable difference (JND) was calculated as the sharpness difference at which the TacTip made the correct judgement on 75 % of trials. (75 % is midpoint between the level of chance, 50 %, and perfect performance, 100 %, and is commonly used as the discrimination threshold in human psychophysics.) For the log curve, the JND was 9.2°, with the upper and lower bounds giving JNDs of 6.5° and 11.8° respectively. There was an apparent anomaly at a difference of 25° which fell below the curve of best fit, but we believe this to be merely an artefact due to noise associated with the experimental set-up.

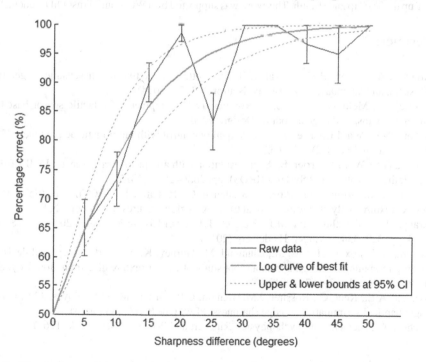

Fig. 3. Results from training on a single exemplar. The red line represents the (actual) percentage of trials on which the TacTip gave the correct judgement for each sharpness difference. Error bars represent standard error of the mean. The blue lines represent the log curves that best fit the data; the solid blue line is the single model that best fits and the dotted lines are the upper and lower 95 % confidence bounds. (Color figure online)

4   Discussion

The results showed that after a single tap on a single training exemplar, the TacTip was sensitive to differences in angle sharpness of 9.2°, which compares to a JND of 8.6° that has previously been found in humans performing the same task after years of tactile experience with edges [7]. To our knowledge this exemplar-generalisation method of training has not been employed before, and holds promise for two reasons. First, it can dramatically cut the amount of training required compared to current alternative techniques such as classification that require training on a large number of separate classes. Second, as generalisation is a fundamental part of human (and other animal) learning it can be used to develop more realistic models of animal perception. Humans are remarkably more sensitive to relative differences in sensory information than absolute values or classes [8], which suggests that learning through exemplar-generalisation may be able to produce more biologically realistic perceptual discrimination than other kinds of learning. We intend to explore this in the future by using this training technique to model human perceptual decision-making and confidence.

Acknowledgements.   We thank Benjamin Ward-Cherrier and Luke Cramphorn for their help setting up the TacTip and stimuli. This work was supported by a Wellcome Trust PhD studentship.

References

1. Grant, R.A., Itskov, P.M., Towal, R.B., Prescott, T.J.: Active touch sensing: finger tips, whiskers, and antennae. Front. Behav. Neurosci. **8**, 50 (2014)
2. Chorley, C., Melhuish, C., Pipe, T., Rossiter, J.: Development of a tactile sensor based on biologically inspired edge encoding. Design (2008)
3. Lepora, N.F., Ward-Cherrier, B.: Tactile quality control with biomimetic active touch. IEEE Robot. Autom. Lett. **1**(2), 646–652 (2016)
4. Lepora, N.F., Ward-Cherrier, B.: Superresolution with an optical tactile sensor. In: IEEE/RSJ on Intelligent Robots and Systems (IROS), pp. 2686–2691 (2015)
5. Winstone, B., Griffiths, G., Pipe, T., Melhuish, C., Rossiter, J.: TACTIP - tactile fingertip device, texture analysis through optical tracking of skin features. In: Lepora, N.F., Mura, A., Krapp, H.G., Verschure, P.F.M.J., Prescott, T.J. (eds.) Living Machines 2013. LNCS, vol. 8064, pp. 323–334. Springer, Heidelberg (2013)
6. Lepora, N.F., Fox, C., Evans, M., Diamond, M., Gurney, K., Prescott, T.: Optimal decision-making in mammals: Insights from a robot study of rodent texture discrimination. Roy. Soc. Interface **9**(72), 1517–1528 (2012)
7. Skinner, A.L., Kent, C., Rossiter, J.M., Benton, C.P., Groen, M.G.M., Noyes, J.M.: On the edge: haptic discrimination of edge sharpness. PLoS One **8**, e73283 (2013)
8. Fechner, G.: Elements of Psychophysics. Rinehart and Winston, New York (1966)

On Rock-and-Roll Effect of Quadruped Locomotion: From Mechanical and Control-Theoretical Viewpoints

Ryoichi Kuratani[✉], Masato Ishikawa, and Yasuhiro Sugimoto

Osaka University, 2-1, Yamadaoka, Suita, Osaka, Japan
kuratani@eom.mech.eng.osaka-u.ac.jp
http://www-eom.mech.eng.osaka-u.ac.jp/web/

Abstract. In this paper, we discuss body-induced motion of quadraped walking, in the presence of rolling contact between its feet and the ground. This paper is based on the authors' previous study, where we had shown that active *"rocking"* of the torso from the left to the right (in the lateral plane) would induce alternative swing of the legs (in the sagittal plane), which result in autonomous forward walking. Now in this paper, we focus on the influence of the rolling contact of the feet and the ground; we point out that the forwading displacement is the sum of *"walking effect"* and *"rolling effect"*, and the rolling effect can be observed even when each pair of the fore legs and the hind legs are bound together. Moreover, in order to extract pure rolling motion, we analyze in detail the model of two hemispheres connected via a spine rod, as an extremely reduced model of quadruped locomotion.

Keywords: Quasi-passive dynamic walking · Rolling locomotion · Non-holonomic system

1 Introduction

Body motion plays an impotant role in quadruped locomotion. Animals with relatively large-sized torso, such as mammals or reptiles, often exhibit quadruped walking associated with synchronized body oscillation.

In particular, the authors have been focusing on the *lateral rocking* of the body, namely, swaying motion of the body from the left to the right. The lateral rocking can be either *the cause* and *the result*; in the authors' previous work on passive-dynamic quadruped walking [4], we developed a quadruped mechanism without any actuator or controller Fig. 1(a), which automatically walks down the slope while the only source of the power was the gravity. The key observation here was that lateral rocking of the body is also automatically induced, with a certain phase gap between the foreleg and the hindleg parts, and the same frequency as that of the walking in the sagittal plane. In the succeeding work, to exchange the cause and the result, we developed a quasi-passive quadruped walking robot [3].

© Springer International Publishing Switzerland 2016
N.F. Lepora et al. (Eds.): Living Machines 2016, LNAI 9793, pp. 503–509, 2016.
DOI: 10.1007/978-3-319-42417-0_54

The robot has two active pendulums, one of which is equipped with the fore-part and the other is equipped with the rear part Fig. 1(b). By actuating the pendulums with an appropriate phase gap, we succeeded in exciting the lateral rocking, and resulting walking locomotion (on an even surface) as well. Through these works we showed that "walking induces the body rocking" and "the body rocking induces walking".

Now we notice that the *feet* play a crucial role in transforming the body motion to the leg motion, and vice versa. Indeed, the surface of the feet are *hemispherical* in the works above, i.e., curvature of the contact point of a foot is always constant. When the body "rocks", it starts to "roll" on the ground thanks to the hemispherical feet.

In this paper, we go into more detail of this rocking and rolling phenomena. First, we explain concept of two spheres vehicle model and derive the kinematic state equation of that in Sect. 2. In Sect. 3, we show periodic control input of Rock-and-Roll locomotion and simulation results.

(a) Overview of Jenkka I : [4]
passive quadruped walker

(b) Overview of duke: [3]
quasi-passive quadruped walker

Fig. 1. Passive and Quasi-passive dynamic walking robot

2   Geometric Setting

2.1   Orientation and Internal Shape

In this subsection, we describe two spheres vehicle model. The sole shape of Fig. 2 is a part of sphere which radius is r, the connecting parts between legs and twistable pole are designed to be same position with the center of each spheres, so this robot (Fig. 2) is regarded as two hemispheres connected by twistable pole (Fig. 3). The coordinate setting of proposal model is shown in Fig. 4.

The hemisphere means a half of sphere, when the hemisphere is between two planes, plane A includes the cross section of the hemisphere, plane B is parallel to plane A, the distance between A and B is r, then the contact point of the hemisphere and plane B is one point and we call it as bottom of the hemisphere in

Fig. 2. Widening distance between legs (Phase difference $\alpha = \pi$[rad])

this paper. In Fig. 4 point O is the bottom of the hemisphere, θ_i, ψ_i are angular variables which show how each hemispheres are inline from the condition in which point O is contact with plane, ϕ is also an angular variable which shows how connecting pole rotate around z-axis. And P is the point in which the sphere is contact with the plane.

The shape of robot feet looks like squares as seen from above, which corresponds to the left model of Fig. 4, it is so called coordinate setting of two wheels vehicle model in which the pole is connected to center of each wheels and input is only revolution speed of wheels. (x_1, y_1) and (x_2, y_2) are the center of each wheels, u_1, u_2 are inputs to revolution speed of each wheels. Wheels of the two wheels vehicle model are replaced with spheres in proposal model, so we call this model as two spheres vehicle model. Throughout this paper, we assume that spheres may roll over the floor, but never slip. All expressed in the inertial coordinate.

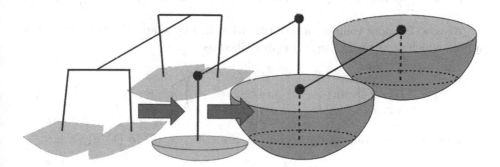

Fig. 3. Model reduction: from the quadruped walker (left) to the two-sphere vehicle model (right)

2.2   Kinematic State Equation of Two Spheres Vehicle Model

Now we derive a reduced kinematic state equation of two spheres vehicle model. First, we consider the kinematic equation of two wheels model and rolling contact condition. Then, we derive the kinematic equation of two spheres vehicle model.

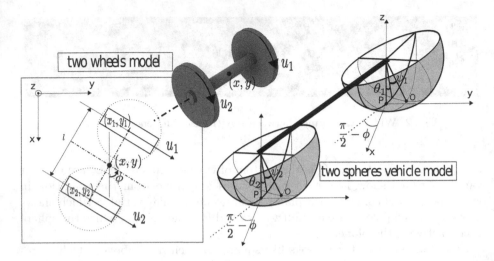

Fig. 4. Coordinate setting of the proposed model

From the coordinate setting of Fig. 4, we obtain the following kinematic equations

$$\dot{x} = \frac{u_1 + u_2}{2}\cos\phi, \quad \dot{y} = \frac{u_1 + u_2}{2}\sin\phi, \quad \dot{\phi} = \frac{u_2 - u_1}{l} \tag{1}$$

$$\dot{x}_1 = \dot{x} - \frac{l}{2}\dot{\phi}\cos\phi, \ \dot{y}_1 = \dot{y} - \frac{l}{2}\dot{\phi}\sin\phi, \ \dot{x}_2 = \dot{x} + \frac{l}{2}\dot{\phi}\cos\phi, \ \dot{y}_2 = \dot{y} + \frac{l}{2}\dot{\phi}\sin\phi \tag{2}$$

We also consider rolling contact condition in Fig. 4, and the velocity of the sphere which center is (x_1, y_1, z_1) is described as

$$\begin{pmatrix} \dot{x}_1 \\ \dot{y}_1 \\ \dot{z}_1 \end{pmatrix} = \begin{pmatrix} \cos\left(\frac{\pi}{2} - \phi\right) & -\sin\left(\frac{\pi}{2} - \phi\right) & 0 \\ \sin\left(\frac{\pi}{2} - \phi\right) & \cos\left(\frac{\pi}{2} - \phi\right) & 0 \\ 0 & 0 & 1 \end{pmatrix} \begin{pmatrix} r\dot{\theta}_1 \\ r\dot{\psi}_1\cos\theta_1 \\ 0 \end{pmatrix} \tag{3}$$

where r is the radius of the sphere [1]. Substituting (3) to (1) (2) we have

$$\begin{cases} u_1\cos\phi = r\dot{\theta}_1\sin\phi - r\dot{\psi}_1\cos\theta_1\cos\phi \\ u_1\sin\phi = r\dot{\theta}_1\cos\phi + r\dot{\psi}_1\cos\theta_1\sin\phi \end{cases} \tag{4}$$

From these equations, we obtain

$$\dot{\theta}_1 = \frac{S(2\phi)}{r}u_1 \tag{5}$$

where $S(x) = \sin x, C(x) = \cos(x)$. We derive following state equation of two spheres vehicle model from equations shown as above

$$
\begin{pmatrix} \dot{x} \\ \dot{y} \\ \dot{\phi} \\ \dot{\theta}_1 \\ \dot{\psi}_1 \\ \dot{\theta}_2 \\ \dot{\psi}_2 \end{pmatrix} = \begin{pmatrix} \frac{1}{2}C\phi \\ \frac{1}{2}S\phi \\ -\frac{1}{l} \\ \frac{S(2\phi)}{r} \\ \frac{1}{r} \\ 0 \\ 0 \end{pmatrix} u_1 + \begin{pmatrix} \frac{1}{2}C\phi \\ \frac{1}{2}S\phi \\ \frac{1}{l} \\ 0 \\ 0 \\ \frac{S(2\phi)}{r} \\ \frac{1}{r} \end{pmatrix} u_2 \rightarrow \dot{\boldsymbol{\xi}} = \boldsymbol{g}_1(\boldsymbol{\xi})u_1 + \boldsymbol{g}_2(\boldsymbol{\xi})u_2 \qquad (6)
$$

where $\xi = (x, y, \phi, \theta_1, \psi_1, \theta_2, \psi_2)^T$.

When we consider the input on holonomy principle, we can understand that the state transfers to the direction of Lie Bracket$[g_1, g_2]$ of g_1, g_2 when with closed curve input on the $U_1 - U_2$ (U_i is time integral of u_i) plane.

3 Control of Rock-and-Roll Locomotion and Simulation Results

The Lie Bracket$[g_1, g_2]$ of g_1, g_2 is calculated as

$$
\boldsymbol{g}_3 := [\boldsymbol{g}_1, \boldsymbol{g}_2](\boldsymbol{\xi}) = \frac{\partial \boldsymbol{g}_2}{\partial \boldsymbol{\xi}} \boldsymbol{g}_1 - \frac{\partial \boldsymbol{g}_1}{\partial \boldsymbol{\xi}} \boldsymbol{g}_2 = \left(\frac{S\phi}{l}, -\frac{C\phi}{l}, 0, -\frac{2C(2\phi)}{rl}, 0, -\frac{2C(2\phi)}{rl}, 0, \right)^T
$$
$$(7)$$

and we can understand ϕ does not change if we calculate higher order Lie Bracket. Then we assume ϕ is small enough, the $\boldsymbol{g}_3$ is

$$
\boldsymbol{g}_3 = \left(0, -\frac{C\phi}{l}, 0, -\frac{2C(2\phi)}{rl}, 0, -\frac{2C(2\phi)}{rl}, 0, \right)^T \qquad (8)
$$

We realize that states transfer in y, θ_1, θ_2 direction with the closed curve input on the $U_1 - U_2$ plane. One of the input is [2]

$$
U_1 = A\sin(\omega t) \qquad (9)
$$
$$
U_2 = A\sin(\omega t - \alpha) \qquad (10)
$$

As a result of simulation, the center point (x, y) on the $x - y$ plane show the following trajectory (Fig. 5(b)). Therefore, we verify that the input derived in the previous section is available.

In the same condition, we obtain following result (Fig. 6) from simulation, where t is simulation time[s], $\alpha = \pi$[rad] and other variables are same with Fig. 5. Figure 6 shows the same phenomenon confirmed in robot experience (Fig. 2), then it is conceivable that the trigger is phase difference in rocking motion corresponding to α.

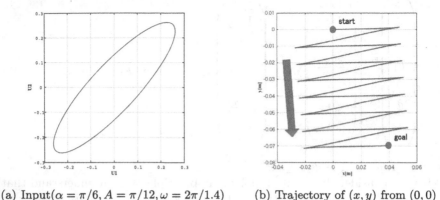

(a) Input($\alpha = \pi/6, A = \pi/12, \omega = 2\pi/1.4$)   (b) Trajectory of (x, y) from $(0, 0)$

Fig. 5. Simulation result of kinematic state equation under the periodic controls

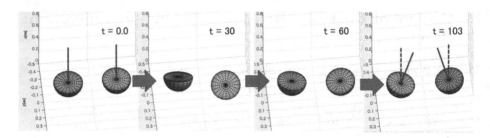

Fig. 6. Stance-widening phenomenon: the distance between the two hemispherical feet increases

4   Conclusion

This paper discussed influence of rocking of the body and rolling about the feet, motivated by quadruped walking. In particular, we proposed a new noholonomic mobile robot and built a new model to explain a phenomenon which could not be given enough explanation by the model proposed in our previous work. The simulation result of the model showed the phenomenon can appear at almost same condition with the real experiment, from which we found the trigger for the phenomenon is the phase difference between each spheres oscillation. In addition, judging from the direction of the vector-field g_3 the model will go down in forwarding motion. And this suggests, quadrupeds get their legs forward not to pull their body but to continue to stand.

In this paper, we have examined this rock-and-roll mechanism through numerical simulations. Future research would include quantitative analysis by comparison between numerical simulation and physical experiments.

References

1. Ishikawa, M., Kobayashi, Y., Kitayoshi, R., Sugie, T.: The surface walker: a hemispherical mobile robot with rolling contact constraints. In: IEEE/RSJ International Conference on Intelligent Robots and Systems, pp. 2446–2451 (2009)
2. Ishikawa, M., Minami, Y., Sugie, T.: Spherical rolling robot: a design and motion planning studies. In: IEEE International Conference on Robotics and Automation, RiverCentre, pp. 9–16 (2010)
3. Kibayashi, T., Sugimoto, Y., Ishikawa, M., Osuka, K., Sankai, Y.: Experiment and analysis of quadrupedal quasi-passive dynamic walking robot Duke. In: International Conference on Intelligent Robots and Systems, pp. 299–304 (2012)
4. Sugimoto, Y., Yoshioka, H., Osuka, K.: Realization and motion analysis of multi-legged passive dynamic walking. In: SICE Annual Conference, pp. 2790–2793 (2010)

Hydromast: A Bioinspired Flow Sensor with Accelerometers

Asko Ristolainen[1(✉)], Jeffrey Andrew Tuhtan[1], Alar Kuusik[2], and Maarja Kruusmaa[1]

[1] Centre for Biorobotics, Tallinn University of Technology,
Akadeemia tee 15a-111, 12618 Tallinn, Estonia
{asko.ristolainen,jeffrey.tuhtan,Maarja.kruusmaa}@ttu.ee
[2] Eliko Competence Centre, Mäealuse 2/1, 12618 Tallinn, Estonia
alar.kuusik@eliko.ee

Abstract. Fish have developed advanced hydrodynamic sensing capabilities using neuromasts, a series of collocated inertial sensors distributed over their body. We have developed the hydromast, an upscaled version of this sensing modality in order to facilitate near bed sensing for aquatic systems. Here we introduce the concept behind this bioinspired flow sensing device as well as the first results from laboratory investigations.

Keywords: Hydromast · Biomimetic · Flow sensing

1 Introduction

Fish experience the surrounding flow using their lateral line and inner-ear sense organs. The lateral line organs consist of linear arrays of neuromasts on the head and along the body, sensitive to changes in the spatial derivatives of the flow field. Neuromasts consist of gelatinous cupula whose deflections correspond to the local motions of the fluid, relative to the motion of the body. Superficial neuromasts located at the surface are sensitive to local changes in the velocity, whereas canal neuromasts, embedded within subcutaneous canals connected by pores to the surrounding flow, act primarily as acceleration detectors. The inner ear responds to changes in the speed of the fish, acting as linear and angular accelerometers, and is sensitive to the average bulk motion of the fish's body within the surrounding flow field [1].

Natural flow fields, such as those occurring in rivers and coastal waters can be described as a complex amalgamation of velocity, vorticity and pressure terms, acting over a wide range of spatiotemporal scales [3]. As such, it is not possible to directly record the complete physical evolution of the flow field with reference to these terms. Instead, our aim is to extract salient, hydrodynamically-relevant features such as the near-bed bulk flow velocity and bed shear stresses by utilizing the fluid-body interaction in natural flow and laboratory investigations. Measurement data can be analysed in both the time and frequency domains to provide a new source of physically-based hydrodynamic measurements. We have shown in previous works that bioinspired measurements based on a fish-shaped lateral line probe consisting of a collocated pressure sensing array is capable of calibrated bulk velocity estimation in a limited range (0–0.5) m/s, including

© Springer International Publishing Switzerland 2016
N.F. Lepora et al. (Eds.): Living Machines 2016, LNAI 9793, pp. 510–517, 2016.
DOI: 10.1007/978-3-319-42417-0_55

large (up to 90°) angular deviations of the sensor body with respect to the freestream flow [4, 5]. Here we introduce a new hydromast device, inspired by the single superficial neuromasts of fish's lateral line for turbulent, flume experiments. The hydromast can be viewed as a bioinspired upscaled inertial measurement device tailored for near-bed studies of complex in-situ flows which are often difficult or impossible obtain using conventional approaches such as acoustic Doppler current profiling (Fig. 1).

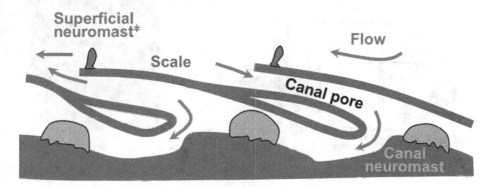

Fig. 1. The lateral line sensory apparatus of a fish, consisting of superficial neuromasts (green) and subdermal canal neuromasts (orange). Illustration after [2]. (Color figure online)

Fishes have the capability to sense local hydrodynamic stimuli close to their bodies using the lateral line modality. We show that an artificial inertial neuromast can be reconstructed and the base biological design can be exploited to measure flow characteristics with accelerometers.

2 Design Concept of the Artificial Neuromast Sensor with Accelerometers

The first approximation to create an upscaled superficial neuromast has been made with a stiff plastic stem of circular cross-section having a 10 mm diameter (see Fig. 2). The stem was fixed onto a silicone membrane (casted from Elite Double 22, Zhermack SpA), which acts similarly to the gelatinous cupula, as a spring element allowing the stem to pivot in the water flow. The stem was fixed trough the membrane with a bolt onto a clamp that held an accelerometer with 10×10 mm printed circuit board. The membrane was fixed between two surfaces in the casing (see Fig. 2).

To remove offsets caused by any slight misalignment of the sensor prototype in the flume, and to measure the stem tilt angles in the casing frame we installed a second stationary accelerometer to the casing close to the moving accelerometer (see Fig. 2). The difference between the base and stem angles from accelerometers thus provides an output signal largely free from any body self-motion due to mounting.

Accelerometers were interfaced with 32-bit ARM microcontroller (ST Microelectronics) over 400 kbps I2C bus. We use the earth gravity vector and arctangent function to calculate the sensor inclination in X, Y coordinates. It is assumed that sensor does

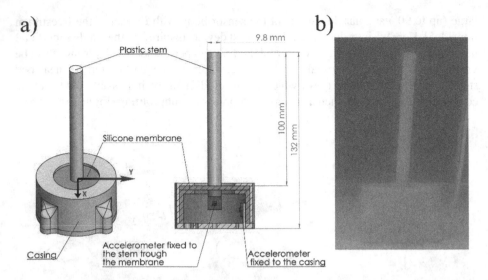

Fig. 2. (a) Schematic of the hydromast prototype and (b) image of the sensor in the flow tunnel

not rotate about the Z axis, and the random acceleration noise caused by water turbulences can be eliminated through the time averaging of the output signal. According to the experimental results more sophisticated sensor inclination measurement methods do not provide large benefits in accuracy over the simple method described. This is because the inertial and magnetic sensor fusion algorithms do not perform well when high frequency acceleration is present.

The accelerometers were connected to the ARM microcontroller using 4 thread thin ear phone wires to minimize disturbance due to wire movement and bending. The accelerometers were coated with Plasti Dip plastic (Plasti Dip International) to make them waterproof.

3   Design Evaluation and First Results

3.1   Experimental Setup

We tested the sensor in a flow tunnel with a working section of 0.5 m × 0.5 m × 1.5 m embedded into a test tank (see [6] for the test tank description). Uniform flow (constant water surface, no change in flowrate over time) in the working section was created with the help of a U-shaped flow straightener and two sequential collimators. An AC motor was used to create the circulation inside the flow tunnel, and the flow speed was calibrated using a digital particle image velocimetry system.

The sensor was fixed onto a metallic plate which was placed in the middle of the bottom in the flow tunnel. We ran experiments at 0.05 m/s speed increments, with maximum speed of 0.5 m/s. The flow was let to stabilize for 30 s and the data was logged for 30 s on each speed.

3.2   Calibration Results

The calibration results showed that the hydromast with the chosen parameters (stem height, diameter, membrane thickness and elasticity) has threshold shift in its behaviour above speeds of the 0.25 m/s. The root mean square (RMS) of the hydromast's stem pivot angles in the water flow (offsets were removed using the data recorded by the secondary accelerometer fixed to the casing from 10 experiments) are plotted in Fig. 3.

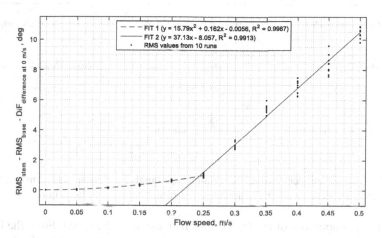

Fig. 3. RMS of the stem tilt angle along the flow direction with removed initial offsets and RMS of the base

At to velocities of up to 0.25 m/s, the hydromast's relation to flow speed was quadratic ($R^2 = 0.9987$) whereas from speeds 0.25 m/s to 0.5 m/s the hydromast responded linearly to increasing flow speed ($R^2 = 0.9913$). This shift is due to the transition to turbulence over the body, where at velocities > 0.25 m/s, larger vortices are suddenly shed from the stem surface, creating a new fluid-body interaction between the skin drag, vortex-induced drag, buoyancy and restoring force by the membrane creating a new fluid-body interaction. The dynamic balance of forces causes a lock-in of the oscillatory motion of the hydromast particular to each flow velocity. This dynamical loading results from the interplay of fluid drag and restoring forces driven by the hydromast's positive buoyancy in conjunction with the membrane stiffness.

The frequency spectra analysis with fast-Fourier-transform (FFT) from the concatenated signal of 10 experiments showed that distinguishable frequencies appear above flow speeds of 0.3 m/s. The experimental data from 10 runs was concatenated to analyse all of the runs as one single dataset. The results from the concatenated data compared with the single runs data showed no significant difference in the FFT spectrum. Before the FFT was performed the casing vibrations were removed from the stem signal by subtracting the casing accelerometer signal from the stem accelerometer signal. The signal was then filtered using median filter and linear trends were removed. The frequency spectra are plotted on Fig. 4.

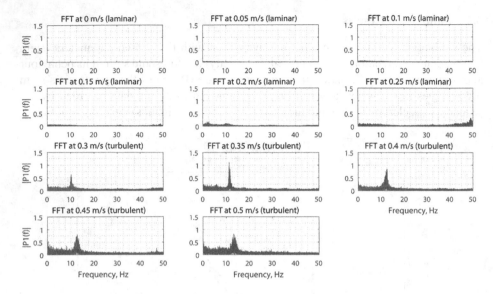

Fig. 4. FFT frequency spectra from the stem-casing signals for flow speeds from 0–0.5 m/s in 0.05 m/s intervals. Peaks arise due to the transition from laminar to turbulent flow.

The mean amplitudes of the frequencies increase with the speed. From the FFT we calculated frequency spectra mean amplitudes which are plotted against the flow speed in Fig. 5. The relation of the mean frequency amplitudes and flow speeds resulted in a quadratic relation with $R^2 = 0.9913$). This trend is due to the size and frequency of vortices shed from the cylinder body, with increasing velocity the vortices become larger and are shed at a proportionally higher rate increasing the force imbalance on the stem and creating larger magnitude fluctuations.

The two flow regimes of the hydromast can also be noted in the standard deviation (STD) values of the vibrations. The stem's vibration standard deviation along flow is increasing around 0.25 m/s with corresponding Reynolds number over 2000. The STD values are plotted of the mean stem angles in the flow are plotted in Fig. 6.

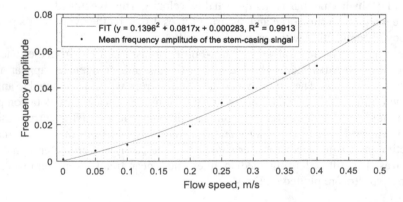

Fig. 5. Mean frequency amplitudes from the stem-casing signal on flow speeds 0–0.5 m/s.

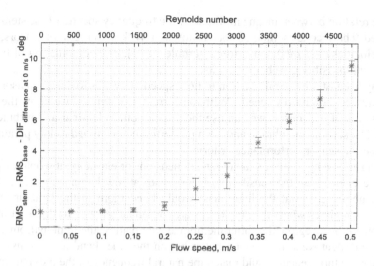

Fig. 6. RMS of the stem angle along the flow corresponding Reynolds numbers and standard deviations, where the highest STD are found during the unstable transition from laminar to turbulent flow (0.25 and 0.30 m/s).

4 Discussion

In contrast to previous work based on bioinspired mammalian tactile whiskers [7] whose sensing ability is based on active detection, relying on the purposeful motion of the sensor within the fluid, our device is based on measurements using passive hydromechanical detection. Passive sensing based on the lateral line sensing modality of fish has the added advantage that knowledge of the instantaneous displacement of the sensor body, relative to the flow, can be directly related to the frequency, amplitude and phase of the flow stimulus [8].

The change in the hydromast's behaviour can be related to the change from near-laminar flow to fully turbulent flow. With a 9.8 mm cylindrical stem the Reynolds numbers indicate that around speeds 0.20 to 0.30 m/s, the flow transition occurs from near-laminar to turbulent (with corresponding Reynolds numbers Re = 1952 to Re = 2928). At the Reynolds numbers between 2000 and 4000 the flow around the cylindrical stem is transient, the flow is therefore in between near-laminar and fully turbulent.

The change from near-laminar to fully turbulent flow region can well be seen in the FFT graphs (see Fig. 4) where distinct frequencies appear at 0.3 m/s. The turbulence can be also noted in the standard deviations of stem vibrations. There is a larger increase of the standard deviation around the speed of 0.25 m/s (see Fig. 6).

The flow speed from hydromast's readings could be detected in two ways. One solution is to use the mean angle of the stem in the flow to estimate flow speed from the sensors. The emergence of clearly distinguishable frequency peaks from the FFT then tells us whether we need to use the linear or quadratic regression of the mean angle of the stem to calculate the flow speed. A second, and complimentary solution is to use the quadratic relation between mean amplitude of the frequency spectra of the stem and the flow speed. The second method has been shown to work with pressure sensors, even at higher velocities (> 1 m/s) and extreme turbulence [5]. The combination of two techniques would possibly give better results.

The hydromast's bilateral reaction to flow speed allows us to design sensors differently for low and high flow speeds. With the current prototype, if we know the characteristic curves of the sensor, we can use the sensor both in laminar and turbulent flow regimes. The sensors membrane and stem properties can be tuned as key parameters in order to define sensor performance criteria.

By altering the hydromast's membrane dimensions or elasticity, it would be possible to increase or decrease the sensibility of the sensor at low laminar flow speeds. With softer membrane material the hydromast would bend more in the laminar and near-laminar flow regimes where the drag forces are caused largely due to skin friction. In the fully turbulent flow region, the softer membrane would allow the light hydromast's stem catch the turbulent wake frequencies better but on the other hand there is also a higher change that the turbulences would match the natural frequency of the moving stem.

The dimensions of the hydromast's stem affect the speed at which the flow around the stems reaches the transition point from laminar to turbulent. Stems with smaller diameters (and smaller Reynolds numbers) would reach the transition point at higher speeds and vice versa in case of larger stem diameters. Thus, even though the fluid-body interactions between the stem and the fully turbulent flow field are complex, the system has only a few simple parameters which can be tuned when developing neuromast-inspired flow sensing devices.

The best suitable sensor combinations will be parameterized in ongoing investigations where we take into account a wider diversity of conditions where the sensors will be applied.

5   Conclusion

In confined laboratory experiments we have shown that a bioinspired hydromast equipped with accelerometers can be used to detect flow speeds up to 0.5 m/s. It is possible to relate the flow speed to the hydromast stem motions both in the near-laminar and turbulent flow regions.

Future research will be focused on the study of how the hydromast behaves at higher flow speeds and how the sensibility is dependent on the hydromast stem and membrane properties.

The areas of interest where we want to apply the hydromast sensors are related to both freshwater (e.g. rivers) and salt water (e.g. sediment transportation on the seabed, current flow speed and direction monitoring in ports etc.) environments where near-bed flow sensing is often difficult, if not impossible to achieve.

The current work was done in the frame of Lakhsmi project (European Union's Horizon 2020 research and innovation program under grant agreement No 635568, www.lakhsmi.eu).

References

1. Kalmijn, A.J.: Hydrodynamic and acoustic field detection. In: Atema, J., Fay, R.R., Popper, A.N., Tavolga, W.N. (eds.) Sensory Biology of Aquatic Animals, pp. 83–130. Springer, New York (1988)
2. McHenry, M.J., Liao, J.C.: The hydrodynamics of flow stimuli. In: Coombs, S., Bleckmann, H., Fay, R.R., Popper, A.N. (eds.) The Lateral Line System, pp. 73–98. Springer, New York (2013)
3. Davidson, P.: Turbulence: An Introduction For Scientists and Engineers. Oxford University Press, USA (2015)
4. Strokina, N., Kämäräinen, J.-K., Tuhtan, J.A., Fuentes-Perez, J.F., Kruusmaa, M.: Joint estimation of bulk flow velocity and angle using a lateral line probe. IEEE Trans. Instrum. Meas. 1–13 (2015)
5. Fuentes-Pérez, J.F., Tuhtan, J.A., Carbonell-baeza, R., Musall, M., Toming, G., Muhammad, N., Kruusmaa, M.: Current velocity estimation using a lateral line probe. Ecol. Eng. **85**, 296–300 (2015)
6. Tuhtan, J.A., Toming, G., Ruuben, T., Kruusmaa, M.: A method to improve instationary force error estimates for undulatory swimmers. Underw. Technol. **33**(3), 141–151 (2016)
7. Rooney, T., Pearson, M.J., Pipe, T.: Measuring the local viscosity and velocity of fluids using a biomimetic tactile whisker. In: Wilson, S.P., Verschure, P.F., Mura, A., Prescott, T.J. (eds.) Living Machines 2015. LNCS, vol. 9222, pp. 75–85. Springer, Heidelberg (2015)
8. van Netten, S., McHenry, M.: The biophysics of the fish lateral line. In: Coombs, S., Bleckmann, H., Fay, R.R., Popper, A.N. (eds.) The Lateral Line System, vol 48, pp. 99–119. Springer, New York (2014)

Developing an Ecosystem for Interactive Electronic Implants

Paul Strohmeier[1]([⊠]), Cedric Honnet[2], and Samppa von Cyborg[3]

[1] University of Copenhagen, Copenhagen, Denmark
paul.strohmeier@gmail.com
[2] Carpe Noctem Labs, San Francisco, USA
cedric@carpenoctem.cc
[3] Von Cyborg Bodyart, Tampere, Finland
sampp@voncyb.org

Abstract. In this work in progress report we present Remora, a system for designing interactive subdermal devices. Remora builds upon methods and technologies developed by body modification artists. The development has so far focussed on battery and power management as well as redundancy features. Remora consists of a series of hardware modules and a corresponding software environment. Remora is designed for body modification artists to design their own implants, but will also be a platform for researchers interested in sub-dermal interaction and hybrid systems. We have so far implemented a prototype device; future work will include in depth evaluation of this device as well as user studies, a graphical development environment and additional hardware and software modules.

Keywords: Implantable devices · Subcutaneous interaction · Human-machine hybrids

1 Introduction

There is growing interest in interacting with devices inside the body in the Human Computer Interaction (HCI) research community, as well as in the Body Modification Community. Implanted devices are anticipated to provide opportunities for sensory augmentation, biometric data collection, data-storage or novel HCI applications [1,2]. This growing interest can be observed looking at workshops at leading HCI conferences[1], various self-experimentation projects, such as the Circadia by Grindhouse Wetware [3], or review papers on this very topic [4].

Implanted devices are not new - pacemakers and dental implants date back to the late 1950s and have since become established in medical practice. Common medical implants include joint replacements, drug delivery systems, cerebral shunts, stents, cardiac monitors and pacemakers, and even brain implants that provide deep brain stimulation to Parkinson patients or treat seizures for epilepsy

---

[1] https://insertables.wordpress.com/.

© Springer International Publishing Switzerland 2016
N.F. Lepora et al. (Eds.): Living Machines 2016, LNAI 9793, pp. 518–525, 2016.
DOI: 10.1007/978-3-319-42417-0_56

patients. Implants are also common for plastic surgery purposes, most notably the breast implant, which dates back to the 1960s.

In parallel to these more established practices, the body modification community has developed methods, practices and technologies of their own, including the development of subdermal implants. These are usually injection molded silicone objects which are implanted under the skin for aesthetic purposes. The manufacturing process of these subdermal implants allows encapsulating objects inside the implant. This enables implantation of unconventional objects, such as the ashes of a loved one or electronic circuitry. There is a trend amongst self-identified grinders or wet-ware hackers to appropriate these methods for the design of implantable electronic devices [3, 4].

Fig. 1. Remora prototype

Inspired by body modification culture, and in a somewhat worried reaction to the design of devices implanted by grinders, we are designing Remora (Fig. 1), an eco-system for the development of implantable devices. This eco-system conforms to the constraints a body modification artist is faced with when producing and implanting an electronic device. Remora is therefore not intended to improve upon, or even compare to the current state of the art in electronic implantable medical devices.

Remora aims to provide a platform with multiple failsafe and redundancy mechanisms around which to design interactive, subdermal devices. We anticipate this to find use in the body modification community, however we also believe that various research communities will benefit from such a platform. For example, Remora would allow research into sub-cutanious sensing [1] to progress within living bodies, and enable novel human computer interaction methods and advanced biometric monitoring.

Remora will consist of a development environment for designing ultra-low power applications, as well as a wireless bootloader with redundancy and self-monitoring functions. This will be paired with hardware module designs which can be combined based on the requirements of a given application. The resulting device would then be encapsulated in parylene C coating and embedded in implant grade silicone. The resulting silicone object can be inserted under the skin using methods established by body modification artists.

2 Related Work

2.1 Posthuman Bodies

Art that challenges the traditional role of the body is becoming succeedingly more mainstream. French artist Orlan has exhibited her plastic surgery performance at the Guggenheim in NYC, while Australian artist Stelarc has become a common reference at human computer interaction conferences. Stelarc's works explore the theme of agency beyond the body. His artworks often involve performances in which control over his limbs is handed to the audience or to random signals generated by internet traffic. He also uses robotic appendixes to expand his body and has experimented with implantation of extra sensory organs [5].

While artists such as Stelarc and Orlan are becoming more mainstream, the body modification community still strongly identifies as a counterculture movement. One of the earliest functional implants common within the body-modification community is a magnet. An implanted magnet vibrates when the implantee is subjected to a varying electromagnetic field. This might occur when walking through security systems or in the close proximity of electric motors or transformers. The first magnetic implants were implanted by Samppa von Cyborg in 2000. Kevin Warwick experimented with direct communication between the nervous system of two implantees. He used a 100 pin micro-array as an electrical interface with the nervous system of each implantee. Nervous impulses were then wirelessly transmitted between them. Warwick was also one of the first to implant RFID tags [6]. RFID and NFC implants have been further popularized by Amal Graafstra [7].

The methods developed by the body modification community have recently been appropriated by self identified grinders, wet-ware- or bio hackers. Most notably Tim Cannon and Grindhouse Wetware have developed a temperature monitoring device and an implantable silicone object illuminated from the inside. While both device types have been implanted, they were since removed due to safety concerns [3].

2.2 Development Environments

There is an ever increasing number of ecosystems and dev-boards that support non-experts in the development of digital devices. Most prominent among them Raspberry Pi and Arduino. Neither however meet the ultra-low power requirements and the compact size required of an implanted device. Other interesting systems include the Bluetooth Low Energy (BLE) development boards by Red-BearLabs[2] or products based on Simblee[3]. These solutions have relatively low power consumption, but the form-factor of the devices by RedBearLabs and the closed source of the Simblee protocol are also problematic. There are various upcoming development kits based around the new nRF52, such as the BMD-300 by Rigado[4] with even lower power consumption. While such modules have

---

[2] http://redbearlab.com/.
[3] https://www.simblee.com/.
[4] https://www.rigado.com/product/bmd-300/.

optimized power consumption, they do not provide the close coupling with wireless charging and battery monitoring which we envision.

3   Design Space

3.1   Implantation Style

Body modification artists typically distinguish between transdermal and subdermal implants. **Transdermal** implants contain a metal base and a threaded stud extending beyond the skin, on which jewelry - for example metal horns - can be mounted. Depending on the design of the transdermal implant, it might have a porous base which increases the strength with which it is fused to the skin. Transdermals developed by Samppa von Cyborg contain a coating to allow tissue to bond with the implant itself, rather than only encapsulating it. **Subdermal** implants are placed underneath the skin. As the skin can close above it, such implants minimize the risk of both infection and rejection. As the silicone is smooth without any porous structures for the skin to latch on to, removal of such an implant is comparably simple.

3.2   Functionality

We define all implants which do not exhibit any behavior as non-functional (some, such as breast implants, may have social functions, but do not exhibit any behavior of their own, so we consider them non-functional). There are two main types of **passive implants** which are gaining increasing popularity: NFC/RFID chips and magnets. These implants transmit and ID or alert the implantee of the presence of a varying electromagnetic field respectively, however they require an external power source. In contrast to these are various explorations into **active powered implants**, most notably the Circadia and North Star by Grindhouse Wetware. While these devices can turn LEDs on and off on demand, or stream temperature data, they have no way of interacting with the implantees body or environment in real time. Remora will support the design of **interactive implants**: implants which can react to the implantees bodies, environment and activities both continuously and in real time.

4   Implementation

4.1   Hardware

Remora consists of a Core and Extensions (Fig. 2). The Core includes the Power Transfer Module, the Battery Module, and the Logic Module. Remoras functionality is determined by the Extension Modules connected to its Core. Such Extension Modules might provide haptic feedback or illuminate the device from the inside. They could collect inertial or biometric data or enable explicit user input. This modular approach was chosen to give developers the greatest possible freedom in developing their own devices. Example devices might include sensory augmentation applications, such as a device which provides a haptic impulse when the implantee is facing north, continuous biometric monitoring, or internal data storage.

The Core consists of a Logic Module, Power Transfer Module and Battery Module (see Fig. 2). Currently we have also implemented an Inertial Sensing Unit as our first Expansion Module.

The **Battery Module** contains the battery and basic circuitry to protect battery life as well as a kill switch mechanism. When this killswitch is activated all circuits are immediately powered off. The killswitch can be triggered in multiple ways: the user can explicitly activate it, the Battery Module can activate it, if the temperature of the battery or the power consumption exceeds a predetermined threshold, and it can be triggered by the Logic Module, if it detects a problem.

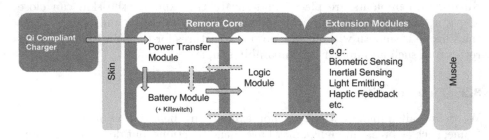

Fig. 2. Overview of a Remora device. Green solid arrows indicate power flow, yellow dashed arrows indicate control. The Logic Module controls extensions and Battery Module, while the Power Transfer Module only has control over the Battery Module. The Battery Module can be removed completely if one is interested in designing a passive device. (Color figure online)

For explicit user killswitch activation, we assume that the user has a secondary passive implant consisting solely of a magnet. For example, if the primary implant is in the left arm, the user might have a magnet in the right hand. By touching the magnet of their right hand to the implant in their left hand, the user has a simple and reliable mechanism of disconnecting the battery from the rest of the circuitry.

The **Power Transfer Module** implements the Qi Compliant Power Transfer standard. This enables users to charge Remora with any off-the-shelf wireless induction charging device. The Qi protocol is implemented using the BQ51050B IC by Texas Instruments. The Power Transfer Module will interrupt the power transfer if: its own temperature rises over a cut-off temperature, if notified of a charging error by the Logic Module, or if it detects the battery to be fully charged. The Qi Compliant standard was chosen because it is currently widely used and therefore a convenient starting point. We anticipate exploring options tailored more specifically to charging an implanted device in the future.

The **Logic Module** is powered by Nordic Semiconductor's nRF52832. This SoC comes with a Cortex M4 processor and various wireles communication options including NFC and a 2.4 GHz radio that supports Bluetooth Low Energy (BLE). The Logic Module supports communication with other devices via UART, SPI and I2C. Additionally the module has a number of GPIOs and 12 bit ADC inputs for interfacing with sensors. The Logic Module monitors its own temperature, the temperature of the Power Transfer Module and the charge state and temperature of the Battery Module. If any of these measures are outside of predetermined levels it can disable wireless charging, activate the killswitch or shut itself down until woken up by explicit user input through the Battery Module.

As any two Core modules can operate with the third module deactivated (See Fig. 2), this setup also allows us to collect debug information about the state of the device in case of suspected problems. The Logic Module can be powered from either the Battery Module or the Wireless Power Module and, if intact, provide debug information via Bluetooth. If the Logic Module has failed, battery health information can be extracted from the Power Transfer Model via the Qi communication interface.

4.2   Software

The software embedded in the nRF52 consists of two parts, the bootloader and the embedded main application. To develop or customize them, we provide the user with a development environment based on the SDK made by Nordic Semiconductors.

The **bootloader** performs two functions: it enables firmware uploads using BLE and it takes over monitoring battery health and power consumption, should there be no functioning firmware (Fig. 3). The firmware is uploaded using a dual bank memory layout. This ensures that only complete and valid images are activated. During dual bank upload the received firmware is stored in an intermediary memory location and, after it has been fully uploaded and verified, is transferred to its intended location. If an error occurs during the transfer, or if the system is unable to validate the new firmware, the previously uploaded firmware continues operating. The bootloader can be triggered through a BLE request, or by activating the killswitch.

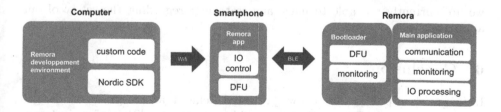

Fig. 3. Schematic overview of Remora software architecture

The **main application** has several tasks, monitoring & redundancy, communication and IO processing. The monitoring & redundancy task is not designed to be modified by the user. It continuously monitors temperature of all modules as well as power drain and battery health. For reliability purpose, these measures are also performed in parallel by dedicated analog circuitry. If either the monitoring & redundancy task or the analog circuitry determines values to be outside of a predetermined range, the killswitch of the Battery Module is activated. The monitoring & redundancy task also uses a watchdog monitoring system to restart the processor, should a software endless loop problem occur. Button-less Device Firmware Update (DFU) is conducted using BLE. BLE is also used to provide actuators with external control signals, or send sensor data to other BLE devices. The IO processing is the most modular part and is designed to be completely customizable. It handles communication with extension modules and runs the user created application.

The development environment is built on top of the development kit provided by Nordic Semiconductors. Users can create custom application using command line arguments and a Makefile, enabling desired extension modules. The development environment optimizes power consumption and memory usage for a given application. It also generates a checksum to avoid transmission errors during the firmware update. Error verification is then performed by the dual bank bootloader. The development environment outputs a zip archive including all required binaries. The archive can then be stored in a cloud services such as Dropbox, to access it from a smartphone app. The smartphone then uploads the firmware using BLE.

5   Future Work

We have implemented the three core modules and the inertial sensing module. We have tested these for basic functionality. Our next steps are to add additional hardware modules and to thoroughly evaluate the existing ones. We intend to optimize power consumption using the Simplicity EnergyAware Profiler at body temperature as well as optimize heat-dissipation using a near infrared camera by Optris[5]. An open question is which battery technology the final platform should target. Using traditional implantable batteries might not be ideal as our power consumption and charging behaviors are different from existing devices. We wish to emphasize that, as we have not conducted any formal evaluation, we are currently not able to make any statement regarding the safety of our approach.

6   Conclusion

In this work in progress report, we presented Remora, an ecosystem for designing implantable devices. Rather than adding to the tradition of mainstream

---

[5] http://silabs.com, http://optris.com.

implantable devices, Remora builds on knowledge and practices collected by body modification artists. Remora is designed for these artists to create their own implants, but will also be a platform for researchers interested in creating hybrid systems. Our design is primarily focused on providing series of monitoring and redundancy features, aimed at protecting the implantee and providing the designer with as much debug information as possible, should an error occur. We have so far designed and implemented a proof-of concept device. Future work will expand upon these and include empirical studies testing the safety and reliability of our designs and evaluating interaction design with and for implanted devices.

References

1. Holz, C., Grossman, T., Fitzmaurice, G., Agur, A.: Implanted user interfaces. In: Proceedings of the 2012 ACM Annual Conference on Human Factors in Computing Systems (2012)
2. Hameed, J., Harrison, I., Gasson, M.N., Warwick, K.: A novel human-machine interface using subdermal magnetic implants. In: 2010 IEEE 9th International Conference on Cyberntic Intelligent Systems (2010)
3. The Half Life of Body Hacking – Motherboard. http://motherboard.vice.com/read/the-half-life-of-body-hacking. Accessed 20 Mar 2016
4. Heffernan, K.J., Vetere, F.: You put what, where? Hobbyist use of insertable devices. In: Proceedings of the 2016 ACM Annual Conference on Human Factors in Computing Systems (2016)
5. Ping Body - Institute for the Unstable Media. http://v2.nl/events/ping-body. Accessed 21 Oct 2015
6. Warwick, K., Ruiz, V.: On linking human and machine brains. Neurocomputing 71(13–15), 2619–2624 (2008)
7. Graafstra, A.: Hands On: How radio-frequency identifion and I got personal. IEEE Spectr. 18, 23 (2007)

Gait Analysis of 6-Legged Robot with Actuator-Equipped Trunk and Insect Inspired Body Structure

Yasuhiro Sugimoto[✉], Yuji Kito, Yuichiro Sueoka, and Koichi Osuka

Department of Mechanical Engineering, Osaka University, Osaka, Japan
{yas,sueoka,osuka}@mech.eng.osaka-u.ac.jp,
kito@dsc.mech.eng.osaka-u.ac.jp

Abstract. Not only 4-legged animals but also 6-legged creatures exhibit a range of gaits. To reveal the mechanism underlying the gait of animals and realize such a gait in a multi-legged robot, it is important to understand the inter-limb coordination. It is possible that the inter-limb coordination is strongly influenced by the structure of trunk to which each leg is connected. In this study, we designed a brand-new insect-inspired trunk mechanism that is equipped with an actuator. Using this trunk, we built a 6-legged robots with passive limbs. We performed walking experiments with the developed robot and analyzed its gait, especially the relationship between the structure of the trunk and the walking velocity. The experimental result shows an insect inspired body structure affects its gait and the walking velocity of 6-legged walking robot.

Keywords: Legged robot · Trunk structure · Gait analysis

1 Introduction

Many terrestrial animals adopt walking as their means of locomotion. It is well known that quadruped animals exhibit specific gaits, such as walking, troting, paceing, depending on their movement speed at which they are moving. Furthermore, not only 4-legged animals but also 6-legged creatures exhibit a range of gaits, such as waving, tripod and tetrapod gaits [13]. And the insects also exhibit the continuum of these gaits [9]. To reveal the mechanisms behind the gaits of animals and realize those gaits in multi-legged robots, it is important to understand the inter-limb coordination. Although a large number of studies have examined the transition of gaits, the inter-limb coordination and its applications to multi-legged robots, many studies have focused on the controller layer, such as the neural network and central pattern generator (CPG) [1,2,6]. It is also interesting to observe the physical layer, that is, the mechanical structure. In particular, it is probable that the inter-limb coordination is strongly influenced by the structure of the trunk because the legs are connects to it mechanically [4,10,12].

To investigate the inter-limb coordination from the viewpoint of the mechanical properties, we focused on passive dynamic walking, studied first by McGeer [7].

© Springer International Publishing Switzerland 2016
N.F. Lepora et al. (Eds.): Living Machines 2016, LNAI 9793, pp. 526–531, 2016.
DOI: 10.1007/978-3-319-42417-0_57

Passive dynamic walking refers to the ability, in a mechanical device, to walk down a shallow slope without using any actuation or sensing. Although the subjects of this research were mostly two-legged robots, some research has addressed multi-legged devices [5, 8]. In this study, to realize level walking and spontaneous inter-limb coordination, we designed a brand-new trunk mechanism incorporating an actuator and built a 6-legged robot using this trunk. The legs of the robot are passive limbs. That is, the legs dose not have any actuators, such that the motions of each leg are instigated by the trunk. As a result, spontaneous inter-limb coordination is expected to occur. In addition, we focused on the structure of an insect's body. Through walking experiments with the developed robot, we analyzed its locomotion, especially the relationship between the structure of the trunk and the walking velocity.

2   Multi Legged Robot

Figure 1 shows the developed 6-legged robot. This robot is based on that developed in our previous study [5]. This robot consists of a series of connected three two-legged elements with two trunks, each incorporating an actuator. Each leg is rigid and has no knee joint. Additionally, each foot is a part of a sphere. Each leg is a passive limb and attached to a free joint at the waist, rotatable about the axial pitch. The rocking motion of each two-legged element is instigated in the lateral plane by the actuator-equipped trunk, with the robot being able to walk by swinging its legs with a rocking motion. There is no direct leg-synchronizing mechanism. Only through the trunk, each element interacts indirectly.

Figure 1(b) shows the developed actuator-equipped trunk. The trunk is equipped with a servo motor and is able to rotate in the directions of the roll and yaw axes. The torque generated by the servo motor leads to rotation in the roll direction. The roll and yaw rotations are coupled because of the gearing mechanism. As a result, rotations in the roll and yaw directions occur at the same time even with only one servo motor. The movement around the pitch axis is decoupled from the roll and yaw axes. The degree of freedom around the pitch axis is given at joints between two-legged elements and the trunk.

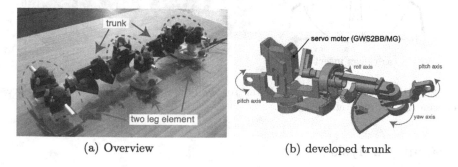

(a) Overview   (b) developed trunk

Fig. 1. Developed 6-legged robot and an developed actuatable trunk.

The developed 6-legged robot has two trunks. The front trunk is located between the front and middle legs while the hind trunk which is located between the middle and hind legs. Inputs to these trunks can be provided individually. We apply the following Eq. (1) as inputs $\phi_f(t)$ and $\phi_h(t)$ to the front and hind trunk servo motors for which A[rad], T[s] and β[rad] are the amplitude, period and phase difference, respectively. We can see that the possible gaits of the robot will change depending on β. In this study, we specifically focused on the phase difference β.

$$\phi_f(t) = A\sin(\frac{2\pi}{T}t), \quad \phi_h(t) = A\sin(\frac{2\pi}{T}t - \beta) \tag{1}$$

Several studies involving an analysis of a multi legged robot have used robots that configured of series-connected two-legged elements, in the same way as our developed robot [5,11]. The structures of the fundamental two-legged elements were almost the same. In contrast, front, middle and hind legs of most insects have different structures. The distance between the landing point of each leg pair is different as shown in Fig. 2(a). Inspired by these interesting insect structures, we implemented a means of adjusting the distances between the left and right leg joints of the robot. By using spacers, the distances can be adjusted. For example, in Fig. 2(b), the distance between the joints of the middle legs is increased.

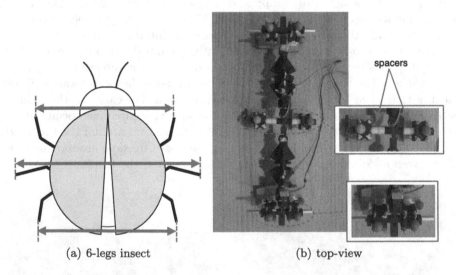

(a) 6-legs insect   (b) top-view

Fig. 2. 6-legs insect and top view of the robot. This robot can change the distances between the left and right leg joint.

3   Gait Analysis

We investigated the walking speed as a function of the input phase difference β. The other values in Eq. (1) are fixed to $A = 20$[deg] and $T = 0.2$[s]. To examine the effect of the body structure, we also enlarged the left-and-right leg joint distances for the front, middle and hind legs. Figure 3 shows the experimental results of our experiments. The vertical and horizontal axes show the walking speed and phase difference β, respectively. Figure 3(a) shows average walking speeds and standard deviations of each phase difference obtained with the normal structure case, that is, when the leg joint distances of the front, middle and hind legs are all the same. Figure 3(b) shows average walking speeds obtained when any of the leg joint distances of the front, middle or hind legs were changed.

The walking speed is almost the same except in the case of $\beta = 0$ and $\beta = 30$ for the normal structure. When $\beta = 0$, the robot almost stopped, that is, it repeatedly stepped in the same location. When $\beta = 60$, the robot attained its maximum value of the walking speed. Figure 4 shows an image of forward walking with $\beta = 180$. Figure 5 shows lateral oscillations θ of each two-legged elements. Because of the particular structure of each two-legged elements, positive θ value means that the right leg is on the ground and left leg is lifted off the ground, and vice versa. It was confirmed that the robot can move forward by wriggling its body. In this case, the robot exhibits a tripod gait. It was also confirmed that the robot could exhibit a wave gait when $\beta = 90$ or $\beta = 270$.

On the other hand, the walking speed varied considerably depending on the value of β when the insects inspired body structure was used. In addition, the transition in the walking speed also varied based on the adjustment location. When the front or hind leg joint distance was increased, the walking speed was a minimum at around $\beta = 180$. However, the walking speed increased to a maximum at around $\beta = 180$ when the middle leg joint distance was increased. In this case, the robot also exhibited a tripod gait when $\beta = 180$ and a wave gait when $\beta = 90$ or $\beta = 270$. It is known that some insects exhibit a transition in their gait; they generate a wave gait at low speeds and then modify their gait to a tetrapod gait at medium speeds [3]. Now, the distance between the middle leg joints of many insects is larger than that of other leg joints as shown in Fig. 2(a). The results of walking experiments seem to exhibit a similar tendency to insect locomotion. A detailed analysis of the relationship between the gaits and the structure will be undertaken as part of our future work.

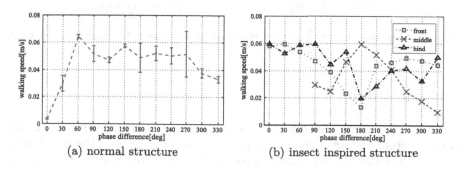

(a) normal structure   (b) insect inspired structure

Fig. 3. Walking speed of 6-legs robot vs. phase difference β

Fig. 4. The walking snapshots of the developed 6-legged robot (phase difference $\beta = 180[\text{deg}]$). In this case, fore and hind leg elements are synchronized and middle leg is antiphase.

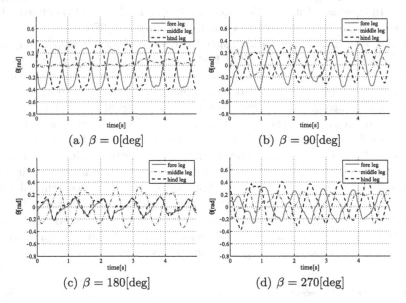

Fig. 5. Lateral oscillation θ of each leg elements (A = 25[deg], T = 0.2[s]) (Color figure online)

4   Conclusion

In this study, we developed a 6-legged walking robot with an insect-inspired body structure and an actuator-equipped trunk for analyzing the relationship between gaits and the structure of a robot. We conducted walking experiments with the developed robot. Our experimental results showed that the walking speed and gait were related to the input phase difference. In addition, it was also verified that the transition in the walking speed changed greatly depending on the body structure.

Acknowledgments. This work was supported by the Grant-in-Aid for Young Scientists (B), No.26870337 and CREST, JST.

References

1. Beer, R.D., Quinn, R.D., Chiel, H.J., Ritzmann, R.E.: Biologically inspired approaches to robotics: what can we learn from insects? Commun. ACM **40**(3), 30–38 (1997)
2. Buschges, A.: Sensory control and organization of neural networks mediating coordination of multisegmental organs for locomotion. J. Neurophysiol. **93**(3), 1127–1135 (2004)
3. Hughes, G.M.: The co-ordination of insect movements. J. Exp. Biol. **29**(2), 267–285 (1952)
4. Iida, F., Gomez, G., Pfeifer, R.: Exploiting body dynamics for controlling a running quadruped robot. In: Proceedings of International Conference on Advanced Robotics (ICAR 2005), pp. 229–235 (2005)
5. Kito, Y., Sueoka, Y., Nakanishi, D., Yasuhiro, S., Ishikawa, M., Wada, T., Osuka, K.: Quadruped passive dynamic walking robot with a new trunk structure inspired by spine In: Proceedings of Dynamic Walking 2014 (2014)
6. Matsuoka, K.: Mechanisms of frequency and pattern control in the neural rhythm generators. Biol. Cybern. **56**(5), 345–353 (1987)
7. McGeer, T.: Passive dynamic walking. Int. J. Robot. Res. **9**(2), 62–82 (1990)
8. Remy, C.D., Buffinton, K., Siegwart, R.: Stability analysis of passive dynamic walking of quadrupeds. Int. J. Robot. Res. **29**(9), 1173–1185 (2009)
9. Schilling, M., Hoinville, T., Schmitz, J., Cruse, H.: Walknet, a bio-inspired controller for hexapod walking. Biol. Cybern. **107**(4), 397–419 (2013)
10. Schilling, M., Paskarbeit, J., Schmitz, J., Schneider, A., Cruse, H.: Grounding an internal body model of a hexapod walker control of curve walking in a biologically inspired robot. In: IEEE International Conference on Intelligent Robots and Systems, pp. 2762–2768 (2012)
11. Sugimoto, Y., Yoshioka, H., Osuka, K.: Realization and motion analysis of multilegged passive dynamic walking. In: Proceedings of SICE Annual Conference 2010, Taipei (2010)
12. Takuma, T., Izawa, R., Inoue, T., Masuda, T.: Mechanical design of a trunk with redundant and viscoelastic joints for rhythmic quadruped locomotion. Adv. Robot. **26**(7), 745–764 (2012)
13. Wilson, D.: Insect walking. Ann. Rev. Entomol. **11**(1), 103–122 (1965)

Quadruped Gait Transition from Walk to Pace to Rotary Gallop by Exploiting Head Movement

Shura Suzuki[1], Dai Owaki[1(✉)], Akira Fukuhara[1,2], and Akio Ishiguro[1,3]

[1] Research Institute of Electrical Communication, Tohoku University,
2-1-1 Katahira, Aoba-ku, Sendai 980-8577, Japan
owaki@riec.tohoku.ac.jp
[2] JSPS, 5-3-1 Kojimachi, Chiyoda-ku, Tokyo 102-0083, Japan
[3] CREST, Japan Science and Technology Agency, 4-1-8 Honcho, Kawaguchi,
Saitama 332-0012, Japan
http://www.cmplx.riec.tohoku.ac.jp

Abstract. The manner in which quadrupeds change their locomotion patterns with change in speed is poorly understood. In this paper, we demonstrate spontaneous gait transition by using a quadruped robot model with a head segment, for which leg coordination can be self-organized through a simple "central pattern generator" (CPG) model with a postural reflex mechanism. Our model effectively makes use of head movement for the gait transition, suggesting that head movement is crucial for the reproduction of the gait transition to high-speed gaits in quadrupeds.

Keywords: Central pattern generator (CPG) · Interlimb coordination · Physical communication · Head movement · Postural reflex

1 Introduction

Quadrupeds exhibit flexible locomotive patterns depending on their locomotion speed, environmental conditions, and animal species [1]. Such locomotor patterns are generated via the coordination of leg movements, i.e. "interlimb coordination". However, the mechanism responsible for interlimb coordination remains unclear. The knowledge of the interlimb coordination underlying quadruped locomotion is essential for understanding the locomotion mechanism in legged animals: furthermore, it will be useful in establishing the fundamental technology for legged robots that can reproduce locomotion similar to that of animals.

Previous neurophysiological experiments involving decerebrate cats [2] have suggested that their locomotive patterns are controlled in part by a distributed neural network, called the "central pattern generator" (CPG), in the spinal cord.

D. Owaki—We acknowledge the support of a JSPS KAKENHI Grant-in-Aid for Young Scientists (A) (25709033).

© Springer International Publishing Switzerland 2016
N.F. Lepora et al. (Eds.): Living Machines 2016, LNAI 9793, pp. 532–539, 2016.
DOI: 10.1007/978-3-319-42417-0_58

These biological findings have prompted many researchers to incorporate *neurally wired* CPG models into legged robots to generate highly adaptive locomotion [3,4]. In contrast, we proposed a novel *non-wired* CPG model consisting of four decoupled oscillators with local force feedback only [5]. Our model explains wide observations of gait patterns in quadrupeds [5], and the predictions based on our results are consistent with biological evidence [6]. However, our model did not sufficiently reproduce high-speed gait patterns, e.g. transverse/rotary gallop [1].

This study aims to reproduce gait transition, especially, in high-speed gait patterns. Thus, we focus on head movements in quadruped locomotion for the following two reasons:

1. In quadrupeds, the mass ratio of the head segment, i.e. head and neck, to the body is around 10 %, whereas the mass ratio of fore or hind legs is around 5.5 % ∼ 6.5 % [7].
2. The relative movement phase relationship between all body segments, e.g. head, neck, trunk, etc. changes according to the gait pattern [8].

These facts suggest that head movements strongly affect the interlimb coordination with an increase in the locomotion speed. Owing to our effort in constructing a quadruped model using a simple CPG [5] with an *additional* postural reflex mechanism, we successfully reproduced gait transition from walk to pace to rotary gallop by using the head movement. Surprisingly, the obtained gait patterns are qualitatively similar to a giraffe's gait patterns [9], suggesting that these animals effectively make use of head and neck movements during gait transitions.

2 Model

2.1 Musculoskeletal Structure

Figure 1 shows an overview of the quadruped robot model constructed in this study. To mainly focus on the effects of the head movement on interlimb coordination, we used a simple musculoskeletal structure of the quadruped robot, which consists of a head, trunk, and four legs. The head and trunk segments are connected via a passive spring that has one degree of freedom (DOF) in the pitch direction only. We implemented two actuators for each leg (2 DOF), in which the rotational actuator drives a leg at the shoulder/hip joint and the linear actuator drives a leg along the leg axial direction, as shown in Fig. 1 right. A phase oscillator was implemented in a leg to generate a rhythmic leg motion during the stance and swing phases according to the oscillator phase ϕ_i (hereafter, $i = 1 \sim 4$, Left fore (LF):1, Left hind (LH):2, Right fore (RF):3, Right hind (RH):4), as explained in Sect. 2.2 in detail. We implemented two types of sensors: pressure sensors on the feet, which can detect ground reaction forces N_i [N] (GRFs) parallel to the leg axis, and an angle sensor on the neck joint, which can detect the displacement ψ [rad] from the equilibrium posture.

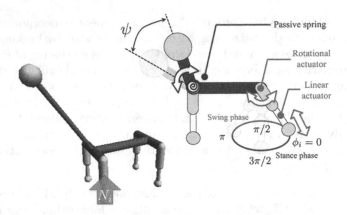

Fig. 1. Quadruped robot model.

2.2 CPG Model with Postural Reflex

We used our CPG model [5] that enables a quadruped to achieve *flexible* interlimb coordination without "neural" communication. We additionally implemented a simple postural reflex mechanism [10] because the whole body posture desta-bilized upon adding the head segment. In the following section, we explain the control scheme of our quadruped robot model.

Each leg is controlled using rotational and linear actuators according to the oscillator phase ϕ_i [rad]. The target angle $\bar{\theta}_i$ [rad] and length $\bar{l}_i$ [m] are described using the oscillator phase as follows:

$$\bar{\theta}_i = -C_{\text{amp}} \cos \phi_i, \tag{1}$$

$$\bar{l}_i = L_0 - L_{\text{amp}} \sin \phi_i, \tag{2}$$

where C_{amp} denotes the amplitude of the target angle, and L_{amp} and L_0 denote the amplitude and offset length of the target length, respectively. Based on the target angle and length, each actuator is controlled using the following equa-tions:

$$\tau_i = -K_r(\theta_i - \bar{\theta}_i), \tag{3}$$

$$F_i = -K_l(l_i - \bar{l}_i), \tag{4}$$

where K_r and K_l are the P gains for each actuator. θ_i and l_i denote the actual angle of the shoulder/hip joint and the actual leg length, respectively. Based on the control scheme, a leg tends to be in the swing phase for $0 < \phi_i \leq \pi$, and in the stance phase for $\pi < \phi_i \leq 2\pi$, as shown in Fig. 1 right.

The dynamics of the phase oscillators are described as follows:

$$\dot{\phi}_i^f = \omega - \sigma N_i^f \cos \phi_i^f - \rho \psi \cos \phi_i^f \tag{5}$$

$$\dot{\phi}_i^h = \omega - \sigma N_i^h \cos \phi_i^h + \rho \psi \cos \phi_i^h \tag{6}$$

where ϕ_i^f and ϕ_i^h denote the phases of the fore and hind leg oscillators, respectively. ω [rad/s] denotes the intrinsic angular velocity of the oscillators. σ [rad/Ns] and ρ [1/s] denote the feedback gains for the second and third terms, respectively. N_i^f and N_i^h [N] denote the GRFs detected by pressure sensors in the fore and hind feet, and ψ [rad] denotes the neck joint angle detected by the angle sensor.

The second term enables interlimb coordination without any neural communication [5]. The physical effect of this term is that a leg remains in the

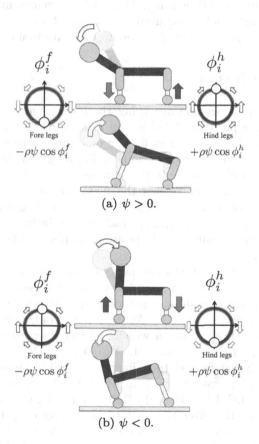

(a) $\psi > 0$.

(b) $\psi < 0$.

Fig. 2. Postural reflex mechanism. The red and blue sections in the phase planes of oscillators ϕ_i^f and ϕ_i^h represent the swing and stance phases, respectively. The green (leg contraction) and orange (leg extension) arrows represent the effect of the postural reflex described by the third term in Eqs. (5) and (6). (Color figure online)

stance ($\sin \phi_i < 0$) phase while supporting the body ($N_i > 0$). The essence of the coordination mechanism based on the local feedback is that the local sensor value N_i includes *global* physical information regarding the other legs, i.e., to what extent the other legs support the body at any given time. Thus, the *physical*-communication-based interlimb coordination can be achieved without direct coupling between oscillators, unlike in most previous CPG models [3,4].

The third term was inspired by the "postural reflex mechanism" [10], which various vertebrates possess as a primitive reflex. The postural reflex tries to maintain an equilibrium position or posture of the body by using visual, vestibular, and deep sensory information. Here, we focused on the "tonic neck reflex" [10], which activates limb muscles according to the neck joint angle. This term provides the following effect: the extension of the fore legs and the contraction of the hind legs enables a posture to be maintained when $\psi > 0$ (Fig. 2(a)), whereas the contraction of the fore legs and the extension of the hind legs enables a posture to be maintained when $\psi < 0$ (Fig. 2(b)). Thus, we can include interlimb coordination mechanism, which enables maintaining the equilibrium posture according to the neck joint angle, thus resulting in postural stabilisation during locomotion.

3   Simulation Result

We conducted dynamic simulations using the Open Dynamics Engine (ODE) by implementing the proposed CPG model in a quadruped robot in the simulation environment. The parameters for the quadruped robot model are shown in Table 1. As shown in the upper graph of Fig. 3, we conducted a gait transition experiment by changing only the parameter ω [rad/s]. The lower graph in Fig. 3 shows the time evolution of the phase differences $\Delta\phi$ [rad] between oscillators during gait transition. The blue, red, and green lines represent $\Delta\phi_{LH-LF}$, $\Delta\phi_{RF-LF}$, and $\Delta\phi_{RH-LF}$ [rad], respectively. The result indicates that we achieved gait transition with spontaneous phase difference modification by changing only ω.

Figure 4 shows the representative gait patterns for (a) $\omega = 8.0$, (b) $\omega = 11.0$, and (c) $\omega = 14.0$. The upper and lower graphs show the time evolution of the neck joint angle ψ [rad] and gait diagrams, where the coloured regions represent the stance phases ($N_i > 0$). At (a) $\omega = 8.0$ [rad/s], our robot exhibited a diagonal-sequence (D-S) walk, in which the feet touched the ground in the following order: LF, LH, RF, and RH. At (b) $\omega = 11.0$ [rad/s], a pace was observed with the ipsilateral feet touching the ground in phase. At (c) $\omega = 14.0$ [rad/s], a rotary gallop was observed, in which the feet touched the ground in the following order: LF, RF, RH, and LH. These results indicate that our robot reproduced gait transition from D-S walk to pace to rotary gallop in response to the locomotion speed. Further, in D-S walk, the neck leaned forward when the LF leg touched the ground, then leaned backward when the LH leg touched the ground, leaned forward again when the RF leg touched the ground, and finally leaned backward when the RH leg touched the ground. In pace, the neck leaned forward at the middle stance phase of the ipsilateral leg pairs, e.g. LF and LH.

Table 1. Parameters for quadruped robot model

Head mass	0.175 [kg]
Total mass	1.38 [kg]
Leg length	0.12 [m]
Shoulder/Hip length	0.12 [m]
Trunk length	0.19 [m]
Neck length	0.20 [m]
Spring constant of neck joint	3.50 [Nm/rad]
σ	0.60 [rad/Ns]
ρ	10.0 [1/s]
C_{amp}	0.1π [rad]
L_{amp}	0.006 [m]
L_0	0.12 [m]
K_r	40.0 [Nm/rad]
K_l	25.0 [N/m]

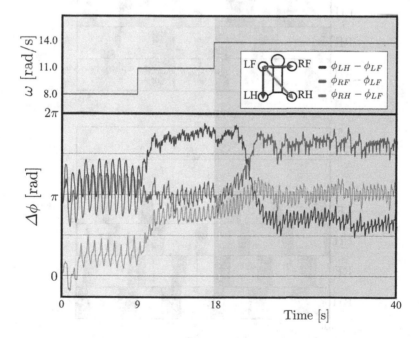

Fig. 3. Gait transition. Top: time profile of ω. Bottom: time evolution of phase differences $\Delta\phi$ between oscillators. (Color figure online)

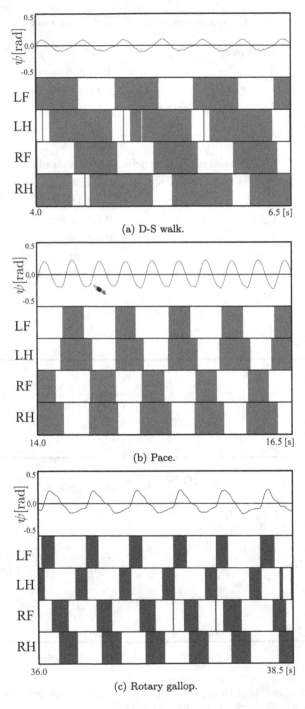

(a) D-S walk.

(b) Pace.

(c) Rotary gallop.

Fig. 4. Gait patterns for $\omega = 8.0, 11.0, 14.0$ [rad/s].

In rotary gallop, the neck leaned forward when the fore legs touched the ground, and it leaned backward when the hind legs touched the ground. The salient effect in rotary gallop is provided by the postural reflex mechanism given by Eqs. (5) and (6).

4 Conclusion and Future Work

To reproduce the gait transition in high-speed gait patterns, we modelled a quadruped robot in a simulation and implemented the proposed CPG model with a postural reflex mechanism. As a result, our robot reproduced gait transitions from D-S walk to pace to rotary gallop in response to the locomotion speed. Surprisingly, the obtained gait patterns are qualitatively similar to a giraffe's gait patterns [9], suggesting that these animals effectively use the head and neck movements during gait transitions. Therefore, our model would contribute to understanding the mechanism underlying giraffes's gait transition. In future, we will clarify the generation mechanism of "transverse gallop", which was not reproduced in this study. Furthermore, we will conduct a theoretical analysis of the gait transitions from low- to middle- to high-speed gait patterns, which seems to be of great interest.

References

1. Muybridge, E.: Animals in Motion. Dover Publications, New York (1957)
2. Shik, M.L., Severin, F.V., Orlovskii, G.N.: Control of walking and running by means of electrical stimulation of the midbrain. Biophysics **11**, 756–765 (1966)
3. Kimura, H., Akiyama, S., Sakurama, K.: Realization of dynamic walking and running of the quadruped using neural oscillator. Auton. Robots **7**, 247–258 (1999)
4. Righetti, L., Ijspeert, A.J.: Pattern generators with sensory feedback for the control of quadruped locomotion. In: Proceedings of ICRA 2008, pp. 819–824 (2008)
5. Owaki, D., Kano, T., Nagasawa, K., Tero, A., Ishiguro, A.: Simple robot suggests physical interlimb communication is essential for quadruped walking. J. R. Soc. Interfac. (2012). doi:10.1098/rsif.2012.0669
6. Duysens, J., Clarac, F., Cruse, H.: Load-regulating mechanisms in gait and posture: comparative aspects. Physiol. Rev. **83**, 83–133 (2000)
7. Buchner, H.H.F., Savelberg, H., Schamhard, H.C., Barneveld, A.: Inertial properties of Dutch warmblood horse. J. Biomech. **30**, 653–658 (1997)
8. Dunbar, D.C., Macpherson, J.M., Simmons, R.W., Zarcades, A.: Stabilization and mobility of the head, neck and trunk in horses during overground locomotion: comparisons with humans and other primates. J. Exp. Biol. **211**, 3889–3907 (2008)
9. Dagg, A.I.: Giraffe: Biology, Behaviour, and Conservation. Cambridge University Press, Cambridge (2014)
10. Magnus, R.: Cameron prize lectures on some results of studies in the physiology of posture. Lancet **208**, 531–536 (1926)

Exploiting Symmetry to Generalize Biomimetic Touch

Benjamin Ward-Cherrier[1,2]([⊠]), Luke Cramphorn[1,2], and Nathan F. Lepora[1,2]

[1] Department of Engineering Mathematics, University of Bristol, Bristol, UK
[2] Bristol Robotics Laboratory, University of Bristol, Bristol, UK
{bw14452,ll14468,n.lepora}@bristol.ac.uk

Abstract. We introduce a method for generalizing tactile features across different orientations and locations, inspired by recent studies demonstrating tactile generalization in humans. This method is applied to two 3d-printed bioinspired optical tactile sensors. Internal pins acting as taxels are arranged with rotational and translational symmetry in these sensors. By rotating or translating a small sample of tactile images, we are able to generalize tactile stimuli to new orientations or locations along the sensor respectively. Applying these generalization methods in combination with active perception leads to the natural formation of a fovea of accurate real tactile data surrounded by moderately less accurate generalized data used to focus the sensor's tactile perception.

Keywords: Tactile sensors · Perceptual generalization

1 Introduction

Recent studies indicate that humans generalize certain tactile stimuli like pressure, roughness or vibrations across the skin's surface [1,2], suggesting that low-level sensory generalization could be an important component of perceptual learning. In biomimetic approaches to touch [3–5], which tend to rely on exhaustive training sampling, applying methods for tactile feature generalization could be both practically beneficial (reducing training time), as well as forming the basis for an investigation of bioinspired perceptual generalization methods.

We introduce here a method for tactile generalization by exploiting sensor symmetries to generate new tactile data. Our method is applied to two optical tactile sensors: a round and flat version of the TacTip [6]. Both sensors optically track pins inspired by papillae in the human fingertip. The rotational and translational symmetry in pin layouts is exploited to generalize tactile features. In both cases, our method moderately increases errors in localization and orientation identification. However these errors can be compensated for by using active perception to relocate the sensor.

Thus we propose that tactile features can be reliably generalized through our method, giving a generalization from known encountered stimuli to novel stimuli from the symmetry of the tactile sensor. Used in conjunction with active perception, the method leads to the separation of the sensor into a fovea of real training data surrounded by coarse generalized data.

© Springer International Publishing Switzerland 2016
N.F. Lepora et al. (Eds.): Living Machines 2016, LNAI 9793, pp. 540–544, 2016.
DOI: 10.1007/978-3-319-42417-0_59

2 Methods

2.1 Hardware

The Round TacTip is a cheap, robust, 3d-printed optical tactile sensor developed at Bristol Robotics Laboratory [6]. It is made up of a 1 mm thick rubber skin, whose inside surface is covered in 127 pins arranged according to a projected hexagonal pattern with 6-fold rotational symmetry and inspired by papillae in the human fingertip (Fig. 1 - left panel). These pins deflect upon object contact, and their deflections in x- and y- directions are tracked by a webcam.

The Flat TacTip's overall structure and design is elongated and based on a modified version of the TacThumb [7]. The Flat TacTip has a rectangular shape and 16-fold approximate translational symmetry, with a total of 80 pins separated by 4 mm from their nearest neighbours (Fig. 1 - right panel).

2.2 Perceptual Generalization

Our method for data generation considers a section of training data, and applies affine transformations to generalise tactile data based on sensor symmetries. The method differs slightly for each sensor, based on the 6-fold rotational symmetry of the round TacTip and the 16-fold translational symmetry of the flat TacTip. Both methods are outlined below:

Orientation. We train the round TacTip to detect edge orientation by rotating over a 360° range in 5° increments. At each orientation the sensor taps down on a fixed edge stimulus. To generalize across orientation, we train only over a sub-sample of edge orientations $(0 - 60°)$ and rotate each set of tactile data gathered 60° around its centre, reassigning taxel identities to obtain the next set of orientations $(60 - 120°)$. By repeating this step 6 times, we are able to obtain tactile data for the whole 360° range, effectively generalising a tactile feature over all possible orientations.

Fig. 1. The round TacTip sensor (left) and the flat TacTip sensor (right). The 6-fold rotational symmetry (round TacTip) and 16-fold translational symmetry (flat TacTip) are exploited to generalize tactile stimuli from a reduced sample.

Location. A cylinder is rolled by the flat TacTip over a range of 28 mm, in increments of 0.5 mm. To generalize data, we train over a 4 mm sub-sample, and apply our translations to the tactile data 6 times to obtain the full 28 mm range. This enables the sensor to generalize a tactile feature across its skin's surface (Fig. 3).

3   Results

3.1   Inspection of Generalized Data

In orientation generalization with the round TacTip, generalized data (Fig. 2 - right panel) looks very similar to the real training data (Fig. 2 - left panel). We would thus predict a strong performance in orientation recognition using our method.

In the location generalization with the flat TacTip, we note that the real data (Fig. 3 - left panel) is not completely symmetrical, likely due to the intrinsic mechanics of the sensor close to its boundaries. The generalized data appears more regular (Fig. 3 - right panel), ignoring the boundary effect and leading us to expect higher localization errors relative to the orientation case.

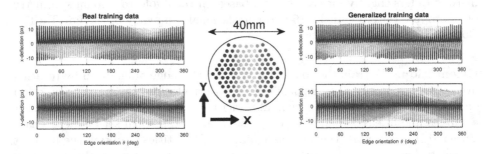

Fig. 2. Round TacTip data: left panel shows real training data, right panel corresponds to generalized data. The middle panel identifies pins according to their position on the sensor skin. (Color figure online)

3.2   Validation of Feature Generalization

Validation of the method is performed with a maximum likelihood probabilistic method. Monte-Carlo sampling of a pre-collected test data set gives us average localization or orientation classification errors for each class. Errors are then compared between real and generalized training data.

For the round TacTip, the real data has an average classification error over all orientation classes of $e_\theta = 0.42°$, whereas generalized data averages $e_\theta = 2.25°$. Errors are increased for the generalized data, but remain within a reasonable range based on the orientation being sampled every 5°.

In the flat TacTip case, using real training data leads to perfect localization ($e_x = 0$ mm), whereas the generalized data errors average $e_x = 1.77$ mm.

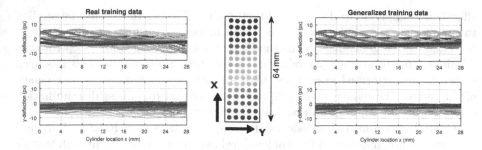

Fig. 3. Flat TacTip data: left panel shows real training data, right panel corresponds to generalized data. The middle panel identifies pins according to their position on the sensor skin. (Color figure online)

The method proposed is contingent on sensors being designed in a symmetrical way. The flat TacTip's higher errors are likely explained by its non-symmetrical mechanical properties, as evidenced in the previous section. Thus, there is a notable increase in errors of edge orientation classification and localization in the generalized data case. To reduce these errors we apply active perception methods to both orientation and location paradigms.

Active perception seeks to re-position the sensor to sample from the accurate real training data. This corresponds to the 0–60° range for orientation (round TacTip), and the 0–4 mm range for location (flat TacTip).

For orientation, the average error for active perception reduces to $e_\theta = 0.17°$ for real data, and for generalized data it becomes $e_\theta = 0.75°$. Thus the use of active perception reduces the disparity between the orientation classification performance of real and generalized training data. This result is expected, since active perception allows the sensor to be repositioned to gather the more precise real training data.

In the location case, errors for localization with the real data remain at $e_x = 0$ mm, whereas with generalized data they are reduced to just $e_x = 0.06$ mm. In this case, the beneficial effect of using active perception is more pronounced which suggests that although the generalized data for the flat TacTip is not very accurate, it is precise enough to effectively reposition the sensor to an area with more accurate tactile data.

The use of our method combined with active perception leads to the appearance of a fovea of real data samples, surrounded by less accurate, generalized data allowing the sensor to focus objects on the fovea.

4   Conclusions and Future Work

We propose here a method for tactile stimuli generalization, implementing low-level perceptual learning with symmetrical sensors. Combining our method with the use of active perception leads to the creation of a foveated sensor which can effectively sample multi-dimensional tactile data. This method could be applied to complex tasks including manipulation in higher dimensions.

The location generalization method could also represent a first step towards mimicking tactile stimuli generalization across the skin's surface. This has been shown to be performed in humans, with an interval discrimination task generalizing across the hand's surface [2] and pressure and roughness discrimination being learned in adjacent and homologous fingers [1].

Overall, our approach considers the shape of the sensor itself to influence perception and implement a form of low-level perceptual learning through feature generalization. This could be linked to the concept of body schema [8] to the extent that knowledge of body positions (or in this case sensor organization) can influence the interpretation of tactile data. Body schema work is gaining traction in robotics [9] and methods that integrate sensor position and shape into perceptual learning are likely to lead to more successful interactions with the robot's environment.

Acknowledgment. We thank Sam Coupland, Gareth Griffiths and Samuel Forbes for their assistance with 3d-printing. BWC was supported by an EPSRC DTP studentship and NL was supported in part by an EPSRC grant on 'Tactile Superresolution Sensing' (EP/M02993X/1).

References

1. Harris, J., Harris, I., Diamond, M.: The topography of tactile learning in humans. J. Neurosci. **21**(3), 1056–1061 (2001)
2. Nagarajan, S., Blake, D., Wright, B., Byl, N., Merzenich, M.: Practice-related improvements in somatosensory interval discrimination are temporally specific but generalize across skin location, hemisphere, and modality. J. Neurosci. **18**(4), 1559–1570 (1998)
3. Lepora, N., Martinez-Hernandez, U., Prescott, T.: Active touch for robust perception under position uncertainty. In: IEEE International Conference on Robotics and Automation (ICRA), pp. 3020–3025 (2013)
4. Lepora, N.: Biomimetic active touch with tactile fingertips and whiskers. IEEE Trans. Haptics **9**(2), 170–183 (2016)
5. Lepora, N., Martinez-Hernandez, U., Evans, M., Natale, L., Metta, G., Prescott, T.: Tactile superresolution and biomimetic hyperacuity. IEEE Trans. Rob. **31**(3), 605–618 (2015)
6. Chorley, C., Melhuish, C., Pipe, T., Rossiter. J.: Development of a tactile sensor based on biologically inspired edge encoding. In: International Conference on Advanced Robotics, ICAR 2009, pp. 1–6. IEEE (2009)
7. Ward-Cherrier, B., Cramphorn, L., Lepora, N.: Tactile manipulation with a tac-thumb integrated on the open-hand m2 gripper. Rob. Autom. Lett. **1**(1), 169–175 (2016)
8. Maravita, A., Spence, C., Driver, J.: Multisensory integration and the body schema: close to hand and within reach. Curr. Biol. **13**(13), R531–R539 (2003)
9. Hoffmann, M., Marques, H., Arieta, A., Sumioka, H., Lungarella, M., Pfeifer, R.: Body schema in robotics: a review. IEEE Trans. Auton. Mental Dev. **2**(4), 304–324 (2010)

Decentralized Control Scheme for Centipede Locomotion Based on Local Reflexes

Kotaro Yasui[1]($\boxtimes$), Takeshi Kano[1], Dai Owaki[1], and Akio Ishiguro[1,2]

[1] Research Institute of Electrical Communication, Tohoku University,
2-1-1 Katahira, Aoba-ku, Sendai 980-8577, Japan
{k.yasui,tkano,owaki,ishiguro}@riec.tohoku.ac.jp
[2] Japan Science and Technology Agency, CREST, 4-1-8 Honcho, Kawaguchi,
Saitama 332-0012, Japan
http://www.cmplx.riec.tohoku.ac.jp/

Abstract. Centipedes exhibit adaptive locomotion via coordination of their numerous legs. In this study, we aimed to clarify the inter-limb coordination mechanism by focusing on autonomous decentralized control. Based on our working hypothesis that physical interaction between legs via the body trunk plays an important role for the inter-limb coordination, we constructed a model wherein each leg is driven by a simple local reflexive mechanism.

Keywords: Centipede locomotion · Inter-limb coordination · Decentralized control · Reflex

1 Introduction

Centipedes move by propagating leg-density waves from the head to tail through the coordination of their numerous legs [1] and can adapt to various circumstances, e.g., changes in walking speed, by appropriately changing the form of the leg density waves in real time [2]. This ability has been honed by a long-time evolutionary process, and there likely exists an ingenious inter-limb coordination mechanism underlying centipede locomotion. Clarification of this mechanism will help develop highly adaptive and redundant robots, as well as provide new biological insights.

Autonomous decentralized control could be the key to understand the inter-limb coordination mechanism in centipede locomotion, and several models based on decentralized control mechanisms have been proposed so far [3,4]. However, they could not truly reproduce the innate behavior of centipedes. This fact indicates that the essential decentralized control mechanism for the inter-limb coordination still remains unclear.

To tackle this problem, here we hypothesized that the physical interaction between legs via the body trunk is essential for the inter-limb coordination. Thus, we constructed a model based on a simple local reflexive mechanism in which physical interaction is fully exploited.

© Springer International Publishing Switzerland 2016
N.F. Lepora et al. (Eds.): Living Machines 2016, LNAI 9793, pp. 545–547, 2016.
DOI: 10.1007/978-3-319-42417-0_60

2   Model

In our previous study, we proposed a simple local reflexive mechanism by using a two-dimensional model in which yaw and roll motion of the body trunk is neglected [5]. However, we consider that three-dimensional motion of the body trunk plays an important role for the inter-limb coordination, and hence, here we extend our previous model to the three-dimensional case.

A schematic of the physical model is shown in Fig. 1. Each leg is composed of a parallel combination of a real-time tunable spring (RTS) and a damper, where RTS is a spring whose natural length can be actively changed [6]. Each leg is connected to the body trunk via a yaw joint to move the leg back and

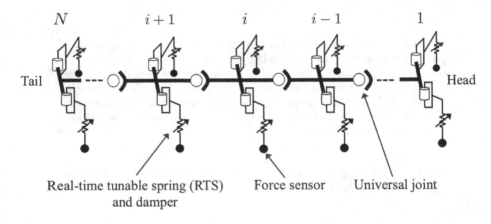

Fig. 1. Schematic of the physical model.

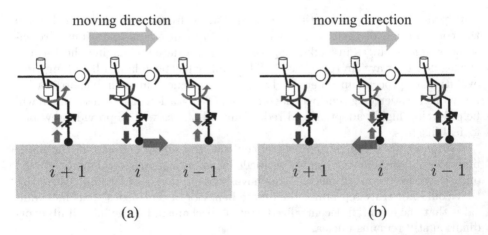

Fig. 2. Effect of the local reflexive mechanism (a) when the right ith leg detects a ground reaction force beneficial for propulsion and (b) when the right ith leg detects a ground reaction force that impedes propulsion. Red arrows denote the horizontal component of ground reaction forces acting on the ith leg tip. (Color figure online)

forth actively; this joint is controlled according to proportional-derivative (PD) control. Passive universal joints are implemented in the body trunk such that the body undulation can be generated through the physical interaction between the legs during locomotion.

The natural lengths of the RTSs and the target angles of the yaw joints are controlled according to the following mechanism. When the ith leg detects a ground reaction force beneficial for propulsion, the ith leg kicks the ground backward and the $i-1$th and $i+1$th legs on the ipsilateral side throw themselves forward with lifting their tips off the ground (Fig. 2(a)). On the other hand, when the ith leg detects a ground reaction force which impedes propulsion, the ith leg throws itself forward with lifting its tip off the ground and the $i-1$th and $i+1$th legs on the ipsilateral side kick the ground backward (Fig. 2(b)). Thus, it is expected that neighboring legs do not interfere with each other such that the body propels itself effectively. Note that neural connection between the right and left legs is not assumed in our model.

3   Conclusion and Future Work

We focused on centipedes and proposed a decentralized control scheme for the inter-limb coordination, wherein a simple local reflexive mechanism was implemented. In the future, we will investigate the validity of the proposed control scheme via simulation.

Acknowledgments. This research was supported by JST CREST. The authors would like to thank Kazuhiko Sakai of Research Institute of Electrical Communication of Tohoku University for helpful discussion.

References

1. Kuroda, S., Kunita, I., Tanaka, Y., Ishiguro, A., Kobayashi, R., Nakagaki, T.: Common mechanics of mode switching in locomotion of limbless and legged animals. J. R. Soc. Interface **11**, 20140205 (2014)
2. Manton, S.M.: The Arthropoda: Habits, Functional Morphology and Evolution. Clarendon Press, Oxford (1977)
3. Inagaki, S., Niwa, T., Suzuki, T.: Follow-the-contact-point gait control of centipede-like multi-legged robot to navigate and walk on uneven terrain. In: 2010 IEEE/RSJ International Conference on Intelligent Robots and Systems (IROS), pp. 5341–5346 (2010)
4. Onat, A., Tsuchiya, K., Tsujita, K.: Decentralized autonomous control of a myriapod locomotion robot. In: Proceedings of the 1st International Conference on Information Technology in Mechatronics, pp. 191–196 (2001)
5. Yasui, K., Sakai, K., Kano, T., Owaki, D., Ishiguro, A.: TEGOTAE-based decentralized control mechanism underlying myriapod locomotion. In: The First International Symposium on Swarm Behavior and Bio-Inspired Robotics (2015)
6. Umedachi, T., Takeda, K., Nakagaki, T., Kobayashi, R., Ishiguro, A.: Fully decentralized control of a soft-bodied robot inspired by true slime mold. Biol. Cybern. **102**(3), 261–269 (2010)

Realization of Snakes' Concertina Locomotion by Using "TEGOTAE-Based Control"

Ryo Yoshizawa[1]([✉]), Takeshi Kano[1], and Akio Ishiguro[1,2]

[1] Research Institute of Electrical Communication, Tohoku University,
2-1-1 Katahira, Aoba-ku, Sendai 980-8577, Japan
{r-yoshi,tkano,ishiguro}@riec.tohoku.ac.jp
[2] JST CREST, 4-1-8 Honcho, Kawaguchi, Saitama 332-0012, Japan
http://www.cmplx.riec.tohoku.ac.jp/

Abstract. Our goal is to develop snake-like robots that can exhibit versatile locomotion patterns in response to the environments like real snakes. Towards this goal, in our other work, we proposed an autonomous decentralized control scheme on the basis of a concept called TEGOTAE, a Japanese concept describing how well a perceived reaction matches an expectation, and then succeeded in reproducing scaffold-based locomotion, a locomotion pattern observed in unstructured environments, via real-world experiments. In this study, we demonstrated via simulations that concertina locomotion observed in narrow spaces can be also reproduced by using the proposed control scheme.

1 Introduction

Snakes can change their locomotion patterns in response to the environments in real time [1,2]. Inspired by this ability, various snake-like robots have been developed thus far [2–4]. However, they could not reproduce the innate behavior of real snakes.

To address this issue, in our other work [5], we proposed an autonomous decentralized control scheme based on TEGOTAE. Then, our snake-like robot implementing the proposed control scheme successfully reproduced scaffold-based locomotion wherein terrain irregularities were actively exploited.

However, applicability of this control scheme to other locomotion patterns is still unclear. In this study, we demonstrated via simulations that concertina locomotion observed in narrow spaces can be reproduced by using the TEGOTAE-based control scheme.

2 Model

Although the musculoskeletal structure of a real snake is complicated, we simply modeled it as shown in Fig. 1. The skeletal system is described by particles

© Springer International Publishing Switzerland 2016
N.F. Lepora et al. (Eds.): Living Machines 2016, LNAI 9793, pp. 548–551, 2016.
DOI: 10.1007/978-3-319-42417-0_61

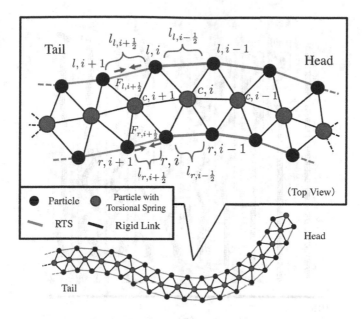

Fig. 1. Schematic of model. (Color figure online)

connected via rigid links. A torsional spring is embedded in each backbone joint. A parallel combination of a damper and a real-time tunable spring (RTS) [6], whose resting length can be changed arbitrarily, is aligned on both sides of the body. The friction coefficient along the body axis is set to be smaller than that along its perpendicular axis, like real snakes. The natural lengths of the RTSs are controlled according to the TEGOTAE-based control scheme, which is essentially the same as the control scheme used in [5].

3 Simulation Results

We performed a simulation in a narrow aisle. The natural lengths of the RTSs for the head part were operated via a keyboard manipulation. The result is shown in Fig. 2(a). The simulated snake exhibited concertina locomotion, *i.e.*, the tail part of the body was first pulled forward with the head part anchored followed by extension of the head part with the tail part anchored, like a real snake (Fig. 2(b)).

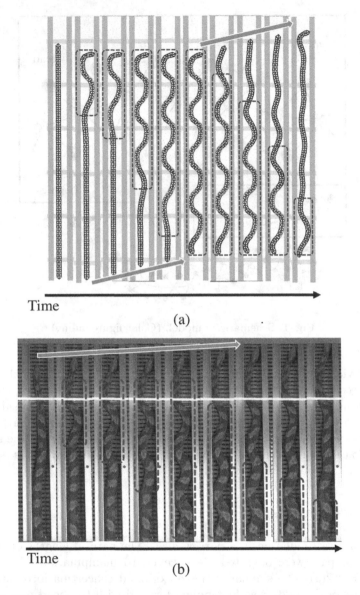

Fig. 2. (a) Simulation result for locomotion in narrow aisle. RTSs generating forces of contraction and expansion derive from feedback are colored by red and green, respectively. Gray areas denote side walls. (b) Photographs of concertina locomotion of a real snake in narrow aisle. (Color figure online)

Acknowledgments. This work was supported in part by NEDO (Core Technology Development Program for Next Generation Robot). The authors would like to thank Dr. Kosuke Inoue of Ibaraki University and Dr. Hisashi Date of University of Tsukuba, and Daiki Nakashima of Tohoku University, for their cooperation.

References

1. Moon, B.R., Gans, C.: Kinematics, muscular activity and propulsion in gopher snakes. J. Exp. Biol. **201**, 2669–2684 (1998)
2. Liljebäck, P., Pettersen, K.Y., Stavdahl, Ø., Gravdahl, J.T.: Snake Robots - Modelling, Mechatronics, and Control: Advances in Industrial Control. Springer, London (2012)
3. Hirose, S.: Biologically Inspired Robots (Snake-like Locomotor and Manipulator). Oxford University Press, Oxford (1993)
4. Rollinson, D., Alwala, K.V., Zevallos, N., Choset, H.: Torque control strategies for snake robots. In: IEEE/RSJ International Conference Intelligent Robots and Systems (IROS), pp. 1093–1099 (2014)
5. Kano, T., Yoshizawa, R., Ishiguro, A.: TEGOTAE-based control scheme for snake-like robots that enables scaffold-based locomotion. In: Lepora, N.F., et al. (eds.) Living Machines 2016. LNCS(LNAI), vol. 9793. Springer, Switzerland (2016)
6. Umedachi, T., Takeda, K., Nakagaki, T., Kobayashi, R., Ishiguro, A.: Fully decentralized control of a soft-bodied robot inspired by true slime mold. Biol. Cybern. **102**, 261–269 (2010)
7. Date, H., Takita, Y.: Adaptive locomotion of a snake like robot based on curvature derivatives. In: IEEE/RSJ International Conference on Intelligent Robots and Systems (IROS), pp. 3554–3559 (2007)

Author Index

Abourachid, Anick 3
Aitken, Jonathan M. 409
Akkus, Ozan 365
Albanese, Ugo 16, 119
Ambrosano, Alessandro 16, 119
Aquilina, Kirsty 393
Arsiwalla, Xerxes D. 389
Asada, Minoru 203

Bai, Guochao 28
Barton, David A.W. 393
Bernabei, Rina 40
Bertrand, Olivier J.N. 167
Blancas, Maria 297, 353, 400
Bosse, Jacob W. 329
Bračun, Drago 319
Breedveld, Paul 307

Cameron, David 297, 353, 409, 413
Camilleri, Daniel 48
Carbonaro, Nicola 58
Chang, Sarah R. 192
Chapin, Katherine J. 365
Charisi, Vicky 297, 353
Chen-Burger, Jessica 131
Cheung, Hugo 409
Chiel, Hillel J. 97, 365
Chua, Adriel 409
Collins, Emily 409, 413
Cominelli, Lorenzo 58, 297, 353
Cramphorn, Luke 418, 485, 540
Cutkosky, Mark R. 288

Daltorio, Kathryn A. 97
Damianou, Andreas 48
Dario, Paolo 341
Davison, Daniel 297, 353
De Rossi, Danilo 58, 297, 353
Desmulliez, Marc P.Y. 71, 131
Dewar, Alex 263
Dodou, Dimitra 307

Eberle, Henry 424
Egelhaaf, Martin 167
Evers, Vanessa 297, 353

Fabre, Remi 227
Falotico, Egidio 16, 119, 341
Fernando, Samuel 297, 353, 413
Fukuhara, Akira 79, 532

Garofalo, Roberto 58, 297, 353
Getsy, Andrew P. 329, 429
Gewaltig, Marc-Oliver 16
Giannaccini, Maria Elena 436
Goda, Masashi 441
Graham, Paul 263
Grechuta, Klaudia 297

Hawley, Emma L. 365
Hayashi, Yoshikatsu 424
Herreros, Ivan 214, 389
Hinkel, Georg 16
Honnet, Cedric 518
Horchler, Andrew D. 97
Horikiri, Shun-ya 472
Hosoda, Koh 467
Howell, Steven 155
Huang, Jiaqi V. 85
Hugel, Vincent 3
Hunt, Alexander J. 144

Inoue, Naoki 467
Ishiguro, Akio 79, 441, 449, 454, 472, 532, 545, 548
Ishihara, Hisashi 203
Ishikawa, Masato 503
Isobe, Yoshihiro 467
Itayama, Susumu 441

Jackson, Harry 48

Kaiser, Jacques 16
Kandhari, Akhil 97

Kano, Takeshi   79, 441, 449, 454, 545, 548
Kent, Christopher   498
Kerdegari, Hamideh   107
Kim, Yeongmi   107
Kirtay, Murat   16, 119
Kito, Yuji   526
Knoll, Alois   16
Kobetic, Rudi   192
Kong, Xianwen   28
Krapp, Holger G.   85
Kruiper, Ruben   131
Kruusmaa, Maarja   510
Kuratani, Ryoichi   503
Kuusik, Alar   510

Laschi, Cecilia   16, 119, 341
Law, James   409
Lawrence, Neil   48
Leonards, Ute   498
Lepora, Nathan F.   393, 418, 436, 485, 498,
    540
Levi, Paul   16
Li, Wei   144
Lindemann, Jens Peter   167
Low, Sock C.   459
Lund, Henrik Hautop   341
Ly, Olivier   227

Maekawa, Koki   467
Maffei, Giovanni   214
Marques-Hueso, Jose   71
Martin, Joshua P.   329
Martínez-Cañada, Pablo   16
Mazzei, Daniele   58, 297, 353
Mealin, Sean   155
Meyer, Hanno Gerd   167
Millings, Abigail   413
Mitchinson, Ben   179
Miyazawa, Sakiko   441
Moore, Roger   297, 353, 413
Morillas, Christian   16
Moulin-Frier, Clément   490

N'Guyen, Steve   227
Nandor, Mark J.   192
Nasuto, Slawomir   424
Ng, Jack Hoy-Gig   71
Nishii, Jun   472

Omedas, Pedro   353
Osuka, Koichi   526
Ota, Nobuyuki   203
Otto, Marc   239
Owaki, Dai   79, 441, 449, 472, 532, 545

Panova, Nedyalka   480
Paskarbeit, Jan   167, 239
Passault, Grégoire   227
Patel, Jill M.   365
Pestell, Nicholas   485
Philippides, Andrew   263
Pieroni, Michael   297, 353
Pirim, Patrick   275
Pope, Morgan T.   288
Power, Jacqueline   40
Prescott, Tony J.   48, 107, 179, 297, 353, 413
Puigbò, Jordi-Ysard   490

Quinn, Roger D.   97, 144, 192, 329, 365, 429

Reidsma, Dennis   297, 353
Ristolainen, Asko   510
Ritzmann, Roy E.   329
Roberts, David L.   155
Roscow, Emma   498
Rouxel, Quentin   227
Ruck, Maximilian   214

Saku, Taro   467
Samuel, Ifor D.W.   480
Sánchez-Fibla, Martí   214
Santos-Pata, Diogo   251
Schilling, Malte   239
Schneider, Axel   167, 239
Sharkey, Amanda   413
Shimizu, Masahiro   467
Škulj, Gašper   319
Sprang, Tim   307
Steadman, Nathan   263
Stokes, Adam   375
Strohmeier, Paul   518
Sueoka, Yuichiro   526
Sugimoto, Yasuhiro   503, 526
Suzuki, Shura   532
Szczecinski, Nicholas S.   144, 329, 429
Szollosy, Michael   413

Tenopala-Carmona, Francisco   480
Thompson, Alexander C.   480
Tognetti, Alessandro   58
Tolu, Silvia   341
Triolo, Ronald J.   192
Tuhtan, Jeffrey Andrew   510

Ulbrich, Stefan   16

van der Meij, Jan   297
van Wijngaarden, Joeri B.G.   459
vander Meij, Jan   353
Vannucci, Lorenzo   16, 119, 341
Verschure, Paul F.M.J.   214, 251, 297, 353,
    389, 400, 459, 490
von Cyborg, Samppa   518
Vouloutsi, Vasiliki   297, 353, 400

Walker, Christopher   263
Wang, Jieyu   28
Wang, Yilin   85
Ward-Cherrier, Benjamin   418, 485, 540
Watson, David E.   71
Webb, Barbara   375
Webster, Victoria A.   365
Wei, Tianqi   375
Whyle, Stuart   436
Wijnen, Frances   297, 353

Yasui, Kotaro   449, 545
Yoshizawa, Ryo   454, 548

Zaraki, Abolfazl   58
Zucca, Riccardo   251, 297, 353, 400
Zucker, George S.   97

Printed in the United States
By Bookmasters